Lecture Notes in Physics

Edited by H. Araki, Kyoto, J. Ehlers, München, K. Hepp, Zürich
R. Kippenhahn, München, H. A. Weidenmüller, Heidelberg
and J. Zittartz, Köln

180

Group Theoretical Methods in Physics

Proceedings of the XIth International Colloquium
Held at Boğaziçi University, Istanbul, Turkey
August 23–28, 1982

Edited by M. Serdaroğlu and E. İnönü

Springer-Verlag
Berlin Heidelberg New York Tokyo 1983

Editors

Meral Serdaroğlu
Erdal İnönü
Physics Department, Boğaziçi University
P.K.2., Bebek Istanbul, Turkey

ISBN 3-540-12291-5 Springer-Verlag Berlin Heidelberg New York Tokyo
ISBN 0-387-12291-5 Springer-Verlag New York Heidelberg Berlin Tokyo

Printing and binding: Beltz Offsetdruck, Hemsbach/Bergstr.
2153/3140-543210

PREFACE

The XIth International Colloquium on Group Theoretical Methods in Physics was held at Boğaziçi University in Istanbul from August 23 to 28, 1982. Like the previous conferences in this series, it brought together physicists, mathematicians and chemists interested in group structures, group representations and their use in the description and understanding of natural phenomena.

The colloquium was divided into nine sessions. Each session began with one or two review talks and continued with short presentations of research results, supplemented in some cases by posters. Following the established pattern, the talks covered most areas of physics, a novelty being the addition of sessions on completely integrable systems and on statistical mechanics.

The proceedings follow the same pattern and are divided into nine chapters, each giving the texts of the invited and review talks and the short communications or posters presented in the session. To avoid delays in publication, we have done practically no editing on the texts. Only in some cases, with the purpose of keeping the book within a reasonable size, did we ask the authors to shorten their texts a little, and we are grateful to them for their gracious cooperation. A few talks unfortunately could not be incorporated in the proceedings, as the texts did not reach us in time.

Participants of previous colloquia will remember that every alternate year since 1978, group theory conferences have provided the setting for the presentation of the Wigner Medal for "outstanding contributions to the understanding of physics through group theory." There was no medal ceremony at the Istanbul meeting, but in order to continue the tradition, we decided to add to these proceedings two appendices related to the Wigner Medal Ceremonies for 1980 and 1982, which were kindly sent to us by Professor A. Böhm, Chairman of the Group Theory Foundation. Appendix I gives the acceptance speech of Professor Gel'fand upon receiving the 1980 medal, and Appendix II contains a summary of the proceedings of the presentation of the Wigner Medal for 1982 to Professor Y. Nee'man, which took place at the University of Texas on November 11, 1982, during the Solway Conference.

It gives us great pleasure to end this brief introduction by expressing our profound gratitude to various organizations and people who made it possible to hold the XIth Group Theory Colloquium in Istanbul and contributed effectively to its success. These include the sponsors of the conference who also provided financial and logistical support:

The International Union of Pure and Applied Physics (IUPAP)
Turkish Physical Society
Boğaziçi University
and the following private Turkish firms, which gave further substantial financial support:

Hisarbank A.Ş.
Türkiye Şişe ve Cam Fabrikaları A.Ş.
Oceanique S.A.
Enka Holding A.Ş.
Nova Ltd.
Burçelik A.Ş.
Hatay Pan Kara Nakliyat A.Ş.
Burla Makina Ticaret A.Ş.
Mekser A.Ş.

We are particularly indebted to the members of the Standing Committee, who chose the conference site and provided continuous moral support to the organizers throughout the conference; the International Advisory Committee members, who helped to organize the sessions, suggesting topics and speakers, and acted as very able chairmen; the President of Boğaziçi University, Professor Ergün Toğrol, who graciously put all the campus facilities at the disposal of the colloquium participants, and finally, our colleagues in the Local Organization Committee and the Physics Department, who spared no effort in ensuring that the colloquium proceeded smoothly and reached a successful conclusion.

February 1983
Istanbul

The Editors
M.Serdaroğlu
E.İnönü

THE INTERNATIONAL STANDING COMMITTEE

H.BACRY	U.E.R. de Luminy, Marseille, France
L.C.BIEDENHARN	Duke University, U.S.A.
J.L.BIRMAN	City College of C.U. New York, U.S.A.
K.BLEULER	University of Bonn, Germany
A.BÖHM	University of Texas, U.S.A.
L.L.BOYLE	University of Kent, England
A.P.CRACKNELL	The University, Dundee, Scotland
G.G.EMCH	University of Rochester, U.S.A.
L.P.HORWITZ	University of Tel-Aviv, Israel
A.JANNER	University of Nijmegen, The Netherlands
B.R.JUDD	Johns Hopkins University, U.S.A.
P.KRAMER	University of Tübingen, Germany
L.MICHEL	IHES, Bures-sur-Yvette, France
M.MOSHINSKY	University of Mexico (UNAM), Mexico
Y.NE'EMAN	Tel-Aviv University, Israel
C.PIRON	University of Geneva, Switzerland
S.STERNBERG	Tel-Aviv University, Israel
P.WINTERNITZ	University of Montreal, Canada
J.A.WOLF	University of California, Berkeley, U.S.A.

THE INTERNATIONAL ADVISORY COMMITTEE

I.BARS	Yale University, U.S.A.
A.O.BARUT	University of Colorado, U.S.A.
A.BÖHM	University of Texas, U.S.A.
D.CHUDNOVSKY	Columbia University, U.S.A.
A.P.CRACKNELL	The University, Dundee, Scotland
F.GÜRSEY	Yale University, U.S.A.
F.IACHELLO	Yale University, U.S.A.
A.JANNER	University of Nijmegen, The Netherlands
B.JULIA	Ecole Normale Supérieure, France
B.KOSTANT	Massachusetts Institute of Technology, U.S.A.
L.MICHEL	IHES, Bures-sur-Yvette, France
M.MOSHINSKY	University of Mexico (UNAM), Mexico
W.NAHM	Max-Planck-Institut, Bonn, Germany
Y.NE'EMAN	Tel-Aviv University, Israel
L.RADICATI	Scuola Normale Superiore Pisa, Italy
H.RÖMER	University of Freiburg, Germany
P.WINTERNITZ	University of Montreal, Canada
M.VERGNE	Massachusetts Institute of Technology, U.S.A.
G.ZUCKERMAN	Yale University, U.S.A.

THE ORGANIZATION COMMITTEE

G.AKTAŞ, M.ARIK, M.HORTAÇSU, E.İNÖNÜ, M.SERDAROĞLU, A.SEVGEN	Boğaziçi University,Turkey
F.GÜRSEY	Yale University,U.S.A.

TABLE OF CONTENTS

SESSION I - GROUP REPRESENTATIONS

SESSION II - COMPLETELY INTEGRABLE SYSTEMS AND GROUP THEORY

NON-COMPACT GROUPS AND IRREDUCIBLE REPRESENTATIONS

Gregg J. Zuckerman*
YALE UNIVERSITY
Mathematics Department
Box 2155 Yale Station
New Haven, CT. 06520
USA

ABSTRACT

We review the modern development of the theory of linear irreducible infinite dimensional representations of non-compact groups. Such representations arise on spaces of states, spaces of observables, spaces of classical fields, etc. We connect the unitarity problem in mathematics to the elimination of ghosts in formalisms using indefinite Hermitian metrics.

We would like in this talk to survey the field of infinite dimensional representations of non compact Lie groups, and the possible connections of this field to physics. We emphasize at first representations on groups on linear spaces, say for example linear spaces of either classical or quantum observables[1]]. Here there is no unitarity hypothesis. Only the space of normalizable physical states carries a positive definite inner product. So we reserve discussion of unitary representations to later in our talk.

§1: Brief history of the "modern" approach to representations of non-compact semisimple Lie groups and their Lie algebras:

~1890: Classification of complex semisimple Lie algebras -- E. Cartan (earlier work by W. Killing).

~1900: Representations of finite groups and finite dimensional associative algebras -- Frobenius, I. Schur, Burnside, Wedderburn.

~1910: Classification of real semisimple Lie algebras; classification

*Supported by NSF Grant #MCS80-05151 and by the Alfred P. Sloan Foundation

of finite dimensional linear representations of semisimple Lie algebras -- E. Cartan.

~ 1920: Invention of abstract algebra: vector spaces, associative algebras, ideals, <u>modules</u>, all possibly infinite dimensional -- D. Hilbert, H. Weyl, E. Noether.

~ 1925: Invention of the matrix form of quantum mechanics -- W. Heisenberg, M. Born, P. Jordan, P. Dirac.

~ 1925: Finite dimensional representations of semisimple Lie groups, compact and (via the unitary trick), non compact -- H. Weyl.

~ 1935: Construction of some special infinite dimensional modules for Lie algebras of type $so(n,2)$, $n = 3$ (deSitter) or $n = 4$ (conformal) -- P. Dirac[2], E. Majorana.

~ 1949: Systematic use of Poincaré's universal enveloping algebra $\mathcal{U}(\mathfrak{g})$ and the Noetherian approach to $\mathcal{U}(\mathfrak{g})$-modules, possibly infinite dimensional -- C. Chevalley, Harish-Chandra[3], (earlier work by E. Wigner, V. Bargmann).

~ 1950 - present: Modern school: Work with a fixed maximal compact subgroup K in real semisimple group G; use Harish-Chandra's dictionary[3]: irreducible linear representations of G correspond to irreducible $(\mathcal{U}(\mathfrak{g}), K)$-modules (no unitarity hypothesis). Construction and classification of <u>all</u> irreducible $(\mathcal{U}(\mathfrak{g}), K)$-modules -- among others, Harish-Chandra, I. M. Gelfand, W. Schmid, R. P. Langlands, E. Stein, A. Knapp, T. Enright, V. Varadarajan, N. Wallach, D. Vogan, A. Beilinson, J. Bernstein, and the speaker (see [4] and references therein).

1925 - present: Progress on the unitary irreducible representations: there is however no general classification theory.

1976: Partial progress on the unitary problem. Classification of <u>all</u> irreducible Hermitian, possibly indefinite, group representations -- R. P. Langlands, W. Schmid, A. Knapp, and the speaker[5].

1982: Open problem: which representations do not have "ghosts", i.e. states of zero or negative norm?

§2. Cohomology and cutting down the size of modules.

First, an educational module concerning module theory:

Start with a representation of a Lie algebra $\mathfrak{g}$ by linear operators on a vector space V. For each X in $\mathfrak{g}$, $\pi(X)$ is the operator representating X. We have $[\pi(X), \pi(Y)] = \pi([X,Y])$.

For each X in g and each v in V we define a product $Xv = \pi(X)v$. This "module product" generalizes and coexists with scalar multiplication by a fixed complex number λ :

(Scalar product) $v \to \lambda v$ (2.1)

(Module product) $v \to Xv$ (2.2)

Knowing the module product is theoretically equivalent to knowing the operators $\pi(X)$ for all X in g. However, we may have an algorithm for computing the module product without having a closed expression for the representing operators.

V equipped with the module product is called a g-module. If we introduce iterated module products, i.e. $v \to X_n \dots X_2 X_1 v$, then we regard V as a $U(g)$-module. In a sense, the $U(g)$-modules are enriched vector spaces.

Philosophy: when the carrier space V, rather than the representation π, is at the center of attention, we should use the language of $U(g)$-modules.

Remarks on bases: One can introduce a basis for g and a basis for V, and write formulas for module products:

$$X_i v_j = \sum_{k=1}^{\infty} \alpha_{ij}^k v_k \tag{2.3}$$

A basic convention of algebraists is that for each i and j, the coefficients α_{ij}^k vanish for all large k, where "large" depends on i and j. Thus, there are no convergence questions. Harish-Chandra's equivalence between group representations and $(U(g), K)$ modules relates therefore a functional analysis problem, where convergence is important, to an algebra problem, where convergence plays no role[3].

Finally, we define the notion of a g-map

$$T: V_1 \to V_2 \tag{2.4}$$

from one g-module to a second. We require that T be a linear operator (not necessarily bounded in any sense), and that for each X in g and each v_1 in V_1,

$$T(Xv_1) = X(Tv_1) \tag{2.5}$$

We can now discuss cohomology: suppose M is a $U(g)$-module and $d: M \to M$ is a g-map such that $d^2 = 0$. Let

$$Z = \text{Ker } d = \{m \text{ in } M \mid dm = 0\},$$

$$B = \text{Im } d = \{m \text{ in } M \mid m = dn \text{ for some } n \text{ in } M\},$$

and $$H(M,d) = Z/B \quad \text{--} \tag{2.6}$$

the quotient module of Z by B, based on the vector space of cosets of Z by B.

Remark: Many of the above concepts in module theory were developed by E. Noether, who influenced Chevalley, who influenced Harish-Chandra, who influenced the whole modern school.

Example 1: The "photon module" for the Lie algebra g of the Poincaré group:

Let C^p be the g module of differential p-forms on space-time. Let $\delta : C^p \to C^{p+1}$ be the exterior derivative -- δ is a g-map. Let $\delta^* : C^p \to C^{p-1}$ be the adjoint to δ relative to the Minkowski metric. In this language, the wave equation f or a p-form ψ reads

$$\Box\psi = (\delta\delta^* + \delta^*\delta)\psi = 0. \tag{2.7}$$

Let C_0^1 be the 1-forms A such that $\Box A = 0$. If A is in C_0^1, so is $\delta^*\delta A$. Moreover, if A is in C_0^1, $(\delta^*\delta)^2 A = -(\delta^* \delta\delta\delta^*)A = 0$.

So, in our example we let M be the g-module C_0^1 and we let d be the g map $\delta^*\delta : M \to M$. Then we can form the cohomology module $H(C_0^1, \delta^*\delta)$. The (normalizable) states in this module correspond to "transverse photons". The module $B = dM = \delta^*\delta\, C_0^1$ consists of pure gauge fields (longitudinal photons).

Example 2: Photon module for the conformal algebra, $g = so(4,2)$[6],[2]:

Let C^p again denote the p-forms on space-time, p = 0,1,2,3,4. We now let

$$\tilde{M} = C^0 \oplus C^1 \oplus C^3 \oplus C^4 \quad , \tag{2.8}$$

and regard $\tilde{M}$ as a module over the conformal algebra, $\tilde{g}$. Let δ again be the exterior derivative, and let $*$ be the Minkowski duality operator (Hodge star) on 2-forms, C^2. We form a "complex" of $\tilde{g}$-modules

$$(0) \to C^0 \xrightarrow{\delta} C^1 \xrightarrow{\delta * \delta} C^3 \xrightarrow{\delta} C^4 \longrightarrow (0); \tag{2.9}$$

the composite of successive $\tilde{g}$-maps above is zero. We now let $\tilde{d}$ be the differential on $\tilde{M}$ which equals δ on C^o, $\delta*\delta$ on C^1, δ on C^3, and 0 on C^4. Then, $H(\tilde{M}, \tilde{d})$ becomes a graded cohomology $\tilde{g}$-module.

$$H^1(\tilde{M}, \tilde{d}) = {}^{\mathrm{Ker}\ \delta_*\delta} /_{\mathrm{Im}\ \delta} \tag{2.10}$$

is what we call the photon module for the conformal algebra. The module Im δ corresponds to pure gauge fields.

§3. We now turn to the problem of Hermitian modules. We say that (M, d) is Hermitian if M carries a nondegenerate Hermitian inner product <,> such that

$$< dm_1, m_2> = < m_1, dm_2> \tag{3.1}$$

for all m_1, m_2 in M; also, < , > is g-invariant:

$$< Xm_1, m_2> + < m_1, Xm_2 > = 0 \tag{3.2}$$

for X in the real Lie algebra g.

If m_1 is in Ker d and m_2 is in Im d, then $<m_1,m_2> = < m_1,dm_3> = <dm_1, m_3> = 0$. It follows that H(M,d) carries a g-invariant Hermitian form induced by the restriction of <,> to Ker d. This induced form on H(M,d) will be nondegenerate under appropriate hypotheses.

Definition: We say that d eliminates ghosts in M if the Hermitian inner product on H(M,d) is proportional to a positive definite inner product, i.e. H(M,d) is unitarizable.

Example 1: Photon module for the Poincaré algebra:

$(C_0^1, \delta^*\delta)$ is Hermitian if we suitably restrict the decay at infinity of our 1-forms. The (normalized) module $H(C_0^1, \delta^*\delta)$ carries the helicity ±1 unitary irreducible representations of the Poincaré group; $\delta^*\delta$ eliminates ghosts, a result going back to Gupta and Bleuler[7].

Example 2: Photon theory for so(4,2):

$$\tilde{M} = C^o \oplus C^1 \oplus C^3 \oplus C^4 \tag{3.3}$$

as above is Hermitian: define the inner product by

$$< w_1,w_2 > = \int_{\text{space-time}} w_1 \wedge \overline{w_2} \tag{3.4}$$

for two differential forms w_1 and w_2 (with suitably restricted decay). Build d as before from δ, $\delta^*\delta$, δ , and 0. Then d is self-adjoint. However, $H(\tilde{M},d)$ is <u>indefinite</u> Hermitian.

Example 3: We work now with a construction in pure mathematics which is currently being investigated by the speaker and other representation theorists[8],[4]. G is a real semisimple group, K is the maximal compact subgroup, and H is a connected closed subgroup of K. We might have $G = SO(4,2)_o$, $K = SO(4) \times SO(2)$, and $H = U(2) \times SO(2)$. We suppose that V is a Hermitian $\mathcal{U}(\mathfrak{g})$-module such that the $\mathfrak{h}$ action on V integrates to a representation of the compact group H.

Let $C^p(K/H; V)$ be the V-valued p-forms on K/H, i.e. in local coordinates such a differential form has coefficients which are V valued functions. By a suitable generalization of the usual exterior calculus, we can define:

(1) An operator for each p,

$$d = C^p(K/H; V) \to C^{p+1}(K/H; V) \tag{3.5}$$

such that $d^2 = 0$.

(2) A Hermitian structure on the sum

$$M = \bigoplus_{p=0}^{\dim K/H} C^p(K/H; V) \tag{3.6}$$

defined by

$$< w_1, w_2 > = \int_{K/H} < w_1 \wedge w_2 >_V \tag{3.7}$$

where $< w_1 \wedge w_2 >_V$ means first wedge than take inner products in V.

By means of the $\mathfrak{g}$-module structure on V we can define a natural $\mathfrak{g}$-module structure on each $C^p(K/H; V)$ (For the experts, we look at the cochain map induced by the $\mathfrak{k}$-module product map $\mathfrak{g} \otimes V \to V$). We finally obtain the structure of a Hermitian differential $\mathfrak{g}$-module on the pair (M,d). We can then form the cohomology module H(M,d), which will be a Hermitian $(\mathcal{U}(\mathfrak{g}), K)$-module, i.e. the $\mathfrak{k}$-action on H(M,d) will integrate to a K representation on V.

Problem: When does the differential d in M eliminate ghosts?
For every semisimple group G, infinitely many examples of Hermitian indefinite $(\mathcal{U}(\mathfrak{g}), H)$ modules lead to differential modules (M,d) in which d eliminates ghosts. Various mathematicians including the speaker have conjectured the unitarity of more general modules of the

type H(M,d), M obtained from V, as above[8]. Whenever we can prove a case of these conjectures, we can conclude :

Cohomology exorcises ghosts.

Howard Garland and the speaker obtained positive results for the unitarity problem in the special case when H(M,d) has a "highest weight vector"[9]. Our method is closely related to an earlier paper by Garland on affine Kac-Moody (loop) algebras[10].

REFERENCES

1] A. Barut, R. Raczka. Theory of Group Representations and Applications. Polish Scientific Publishers, Warsaw (1977).

2] Dirac, P.A.M., Wave equations in conformal space. Annals of Math., Vol. 37 (1936), pp. 429-442.

3] Harish-Chandra, Representations of semisimple Lie groups II, Trans. Amer. Math. Soc. 76 (1954), 26-65.

4] D. Vogan, Representations of real reductive Lie groups. Progress in Mathematics Series. Birkhauser, Boston (1981).

5] A. Knapp, G. J. Zuckerman. Classification theorems for representations of semisimple Lie groups, Non-Commutative Harmonic Analysis, Lecture Notes in Math., vol. 587, Springer-Verlag, New York, (1977), pp. 138-159.

6] G. J. Zuckerman, Induced representations and quantum fields. To appear in "Group theory and its applications to Physics and Mathematical Physics," proceedings of a conference held at University of Chicago, July 6-16, 1982.

7] Davydov, A.S., Quantum Mechanics. NEU Press Technical Translation Series, (1966).

8] D. Vogan, G. J. Zuckerman, Unitary representations with non-zero cohomology. Preprint (1982).

9] H. Garland, G. J. Zuckerman, On unitarizable highest weight modules of Hermitian pairs. Journal of the Faculty of Science, University of Tokyo, Sec. I A. Vol. 28, No. 3, pp. 877-889, (1982).

10] H. Garland, The arithmetic theory of loop algebras, Journal of Algebra 53 (1978), 480-551.

DUALITY THEOREMS IN CONFORMAL GEOMETRY

Hubert Goldschmidt
THE INSTITUTE FOR ADVANCED STUDY
Princeton, New Jersey 08540
U.S.A.

ABSTRACT

For conformally flat Riemannian manifolds of dimension $n \geq 3$, we describe an explicit resolution of the sheaf Θ_c of conformal Killing vector fields which is formally self-adjoint, and we deduce a duality theorem for the cohomology of X with values in Θ_c.

Let X be a manifold of dimension $n \geq 3$, whose tangent and cotangent bundles we denote by T and T^* respectively. Let $\Lambda^k T^*$ and $S^k T^*$ be the k-th exterior and symmetric powers of T^*. We shall denote by $\mathcal{E}$ the sheaf of sections of a vector bundle E over X and by $C^\infty(E)$ the space of sections of E over X.

Two Riemannian metrics g, g' on X are conformally equivalent if there exists a real-valued differentiable function u on X such that $g' = e^u g$; a conformal structure on X is a conformal class of metrics. We now fix a metric g on X. We say that the Riemannian manifold (X,g) is conformally flat if, for every $x \in X$, there is a diffeomorphism φ of a neighborhood U of x onto an open subset of $\mathbb{R}^n$ and a real-valued function u on U such that

$$\varphi^* g' = e^u g ,$$

where g' is the Euclidean metric on $\mathbb{R}^n$.

Let

$$\mu : \Lambda^{j+1}T^* \otimes \Lambda^2 T^* \longrightarrow \Lambda^{j+2}T^* \otimes T^* ,$$

$$\mathrm{Tr}^j : \Lambda^{j+1}T^* \otimes T^* \otimes T^* \longrightarrow \Lambda^j T^* \otimes T^*$$

be the mappings defined by

$$\mu(u)(\xi_1,\dots,\xi_{j+2},\eta) = \sum_{\ell=1}^{j+2}(-1)^{\ell+1}u(\xi_1,\dots,\hat{\xi}_\ell,\dots,\xi_{j+2},\xi_\ell,\eta) ,$$

$$(\mathrm{Tr}^j v)(\xi_1,\dots,\xi_j,\eta) = \sum_{i=1}^{n} v(t_i,\xi_1,\dots,\xi_j,t_i,\eta) ,$$

where $u \in (\Lambda^{j+1}T^* \otimes \Lambda^2 T^*)_x$, $v \in (\Lambda^{j+1}T^* \otimes T^* \otimes T^*)_x$ and $\xi_1,\dots,\xi_{j+2},\eta \in T_x$, with $x \in X$, and where $t_1,\dots,t_n$ is an orthonormal basis of T_x . We denote by E_j the kernel of μ and by E_j^0 the kernel of the restriction of Tr^j to E_j . We have

(1) $$E_{n-2}^0 = 0 \quad , \quad E_{n-1}^0 = 0 .$$

The vector bundles E_j are endowed with scalar products induced by the metric g . We denote by ρ_j the orthogonal projection of E_j onto E_j^0 .

We denote by ∇ the Levi-Civita connection of g . The curvature tensor R of (X,g) is the section of $\Lambda^2T^* \otimes \Lambda^2T^*$ determined by

$$R(\xi_1,\xi_2,\xi_3,\xi_4) = g((\nabla_{\xi_1}\nabla_{\xi_2} - \nabla_{\xi_2}\nabla_{\xi_1} - \nabla_{[\xi_1,\xi_2]})\xi_3,\xi_4) ,$$

for all vector fields ξ_1,ξ_2,ξ_3,ξ_4 on X . According to the Bianchi identity, R is a section of E_1 and the Weyl tensor W of (X,g) is the section $\rho_1 R$ of E_1^0 . If $n \geq 4$, a classic result of H. Weyl asserts that the Riemannian manifold (X,g) is conformally flat if and only if its Weyl tensor W vanishes. If $n = 3$, according to (1), W vanishes identically; in this case, the conformal flatness of (X,g) is equivalent to the vanishing of a conformal invariant, the Schouten tensor.

A vector field ξ on X is a conformal Killing vector field if there is a real-valued function f on X such that

(2) $$\mathcal{L}_\xi g = fg ,$$

where $\mathcal{L}_\xi g$ is the Lie derivative of g along ξ. If $\operatorname{Tr} h$ denotes the trace of a symmetric 2-form h, equation (2) is equivalent to

$$(3) \qquad D_0^c \xi = \mathcal{L}_\xi g - \frac{1}{n} \operatorname{Tr} (\mathcal{L}_\xi g) = 0 \quad .$$

In studying deformations of the conformal structure of (X,g), we are led to interpret the cohomology groups $H^j(X,\Theta_c)$ of X with values in the sheaf Θ_c of conformal Killing vector fields on X. In fact, one of the problems consists in finding conditions on a deformation of g in order that it be conformally rigid: if g_t is a one-parameter family of metrics on X, with $g_0 = g$, when does there exist one-parameter families φ_t of diffeomorphisms of X and u_t of real-valued functions on X such that $\varphi_0 = \mathrm{id}$ and $u_0 \equiv 0$ and satisfying the relation

$$\varphi_t^* g = e^{u_t} g_t \qquad ?$$

Differentiating the above equality with respect to t at $t = 0$, we obtain the equation

$$(4) \qquad \mathcal{L}_\xi g - ug = h \quad ,$$

where ξ is the vector field $d\varphi_t/dt_{|t=0}$, and u is the function $du_t/dt_{|t=0}$ on X, and where h is the infinitesimal deformation, the symmetric 2-form $dg_t/dt_{|t=0}$. We may always assume that the trace of h vanishes, and in this case (4) is equivalent to the inhomogeneous equation

$$(5) \qquad D_0^c \xi = h \quad ;$$

we now wish to express the compatibility conditions on h for the solvability of the equation (5).

If h is a metric on X, we denote by $\mathcal{W}(h)$ the Weyl tensor of h ; the linearization of the non-linear differential operator $\mathcal{W}$ along g

$$P_1 = \mathcal{W}'_g : S^2\mathcal{T}^* \longrightarrow \mathcal{E}_1$$

is the second-order linear differential operator defined by

$$\mathcal{W}'_g(h) = \frac{d}{dt}\mathcal{W}(g+th)_{|t=0} \quad ,$$

for $h \in S^2\mathcal{T}^*$. We denote by $S_0^2T^*$ the sub-bundle of S^2T^* consisting of those symmetric 2-forms with zero trace. If $n \geq 4$ and $W = 0$, the following theorem shows that the compatibility condition on h for the local solvability of (5) is given by $\mathcal{W}'_g$.

<u>Theorem</u> 1([2]). If $n \geq 4$ and $W = 0$, the sequence

$$0 \longrightarrow \Theta_c \longrightarrow \mathcal{T} \xrightarrow{D_0^c} S_0^2\mathcal{T}^* \xrightarrow{\mathcal{W}'_g} \mathcal{E}_1$$

is exact.

In fact, when $W = 0$, the operator $\mathcal{W}'_g$ takes its values in E_1^0 . From Theorem 1, we obtain a canonical isomorphism

$$H^1(X,\Theta_c) \simeq \frac{\mathrm{Ker}\{\mathcal{W}'_g : C^\infty(S_0^2T^*) \longrightarrow C^\infty(E_1^0)\}}{D_0^c C^\infty(T)} \quad ,$$

and the interpretation of $H^1(X,\Theta_c)$ as the space of infinitesimal deformations of the conformal structure of (X,g) . If (X,g) is conformally flat, we can express the higher cohomology groups in a similar way; in [2], we extend the sequence of Theorem 1 and construct a resolution of Θ_c .

We also denote by ∇ the connection on Λ^kT^* induced by the Levi-Civita connection of g and consider the first-order differential operator

(6) $$\nabla^\wedge : \Lambda^j\mathcal{T}^* \otimes \Lambda^k\mathcal{T}^* \longrightarrow \Lambda^{j+1}\mathcal{T}^* \otimes \Lambda^k\mathcal{T}^*$$

determined by

$$\nabla^{\wedge} (\omega \otimes \alpha) = d\omega \otimes \alpha + (-1)^j \omega \wedge \nabla \alpha \quad ,$$

for $\omega \in \Lambda^j \mathcal{R}^*$, $\alpha \in \Lambda^k \mathcal{R}^*$. For $k = 2$, one verifies easily that $\nabla^{\wedge} (\mathcal{E}_{j-1}) \subset \mathcal{E}_j$. We set

$$P_j = \rho_j \nabla^{\wedge} : \mathcal{E}^0_{j-1} \longrightarrow \mathcal{E}^0_j \ ,$$

for $2 \leq j \leq n-3$. We also consider the first-order differential operator

$$P_{n-1} = \nabla^{\wedge} \cdot \mathrm{Tr}^{n-1} : \Lambda^n \mathcal{R}^* \otimes S^2_0 \mathcal{R}^* \longrightarrow \Lambda^n \mathcal{R}^* \otimes \mathcal{R}^* \ ,$$

where $\nabla^{\wedge}$ is the operator (6) with $j = n-1$ and $k = 1$.

For the remainder of this paper, we assume that (X,g) is conformally flat, with $n \geq 4$. We set

$$F_0 = T \ , \qquad F_1 = S^2_0 T^* \ ,$$

$$F_j = E^0_{j-1} \ , \qquad \text{for} \quad 2 \leq j \leq n-2 \ ,$$

$$F_{n-1} = \Lambda^n T^* \otimes S^2_0 T^* \ , \qquad F_n = \Lambda^n T^* \otimes T^* \ ,$$

and $P_0 = \frac{1}{2} D^c_0$; in [2], we define a second-order linear differential operator

$$P_{n-2} : \mathcal{F}_{n-2} \longrightarrow \mathcal{F}_{n-1}$$

in terms of ∇ , g and the Ricci curvature of (X,g) , and prove the following result:

<u>Theorem</u> 2. If $n \geq 4$ and $W = 0$, the sequence

$$(7) \quad 0 \longrightarrow \Theta_c \longrightarrow \mathcal{F}_0 \xrightarrow{P_0} \mathcal{F}_1 \xrightarrow{P_1} \mathcal{F}_2 \longrightarrow \cdots \longrightarrow \mathcal{F}_n \longrightarrow 0$$

is an exact elliptic complex.

Now assume that X is oriented and let Ω be the volume form of (X,g). Let $g^\flat : T \longrightarrow T^*$ and $g^\# : T^* \longrightarrow T$ be the isomorphisms determined by g. We define isomorphisms $* : F_j \longrightarrow F_{n-j}$, for $j = 0,1,n-1,n$, by

$$*\xi = \Omega \otimes g^\flat(\xi) \;, \qquad *(\Omega \otimes \alpha) = g^\#(\alpha) \;,$$

$$*h = \Omega \otimes h \;, \qquad *(\Omega \otimes h) = (-1)^{n-1}h \;,$$

for $\xi \in T$, $\alpha \in T^*$ and $h \in S_0^2T^*$. Consider the isomorphism

$$* : \Lambda^{j+1}T^* \otimes \Lambda^2T^* \longrightarrow \Lambda^{n-j-1}T^* \otimes \Lambda^2T^*$$

sending $\alpha \otimes \beta$ onto $(*\alpha) \otimes \beta$; by restriction, this mapping induces an isomorphism

$$* : E_j^0 \longrightarrow E_{n-j-2}^0 \;,$$

for $1 \leq j \leq n-3$. Thus we have isomorphisms $* : F_j \longrightarrow F_{n-j}$, for $0 \leq j \leq n$, such that

$$**u = (-1)^{nj+j}u \;,$$

for $u \in F_j$ (see [3]). The vector bundles F_j are endowed with scalar products $(\;,\;)$ induced by g. For $0 \leq j \leq n$, there is a unique morphism

$$F_j \otimes F_{n-j} \longrightarrow \Lambda^nT^* \;,$$

sending $u \otimes v$ onto $\langle u,v \rangle$, such that

$$\langle u,*v \rangle = (u,v)\,\Omega \;,$$

for $u,v \in F_j$. Then

$$\langle u,v \rangle = (-1)^{nj+j} \langle v,u \rangle \;,$$

for $u \in F_j$, $v \in F_{n-j}$, and in [3] we verify the following:

<u>Proposition</u> 1. If X is oriented, for $0 \leq j \leq n-1$, there is a differential operator

$$\alpha_j : \mathcal{F}_j \oplus \mathcal{F}_{n-j} \longrightarrow \Lambda^{n-1}\mathcal{R}^*$$

of order k_j-1, where k_j is the order of P_j, such that

$$\langle P_j u, v\rangle + (-1)^j \langle u, P_{n-j-1} v\rangle = d\alpha_j(u \oplus v) ,$$

for all $u \in \mathcal{F}_j$, $v \in \mathcal{F}_{n-j-1}$. Moreover, the formal adjoint $P_j^* : \mathcal{F}_{j+1} \longrightarrow \mathcal{F}_j$ of P_j is equal to $(-1)^{nj+1} * P_{n-j-1} *$.

The last assertion of this proposition implies that the sequence (7) is formally self-adjoint. We denote by H^j the cohomology of the sequence

$$0 \longrightarrow C^\infty(F_0) \xrightarrow{P_0} C^\infty(F_1) \xrightarrow{P_1} C^\infty(F_2) \longrightarrow \cdots \longrightarrow C^\infty(F_n) \longrightarrow 0$$

at $C^\infty(F_j)$. Theorem 1 gives us a canonical isomorphism

$$H^j(X,\Theta_c) \simeq H^j .$$

If X is compact, according to Proposition 1 and Stokes' formula the mapping

$$H^j \otimes H^{n-j} \longrightarrow \mathbb{R}$$

sending $[u] \otimes [v]$ into

$$\int_X \langle u,v\rangle ,$$

where $[u]$, $[v]$ are the classes of $u \in C^\infty(F_j)$ and $v \in C^\infty(F_{n-j})$, satisfying $P_j u = 0$, $P_{n-j} v = 0$, in H^j and H^{n-j} respectively, is well-defined. Thus we obtain a linear mapping

$$\text{(8)} \qquad H^j(X,\Theta_c) \otimes H^{n-j}(X,\Theta_c) \longrightarrow \mathbb{R}$$

that depends only on the conformal class of g . Using Hodge theory, in [3] we deduce the following duality result:

Theorem 3. If (X,g) is a conformally flat, compact oriented Riemannian manifold of dimension $n \geq 4$, the pairing (8) is non-degenerate.

Remark. If $n = 3$ and (X,g) is conformally flat, a resolution of Θ_c similar to (7) is constructed in [2], and in [3] we prove the analogue of Theorem 3 in this case.

References

[1] J. Gasqui, H. Goldschmidt, Théorēmes de dualité en géométrie conforme, C. R. Acad. Sci. Paris, vol. 294, Série I, (1982), 99-102; 201-203.

[2] J. Gasqui, H. Goldschmidt, Déformations infinitésimales des structures conformes plates. I (to appear).

[3] J. Gasqui, H. Goldschmidt, Déformations infinitésimales des structures conformes plates. II (to appear).

COVARIANT DIFFERENTIAL OPERATORS

Michael Harris*
Brandeis University, Waltham, Mass. 02254, U.S.A

and

Hans Plesner Jakobsen
Mathematics Institute, Universitetsparken 5
2100 Copenhagen Ø, Denmark

Introduction

In this article, results on covariant differential operators[1)] are generalized and then used to discuss in detail the situations in which a constant coefficient holomorphic differential operator D, on a Hermitian symmetric space G/K, satisfies a covariance property $DU_1 = U_2D$ for two holomorphically induced (highest weight) representations, U_1 and U_2, of G. The results generalize and give new proofs of previously obtained results on covariant differential operators by a number of authors. See Ref. 1 for further details. We stress that the method presented here is completely intrinsic. In fact, the operator D is determined by the K-types that are missing from the representation U_1.

To illustrate the generality of our approach, and at the same time cover several series of interesting groups (e.g. $Sp(n,\mathbb{R})$, $SO_0(2,n)$, and $SU(n,n)$; $n = 1,\ldots$) we have chosen to formulate the results for an arbitrary Hermitian symmetric space (of the non-compact type). Based on results in Ref. 2, the results are then specialized to $G = SU(n,n)$, and detailed results for the cases in which U_1 is unitary are furnished. (See also Ref. 3.) Note in this context that a highest weight representation corresponds to a positive

*Partially supported by NSF grant MCS77 - 04951

energy representation. Our main motivation for studying $G = SU(n,n)$ in detail is, naturally, that for $n = 2$, G is the conformal group. See Ref.'s 4 and 5, and references cited therein for additional background. Thus, among our operators D are the wave operator, the Dirac operator, and a covariant operator corresponding to Maxwell's equations. In fact, for the conformal group we describe completely the set of holomorphic differential operators D that can satisfy a covariance property as above, with U_1 unitary. Observe that since the extended Poincaré group is a subgroup of $SU(2,2)$ we obtain analogous results for this group. We finally remark that due to the holomorphic nature of the spaces and representations involved, it is possible to formulate the results either on the generalized upper half plane, whose Shilov boundary is Minkowski space, or on the generalized unit disk, whose Shilov boundary is a local version of the Segal cosmos. From the point of view of the present work, the latter is the natural.

We wish to thank I.E. Segal, B. Speh, M. Vergne, and B. Ørsted for discussions during, in particular, the initial phases of our study.

1. Notation and the Basic Objects.

1.1. Let $\mathfrak{g}$ be a simple Lie algebra over $\mathbb{R}$ and $\mathfrak{g} = \mathfrak{k} + \mathfrak{p}$ a Cartan decomposition of $\mathfrak{g}$. Throughout we assume that $\mathfrak{k}$ has a non-empty center η; in this case $\eta = \mathbb{R}\cdot h_0$ for an $h_0 \in \eta$ whose eigenvalues under the adjoint action on $\mathfrak{p}^{\mathbb{C}}$ are $\pm i$. Let

$$\mathfrak{p}^{\pm} = \{z \in \mathfrak{p}^{\mathbb{C}} \mid [h_0, z] = \pm iz\}.$$

$\mathfrak{p}^{+}$ and $\mathfrak{p}^{-}$ are abelian subalgebras.

1.2. $G^{\mathbb{C}}$ denotes the simply connected Lie group with Lie algebra $\mathfrak{g}^{\mathbb{C}}$, where $\mathfrak{g}^{\mathbb{C}}$ is the complexification of $\mathfrak{g}$. $K^{\mathbb{C}}$, G, and K are the connected subgroups of $G^{\mathbb{C}}$ with Lie algebras $\mathfrak{k}^{\mathbb{C}}$, $\mathfrak{g}$, and $\mathfrak{k}$, respectively.

1.3. Under the above assumptions G/K is a Hermitian symmetric space (of the non-compact type).

1.4 In an open submanifold of $G^{\mathbb{C}}$, containing G, one has a unique factorization

$$g = (\exp z^+(g))(k(g))(\exp z^-(g)) ,$$

where $z^{\pm}(g) \in p^{\pm}$, and $k(g) \in K^{\mathbb{C}}$. The mapping $g \to z^+(g)$ identifies G/K with a bounded domain $\mathcal{D}$ of p^+; the so-called Harish-Chandra realization of G/K [6)]

1.5 When we specialize to $SU(n,n)$, the letters $a,b,c,d,z,w,$ etc. will denote elements of $M(n,\mathbb{C})$; the set of $n\times n$ complex matrices. It is convenient to consider two realizations of $SU(n,n)$;

$$SU(n,n)_H = \left\{g = \begin{pmatrix} a & b \\ c & d \end{pmatrix} \in S\ell(2n,\mathbb{C}) \mid g^*\begin{pmatrix} 0 & 1 \\ -1 & 0 \end{pmatrix}g = \begin{pmatrix} 0 & 1 \\ -1 & 0 \end{pmatrix}\right\}$$

and

$$SU(n,n)_B = \left\{g = \begin{pmatrix} a & b \\ c & d \end{pmatrix} \in S\ell(2n,\mathbb{C}) \mid g^*\begin{pmatrix} 1 & 0 \\ 0 & -1 \end{pmatrix}g = \begin{pmatrix} 1 & 0 \\ 0 & -1 \end{pmatrix}\right\} .$$

1.6 $SU(n,n)_H^{\mathbb{C}} = SU(n,n)_B^{\mathbb{C}}$. Up to at most multiplication by a complex number of modulus one, $c = \frac{1}{\sqrt{2}}\begin{pmatrix} i & 1 \\ -i & 1 \end{pmatrix} \in SU(n,n)_B^{\mathbb{C}}$, and the map $g \to c^{-1}gc$ is an isomorphism of $SU(n,n)_B$ onto $SU(n,n)_H$.

1.7 For $G = SU(n,n)_B$ the natural choice of K is

$$K = \left\{\begin{pmatrix} u & 0 \\ 0 & v \end{pmatrix} \mid u,v \in U(n) , \det(u\cdot v) = 1\right\} .$$

We choose $h_0 = \begin{pmatrix} \frac{i}{2} & 0 \\ 0 & -\frac{i}{2} \end{pmatrix}$; then $p^+ = \left\{\begin{pmatrix} 0 & z \\ 0 & 0 \end{pmatrix} \mid z \in M(n,\mathbb{C})\right\}$,

and $p^- = \left\{\begin{pmatrix} 0 & 0 \\ w & 0 \end{pmatrix} \mid w \in M(n,\mathbb{C})\right\}$.

When $n = 2$, h_0 is the generator of the Segal time-translation subgroup of K. The domain $\mathcal{D} \subset p^+$ is denoted by B and is equal to the generalized unit disk;

$$B = \{z \in M(n,\mathbb{C}) \mid z^*z < 1\} .$$

1.8 If $x \in g$ we denote by $r(x)$ and $\ell(x)$ the differential operators which act on C^∞ functions on G by

$$(r(x)f)(g) = \frac{d}{dt}\Big|_{t=0} f(g \exp tx) ,$$

and

$$(\ell(x)f)(g) = \frac{d}{dt}\Big|_{t=0} f(\exp(-tx)g) ,$$

respectively. The maps r and ℓ are homomorphisms of $\mathfrak{g}$ and thus have unique extensions to complex homomorphisms of $U(\mathfrak{g}^{\mathbb{C}})$; the universal enveloping algebra of $\mathfrak{g}^{\mathbb{C}}$. We maintain the notation r and ℓ for these extensions.

1.9 Since $\mathfrak{p}^-$ is K-invariant we may define the set $\mathcal{O}(\mathcal{D})$ of holomorphic functions on $\mathcal{D}$ by

$$\mathcal{O}(\mathcal{D}) = \{\text{analytic functions } \varphi \text{ on } G/K \mid r(x)\varphi = 0 \text{ for all } x \in \mathfrak{p}^-\} .$$

Likewise, the set $A(\mathcal{D})$ of anti-holomorphic functions on $\mathcal{D}$ may be identified with

$$A(\mathcal{D}) = \{\text{analytic functions } \varphi \text{ on } G/K \mid r(x)\varphi = 0 \text{ for all } x \in \mathfrak{p}^+\} .$$

1.10 Let τ denote a unitary finite-dimensional representation of K on a vector space V_τ. Define

$$\mathcal{O}(G,K,V_\tau) = \{\text{analytic functions } \varphi\colon G \to V_\tau \mid \forall k \in K : \varphi(g\cdot k) = \tau(k)^{-1}\varphi(g) , \text{ and } \forall x \in \mathfrak{p}^- : r(x)\varphi = 0\} .$$

We extend τ to a holomorphic representation of $K^{\mathbb{C}}$ and define an isomorphism P of $\mathcal{O}(G,K,V_\tau)$ onto $\mathcal{O}(\mathcal{D},V_\tau)$, the set of holomorphic functions on $\mathcal{D}$ with values in V_τ , by $(P\varphi)(g) = \tau(k(g))\varphi(g)$. ($k(g)$ as in (1.4).)

1.11 The action U of G on $C^\infty(G)$ from the left;

$$(U(g)\varphi)(g_0) = \varphi(g^{-1}g_0) ,$$

preserves $\mathcal{O}(G,K,V_\tau)$. Through the isomorphism P we thus get a representation U of G on $\mathcal{O}(\mathcal{D},V_\tau)$. This representation has the form

$$(U_\tau(g)f)(z) = J_\tau(g^{-1},z)^{-1}f(g^{-1}z) \ .$$

Observe that the origin 0 in p^+ belongs to $\mathcal{D}$, and corresponds to $\underline{e}$ in G/K . If $z = g_0 \cdot 0$,

$$J_\tau(g,z) = \tau(k(gg_0))\tau(k(g_0))^{-1} \ ,$$

and we have

$$J_\tau(g_1g_2,z) = J_\tau(g_1,q_2\cdot z)J_\tau(g_2,z) \ ; \ g_1,g_2 \in G \ , \ z \in \mathcal{D} \ ,$$

and $J_\tau(k,0) = \tau(k)$; $k \in K$. In particular, for $k \in K$, $(U_\tau(k)f)(z) = \tau(k)f(k^{-1}z)$, and $k^{-1}z$ is the adjoint action of k^{-1} on $z \in \mathcal{D} \subset p^+$. Also observe that for $f \in \mathcal{O}(\mathcal{D},V_\tau)$, the function φ defined by

$$\varphi(g) = J_\tau(g,0)^{-1}f(g\cdot 0)$$

belongs to $\mathcal{O}(G,K,V_\tau)$, and $p\varphi = f$.

1.12 If τ is irreducible,

$$(U_\tau(\exp(\theta\cdot h_0))f)(z) = e^{im_\tau\cdot\theta} f(e^{-i\theta}\cdot z) \ ,$$

where $m_\tau \in \mathbb{Z}$ is determined by τ . Let $\mathcal{O}^K(\mathcal{D},V_\tau)$ denote the space of functions in $\mathcal{O}(\mathcal{D},V_\tau)$ that transform under $U_\tau|_K$ according to irreducible unitary representations. Since $0 \in \mathcal{D}$, it follows by power series expansion around that point that $\mathcal{O}^K(\mathcal{D},V_\tau)$ consists of polymomials.

1.13 The infinitesimal action dU_τ of U_τ on $\mathcal{O}(\mathcal{D},V_\tau)$ is described in Ref. 7, and in the appendix.

1.14 Consider U(n) ; the group of unitary n×n matrices, along with its diagonal subgroup T ;

$$T = \left\{\begin{pmatrix} d_1 & & 0 \\ & \ddots & \\ 0 & & d_n \end{pmatrix} \mid d_i \in \mathbb{C} \ , \text{ and } |d_i| = 1 \ ; \ i = 1,\dots,n\right\}$$

The unitary irreducible representations $\hat{U(n)}$ of $U(n)$ are parametrized by n-tuples $(n_1,\ldots,n_n)$ of integers with $n_1 \geq n_2 \geq \ldots \geq n_n$. To a given such tuple there is a finite-dimensional vector space V_τ, an irreducible unitary representation τ of $U(n)$ on V_τ, and a unique (up to multiplication by a non-zero complex number) vector $e_0 \in V_\tau$ such that

$$\tau\left(\begin{pmatrix} d_1 & & 0 \\ & \ddots & \\ 0 & & d_n \end{pmatrix}\right)e_0 = d_1^{n_1} \ldots d_n^{n_n} \cdot e_0 ,$$

and such that for any other non-zero vector $e \in V_\tau$ for which

$$\tau\left(\begin{pmatrix} d_1 & & 0 \\ & \ddots & \\ 0 & & d_n \end{pmatrix}\right) e = d_1^{m_1} \ldots d_n^{m_n} \cdot e ,$$

$n_i > m_i$ at the first place where $(m_1,\ldots,m_n)$ differs from $(n_1,\ldots,n_n)$. Conversely, any unitary irreducible representation τ of $U(n)$ is of the form $\tau = \tau(n_1,\ldots,n_n)$ for a unique such tuple. We let

$$\hat{U(n)}_h = \left\{\tau(n_1,\ldots,n_n) \mid n_1 \geq \ldots \geq n_n \geq 0\right\}$$

and observe that any $\tau \in \hat{U(n)}$ can be written as a tensor product of an element from $\hat{U(n)}_h$ with a representation $u \to (\det u)^k$ for some $k \in \mathbb{Z}$.

The Young diagram corresponding to an element of $\hat{U(n)}_h$ consists of n rows, where the i'th row contains n_i boxes. In this realization one anti-symmetrizes vertically, e.g. $u \to \det u$ is represented by a single column (and corresponds to $(1,\ldots,1)$).

1.15 The unitary irreducible representations of the group K in (1.7) are of the form

$$\tau : (u,v) \to (\det v)^k \tau_1(u) \otimes \tau_2(v) ,$$

where $\tau_2 \in \hat{U(n)}_h$, and τ_1 is the contragredient (dual) of an element of $\hat{U(n)}_h$ (cf. (2.1)). We let

$$\delta_k(u,v) = (\det v)^k .$$

1.16 The representation of $SU(n,n)_B$ resulting, along the lines of (1.10), from a such τ ("holomorphically induced") has the form

$$(U_\tau(g)f)(z) = \det(cz+d)^{-k}\tau_1(a-(g^{-1}z)c)^{-1} \otimes \tau_2(cz+d)^{-1}f(g^{-1}z),$$

where $z \in B$, $g^{-1} = \begin{pmatrix} a & b \\ c & d \end{pmatrix} \in SU(n,n)_B$, $g^{-1}z = \frac{az+b}{cz+d}$,

and f is a holomorphic function on B with values in V_τ .

1.17 Observe that we tacitly avoid the simpler expression (a^*+zb^*) for the term $(a-(g^{-1}z)c)^{-1}$ in (1.16).

1.18 Let

$$H = \left\{ z \in M(n,\mathbb{C}) \mid \frac{z-z^*}{2i} > 0 \right\} .$$

The Cayley transform C_0 ,

$$c_0(z) = \frac{1+iz}{1-iz}$$

is a biholomorphic map from H onto B . More generally, with c as in (1.6),

$$g \to U_\tau(c)^{-1}U_\tau(g)U_\tau(c) = U_\tau(c^{-1}gc)$$

is a representation of $SU(n,n)_B$ on the space of holomorphic V_τ-valued functions on H . Here,

$$(U_\tau(c)f)(z) =$$
$$\det\left(\frac{1}{\sqrt{2}} (z+1)\right)^{-k} \tau_1\left(\frac{i}{\sqrt{2}} (z+1)\right) \otimes \tau_2\left(\frac{1}{\sqrt{2}} (z+1)\right)^{-1} f\left(\frac{z-1}{i(z+1)}\right) .$$

Phrased differently, the representation of $SU(n,n)_H$ on the space of V_τ-valued holomorphic functions on H also given by (1.16), but where now $g^{-1} = \begin{pmatrix} a & b \\ c & d \end{pmatrix} \in SU(n,n)_H$, and $z \in H$, is equivalent to that of (1.16). It is a key point that this equivalence is given by an element of the complexified group. Also observe that in the $SU(n,n)_H$-version,

$(a-(g^{-1}z)c)^{-1} = zc^* + d^*$.

1.19 When $n = 2$, the Shilov boundary of H is Minkowski space, whereas the Shilov boundary of B is $U(2)$, and $\widetilde{U(2)} = \mathbb{R} \times S^3$ is the Segal cosmos.

2. Differential Operators.

2.1 In the following the symbol $(.,.)$ will always denote the bilinear pairing between a vector space and its dual, and τ' always denotes the contragredient representation to τ . We recall that $(\tau'(k)v',v) = (v',\tau(k^{-1})v)$ for all $v \in V_\tau$, $v' \in V_{\tau'} = V_\tau'$ and $k \in K$. Clearly , $\tau'' = \tau$.

2.2 K acts on p^+ through the restriction of the adjoint representation. We extend this action to $S(p^+)$; the symmetric algebra over p^+ , in the obvious way. Since p^+ is commutative , $\mathcal{U}(p^+) = S(p^+)$. Let τ be a unitary finite dimensional representation of K on V_τ and consider the natural representation of K on $S(p^+) \otimes V_{\tau'}$. In particular, let A be a finite-dimensional K-invariant subspace, and denote the corresponding representation by τ_A' . A differential operator $D_A : C^\infty(G,V_\tau) \to C^\infty(G,V_{\tau_A})$ is defined componentwise as follows: For $\Phi \in C^\infty(G,V_\tau)$, and $\sum_{i=1}^N q_i \otimes v_i' \in A(q_i \in S(p^+)$, and $v_i' \in V_\tau'$; $i = 1,\ldots,N)$,

$$(D_A\Phi , \sum_{i=1}^N q_i \otimes v_i')(g) = \sum_{i=1}^N (r(q_i)(\Phi,v_i'))(g) .$$

2.3 The following is immediate.

2.3 Lemma $(D_A\Phi)(g\cdot k) = \tau_A(k^{-1})(D_A\Phi)(g)$.

2.4 Corresponding to D_A we obtain a differential operator $\delta_A : C^\infty(\mathcal{D},V_\tau) \to C^\infty(\mathcal{D},V_{\tau_A})$;

$$(\delta_A f)(z) = J_{\tau_A}(g,o)(D_A(J_\tau(\cdot,o)^{-1}f((\cdot)o)))(g) ,$$

where $g\cdot o = z$. Since D_A and δ_A are obtained from the right action, whereas U_τ comes from the left, we get

2.4 Proposition $\delta_A U_\tau = U_{\tau_A} \delta_A$.

Here , U_τ and U_{τ_A} are representations of G on $C^\infty(\mathcal{D},V_\tau)$ and $C^\infty(\mathcal{D},V_{\tau_A})$, respectively, obtained in analogy with (1.11) by omitting the holomorphy requirement. Naturally, these representations preserve the respective subspaces of holomorphic functions, but the crucial point is, that, in general , δ_A does not map $\mathcal{O}(\mathcal{D},V_\tau)$ into $\mathcal{O}(\mathcal{D},V_{\tau_A})$. The main topic of this article is to investigate the situations in which δ_A does preserve holomorphy.

2.5 The $r(q_i)$'s in (2.2) are sums of terms $r(p_1)r(p_2)\ldots r(p_s)$; $p_1,p_2,\ldots,p_s \in p^+$. It thus follows from (1.9) that δ_A annihilates the anti-holomorphic functions. Hence,

2.5 Proposition. δ_A is a polynomial in the holomorphic tangent vectors. The coefficients are C^∞ functions from $\mathcal{D}$ to $\mathrm{Hom}(V_\tau,V_{\tau_A})$.

2.6 2.6 Corollary. If δ_A maps holomorphic functions to holomorphic functions, it is a holomorphic differential operator with constant coefficients.

3. Modules.

3.1 With dU_τ as in (1.13), we introduce the space

$$W_\tau = \{dU_\tau(u)\cdot v \mid u \in \mathcal{U}(g^{\mathbb{C}}) \; , \; v \in V_\tau\} \; .$$

W_τ is a space of polynomials on $\mathcal{D}$ with values in V_τ , and is clearly invariant under dU_τ . Moreover, any dU_τ-invariant subspace of $\mathcal{O}^K(\mathcal{D},V_\tau)$ (1.12) contains W_τ , as follows by considering the action of $dU_\tau(x)$ for $x \in p^+$.

(W_τ is the irreducible quotient of $\mathcal{U}(g^{\mathbb{C}}) \otimes_{\mathcal{U}(k^{\mathbb{C}} \oplus p^+)} V_\tau$.)

3.2 Consider the module $E_{\tau'} = \mathcal{U}(\mathfrak{g}^{\mathbb{C}}) \otimes_{\mathcal{U}(\mathfrak{k}^{\mathbb{C}} \oplus \mathfrak{p}^-)} V_{\tau'}$.
Since $\mathcal{U}(\mathfrak{g}^{\mathbb{C}}) = \mathcal{U}(\mathfrak{p}^+)\mathcal{U}(\mathfrak{k}^{\mathbb{C}})\mathcal{U}(\mathfrak{p}^-)$, $E_{\tau'}$, as a $\mathcal{U}(\mathfrak{k}^{\mathbb{C}})$-module, is equal to $S(\mathfrak{p}^+) \otimes V_{\tau'}$. The action of $\mathcal{U}(\mathfrak{g}^{\mathbb{C}})$ on $E_{\tau'}$ is the natural; from the left.

3.3 $B(\cdot,\cdot)$ denotes the complex bilinear form on $\mathfrak{g}^{\mathbb{C}}$ obtained from the Killing form on $\mathfrak{g}$. Through B we get a canonical identification of $S(\mathfrak{p}^+) \otimes V_{\tau'}$ with $P(\mathfrak{p}^-) \otimes V_{\tau'}$; the space of $V_{\tau'}$-valued polynomials on $\mathfrak{p}^-$: For $z_i \in \mathfrak{p}^+$, $i = 1,\ldots,r$, and $v' \in V_{\tau'}$, $q \in P(\mathfrak{p}^-) \otimes V_{\tau'}$ is defined by

$$q(w) = B(z_1,w)\cdot\ldots\cdot B(z_r,w)\cdot v' .$$

The full description (as in several situations in the sequel) follows by linearity. By (3.2) this sets up a linear isomorphism between $E_{\tau'}$ and $P(\mathfrak{p}^-) \otimes V_{\tau'}$, and in this way we get an action $dL_{\tau'}$ of $\mathcal{U}(\mathfrak{g}^{\mathbb{C}})$ on the latter space corresponding to the left action on the former.

3.4 Analogously, $S(\mathfrak{p}^-) \otimes V_\tau$ may be identified with $P(\mathfrak{p}^+) \otimes V_\tau$ by

$$p(z) = B(z,w_1)\cdot\ldots\cdot B(z,w_s)\cdot v .$$

for $w_i \in \mathfrak{p}^-$, $i = 1,\ldots,s$,and $v \in V_\tau$. Recall that $P(\mathfrak{p}^+) \otimes V_\tau = \mathcal{O}^K(\mathcal{D},V_\tau)$ is the space of K-finite vectors for the representation U_τ of G on $\mathcal{O}(\mathcal{D},V_\tau)$. In what follows, dU_τ denotes the action (1.13) of $\mathcal{U}(\mathfrak{g}^{\mathbb{C}})$ on $P(\mathfrak{p}^+) \otimes V_\tau$.

3.5 Define a pairing of $P(\mathfrak{p}^-) \otimes V_{\tau'}$ with $P(\mathfrak{p}^+) \otimes V_\tau$ by

$$(q,p) = \Big(\sum_{\sigma\in S_n} \prod_i B(z_i,w_{\sigma(i)})\Big)(v',v)\,\delta_{r,s}$$

for q as in (3.3), and p as in (3.4). This pairing clearly places the two spaces in duality.

3.6 3.6. Proposition. For q as in (3.3), and p as in (3.4),

$$(q,p) = (v', (\delta(z_1) \ldots \delta(z_r)p)(0)) .$$

Proof. This follows from the definition of δ (1.13). □

Thus, the space $\mathcal{P}(p^-) \otimes V_{\tau'}$ coincides with the space of $V_{\tau'}$-valued holomorphic constant coefficient differential operators $q(\frac{\partial}{\partial z})$ on $\mathcal{D}$, and the pairing in (3.5) is given by $(q,p) = \left(q(\frac{\partial}{\partial z}) , p(\cdot)\right)(0)$.

3.7 3.7 Proposition. $dL'_{\tau'} = dU_\tau$.

4. Covariance and Missing K-Types.

4.1 Let τ_1 and τ_2 be finite dimensional unitary representations of K, and let $D : \mathcal{O}(\mathcal{D},V_{\tau_1}) \to \mathcal{O}(\mathcal{D},V_{\tau_2})$ be a constant coefficient holomorphic differential operator. If

$$U_{\tau_2} D = D U_{\tau_1} ,$$

U_{τ_1} leaves invariant the space $\{\varphi \in \mathcal{O}(\mathcal{D},V_{\tau_1}) \mid D\varphi = 0\}$. It follows that W_{τ_1} is strictly smaller than $\mathcal{O}^K(\mathcal{D},V_{\tau_1})$, i.e. some K-types are missing from W_{τ_1} .

4.2 Consider an arbitrary irreducible unitary representation τ of K, and define

$W_\tau^o = \{q \in \mathcal{P}(p^-) \otimes V_{\tau'} \mid (q,p) = 0 \text{ for all } p \in W_\tau\}$. Suppose that $W_\tau \neq \mathcal{P}(p^+) \otimes V_\tau = S(p^-) \otimes V_\tau$ or, equivalently, that $W_\tau^o \neq 0$. Since K preserves the degree of homogeneity, the degree of a K-irreducible subspace of W_τ^o is well-defined. Let A be a K-irreducible subspace of lowest degree in W_τ^o . Since W_τ^o clearly does not contain any constants, this degree is greater than or equal to 1 .

4.2 Proposition. Under the above hypothesis, δ_A is a holomorphic constant coefficient differential operator.

5. SU(n,n) .

5.1 We write the elements of B (or H) as n×n matrices $z = [z_{ij}]$, and we write $\frac{\partial}{\partial z}$ for the $n\times n$ matrix whose (i,j)th entry is $\frac{\partial}{\partial z_{ij}}$. If a subspace V of $\mathcal{O}^K(B,V_\tau)$ transforms according to a representation $\tau_1 \in \hat{K}$, the space $V' = \{q(\frac{\partial}{\partial z}) \mid q(z) = \bar{p}(\bar{z})$, and $p \in V\}$ transforms according to $\tau_1{}'$. Here, the bar in $\bar{p}$ refers to a conjugation in V .

5.2 A subspace V_{τ_1} of $\mathcal{O}^K(B,V_\tau)$ transforms according to $\tau_1 \in \hat{K}$ if and only if there exists a nontrivial linear map $T : V_{\tau_1} \to \mathcal{O}^K(B,V_\tau)$ such that

$$\forall z \in B \text{ , } \forall (u,v) \in K \text{ , and } \forall x \in V_{\tau_1} :$$
$$(T\tau_1(u,v)x)(z) = \tau(u,v)(Tx)(u^{-1}zv) \text{ .}$$

It follows that there must exist an $x \in V_{\tau_1}$ such that $(Tx)(1) \neq 0$, where 1 denotes the unit $n\times n$ matrix. Extend τ_1 and τ to $U(n)\times U(n)$ in such a manner that the intertwining relation remains valid. With these extensions,

$$(Tx)(z) = \tau(1,z)^{-1}(T\tau_1(1,z)x)(1) \text{ ,}$$

since a polynomial on B is completely determined by its values on $U(n)$. Observe that $Mx = (Tx)(1)$ is a nontrivial linear map from V_{τ_1} to V_τ , and

$$\forall u \in U(n) : M\tau_1(u,u) = \tau(u,u)M \text{ .}$$

5.3 In the notation of (1.14), (1.15), and (1.16), U_τ is unitary if and only if

$$\tau = \tau_1(0,0,\ldots,-m_1,\ldots,-m_j) \otimes \tau_2(n_1,\ldots,n_i,0,\ldots,0) \otimes \delta_k \text{ ,}$$

and $i+j \leq k$ (Ref. 2, Proposition 4.5.) . $W_\tau{}^{\circ} \neq 0$ if and only if

$$k = i + j + r \text{ ; } r = 0,\ldots,\min(k-i-1,k-j-1) \text{ ,}$$

and in this case, the missing K-type of the lowest degree $(r+1)$ is

$$\tau = \tau_1(0,0,\ldots,\underbrace{-1,\ldots,-1}_{r+1},-m_1,\ldots,-m_j) \otimes$$
$$\tau_2(n_1,\ldots,n_i,\underbrace{1,\ldots,1}_{r+1},0,\ldots,0) \otimes \delta_k ,$$

as follows easily from the results in Ref. 2 by tensoring with the representation U_τ in which τ_1 and τ_2 are the trivial representation, and $k = 1$. It should be mentioned that for the representations U_τ ; $\tau = \delta_k$, the result is due to Wallach [8]. See also Ref's 9 and 10. We finally remark that it may be shown that if U_{τ_1} is unitary, any holomorphic differential operator D ; $U_{\tau_2} D = D U_{\tau_1}$, factors through a covariant operator constructed from the massing K-type of lowest degree. For SU(2,2), this is obvious.

5.4 When $G = SU(2,2)$, the result may be summarized as follows: The covariance relations $U_{\tau_2} D = D U_{\tau_1}$, written for simplicity as $[\tau_2]D = D[\tau_1]$, in which U_{τ_1} is unitary, are of the form $(m,n \in \mathbb{N})$

$$[(0,-(m-1)) \otimes (1,0) \otimes \delta_2]D = D[(0,-m) \otimes (0,0) \otimes \delta_1] ,$$
$$[(0,-1) \otimes (n-1,0) \otimes \delta_2]D = D[(0,0) \otimes (n,0) \otimes \delta_1] ,$$
$$[(0,-(m-1)) \otimes (n-1,0) \otimes \delta_4]D = D[(0,-m) \otimes (n,0) \otimes \delta_2]$$

and $[\delta_3]D = D[\delta_1]$. Thus, the longest chains of differential operators have lenght 2, and originates at representations of the form $[(0,-m) \otimes (0,0) \otimes \delta_1]$; $m \geqq 1$, or $[(0,0) \otimes (n,0) \otimes \delta_1]$; $n \geqq 1$.

5.5 The operators D of the preceding paragraph can be explicitly described by means of (5.1) and (5.2).

Ex. 1. $$[\delta_3]D_1 = D_1[\delta_1] .$$

The missing K-type of the lowest degree is $\delta_3(u,v) = \det v^3$. The operator M of (5.2) can clearly be taken as the identity operator on $V = \mathbb{C}$, and the extensions of δ_3 and δ_1 to $U(2)\times U(2)$, $\tilde{\delta}_3$ and $\tilde{\delta}_1$, must satisfy

$$\tilde{\delta}_3(u,v) = \det u^{\beta} \det v^{\beta+3} ,$$

and

$$\tilde{\delta}_1(u,v) = \det u^{\beta+1} \det v^{\beta+2}$$

for some $\beta \in \mathbb{Z}$. It follows that the one-dimensional subspace V_{δ_3} of $\mathcal{O}^K(B,V_{\delta_1})$ is generated by the polynomial $z \to \det z = z_1z_4 - z_2z_3$. Thus,

$$D_1 = \frac{\partial}{\partial z_4}\frac{\partial}{\partial z_1} - \frac{\partial}{\partial z_2}\frac{\partial}{\partial z_3} .$$

Ex. 2. $[(0,-1)\otimes(0,0)\otimes\delta_2]D_2 = D_2[(0,0)\otimes(1,0)\otimes\delta_1]$.

It follows easily that the space V of polynomials corresponding to $[(0,-1)\otimes(0,0)\otimes\delta_2]$ is

$$V = \left\{\begin{pmatrix} z_4 & -z_2 \\ -z_3 & z_1\end{pmatrix}x \mid x \in \mathbb{C}^2\right\} .$$

Observe that $\tilde{z} = \begin{pmatrix} z_4 & -z_2 \\ -z_3 & z_1\end{pmatrix}$ is the co-factor of z , i.e. $z\cdot\tilde{z} = \det z$. According to (5.1), as a basis $\{e_1,e_2\}$ of the contragredient representation we may take $e_1 = \left(\frac{\partial}{\partial z_4}, -\frac{\partial}{\partial z_3}\right)$, and $e_2 = \left(\frac{\partial}{\partial z_2}, -\frac{\partial}{\partial z_1}\right)$. Thus,

$$D_2 = \begin{bmatrix} \frac{\partial}{\partial z_4} & -\frac{\partial}{\partial z_3} \\ -\frac{\partial}{\partial z_2} & \frac{\partial}{\partial z_1}\end{bmatrix}$$

Naturally, the coordinates on B ;

$$z = \begin{bmatrix} z_0 + z_1 & z_2 + iz_3 \\ z_2 - iz_3 & z_0 - z_1\end{bmatrix} ,$$

lead to the more conventional forms of D_1 and D_2 (scaled appropriately):

$$D_1 = \frac{\partial^2}{\partial z_0^2} - \frac{\partial^2}{\partial z_1^2} - \frac{\partial^2}{\partial z_2^2} - \frac{\partial^2}{\partial z_3^2} ,$$

and

$$D_2 = \begin{bmatrix} \frac{\partial}{\partial z_0} - \frac{\partial}{\partial z_1} & -\frac{\partial}{\partial z_2} - i\frac{\partial}{\partial z_3} \\ -\frac{\partial}{\partial z_2} + i\frac{\partial}{\partial z_3} & \frac{\partial}{\partial z_0} + \frac{\partial}{\partial z_1} \end{bmatrix} .$$

In this connection observe that the other half of the Dirac operator comes form the D_3 of

$$[(0,0) \otimes (1,0) \otimes \delta_2]D_3 = D_3[(0,-1) \otimes (0,0) \otimes \delta_1] .$$

5.6 The results for $G = SU(n,n)$ have so far been formulated in terms of holomorphic functions and holomorphic differential operators on the generalized unit disk, B . The representations involved have been those of (1.16). Specifically, $G = SU(n,n)_B$. However, the results may equally well be formulated on the generalized upper half plane, H : First observe that H and B are contained in $M(n,\mathbb{C})$, and $H \cap B$ is non-empty and open. Therefore it makes sense to compare constant coefficient holomorphic differential operators on B and H . We say that D_B is the same as D_H if $_BD_H = D_B$ on $B \cap H$. For a fixed $\tau \in \hat{K}$, we write $U_\tau{}^B$ for the representation of $G = SU(n,n)_B$ given by (1.16), and $U_\tau{}^H$ for the corresponding representation of $G = SU(n,n)_H$ given by (1.18). We can now formulate the transition to H .

5.6 Proposition. If a holomorphic differential operator D_B on B satisfies: $\forall g \in SU(n,n)_B : U_{\tau_2}{}^B(g)D_B = D_B U_{\tau_1}{}^B$, the same operator D_H on H satisfies: $\forall g \in SU(n,n)_H$: $U_{\tau_2}{}^H(g)D_H = D_H U_{\tau_1}{}^H(g)$.

REFERENCES

1. M. Harris and H.P. Jakobsen, Singular holomorphic representations and singular modular forms, Math. Ann. 259, 227-244 (1982).

2. H.P. Jakobsen, On singular holomorphic representations, Invent. Math. 62, 67-78 (1980).

3. H.P. Jakobsen, Hermitian symmetric spaces and their unitary highest weight modules, preprint (1981).

4. H.P. Jakobsen, B. Ørsted, I.E. Segal, B. Speh and M. Vergne, Symmetry and causality properties of physical fields, Proc. Natl. Acad. Sci. USA 75, 1609-1611 (1978).

5. I.E. Segal, H.P. Jakobsen, B. Ørsted, S.M. Paneitz and B. Speh, Covariant chronogeometry and extreme distances: Elementary particles, Proc. Natl. Acad. Sci. USA 78, 5261-5265 (1981).

6. S. Helgason, Differential Geometry and Symmetric Spaces, New York: Academic Press (1962).

7. H.P. Jakobsen and M. Vergne, Restrictions and expansions of holomorphic representations, J. Functional Analysis 34, 29-53 (1979).

8. N. Wallach, Analytic continuation of the discrete series II, Trans. Amer. Math. Soc. 251, 19-37 (1979).

9. H. Rossi and M. Vergne, Analytic continuation of the holomorphic discrete series of a semi-simple Lie group, Acta Math. 136, 1-59 (1976).

10. M. Kashiwara and M. Vergne, On the Segal-Shale-Weil representation and harmonic polynomials, Invent. Math. 44, 1-47 (1978).

Appendix

Proofs and further details.

A.1 The infinitesimal action dU_τ of U_τ on $\mathcal{O}(\mathcal{D},V_\tau)$ may be described as follows: Let

$$(\delta(z_0)f)(z) = \frac{d}{dt}\Big|_{t=0} f(z+tz_0) ,$$

for $z_0, z \in \mathfrak{p}^+$, and $f \in C^\infty(\mathfrak{p}^+)$. Then, for $f \in \mathcal{O}(\mathcal{D},V_\tau)$,

$$(dU_\tau(x)f)(z) = -(\delta(x)f)(z) \qquad \text{for } x \in \mathfrak{p}^+ ,$$

$$(dU_\tau(x)f)(z) = d\tau(x)f(z) - (\delta([x,z])f)(z) \qquad \text{for } x \in \mathfrak{k}^{\mathbb{C}} ,$$

and

$$(dU_\tau(x)f)(z) = d\tau([x,z])f(z) - \frac{1}{2}(\delta([[x,z],z])f)(z) \qquad \text{for } x \in \mathfrak{p}^- .$$

A.2 Proof of Corollary 2.6. By assumption, the coefficients of δ_A are holomorphic. It follows from (2.4) that for all $x \in \mathfrak{g}^{\mathbb{C}}$, $\delta_A dU_\tau(x) = dU_{\tau_A}(x)\delta_A$. Let $f \in \mathcal{O}(\mathcal{D},V_\tau)$, and $x \in \mathfrak{p}^+$. Since $\delta_A f$ is holomorphic, (A.1) implies that $\delta_A\delta(x)f = \delta(x)\delta_A f$. □

A.3 Proof of Proposition 3.7. We must show that $(dL_{\tau'}(x)q,p) = -(q,dU_\tau(x)p)$ for all $x \in \mathfrak{g}^{\mathbb{C}}$, $q \in \mathcal{P}(\mathfrak{p}^-) \otimes V_{\tau'}$, and $p \in \mathcal{P}(\mathfrak{p}^+) \otimes V_\tau$. For $x \in \mathfrak{k}^{\mathbb{C}}$, this is trivially true. For $x \in \mathfrak{p}^+$, the equality follows from Proposition 3.6 since $(dL_{\tau'}(x)q)(w) = B(x,w)q(w)$, and $dU_\tau(x)p = -\delta(x)p$. Finally, let $x \in \mathfrak{p}^-$ and take q as in (3.3). By (3.2) and (3.3), $dL_{\tau'}(x)q$ corresponds to the element $x\cdot z_1\cdot\ldots\cdot z_r\cdot v'$ of $\mathcal{U}(\mathfrak{g}^{\mathbb{C}}) \otimes_{\mathcal{U}(\mathfrak{k}^{\mathbb{C}}\oplus\mathfrak{p}^-)} V_{\tau'}$. Now,

$$x\cdot z_1\cdot\ldots\cdot z_r\cdot v' = [x,z_1\cdot\ldots\cdot z_r]\cdot v'$$

$$= \sum_{i=1}^{r} z_1\cdot\ldots\cdot z_{i-1}[x,z_i]z_{i+1}\cdot\ldots\cdot z_r\cdot v'$$

$$= \sum_{i=1}^{r} \sum_{j=i+1}^{r} [[x,z_i],z_j]z_1 \cdot \ldots \cdot \hat{z}_i \cdot \ldots \cdot \hat{z}_j \cdot \ldots \cdot z_r \cdot v'$$

$$+ \sum_{i=1}^{r} z_1 \cdot \ldots \cdot \hat{z}_i \cdot \ldots \cdot z_r \, d\tau'([x,z_i])v' \; .$$

Recall that p^+ is commutative. Thus,

$$(dL_{\tau'}(x)q)(w)$$

$$= \sum_{i=1}^{r} B(z_1,w) \cdot \ldots \cdot B(\hat{z_i},w) \cdot \ldots \cdot B(z_r,w)d\tau'([x,z_i]) \cdot v'$$

$$+ \sum_{i=1}^{r} \sum_{j=i+1}^{r} B([[x,z_i],z_j],w)B(z_1,w) \cdot \ldots \cdot B(\hat{z_i},w) \cdot \ldots$$

$$\cdot B(\hat{z_j},w) \cdot \ldots \cdot B(z_r,w) \cdot v' \; .$$

According to (A.1), to complete the proof we must show that

$$(dL_{\tau'}(x)q,p)$$

$$= (\delta(z_1) \cdot \ldots \cdot \delta(z_r)(v_i' - d\tau([x,\cdot])p(\cdot) + \tfrac{1}{2}\delta([[x,\cdot],\cdot])$$

$$p(\cdot)))(0) \; .$$

This is now immediate, since by (3.6) we are evaluating at 0 hence must differentiate the terms $-d\tau([x,\cdot])$ and $\frac{1}{2}\delta([[x,\cdot],\cdot])$ appropriately. Observe that the commutativity of p^+ together with the Jacobi identity imply that for any polynomial $\tilde{p}$ on p^+,

$$(\delta(z_i)\delta(z_j)(\tfrac{1}{2}\delta([[x,\cdot],\cdot])\tilde{p}(\cdot)))(0) = (\delta([[x,z_i],z_j])$$

$$\tilde{p}(\cdot))(0) \; .$$

□

A.4 Proof of Proposition 4.2. By Corollary 2.6, it is sufficient to prove that $\delta_A f \in \mathcal{O}(\mathcal{D},V_{\tau_A})$ for all $f \in \mathcal{O}(\mathcal{D},V_\tau)$. Equivalently, it is sufficient to prove that if $\tilde{f} \in \mathcal{O}(G,K,V_\tau)$, then $D_A\tilde{f} \in \mathcal{O}(G,K,V_{\tau_A})$. Thus, by (2.2), we must demonstrate that

$$r(x)\left(\sum_{i=1}^{N}(r(q_i)(\tilde{f},v_i'))(\cdot)\right) = 0$$

for all $x \in p^-$, and all elements $\sum_{i=1}^{N} q_i \otimes v_i'$ of A, provided that $r(x)\tilde{f} = 0$ for all $x \in p^-$. First notice that it follows from the assumption on $\tilde{f}$ that

$r(x)\left(\sum_{i=1}^{N} r(q_i)(\tilde{f},v_i')\right) = \sum_{i=1}^{N} r([x,q_i])(\tilde{f},v_i')$. Then observe that since W_τ is invariant under dU_τ, W_τ^o is invariant under $dU_{\tau'}$, which, by (3.7), is equal to $dL_{\tau'}$. Thus, $\sum_{i=1}^{N} r(x)r(q_i) \otimes v_i' = \sum_{i=1}^{N} r([x,q_i]) \otimes v_i'$ belongs to W_τ^o and hence is zero, since it is of degree one less than that of A. □

A.5 Proof of Proposition 5.6. By (1.18), it is clearly sufficient to prove that for all $f \in \mathcal{O}(B,V_{\tau_1})$, and all $z \in H$,

$$(D_H(U_{\tau_1}(c^{-1})f))(z) = (U_{\tau_2}(c^{-1})D_B f)(z) .$$

According to (1.6), we may assume that $c^{-1} = \exp(x)$ for an $x \in g^{\mathbb{C}}$. Clearly there exists a non-empty open set $R \subseteq B \cap H$, and an open set $S \subseteq \mathbb{C}$, whose intersection $]a,b[$ with $\mathbb{R}$ contains 0, such that the function $(y,z) \to (U_{\tau_1}(\exp yx)f)(z)$ is well-defined on $S \times R$. Naturally, in this case it is analytic. It is no essential loss of generality to assume that $[0,1] \subset]a,b[$. (Just repeat the argument a finite number of times.) Since $U_{\tau_1}(c^{-1})f = \sum_{i=0}^{\infty} \frac{1}{n!} dU_{\tau_1}(x)^n f$ on R, and $D_H dU_{\tau_1}(x) = dU_{\tau_2}(x)D_H$ on $B \cap H$, it follows that $D_H U_{\tau_1}(c^{-1})f$ and $U_{\tau_2}(c^{-1})D_B f$ agree on R, and hence everywhere on H.

□

GENERALIZED YOUNG TABLEAUX AND WEIGHT MULTIPLICITY FOR CLASSICAL LIE GROUPS

G.Girardi*, A.Sciarrino** and P.Sorba*

*LAPP, BP909 Annecy-le-Vieux, Cedex 74019, FRANCE
**Istituto di Fisica Teorica, 80125 Napoli, ITALY

The irreducible representations (IR) of a classical Lie group can be characterized by a n-ple of integers (or half-integers for spinorial representations of SO(n)), [m]; the highest weight. Each vector of the IR's space can be labelled by a weight, but, generally, in each IR many states are labelled by the same weight. Gilmore[1] has given an algebraic method to build up the weight space. R.King[2] has given a solution to the problem of computing the weight multiplicity (w.m.) using branching rules associated to the subgroups chain. Computer's methods[3] have been also deviced for unitary and orthogonal groups. Here we sketch a method to compute the w.m. based on the use of generalized Young tableaux (GYTs) introduced by the authors[4-5] to develop an algorithm to perform Kronecker product of orthogonal and symplectic groups. Our method applies to all classical Lie groups, due to lack of space we discuss briefly only U(n) and SO(2n) groups. The general method and details will be published elsewhere.

To any ordered n-ple of integers $[\mu]$, ($\mu_i \geq \mu_{i+1}, \mu_i = m_i$ or $m_i - 1/2$) we associate a GYT[4] which has μ_i boxes to the right of a vertical line in the i-th row if $\mu_i > 0$ or $-\mu_i$ boxes to the left if $\mu_i < 0$. To any GYT corresponds a set of states of an IR which are labelled by the following weights:

1- the $[\mu]$ n-ple and its permutation;

2- the n-ples obtained as it follows and their permutations

a) $\mu_i \geq 0$. Label lexically the boxes of the rows, shift the boxes of the i-th row down in all the possible ways such that if l_i is the number of -say- b boxes in the first i rows at least l_i a boxes are present in the first i-1 rows and consider all the ordered n-ples or GYTs so obtained with their multiplicity;

b) $\mu_i \gtrless 0$. Add to GYT $-\mu_n > 0$ boxes in each row, proceed as in a) and at the end decrease each row by $-\mu_n$ boxes.

A product between two GYTs can be defined[4] which reduces to the usual Young tableaux product if no negative boxes are present.

U(n) groups

The states of IR $[m]$ are given by the GYT $[\mu]$ following prescriptions 1 and 2.a.

SO(2n) groups

The states of IR $[m]$ are given by:

1- proceed as for GYT $[\mu]$ of U(n);

2- consider the change of sign of an even number of labels in the n-ples obtained following 1), without the permutations, getting n-ples $[\lambda]$, $\Sigma_i \lambda_i = \Sigma_i \mu_i - 2K \geq 0$. For any k, order the n-ples and pick-up the "highest" one and consider the associated GYT, say $[\lambda]_M$. Take off from the set of n-ples those which are contained in GYT$[\lambda]_M$ according to 2.b and associate to the remaining "highest" one the corresponding GYT and so on;

3- do the product $L_n^{2K}(\{\alpha_i\}) \otimes [\mu]$ where L_n^{2K} are negative GYTs of the form:

$$L_n^{2K}(\{\alpha_i\}) = \left[0,\ldots,-\alpha_i,-\alpha_i,\ldots,-\alpha_1,-\alpha_1\right]; \ \Sigma_i \alpha_i = K, \ \alpha_i \text{ int.}$$

and take in the product only the positive GYTs and proceed on them as in 1) and 2).

All the positive n-ples found in 1)-3) label, apart from a permutation and a change of sign of an even number of labels-Weyl symmetries- the states of IR $[m]$ of SO(2n) with the obtained multiplicity.

REFERENCES

1- R.Gilmore, J.Math.Phys. 11, 513, (1970).

2- R.King, Proceeding of the Conference "Group theoretical Methods in Physics", Nijmegen (1975), ed. A.Janner, J.Janssen and M.Boon, Springer Verlag.

3- V.Amar, U.Dozzio and C.Oleari, "An algorithm for computing inner multiplicities of orthogonal groups", Instituto di Fisica Preprint, Parma (1981) and references therein.

4- G.Girardi, A.Sciarrino and P.Sorba, J.Phys.A:Math.Gen. 15, 1119 (1982).

5- G.Girardi, A.Sciarrino and P.Sorba, LAPP-TH-46 (1981) Preprint, preliminary version.

G(2) ⊃ SU(2) x SU(2) SHIFT OPERATORS AND SCALARS

J.W.B.Hughes
Department of Applied Mathematics, Queen Mary College,
University of London, Mile End Road, London E1 4NS, U.K.

We report here some preliminary results on the application of shift operator techniques to the problem of G(2) ⊃ SU(2) x SU(2). Similar techniques have already been used by the Author[1] to treat the SO(5) ⊃ SU(2) x SU(2) problem, and use the generalisation to SU(2) x SU(2) of the SU(2) shift operators developed by Hughes and Yadegar[2]. A basis for G(2) can be chosen to consist of the $SU(2)^j \times SU(2)^k$ generators $\{j_o, j_\pm; k_o, k_\pm\}$ together with the elements of an eight dimensional tensor representation $R^{[3/2,\ 1/2]}$ of $SU(2)^j \times SU(2)^k$ satisfying the commutation relations (where $\mu = \pm 1/2,\ \pm 3/2$, $\nu = \pm 1/2$):

$$[j_o, R_{\mu\nu}] = \mu R_{\mu\nu}, [j_\pm, R_{\mp 1/2\ \nu}] = 2\, R_{\pm 1/2 \nu},$$

$$[j_\pm, R_{\pm 1/2 \nu}] = \sqrt{3}\, R_{\pm 3/2 \nu}, \ [j_\pm, R_{\mp 3/2 \nu}] = \sqrt{3}\, R_{\mp 1/2\ \nu}$$

$$[k_o, R_{\mu\nu}] = \nu R_{\mu\nu}, \ [k_\pm, R_{\mu \mp 1/2}] = R_{\mu \pm 1/2}\ . \qquad (1)$$

The non-vanishing commutators of the $\{R_{\mu\nu}\}$ amongst themselves are

$$[R_{\pm 3/2\ \pm 1/2}\ , R_{\mp 1/2\ \mp 1/2}] = -[R_{\pm 3/2\ \mp 1/2}\ , R_{\mp 1/2\ \pm 1/2}] = -\frac{1}{2\sqrt{3}}\, j_\pm,$$

$$[R_{\pm 1/2\ 1/2}\ , R_{\pm 1/2\ -1/2}] = \pm 1/3\ j_\pm,$$

$$[R_{3/2\ \pm 1/2}\ , R_{-3/2\ \pm 1/2}] = -[R_{1/2\ \pm 1/2}\ , R_{-1/2\ \pm 1/2}] = \mp 1/2\ k_\pm,$$

$$[R_{3/2\ \pm 1/2}\ , R_{-3/2\ \ 1/2}] = 1/2(k_o \pm j_o), \ [R_{1/2\ \pm 1/2}\ , R_{-1/2 \mp 1/2}] =$$

$$= -1/6\ (3k_o \pm j_o). \qquad (2)$$

Denote the eigenvalues of the $SU(2)^j \times SU(2)^k$ Casimirs J^2 and K^2 by $j(j+1)$ and $k(k+1)$, and those of j_o and k_o by m and n. Then the basis states for an irreducable representation of G(2) can be labelled by $|\alpha; j, m; k, n\rangle$, where α denotes a set of missing state labels. General theory then tells one that one may construct eight shift operators $A^{(\mu\nu)}_{(jk)}$ which when acting upon $|\alpha; j,m; k,n\rangle$ shift the values of j and k by μ and ν, respectively, where $\mu = \pm 3/2$, $\pm 1/2$ and $\nu = \pm 1/2$. The form of $A^{(1/2\ 1/2)}_{(jk)}$ is given by

$$
\begin{aligned}
&\left[(j + m + 1)(k + n + 1)\right]^{1/2} A^{(1/2\ 1/2)}_{(jk)} = \\
&-(k + n + 1)\, R_{-3/2\ 1/2}\, j_+^2 - \frac{1}{\sqrt{3}}(j + 3m + 3)(k + n + 1) R_{-1/2\ 1/2}\, j_+ \\
&+ \frac{1}{\sqrt{3}}(j-3m)(j+m+1)(k+n+1) R_{1/2\ 1/2} + (j+m+1)(k+n+1) R_{3/2\ 1/2}\, j_- \\
&- R_{-3/2\ -1/2}\, j_+^2 k_+ - \frac{1}{\sqrt{3}}(j+3m+3)\, R_{-1/2\ -1/2}\, j_+ k_+ \\
&+ \frac{1}{\sqrt{3}}(j-3m)(j+m+1) R_{1/2\ -1/2}\, k_+ + (j+m+1) R_{3/2\ -1/2}\, j_- k_+
\end{aligned}
\qquad (3)
$$

The other shift operators have similar forms but space does not permit us to give them here.

Now the product operator $A^{(\frac{1}{2}\ \frac{1}{2})}\ A^{(\frac{1}{2}\ -\frac{1}{2})}$ shifts j and k by zero and must therefore be expressible in terms of SU(2) x SU(2) scalars, i.e. operators which commute with the j's and k's. There are four algebraically independent scalars as well as two invariants, and these may be extracted from products such as the above. $A^{(\frac{1}{2}\ \frac{1}{2})}\ A^{(\frac{1}{2}\ -\frac{1}{2})}$ in fact contains three scalars of second order in the R's : Z_1 is of order 1 in the j's and k's, Z_2 has order 3 in the j's and 1 in the k's, and Z_3 is of order 2 in the j's, so Z_1 and Z_3 have total order 4 and Z_2 has total order 6. Their forms are too lengthy to give here but

$$
\begin{aligned}
A^{(\frac{1}{2}\ \frac{1}{2})}_{(j-\frac{1}{2}\ k-\frac{1}{2})}\ A^{(-\frac{1}{2}\ -\frac{1}{2})}_{(j\ k)} &= jZ_1 + Z_2 + k(j-2)\, Z_3 + 2jk(j^2-1) I_2 - kj(j+1)(k^2(j-1) \\
&\quad -1/3\ k(j+1) + 1/9(3j^3 - 4j^2 - 23j + 2)).
\end{aligned}
\qquad (4)
$$

I_2 is the second order invariant given by

$$
\begin{aligned}
I_2 = {} & R_{3/2\ 1/2} R_{-3/2\ -1/2} - R_{3/2\ -1/2} R_{-3/2\ 1/2} - R_{1/2\ 1/2} R_{-1/2\ -1/2} + \\
& + R_{1/2\ -1/2} R_{-1/2\ 1/2} + 1/6\ J^2 + 1/2\ K^2 - 2/3\ j_o .
\end{aligned}
\qquad (5)
$$

A full analysis of the representations of G(2) requires the use of relations such as above, but these on their own are insufficient since one needs also to use relations involving the 6th order invariant I_6, which is of order 6 in the R's, and the 6th order scalar Z_4 which is of order 4 in the R's and 2 in the j's. I_6 is known (private communication by H. De Meyer) and Z_4 is known to the Author. One may obtain this using SU(3) type shift operators[3,4]: $R^{[3/2,\ 1/2]} \times R^{[3/2,\ 1/2]}$ contains an irreducible $q^{[2,0]}$ component, so one obtains SU(3) type operators $\{q_o, q_{\pm1}, q_{\pm2}\}$ of order 2 in the R's which commute with the k's. One may then construct shift operators Q_j^β ($\beta = 0,\pm1,\pm2$) similar to those constructed by Hughes[3,4]. Q^o is closely related to Z_3, and $Q^{+1}Q^{-1}$, which is of order 4 in the R's, contains both Z_3^2, Z_1, Z_2 and Z_4. This expression has already been obtained by the Author but is very lengthy. In order to introduce I_6 the most efficient way is probably to look at the $p^{[1/2,\ 1/2]}$ component of $R^{[3/2,\ 1/2]} \times R^{[3/2,\ 1/2]} \times R^{[3/2,\ 1/2]}$. This enables one to form shift operators $P^{(\alpha \atop \beta)}$ ($\alpha,\beta = 1/2$) of third order in the R's; $P^{(\frac{1}{2}\ \frac{1}{2})}$ $P^{(-\frac{1}{2}\ -\frac{1}{2})}$ will contain the invariant I_6. Another method is to use the $q^{[2,0]}$ and form the shift operator product $Q^{+2}Q^{-1}Q^{-1}$, but this is probably more tedious. This task has not yet been completed.

The ultimate aim is to obtain state labelling operators for an analysis of representations of G(2) $\supset$ SU(2) x SU(2). In addition to I_6, I_2, K^2, K_o, J^2, j_o one needs two further commuting scalars. It is hoped that these may be formed from linear combinations of Z_1, Z_2, Z_3 and Z_4. Computer programs developed by H. De Meyer and G. Vanden Berghe are being used to facilitate these computations, and a fuller version of the results reported here will be published in due course.

References

1. J.W.B.Hughes, SO(4) Shift Operators and Representations of SO(5), J.Math.Phys. in press.
2. J.W.B.Hughes and J.Yadegar, J.Math.Phys. 19, 2068 (1978).
3. J.W.B.Hughes, J.Phys.A: Math.Nucl.Gen. 6, 48 (1973).
4. J.W.B.Hughes, J.Phys.A: Math.Nucl.Gen. 6, 281 (1973).

GENERALISED YOUNG TABLEAUX FOR LIE ALGEBRAS AND SUPERALGEBRAS

Ronald C.King
Mathematics Department, University of Southampton, Southampton, S09 5NH England

ABSTRACT

Young diagrams and tableaux are defined for covariant tensor irreducible representations of U(m) and are then generalised to cover the case of mixed tensors. The extension of these techniques to U(m,n) is described, including a discussion of characters, supercharacters, branching rules and Kronecker products.

1- YOUNG TABLEAUX FOR U(m)

Each covariant tensor irreducible representation $\{\lambda\}$ of U(m) may be labelled by means of a partition $\lambda = (\lambda_1, \lambda_2,\ldots,\lambda_p)$ with $p \leq m$. The partition λ of ℓ also specifies a Young diagram F^λ consisting of boxes arranged in left-adjusted rows of lengths λ_i for $i = 1,2,\ldots,p$. The partition $\tilde{\lambda}$, conjugate to λ, specifies $F^{\tilde{\lambda}}$, obtained from F^λ by interchanging rows and columns. Hence the length of the jth column of F^λ is $\tilde{\lambda}_j$ for $j = 1,2,\ldots,\lambda_1$.

Each box of F^λ is associated with a tensor index. The rows and columns signifiy symmetrisation and antisymmetrisation of the indices. It follows that the basis states of $\{\lambda\}$ may be enumerated by constructing all standard Young tableaux formed by inserting entries chosen from the set $N = \{1,2,\ldots,m\}$ into each of the boxes of F^λ in such a way that the entries are non-decreasing across rows and strictly increasing down columns.

Each tableaux T^λ defines a weight vector $\underline{w}$ with components $w_i = n_i$ for $i = 1,2,\ldots,m$ where n_i is the number of entries i in T^λ. The character of $\{\lambda\}$ is then given by

$$\chi^{\lambda}(\underline{\phi}) = \sum_{\underline{w}} M^{\lambda}_{\underline{w}} \exp (i\underline{w}.\underline{\phi}) \tag{1}$$

where $M^{\lambda}_{\underline{w}}$, the weight multiplicity, is the number of distinct tableaux T^{λ} corresponding to the same weight vector $\underline{w}$. The highest weight vector of $\{\lambda\}$ is simply $\Delta = (\lambda_1, \lambda_2,...,\lambda_p, 0,0,...,0)$. The dimension $d_m\{\lambda\}$ of $\{\lambda\}$ is given by the famous hook length formula.

Kronecker products of U(m) may be decomposed into irreducible constituents by means of the formula

$$\{\sigma\} \times \{\tau\} = \{\sigma.\tau\} = \Sigma\, m^{\lambda}_{\sigma\tau}\{\lambda\} \tag{2}$$

where the coefficients are determined by the Littlewood-Richardson rule. These same coefficients govern the restriction from U(m+n) to U(m) x U(n) for which the branching rule takes the form

$$\{\lambda\} \to \sum_{\tau} \{\lambda/\tau\}\ , \{\tau\} = \sum_{\sigma,\tau} m^{\lambda}_{\sigma\tau}\{\sigma\}\ \{\tau\}\ . \tag{3}$$

2- GENERALISED YOUNG TABLEAUX FOR U(m)

More generally each mixed tensor irreducible representation $\{\bar{\nu};\mu\}$ may conveniently be labelled by means of a pair of partitions μ and ν into p and q parts with $p+q \leq m$. This pair of partitions specifies a generalised Young diagram $F^{\nu;\mu}$ consisting of rows of dotted boxes of lengths ν_j for $j = 1,2,...,q$ right-adjusted to a vertical line, and rows of boxes of lengths μ_i for $i = 1,2,...,p$ left-adjusted to the same line. Boxes and dotted boxes are associated with covariant and contravariant indices respectively, appropriately symmetrised.

The basis states may be enumerated by constructing all generalised standard tableaux $T^{\bar{\nu};\mu}$ formed by inserting positive and negative entries chosen from the sets $N = \{1,2,...,m\}$ and $\bar{N} = \{\bar{1},\bar{2},...,\bar{m}\}$ into each box and each dotted box of $F^{\bar{\nu};\mu}$ respectively, in such a way that the entries are non-decreasing across rows, are strictly increasing in magnitude down columns and $r_i+r_{\bar{i}} \leq i$ for $i = 1,2,...,m$ where r_i and $r_{\bar{i}}$ are the lowest rows containing i and $\bar{i}$ respectively. An entry $\bar{i}$ is to be interpreted as $-i$, and the corresponding weight vector $\underline{w}$ has components $w_i = n_i-n_{\bar{i}}$ for $\underline{i} = 1,2,...,m$, where n_i and $n_{\bar{i}}$ are the number of entries i and $\bar{i}$ in $T^{\bar{\nu};\mu}$. The highest weight vector of $\{\bar{\nu};\mu\}$ is thus

$\Delta = (\mu_1,\mu_2,\ldots,\mu_p,0,0,\ldots,0,-\nu_q,\ldots,-\nu_2,-\nu_1)$, and a dimension formula is known.

Kronecker products are decomposed through the use of the rule

$$\{\bar{\nu};\mu\} \times \{\bar{\lambda};\rho\} = \sum_{\sigma,\tau} \{\overline{\nu/\sigma \cdot \lambda/\tau} \; ; \mu/\tau \cdot \rho/\sigma\} \tag{4}$$

and the restriction from U(m+n) to U(m) x U(n) is governed by

$$\{\bar{\nu};\mu\} \to \sum_{\rho\sigma\tau}\{\overline{\nu/\sigma}; \mu/\tau\}, \{\overline{\sigma/\rho}; \tau/\rho\} \tag{5}$$

In the case of SU(m), mixed tensor representations $\{\bar{\nu};\mu\}$ remain irreducible but are equivalent to covariant tensor representations $\{\lambda\}$ where $\Delta = (\lambda_1,\lambda_2,\ldots,\lambda_{n-1},0) = (\nu_1+\mu_1,\nu_1+\mu_2\ \ldots,\nu_1-\nu_3,\nu_1-\nu_2,0)$.

3- YOUNG TABLEAUX FOR U(m,n)

The Lie superalgebra U(m,n) also has covariant tensor irreducible representations $\{\lambda\}$ labelled by means of λ with p parts but now there is no longer any limit on p. Once again the Young diagram F^λ specifies the symmetry properties of the tensors which are now graded, with components in V_0 or V_1 according as the number of tensor indices taking values in the set $N_1 = \{m+1, m+2,\ldots,m+n\}$ rather than the set $N_0 = \{1,2,\ldots,m\}$ is even or odd.

The basis states of $\{\lambda\}$ may be enumerated for U(m,n) by constructing semistandard Young tableaux S^λ. These are formed by inserting entries chosen from the sets N_0 and N_1 into each of the boxes of F^λ in such a way that entries from N_0 are non-decreasing across rows and strictly increasing down columns, whilst those from N_1 are strictly increasing across rows and non-decreasing down columns. In both rows and columns entries from N_0 precede those from N_1.

One immediate consequence of these rules is that the representation $\{\lambda\}$ of U(m,n) and the corresponding basis states specified by S^λ exist if and only if $\lambda_{m+1} \le n$, or equivalently $\tilde{\lambda}_{n+1} \le m$. The weight vector $\underline{w}$ corresponding to S^λ has components $w_i = n_i$ for $i = 1,2,\ldots, m+n$, where n_i is the number of entries i in S^λ. Weight multiplicities may be found by enumerating all distinct S^λ and the corresponding character and supercharacter of $\{\lambda\}$ are given by

$$\chi^{\lambda}(\underline{\phi}) = \sum_{\underline{w}} M_{\underline{w}}^{\lambda} \exp(i\underline{w}\cdot\underline{\phi}) \text{ and } \tilde{\chi}^{\lambda}(\underline{\phi}) = \sum_{\underline{w}} M_{\underline{w}}^{\lambda}(-1)^{\eta_{\underline{w}}} \exp(i\underline{w}\cdot\underline{\phi}) \tag{6}$$

respectively, where $\eta_{\underline{w}} = w_{m+1} + w_{m+2} + \dots + w_{m+n}$. The highest weight vector of $\{\lambda\}$ is given by $\Delta = (\lambda_1,\lambda_2,\dots,\lambda_m,\tilde{\lambda}_1-m,\tilde{\lambda}_2-m,\dots,\tilde{\lambda}_s-m,0,0,.,0)$ with $s = \lambda_{m+1}$.

Remarkably Kronecker products in U(m,n) may be decomposed, in the case of covariant tensor irreducible representations, by the use once more of (2). The branching rule for the restriction from U(m,n) to U(m) x U(n) takes the form

$$\{\lambda\} \to \sum_{\tau} \{\lambda/\tau\}, \{\tilde{\tau}\} = \sum_{\sigma,\tau} m_{\sigma\tau}^{\lambda} \{\sigma\}, \{\tilde{\tau}\} \tag{7}$$

The crucial appearance of $\tilde{\tau}$ rather than τ, is necessitated by the fact that symmetrisation and antisymmetrisation operators have their roles reversed in V_1 as compared to V_0.

The dimension and superdimension of $\{\lambda\}$ are given by

$$d_{m,n}\{\lambda\} = \sum_{\tau} d_m\{\lambda/\tau\}\, d_n\{\tilde{\tau}\} \text{ and } \tilde{d}_{m,n}\{\lambda\} = d_{m-n}\{\lambda\} \tag{8}$$

In general a representation of U(m,n) may contain irreducible constituents into which it cannot be decomposed as a direct sum. Irreducible representations which appear only as direct summands are said to be typical. The representation $\{\lambda\}$ is typical if and only if $\lambda_m \geq n$, or equivalently $\tilde{\lambda}_n \geq m$. These conditions imply that F^{λ} contains an mxn rectangle of boxes in the upper lefthand corner. In such a case

$$d_{m,n}\{\lambda\} = 2^{mn}\, d_m\{\mu\}\, d_n\{\tilde{\nu}\} \tag{9}$$

where $\mu_i = \lambda_i - n$ for $i = 1,2,\dots,m$ and $\tilde{\nu}_j = \tilde{\lambda}_j - m$ for $j = 1,2,\dots,n$.

4- GENERALISED YOUNG TABLEAUX FOR U(m,n)

Even for SU(m,n) this is not enough. Mixed tensor irreducible representations, $\{\bar{\nu};\mu\}$ exist which are not equivalent to covariant tensor representations. To these correspond generalised semistandard Young tableaux $S^{\bar{\nu};\mu}$ formed by inserting positive and negative entries from N_0 and N_1 and from $\bar{N}_0$ and $\bar{N}_1$, respectively, into the boxes and dotted boxes of $F^{\bar{\nu};\mu}$. Entries from N_0 and $\bar{N}_0$ are non-decreasing across rows and strictly increasing in magnitude down columns, whilst those from N_1 and $\bar{N}_1$ are strictly increasing across rows and non-decreasing in magnitude down columns. Entries from N_0 and $\bar{N}_0$ precede those from N_1 and $\bar{N}_1$ in both rows and columns. The additional constraint $r_i+r_{\bar{i}} \leq i$ still applies for i in N_0 but not, in this form, for i in N_1. Indeed the required constraint is not known at present.

Fortunately the Kronecker product rule for U(m,n) is still (4), and the inverse of (4) in the case $\mu=\lambda=0$ may be used to express $\{\bar{\nu};\rho\}$ in terms of co- and contravariant representations so that the allowed semistandard Young tableaux $S^{\bar{\nu};\rho}$ may be determined and weight vectors and their multiplicities calculated.

For the large class of representations $\{\bar{\nu};\mu)$ of U(m,n) such that $p+t\leq m$ or $q+s\leq n$, where $p = \tilde{\mu}_1$, $q = \nu_1$, $s = \mu_{m+1}$ and $t = \tilde{\nu}_{n+1}$, the highest weight is $\Delta = (\mu_1,\mu_2,...,\mu_p,0,...,0,n-\nu_t,...,n-\nu_2,n-\nu_1,-\tilde{\nu}_n,...,-\tilde{\nu}_2,-\tilde{\nu}_1)$ or $(\mu_1,\mu_2,...,\mu_m,\tilde{\mu}_1-m,\tilde{\mu}_2-m,...,\tilde{\mu}_s-m,0,...,0,-\tilde{\nu}_q,...,-\tilde{\nu}_2,-\tilde{\nu}_1)$. Such a representation is typical if and only if $\mu_i+m-i \neq \tilde{\nu}_j+n-j$ for $1\leq i\leq m-t$ and $1\leq j\leq n-s$. In this case the superdimension is zero, but in general

$$\tilde{d}_{m,n}\{\bar{\nu};\mu\} = d_{m-n}\{\bar{\nu};\mu\} \tag{10}$$

The restriction from U(m,n) to U(m)xU(n) is governed by the rule

$$\{\bar{\nu};\mu\} \to \sum_{\rho\sigma\tau} \{\overline{\nu/\sigma};\mu/\tau\}, \{\overline{\tilde{\sigma}/\rho};\tilde{\tau}/\rho\} \tag{11}$$

and this may be used to calculate the dimension of $\{\bar{\nu};\mu\}$. In the case of the typical representations already mentioned

$$d_{m,n}\{\bar{\nu};\mu\} = 2^{mn}\, d_m\{\bar{\kappa};\lambda\}\, d_n\{\bar{\sigma};\tau\} \tag{12}$$

with $\{\bar{\kappa};\lambda\}$ and $\{\bar{\sigma};\tau\}$ determined by the first m and last n components of Δ.

5- YOUNG DIAGRAMS FOR O(m), Sp(m), OSp(m,n) and SpO(m,n)

Covariant tensor irreducible representations $[\lambda]$, of both O(m) and OSp(m,n), and <λ> of both Sp(m) and SpO(m,n) may also be labelled by partitions λ. In the case of O(m) and Sp(m) Young tableaux may be used to specify basis states and determine weight vectors, their multiplicities and characters. The task of extending these ideas to OSp(m,n) and SpO(m,n) remains to be completed. However the branching rules

$$\{\lambda\} \to [\lambda/D] = [\lambda] + [\lambda/2] + [\lambda/4] + [\lambda/2^2] + \ldots \quad (13)$$

$$\{\lambda\} \to \langle\lambda/B\rangle = \langle\lambda\rangle + \langle\lambda/1^2\rangle + \langle\lambda/1^4\rangle + \langle\lambda/2^2\rangle + \ldots \quad (14)$$

apply not only to the restrictions from U(m) to O(m) and from U(m) to Sp(m) but also to those from U(m,n) to OSp(m,n) and from U(m,n) to SpO(m,n). In addition the Kronecker product rules

$$[\lambda]\times[\mu] = \sum_{\rho}[\lambda/\rho\cdot\mu/\rho] \text{ and } \langle\lambda\rangle\times\langle\mu\rangle = \sum_{\rho}\langle\lambda/\rho\cdot\mu/\rho\rangle \quad (15)$$

apply to both the relevant algebras and superalgebras. New formulae are the branching rules for OSp(m,n) to O(m) Sp(n) and SpO(m,n) to Sp(m) O(n) which take the form

$$[\lambda] \to \sum_{\rho}[\lambda/\rho] , \langle\tilde{\rho}/B\rangle \text{ and } \langle\lambda\rangle \to \sum_{\rho}\langle\lambda/\rho\rangle , [\tilde{\rho}/D] \quad (16)$$

Dimensions and superdimensions may be calculated from these results. In particular it may be verified that

$$\tilde{d}_{m,n}[\lambda] = d_{m-n}[\lambda] \text{ and } \tilde{d}_{m,n}\langle\lambda\rangle = d_{m-n}\langle\lambda\rangle \quad (17)$$

A necessary condition that $[\lambda]$ is a typical representation of OSp(2m+1,2n) or OSp(2m,2n) is $\lambda_m \geq n$, just as for U(m,n). In such a case

$$d_{2m+1,2n}[\lambda] = 2^{2mn}\, d_{2m+1}[\mu]\, d_{2n+1}[\tilde{\nu}] \quad (18)$$

$$d_{2m,2n}[\lambda] = 2^{2mn}\, d_{2m}[\mu]\, d_{2n}\langle\tilde{\nu}\rangle \quad (19)$$

where the notation is that of (9).

Similar formulae may be written down for the dimension of a typical representation $\langle\lambda\rangle$, with $\lambda_n \geq m$, of SpO(2n,2m+1) or SpO(2n,2m). Alternatively use may be made of the isomorphism between SpO(k,ℓ) and OSp(ℓ,k) which implies $\langle\lambda\rangle = [\tilde{\lambda}]$.

This work leans heavily on that of Dondi and Jarvis[1], who introduced the notion of graded tensors; of Balantekin and Bars[2], who extended the author's[3] use of generalised Young diagrams from Lie algebras to superalgebras; of Berele and Regev[4], who independently introduced semistandard Young tableaux; and on earlier work[5-7] on Young diagrams and tableaux, branching rules and dimension formulae for Lie algebras. The work on branching rules, in particular, provides via the special case of (16) for SpO(2n,1) an easy way to show that the character $\langle\lambda\rangle$ of this superalgebra is identical with the character $[\lambda]$ of the algebra O(2k+1), as pointed out by Rittenberg and Scheunert[8]. Finally mention must be made of the definitive work of Kac[9] on whose results many of the formulae of this paper depend.

REFERENCES

1- P.H.Dondi and P.D.Jarvis, J.Phys. A14 (1981) 547-563.

2- A.B.Balantekin and I.Bars, J.Math.Phys. 22 (1981) 1810-1818.

3- R.C.King, J. Math. Phys. 11 (1970) 280-294.

4- A.Berele and A.Regev, Preprint: Weizmann Inst. of Sci. Rehovot, Rehovot, Israel.

5- R.C.King in Lecture Notes in Physics (New York: Springer) 50 (1975) 490-499.

6- R.C.King, J. Phys. A8 (1975) 429-449.

7- N.El Samra and R.C.King, J. Phys. A12 (1979) 2317-2328.

8- V.Rittenberg and M.Scheunert, Preprint: TH3035 CERN

9- V.Kac in Lecture Notes in Maths. (New York: Springer) 676 (1978) 597-626.

THE DeSITTER SYMMETRY OF THE DIRAC EQUATION

P. Moylan
Center for Particle Theory and Physics Department
University of Texas; Austin, Texas 78712
USA

ABSTRACT

The SO(4,1) symmetry of the Dirac equation is constructed. It is realized on the full space of solutions of the Dirac equation. Generalizations and possible physical implications are noted.

1. Introduction

The invariance of the Dirac equation under the restricted Poincaré group is well-known.[1] There is, however, another much less well-known symmetry of the Dirac equation, namely the DeSitter group, SO(4,1). Here we provide an explicit construction of a representation of SO(4,1) operating on solutions of the Dirac eqation; the representation is a generalization to include spin of the sort of projective representations utilized by V. Bargmann in his study of the Lorentz groups.[2]

First we recall some elementary notions of the Dirac theory and the Poincaré group. We then discuss projective transformations on the mass hyperboloid, $T_3 = \{p_\mu | p^\mu p_\mu = m^2\}$, and describe a projective representation of the DeSitter group, from which we construct the representation of SO(4,1) on the space of solutions to the Dirac equation. Finally we present some concluding remarks. The conventions in this paper are the same as in ref. [1]. $\overline{SO_o(4,1)}$ and $\overline{P}$ denote the simply connected covering groups of the DeSitter and Poincaré groups, respectively; the representation which we construct is, of course, a representation of the covering group of SO(4,1).

2. The Dirac Equation

The Dirac equation for a spin $\frac{1}{2}$ particle of mass m is in momentum space:

$$\gamma_\mu P^\mu \psi(p,\xi) = m\psi(p,\xi) \qquad (\{\gamma^\mu,\gamma^\nu\} = 2g^{\mu\nu}) \tag{1}$$

Let R be the space of all $\psi(p,\xi)$ which satisfy eqn. (1). Consider the following inner product on R[3]

$$(\phi,\psi) = \int_{T_3} d\Omega_{T_3} \sum_\xi \phi^*(p,\xi)\gamma^o\psi(p,\xi) \qquad (d\Omega_{T_3} = \frac{1}{2|p_o|} d^3p) \tag{2}$$

We have the result that $R = H(m,\frac{1}{2};+) \boxplus H(m,\frac{1}{2};-)$, where $H(m,\frac{1}{2};+)$ and $H(m,\frac{1}{2};-)$ are the positive and negative energy irreducible representation spaces (UIR's) of $\overline{P}$, respectively ($\boxplus$ means direct sum).

Note that $i\Gamma_\mu$ and $S_{\mu\nu}$ generate an $\overline{SO_o(4,1)}_{i\Gamma_\mu,S_{\mu\nu}}$ where $\Gamma_\mu = \frac{1}{2}\gamma_\mu$ and $S_{\mu\nu} = \frac{i}{4}[\gamma_\mu,\gamma_\nu]$.[4] In the following we find an $\overline{SO_o(4,1)}$ symmetry group which has this $\overline{SO_o(4,1)}_{i\Gamma_\mu,S_{\mu\nu}}$ for its spin part.

3. A Non-unitary Representation of $SO_o(4,1)$ on $L^2(T_3)$

In order to construct the DeSitter symmetry we must describe a non-unitary projective representation of $SO_o(4,1)$. Suppose X^a (a=1,...5) is a light-like vector in 5 dimensions, $X^aX_a = X_o^2 - X_1^2 - X_2^2 - X_3^2 - X_5^2 = 0$. All transformations Λ of O(4,1) leave this this length invariant i.e. $\Lambda^a{}_bX^b = Y^a \quad (Y^aY_a = 0)$. To each vector X^a, we may associate a vector p^μ on T_3 as: $\frac{p^\mu}{m} = \frac{X^\mu}{X^5}$. The linear transformation $\Lambda^b{}_aX^b = Y^a$ corresponds to the following projective transformation on T_3

$$p^{\mu\prime} = m\frac{Y^\mu}{Y^5} = m\frac{(\Lambda X)^\mu}{(\Lambda X)^5} \tag{3}$$

A bounded, continuous representation of $SO_o(4,1)$ on $L^2(T_3)$ is provided by the operators $T(\Lambda)$ for each $\Lambda \in SO_o(4,1)$:[5]

$$[T(\Lambda)\psi](p^\mu) = \frac{1}{[(\Lambda^{-1})^5{}_5 + (\Lambda^{-1})^5{}_\mu\frac{p^\mu}{m}]^{1+i\rho}}\psi\{m\frac{(\Lambda^{-1})^\mu{}_5 + (\Lambda^{-1})^\mu{}_\nu\frac{p^\nu}{m}}{(\Lambda^{-1})^5{}_5 + (\Lambda^{-1})^5{}_\nu\frac{p^\nu}{m}}\} \tag{4}$$

It is necessary to compute the action of the infinitesmal generators of rotations in the a-b planes of projective space. If I^{ab} is an infinitesmal rotation, the essentially-self-adjoint (esa) generator is:

$$-iM^{ab(o)}\psi(p) = \frac{d}{d\omega}\{T(e^{\omega I^{ab}})\psi\}(p)\Big|_{\omega=0} \quad (\psi \in L^2(T_3)) \tag{5}$$

The results are[4]

$$M_{\mu\nu}^{(o)} = \{Q_\mu P_\nu - Q_\nu P_\mu\}\cdot I = M_{\mu\nu}\cdot I; \quad M_{5\mu}^{(o)} = \{-\frac{1}{\lambda}B_\mu^{(o)} - \frac{i}{2m}P_\mu\}\cdot I \tag{6}$$

$$Q^\mu = -i\frac{\partial}{\partial p_\mu}; \quad \frac{1}{\lambda}B_\mu^{(o)} = [\frac{1}{2m}\{P^\rho,M_{\rho\mu}\} + \frac{\rho}{m}P_\mu]\ ; \quad I = \begin{vmatrix} 1 & 0 \\ 0 & 1 \end{vmatrix}$$

The factor I is necessary in order that the $M_{5\mu}^{(o)}$ be esa.[6] It comes from the following considerators: $L^2(T_3) \cong L^2(T_3^+) \boxplus L^2(T_3^-)$ as Hilbert spaces, and a vector $\psi \in L^2(T_3)$ may be written as ψ_i (i=1,2) with $\psi_1 \in L^2(T_3^+)$ and $\psi_2 \in L^2(T_3^-)$. The I specifies the operators $M^{(o)}_{\mu\nu}$ and $M^{(o)}_{5\mu}$ as the direct sum of operators on $L^2(T_3^+)$ and $L^2(T_3^-)$.

4. The DeSitter Symmetry of the Dirac Equation

Now we are able to define a representation of $\overline{SO_o(4,1)}$ on $L^2(T_3)\times C^4$ as follows:

$$[U(\Lambda)\psi](p^\mu,\xi') = e^{-i(\frac{\omega^{\mu\nu}}{2} S_{\mu\nu} + i\omega^\rho \Gamma_\rho)^{\xi'\xi}} [T(\Lambda)\psi](p^\mu,\xi) \quad (7)$$

Here $\omega^{\mu\nu}$ and $\omega^\rho = \omega^{5\rho}$ are the 10 parameters of the $\overline{SO_o(4,1)}$ transformation. To prove this is a representation we must first show that it leaves invariant the space of solutions of the Dirac equation, R. In order to do this we use the following operator identity, which is valid acting on any $\psi \in R$:

$$i\Gamma_\mu = -\frac{1}{2m}\{P^\rho, S_{\rho\mu}\} + \frac{i}{2m} P_\mu \quad (8)$$

Using (4) and (6) we can rewrite (7) as

$$[U(\Lambda)\psi](p^\mu,\xi') = e^{-i\{\frac{\omega^{\mu\nu}}{2} L_{\mu\nu}\cdot I - \omega^\rho \frac{1}{\lambda} B^{(o)}_\rho \cdot I + i\omega^\rho \Gamma_\rho - \frac{i}{2m}\omega^\rho P_\rho \cdot I\}} \psi(p^\mu,\xi) \quad (9)$$

on any ψ for which it makes sense ($L_{\mu\nu} = M_{\mu\nu} + S_{\mu\nu}$). Expanding this exponential as a formal power series we can repeatedly use (8) and $L_{\mu\nu}\psi \in R$ and $B_\mu \psi \in R$ if $\psi \in R$ ($B_\mu = B^{(o)}_\mu + \frac{\lambda}{2m}\{P^\rho, S_{\rho\mu}\}$ and $[B_\mu, \gamma^\rho P_\rho] = 0$, so that $B_\mu \psi \in R$ if $\psi \in R$) to obtain the following result

$$[U(\Lambda)\psi](p^\mu,\xi') = e^{-i\{\frac{\omega^{\mu\nu}}{2} L_{\mu\nu} - \omega^\rho \frac{1}{\lambda} B_\rho\}\cdot I} \psi(p^\mu,\xi) \quad (10)$$

The invariance of R is a direct consequence of (10), since $L_{\mu\nu}$ and B_μ both commute with the Dirac equation. The representation property follows from (7) along with the facts that the $T(\Lambda)$'s are a representation, and the $S_{\mu\nu}$'s and $i\Gamma_\mu$'s generate a representation. It can be proven that this representation is irreducible and unitary.[4)]

We can carry this representation of $\overline{SO_o(4,1)}$ over into $\bar{R} = H(m,\frac{1}{2};+) \boxplus H(m,\frac{1}{2};+)$ by defining the new representation to be the image of $U(\Lambda)$ under the unitary map which sends the pair of canonical basis vectors $\begin{pmatrix} |p,s_3;+\rangle \\ |p,s_3;-\rangle \end{pmatrix}$ into $\begin{pmatrix} |p,s_3;+\rangle \\ |p,s_3;+\rangle \end{pmatrix}$. In this way we eliminate the troublesome negative energy states. If we choose $\rho = \frac{m}{\lambda}$, then we can verify that our representation is the principal series representation with parameters $\rho = \frac{m}{\lambda}$ and $s = \frac{1}{2}$.[4)]

5. Conclusions

The above construction generalizes to yield an arbitrary principal series representation of $\overline{SO_o(4,1)}$ on the space of solutions of the Bargmann-Wigner equations for a particle of spin s and mass m.[4)] This method can be generalized to other SO(p,q) groups; it should be especially attractive to physicists since it uses concepts well-known to them.

Is there any physical interpretation of the B_μ's-- the DeSitter boosts?[7)] If they correspond to observables then the space of physical states must be R since on this space the B_μ's are esa. For fermions this doubling of states can be associated with the antiparticle. However, there must exist a superselection rule which prevents arbitrary superpositions of states from the two H(m,s;+)'s. In this connection, we note that under the Inönü-Wigner group contraction process,[8)] in which $\lambda \to 0$, the $\overline{SO_o(4,1)}$ representation goes over into two UIR's of $\overline{P}$.[9)]

References

1. J.D. Bjorken and S.D. Drell, Relativistic Quantum Mechanics, McGraw-Hill (1964).
2. V. Bargmann, Ann. Math., 48, 568 (1947).
3. V. Bargmann, E.P. Wigner, Proc. Nat. Acad. Sci. 34,5, 211-233 (1946).
4. P. Moylan, JMP (to be published) and Dissertation, UT Austin.
5. H. Abarbanel in Studies in Math. Phys., eds. E. Lieb, B. Simon & A. S. Wightman, Princeton (1976).
6. N. Mukunda, JMP, 1 (9), (1968).
7. See the article in these Proceedings by A. Böhm on group contraction. There he shows that the B_i's go over into λQ_i (i=1,2,3), in the contraction of the Poincaŕe group into the extended Euclidean group. (Q_i is the usual position operator in Euclidean space.)
8. E. Inönü, E.P. Wigner, Proc. National Acad. Sci. 39, 50, (1953).
9. A. Böhm, in Studies in Math. Phys, ed. Barut, Reidel Publ. (1973). See also W. Drechsler, JMP, 18, 7 ,1363 (1977). He shows that the quotient space SO(4,1)/SO(3,1) goes into two copies of Minkowski space. However his SO(4,1) realization is different than the one considered here.

CASIMIR OPERATORS FOR INHOMOGENEOUS CLASSICAL GROUPS

Marcel Perroud
Département de mathématiques appliquées
Ecole Polytechnique de Montréal
Montréal H3C 3A7 Canada

ABSTRACT

The Casimir operators of the following groups are explicitely constructed: $\mathbb{R}^n \square G\ell(n,\mathbb{R})$, $\mathbb{R}^n \square S\ell(n,\mathbb{R})$, $\mathbb{R}^n \square O(p,q)$, $\mathbb{C}^n \square U(p,q)$, $\mathbb{R}^{2n} \square Sp(n,\mathbb{R})$, $\mathbb{C}^n \square G\ell(n,\mathbb{C})$, $\mathbb{C}^n \square S\ell(n,\mathbb{C})$, $\mathbb{C}^n \square S\ell(n,\mathbb{C})$, $\mathbb{C}^n \square O(n,\mathbb{C})$, $\mathbb{C}^{2n} \square Sp(n,\mathbb{C})$, $\mathbb{H}^n \square U^*(2n)$, $\mathbb{H}^n \square SU^*(2n)$, $\mathbb{H}^n \square O^*(2n)$, $\mathbb{H}^n \square Sp(p,q)$. The method is based on a particular fibre bundle structure of the generic orbits generated by the co-adjoint representation of a semi-direct product.

In a recent series of papers[1)], the number of Casimir operators of the groups $\mathbb{R}^n \square G\ell(n,\mathbb{R})$, $\mathbb{R}^n \square S\ell(n,\mathbb{R})$, $\mathbb{R}^n \square O(p,q)$, $\mathbb{C}^n \square U(p,q)$ and $\mathbb{C}^n \square SU(p,q)$ have been computed. In this contribution we outline a method for constructing explicitely the Casimir operators of all classical inhomogeneous groups, details can be found elsewhere[2)].

It is well known that the Casimir operators are in a 1-1 correspondance with the invariant polynomial functions on the dual space of the Lie algebra of the group. On the other hand, it is clear that the generic orbits of the co-adjoint action of the group are contained in the simultaneous level sets of these invariant polynomial functions (if any). The method consists of extracting these polynomials from the knowledge of the orbits.

Let A be a real vector space, $G \subseteq G\ell(V)$ a classical group, $H = A \square G$ the natural semi-direct product and $h^* = A^* \times g^*$ the dual space of the Lie algebra $h = A \times g$ with the scalar product

$$<(p,M), (\xi,X)> = <p,\xi>_1 + <M,X>_2 .$$

The co-adjoint action of H can be written in the form

$$Ad^*(a,g)(p,M) = (g^*p,\ g^*Mg^{*-1} + L(g^*p,a))$$

where $<g^*p,\xi>_1 = <p,g^{-1}\xi>_1$ and $L(g^*p,a)$ is a well defined bilinear function.

It follows from this particular action that the generic Ad*-orbits $O_H(p_0, M_0)$ are provided with a natural fibre bundle structure $O_H(p_0, M_0) \to O_H(p_0) \subseteq A^*$ with typical fibre

$$O_{H(p_0)}(M_0) \simeq O_{G(p_0)}(i(p_0)^* M_0) \times W(p_0)$$

where $H(p_0) = A \square G(p_0)$ is the stabilizer of $p_0 \in A^*$, $i(p_0): g(p_0) \hookrightarrow g$ is the inclusion mapping and $W(p_0)$ is some vector subspace of g^*. Hence $O_H(p_0, M_0)$ is completely characterized by the orbit $O_H(p_0) \subseteq A^*$ and by the orbit $O_{G(p_0)}(i(p_0)^* M_0) \subseteq g(p_0)^*$. Then, it follows from this result that the invariant polynomial functions on h^* are of the two following kinds.

1) $\psi(p)$, invariant polynomial function on A^*.

2) $\psi(p,M) = \phi(i(p_0)^*(\Lambda(p)^{*-1} M \Lambda(p)^*))$, where ϕ is an invariant function on $g^*(p_0)$ and $\Lambda(p): p_0 \mapsto p$ is a "boost".

<u>Some results</u>

1) $\mathbb{R}^n \square G\ell(n,\mathbb{R})$: no invariant function

2) $\mathbb{R}^n \square S\ell(n,\mathbb{R})$: $\psi(p,M) = \det(p, Mp, \ldots, M^{n-1}p)$

3) $\mathbb{R}^n \square O(p,q)$: $\psi_0(p) \equiv \nu = p^t I_{p,q} p$

$\psi_k(p,M) = C_{2k}((\nu I_n - pp^t I_{p,q})M)$, $k = 1,\ldots,[\frac{n-1}{2}]$

4) $\mathbb{C}^n \square U(p,q)$: $\psi_0(p) \equiv \nu = p^\dagger I_{p,q} p$

$\psi_k(p,M) = C_k(\sqrt{-1}(\nu I_n - pp^\dagger I_{p,q})M)$, $k = 1,\ldots,n-1$

5) $\mathbb{R}^{2n} \square Sp(n,\mathbb{R})$: $\psi_0(p,M) \equiv \nu = -p^t J_n M p$

$\psi_k(p,M) = C_{2k}(\nu M + (Mp)(Mp)^t J_n)$, $k = 1,\ldots,n-1$

where

$$\det(X - \lambda I_n) = \sum_{\ell=0}^{n} C_\ell(X)\lambda^{n-\ell}, \; I_{p,q} = -I_p \oplus I_q, \; J_n = J \oplus \ldots \oplus J, \; J = \begin{bmatrix} 0 & -1 \\ 1 & 0 \end{bmatrix}.$$

1) A.P. Deemicev, N.F. Nelipa, Vestnik Moskov Univ. Ser. III, Fiz. Astronom. <u>21</u>, no. 2, 3, 7; ibid no. 4, 23; ibid no. 4, 27 (1980).

2) M. Perroud, The fundamental invariants of inhomogeneous classical groups, preprint 1982 (submitted to the J. Math. Phys.).

CONSTRUCTION AND UNITARY REPRESENTATIONS OF THE NON - COMPACT GROUPS OF SUPERGRAVITY

Cihan Saçlıoğlu

Physics Department, Boğaziçi Univ., Bebek, Istanbul, Turkey

1- INTRODUCTION

It has been observed[1,2,3] that the bosonic sectors of the N=4-8 extended supergravity (ESG) theories exhibit certain non-compact symmetries on the mass shell. These involve linear transformations between the vector field strengths and their duals on the one hand and a nonlinear realization over the scalar fields on the other. The unitary representations of these groups are of interest as the bound state spectra of similar models in two dimensions fall into such multiplets[4]. The aim of this note is to outline a simple mathematical formalism employing annihilation and creation operators (acting on a Fock-type space), in terms of which the generators and unitary representations of the aforementioned non-compact groups can be constructed. More details can be found in references 5 and 6.

2- THE CONSTRUCTION

The passage from the generators of U(N) to those of the bigger noncompact groups of ESG is an extension of a well-known construction[7,8] giving Sp(2n,R) and SO*(2n) from U(n). We shall first review these familiar cases:

a) $U(n) \to Sp(2n,R)$ and $SO^*(2n)$:

Let $a_i(K)$, $b_i(K)$ ($a_i^+(K) \equiv a^i(K)$, $b_i^+(K) \equiv b^i(K)$) be two sets of annihilation (creation) operators transforming as the $\tilde{n}(\tilde{n}^*)$ representation of U(n). The index i runs from 1 to n, while K simply reproduces additional copies of the same operators obeying

$$[\, a_i(K),\, a^j(L)] = [\, b_i(K),\, b^j(L)\,] = \delta_{KL}\delta_i^j \tag{1}$$

with all other commutators vanishing. The U(n) generators are

$$I^m_n = \vec{a}^m.\vec{a}_n + \vec{b}_n.\vec{b}^m \qquad (2)$$

where the dot product denotes a sum over K.

One can augment I^m_n by symmetric and antisymmetric combinations of diboson operators:

$$S_{ij} = \vec{a}_i.\vec{b}_j + \vec{a}_j.\vec{b}_i \quad ; \quad S^{kl} = \vec{a}^k.\vec{b}^l + \vec{a}^l.\vec{b}^k \qquad (3.a)$$

$$A_{ij} = \vec{a}_i.\vec{b}_j - \vec{a}_j.\vec{b}_i \quad ; \quad A^{kl} = \vec{a}^k.\vec{b}^l - \vec{a}^l.\vec{b}^k \qquad (3.b)$$

The sets (I^m_n, S_{ij}, S^{kl}) and (I^m_n, A_{ij}, A^{kl}) generate the algebras of Sp(2n,R) and SO*(2n) respectively. The non-compactness originates from the fact that the dibosons, when used as ladder operators, create infinite dimensional unitary representations.

b) The algebras of ESG:

We now introduce operators $a_{ij}(K) = -a_{ji}(K)$, etc., transforming as $\widetilde{n(n-1)/2}$ under U(n). These satisfy

$$[a_{ij}(K), a^{kl}(L)] = [b_{ij}(K), b^{kl}(L)] = \delta_{KL}(\delta^k_i\delta^l_j - \delta^l_i\delta^k_j) \qquad (4)$$

The U(n) generators are written as

$$I^m_n = \vec{a}^{mk}.\vec{a}_{nk} + \vec{b}_{nk}.\vec{b}^{mk} \qquad (5)$$

The cases n=2 and n=3 revert to the U(1) → Sp(2,R) and U(3) → Sp(6,R) or SO(6) when dibosons are added. On the other hand, for $4 \leq n \leq 8$, it is only with the dibosons $A_{i_1i_2i_3i_4} = \varepsilon_{i_1i_2i_3i_4jklm}\vec{a}^{jk}.\vec{b}^{lm}$ (using as many of the indices $i_1i_2i_3i_4$ as required) and their conjugates that one obtains new algebras. In particular, starting from the compact algebras U(4), U(5), U(6) and SU(8), the new non-compact extensions are precisely the ESG series SU(4)XSU(1,1), SU(5,1), SO*(12) and $E_{7(+7)}$. Interestingly, the construction does not give any other solutions for n > 8. Note also that A_{ijkl} is real and 70-dimensional for n=8; thus a single 70 without its conjugate and the 63 of SU(8) rather than U(8) add up to $E_{7(+7)}$.

3- SOME INFINITE DIMENSIONAL REPRESENTATIONS

All of the above generators except those of $E_{7(+7)}$ decompose under the respective maximal compact subgroups in the same way: The compact generators form a set L_o including a U(1) operator. The remaining noncompact generators L_+, L_- are conjugates with opposing U(1) charges. When, in addition, the $L_+(L_-)$ generators do not involve admixtures of annihilation (creation) operators, one can construct infinite dimensional unitary irreps by the following method.

Choose an initial state $|\psi_I\rangle$ which is annihilated by the operators in the L_- space, i.e.,

$$L_-|\psi_I\rangle = 0 \tag{6}$$

Then the infinite set

$$|\Psi\rangle = (1 \oplus L_+ \oplus L_+L_+ \oplus \ldots \oplus L_+^k \quad \ldots)|\psi_I\rangle \tag{7}$$

constitutes a unitary representation.

Proof: Write the quadratic and higher Casimir operators with L_-'s standing on the right. Let the nth order Casimir operator C_n have the eigenvalue λ_n when applied on $|\psi_I\rangle$. Because of (6) and the property $[C_n, L_+^k] = 0$, this is also the eigenvalue of C_n on $|\psi_I\rangle$. Thus all these operators are diagonalized on $|\Psi\rangle$, proving its irreducibility.

As an example we treat SO*(12)/U(6) with operators as in section 2.a. For simplicity, let $a^i(K)$, $b^i(K)$ represent the Young tableau (1,0,0,0,0,0) with one box in the first row and none in the others. The vacuum $|\Omega\rangle$ is a product of all the vacua annihilated by $a_i(K)$ and $b_i(K)$ for all i and K. Then an initial state $|\psi_I\rangle$ satisfying (6) is given by any linear combination of states such as

$$\left[a^i(1)\right]^{k_1} \left[a^j(2)\right]^{k_2} \ldots \left[a^m(L)\right]^{k_L} |\Omega\rangle \tag{8}$$

with similar states built from b^+'s. The role of the extra index is now apparent: it allows us to arbitrarily enlarge the possible set of $|\psi_I\rangle$'s. To make $|\psi_I\rangle$ irreducible, one can apply a suitable projection operator on (8). For definiteness, let us examine the simplest case K=1. Then the operators $(L^+)^k \sim (a_i b_j - a_j b_i)^k$ transform as (k,k,0,0,0,0), leading to the $|\Psi\rangle$

$$|\Psi\rangle = (\Sigma_{k=0}^{\infty} \oplus (k,k,0,0,0,0)) \oplus |\psi_I\rangle \qquad (9)$$

4- CONCLUDING REMARKS

Our construction of the ESG algebras shares only one input with these theories; namely the use of boson operators a_{ij} in the $\widetilde{n(n-1)/2}$ representation of U(n). We have seen that this suffices to determine correctly the series of non-compact groups of ESG. In particular, the non-compactness is seen to be a direct consequence of having diboson operators.

Our method for constructing irreps depends crucially on the existence of a triply graded structure $L_+ \oplus L_o \oplus L_-$. In addition, the L_- generators should consist of bilinears in the annihilation operators only. These conditions do not hold for $E_{7(+7)}$, which decomposes into $\tilde{63} + \tilde{70}$ (real) under SU(8). Correspondingly, the representations obtained by applying the $\tilde{70}$ repeatedly on a given initial irrep of SU(8) are infinitely reducible.

For a discussion of how our work relates to previous work in mathematical literature[9,10], we refer the reader to ref.6.

5- REFERENCES

1- E.Cremmer and B.Julia, Phys. Lett. 80B(1978)48; Nucl. Phys. B159 (1979) 141.

2- E.Cremmer and J.Scherk and S.Ferrara, Phys. Lett. 74B(1978)61).

3- S.Ferrara,J.Scherk and B.Zumino, Nucl. Phys. B172(1977)393.

4- H.Haber, I.Hinchcliffe and E.Rabinovici, Nucl. Phys. B172(1980)458.

5- M.Günaydin and C.Saçlioğlu, Phys. Lett. 108B(1982)180.

6- M.Günaydin and C.Saçlioğlu, Comm. Math. Phys. (in press).

7- H.Lipkin, "Lie Groups for pedestrians", p.95, North-Holland Amsterdam (1965)

8- Y.Nambu, Lectures given at the International Summer School in High Energy Physics, Erice, Italy (1972), edited by A.Zichichi.

9- M.Kashiwara and M.Vergne, Inventiones Math. 44(1978)1.

10- R.Howe, "Classical Invariant Theory", "Transcending Classical Invariant Theory", unpublished Yale University preprints.

A NEW LOOK AT GROUP ORTHOGONALITY RELATIONS

Ronald Shaw

Applied Mathematics, Hull University, ENGLAND.

ABSTRACT Group orthogonality relations are presented in various coordinate-free, and possibly new, guises; it is not assumed that the ground field is algebraically closed. One of the more unlikely guises is used to give a basis-free proof of the "generalized Frobenius-Schur criterion" for the Wigner type of a corepresentation.

1. Notation Let D be an irrep., of a finite (or compact) group H, carried by a vector space X of dimension n over a field F, where char $F \nmid |H|$. The vector space $W = L(X,X)$ is equipped with a non-degenerate orthogonal geometry by the scalar product

$$\langle A, B\rangle = \mathrm{tr}(AB), \quad A,B \in W. \tag{1}$$

Subspaces of W include the enveloping algebra $\underline{A}$ of D, and the commutant $\underline{B} = [\underline{A}]$ of $\underline{A}$ (or of D). By Schur's lemma, $\underline{B}$ is a division algebra over F, and, by Wedderburn's result, $[\underline{B}] = \underline{A}$. One can prove [1), 2)] that $\underline{A}$ and $\underline{B}$ are nonsingular subspaces. Let $\{J_\mu\}$, $\{J^\mu\}$ be a pair of dual bases for $\underline{B}$:

$$\langle J_\mu, J^\nu\rangle = \delta_\mu^\nu, \quad \mu, \nu = 1, 2, \ldots d = \dim_F \underline{B} \tag{2}$$

Linear operators $W \to W$ include the dyad $|A\rangle\langle B| : C \mapsto \langle B,C\rangle A$, the perp. projections $E_{\underline{A}}$, $E_{\underline{B}}$ onto $\underline{A}$, $\underline{B}$, and, for $A \in GL(X)$, $\mathrm{Ad}A : B \mapsto ABA^{-1}$.

2. Orthogonality relations The O.R.'s involving D inhabit the space $\otimes^2_2(X) = X \otimes X \otimes X' \otimes X'$, X' = dual of X, and will be exhibited below in 3 equivalent guises. In (3), (4) we view $\otimes^2_2(X)$ as $L(W,W)$, and in (5) as $L(\otimes^2 X, \otimes^2 X)$. In (5), τ denotes the flip operator $x \otimes y \mapsto y \otimes x$, $x,y \in X$.

Theorem (i) $\mathcal{A}v_{h\in H} \mathrm{Ad}\, D(h) = E_{\underline{B}}$, (3)

(ii) $\mathcal{A}v\, |D(h)\rangle\langle \bar{D}(h)^{-1}| = m^{-1} E_{\underline{A}}$, $(m = n/d)$, (4)

(iii) $\mathcal{A}v\, D(h) \otimes D(h)^{-1} = \sum_{\mu=1}^{d} (J_\mu \otimes J^\mu) \circ \tau$. (5)

Proof The usual approach to O.R.'s yields (3). See Shaw[2)] for (4),(5).

Remark If $F = \mathbb{R}$, then d = 1, 2 or 4, since (Frobenius) $\underline{B}$ must be isomorphic to $\mathbb{R}$, $\mathbb{C}$ or $\mathbb{H}$. All 3 possibilities do occur.

Remark If $F = \mathbb{C}$ (or is algebraically closed), then (Schur) $\underline{B}$ consists of scalar multiples of the identity I_X, d = 1; also (Burnside) $\underline{A} = W$ and so $E_{\underline{A}} = I_W$. Noting $\langle I_X, I_X\rangle = n$, the O.R.'s now read

$$\mathrm{Av}\ \mathrm{AdD}(h) = n^{-1}|I_X\rangle\langle I_X| \qquad (3')$$
$$\mathrm{Av}\ |D(h)\rangle\langle D(h)^{-1}| = n^{-1} I_W \qquad (4')$$
$$\mathrm{Av}\ D(h) \otimes D(h)^{-1} = n^{-1}\tau \qquad (5')$$

Remark The usual approach to O.R.'s cannot directly yield (4'),(5') - note e.g. that $h \mapsto D(h) \otimes D(h)^{-1}$ is not a representation! For a direct proof of (5'), see Shaw[2].

3. Application of (5') Let H be of index 2 in $G: G = H \cup aH$. If D is a unitary irrep. of H, then so is E, where $E(h) = D(a^{-1}ha)$. Suppose there exists $K \neq 0$ which antilinearly intertwines E with D:

$$KE(h)K^{-1} = D(h),\ h\in H,\ K \in AL(X,X). \qquad (6)$$

Using Schur, we may assume K is antiunitary: $(Kx,Ky) = (y,x)$, and prove that it satisfies

$$K^2 = \epsilon D(a^2),\ \epsilon = \pm 1. \qquad (7)$$

The irreducible corepresentation of G arising from D (see e.g. Theorem B of Shaw and Lever[3]) is of Wigner type I or II according as $\epsilon = 1$ or -1. If we define a bilinear form x.y on X by

$$x.y = (Kx,y), \qquad (8)$$

it is clearly nondegenerate, and, using (6),(7),(8), satisfies:

(Invariance property) $E(h)x.D(h)y = x.y,\ h\in H,$ (9)

(Symmetry property) $x.y = \epsilon y.D(a^2)x.$ (10)

Theorem (Criterion for Wigner type) $\mathrm{Av}_{h\in H} D((ah)^2) = n^{-1}\epsilon I_X$; thus the character χ of D satisfies $\mathrm{Av}_{h\in H}\chi((ah)^2) = \epsilon$.

Proof Use (5'),(9),(10), together with $a^2.a^{-1}ha.h = (ah)^2$:

$$n^{-1}x.y = \mathrm{Av}\ D(h)y.D(h)^{-1}x = \mathrm{Av}\ E(h)D(h)y.x$$
$$= \epsilon\,\mathrm{Av}\ x.D(a^2)E(h)D(h)y = \epsilon\,\mathrm{Av}\ x.D((ah)^2)y.$$

Remark A simpler proof yields $\mathrm{Av}\ D((ah)^2) = 0$ for Wigner type III.

Remark The Frobenius-Schur criterion for the reality type of D is obtained as the special case $G = H \times Z_2$, whence $D = E$, $D((ah)^2) = D(h^2)$, $D(a^2) = I$.

Remark For σ-coreps of G we need $E(h) = \{\bar\sigma(h,a)/\bar\sigma(a,a^{-1}ha)\}D(a^{-1}ha)$ and $K^2 = \epsilon\sigma(a,a)D(a^2)$, as in Theorem B of [3]. Then $x.y = \epsilon\sigma(a,a)y.D(a^2)x$, and the above proof yields the result $\mathrm{Av}_{b\in aH}\sigma(b,b)D(b^2) = n^{-1}\epsilon I_X$.

1) R.Shaw, Linear algebra and group representations, Acad. Press 1982
2) R.Shaw, Multilinear algebra and group representations, Acad. Press 1982.
3) R.Shaw, J. Lever, Commun. Math. Phys. 38, 257-277, (1974).

LIE SUPERGROUPS AND GRADED LIE GROUPS

B.R.Sitaram,

Physics Department,
Delhi University, Delhi-110007,
India

We study in this communication, the basic differences in the properties of Lie supergroups and graded Lie groups with regard to uniqueness, representations and classification.

A central problem in the study of graded Lie algebras (GLA)[1] is the construction of a structure which would in some sense correspond to the GLA as the Lie group corresponds to the Lie algebra. Two structures of this type reported in the past, are Lie supergroups (LSG)[2] and graded Lie groups (GLG)[3]. We wish to study here the basic difference between the two structures.

Briefly, the constructions of LSG's and GLG's proceed as follows: let $g = g_o + g_1$ be a Z_2 graded Lie algebra over K. Let V be a K-commutative graded algebra. Let g' be the tensor product of V and g over K and define a commutator in g' by $[(\sigma,X), (\tau,Y)] = (\sigma\tau, [X,Y])$, $\sigma, \tau \epsilon V$, $X,Y \epsilon g$. g' is a Lie algebra under this bracket. Let G' be the connected, simply connected Lie group corresponding to g'. Then, G' is the LSG corresponding to g with parameters in V.

On the other hand, let G be the connected, simply connected Lie group corresponding to g_o, let K(G) be its K group ring and let U(g) be the Universal algebra over g; define,

$$E(G,g) = K(G) \# U(g),$$

with a K linear associative product defined by

$$(S,X)\ (t,Y) = (St,\ X.SYS^{-1}),\quad S,t\epsilon\ G, X,Y\epsilon U(g).$$

Then E(G,g) is said to be the GLG corresponding to g.

It is worthwhile noticing that G' is a Lie group while E(G,g) has the structure of a Hopf algebra.

We first consider the problem of uniqueness. It is clear that the structure of a LSG depends on both the GLA and the parameter space, V. If we restrict ourselves to a unique parameter space for each GLA, the map $g \to G'$ would be well defined. The inverse map $G' \to g$ need not, however, be well defined. In fact, let g(a) be the GLA over two generators J_o and J_1 with $|J_o| = 0$, $|J_1| = 1$.

$$[J_1, J_1] = a\, J_o$$

It is clear that g(0) is not isomorphic to g(1). Let V be the algebra generated by 1 and τ, $\tau^2 = 0$. Then g'(a), for all values of a is the abelian Lie algebra over two generators. Hence, the map $g \to g'$ is surjective but not injective.

On the other hand, given g,E(G,g) is uniquely determined and vice-versa. In fact, g can be recovered from E(G,g) by considering the primitive elements in E(G,g). Thus the map $g \to E(G,g)$ is invertible.

We now come to the question of representations. Let σ: $g \to h$ be a homomorphism. It is easy to see that, by K linearity, σ defines a homomorphism of Lie algebras σ': $g' \to h'$ and hence, a Lie group homomorphism $\Sigma' : G' \to H'$. Assume now that Σ': $G' \to H'$, or equivalent, σ': $g' \to h'$ is given. Since g' is ungraded, there may not exist any homomorphism σ: $g \to h$ which induces σ'. For example, let g = h = g(1) and let σ' be the map defined by $J_o \leftrightarrow \tau J_1 . \sigma'$ is obviously an isomorphism of Lie algebras, but cannot be induced by a GLA homomorphism as this map would have to change the Z_2 degree. Thus the map $Hom(g,h) \to Hom(G',H')$ is not surjective.

The situation is better, however, in the case of GLG's. We have proved elsewhere[4] that there is a natural invertible map $Hom(g,h) \to Hom(E(G,g),E(H,h))$. In fact σ: $g \to h$ induces as usual σ_o: $g_o \to h_o$ and hence the map exp σ_o: $G \to H$; and K(exp σ_o): $K(G) \to K(H)$. Also σ induces as usual, $U(\sigma)$: $U(g) \to U(h)$. combining, we get a map $E(G,g) \to E(H,h)$. In effect, we can summarize the above by stating that there exists an invertible functor, Kos, from the category of GLA's to the category of GLG's.

We now come to the question of classifications. It is natural in the case of LSG's to take over the classification theory of Lie groups in toto. However, it is easy to see that the passage from GLA to the LSG does not necessarily respect classification. In fact considering the g(a) defined above, it is obvious that g(0) is abelian, while g(1) is nilpotent. In both cases G' as defined above is abelian.

Once again, the results for GLG's are superior. We have shown[5] that it is possible to give a classification theory for GLG's which is natural in the sense that (1) the theory reduces to the theory for Lie groups whenever g is trivially graded; (2) Kos preserves classification i.e., E(G,g) is for example semisimple iff g is semisimple. A further result is that the theory is non trivial in the sense that there exist (for example) semisimple and solvable GLG's and no semisimple GLG is solvable and viceversa. Finally, it is possible to prove the counterpart in the case of GLG's of many of the standard results of Lie groups classification theory: e.g., nilpotent → completely → solvable → solvable. All these results basically stem from the fact that the functor Kos is exact.

In view of the above results, we believe that the theory of GLG's is mathematically superior to the theory of LSG's, A price one has to pay for this is the lack of a group structure on a GLG; we believe however that this is a small price to pay in pay in view of the natural elegance of Kostant's work.

ACKNOWLEDGEMENTS

I am deeply thankful to the organisers of the Colloquium for giving me an opportunity to present my paper at the Colloquium.

REFERENCES

1- M.Scheunert, "The theory of Lie superalgebras", Springer Lecture Notes in Maths, 716 (1979).

2- F.Berezin & G.I.Kac, Mat, Sb. USSR, 11, 311 (1970): A.Rogers, J. Math. Phys, 22 939 (1981).

3- B.Kostant in "Differential Geometric Methods in Phys. Bonn, 1977" ed, K.Bleuler & A Reetz, Springer Lecture Notes in Maths, 570, (1977).

4- B.R.Sitaram & K.C.Tripathy, "Representations of graded Lie groups" J.Math.Phys. (to appear).

5- B.R.Sitaram & K.C.Tripathy, "Representation of graded Lie groups II" submitted to J.Maths.Phys.

THE INVARIANTS OF THE NONDEGENERATE REPRESENTATIONS OF THE GROUP OF THE PSEUDO-ORTHOGONAL MATRICES SO(P,1)

Yi.A.Verdiyev

Institute of Physics of the Academy of Sciences
of the AzSSR, Baku 73, U.S.S.R.

ABSTRACT

The invariants of the nondegenerate representations of the group of pseudo-orthogonal matrices SO(p,1) are constructed.

1- The representation $T_\chi(g)$, $\chi = (\sigma,L)$, $g\varepsilon\, SO(p,1)$ can be realized on the space D vector-function F(x) on the cone $[x,x] = x_1^2+\ldots+x_p^2-x_{p+1}^2=0$, $x_{p+1}>0$, of the **homogeneity** degree, whose values belong to the space D_L of the representation $D^{(L)}(r)$, $r\varepsilon SO(p,1)$

$$T_\chi(g)F(x) = D^{(L)T}(r(x,g)F(xg)$$

In the spherical coordinate system

$$F(x) = e^{\sigma\alpha}f(\xi)$$

$$T_\chi(a)f(\xi) = e^{\sigma(\alpha_a-\alpha)}f(\xi_a),\quad a = g_{p,p+1}\varepsilon SO(1,1)$$

$$T_\chi(k_o)f(k) = f(k\, k_o),\quad k,k_o\varepsilon SO(p)$$

$$f(rk) = D^{(L)T}(r)f(k)$$

$$\xi = \dot{\xi}h(\xi),\quad \dot{\xi} = (0,\ldots,0,1,0),\quad k = rh(\xi),\quad r\varepsilon SO(p-1)$$

The bilinear invariant functionals for the pair of representations $\chi_1 = (\sigma_1,L_1)$ and $\chi_2 = (\sigma_2,L_2)$ exist in the cases:

a) when $\tilde{L}_2 = L_1$, $\sigma_1+\sigma_2+p-1 = 0$. In this case

$$B(f_1,f_2) = \int\langle f_1(k),\ f_2(k)\rangle\ dk$$

b) when $L_1 = L_2 = L$, $\sigma_1 = \sigma_2 = \sigma$. In this case we have

$$B(f_1,f_2) = \int < f_1(k_1),\ K^{(2)}(k_1,k_2) f_2(k_2) > dk_1 dk_2$$

$$K^{(2)}(k) = D^{(L)}((rr')^{-1}) D^{(L)}(I_s)(1-\dot{\xi}k\dot{\xi})^{-p-\sigma+1},\quad k = rg_{12}(\Psi)r',$$

where I_s is reflection of the first p-1 axes.

The kernel of a three-linear functional

$$P(f_1,f_2,f_3) = \int < \prod_{i=1}^{3} dk_i f_i(k_i) K^{(3)}(k_1,k_2,k_3) >,$$

$$f_i(k) \varepsilon D_{\chi_i} \quad \chi_i = (\sigma_i, L_i),\ i = 1,2,3,$$

satisfies the conditions

$$K^{(3)}(k_1k_o, k_2k_o, k_3k_o) = K^{(3)}(k_1,k_2,k_3),$$

$$K^{(3)}(r_1k_1, r_2k_2, r_3k_3) = \prod_{i=1}^{3} \otimes D^{(L_i)}(r_i^{-1}) K^{(3)}(k_1,k_2,k_3).$$

The simultaneous solution of these conditions can be represented in the form

$$K^{(3)}(k_1,k_2,k_3) = \sum_{L_{12},L_{23},L_{13}} A(L_{12},L_{23},L_{13}) < C^{L_1,L_{13},L_{12}}$$

$$K^{(12)}(k_1k_2^{-1}) C^{L_2,L_{12},L_{23}} K^{(23)}(k_2k_3^{-1}) C^{L_3,L_{23},L_{13}} K^{(13)}(k_3k_1^{-1}) >,$$

where

$$K^{(ij)}(g_{12}(\Psi)) = (1 - \dot{\xi}k_i k_j^{-1}\dot{\xi})^{-a_{ij}}$$

$$2a_{12} = \sigma_1-\sigma_2+\sigma_3-p+1,\quad 2a_{23} = -\sigma_1+\sigma_2+\sigma_3-p+1,\quad 2a_{13} = \sigma_1-\sigma_2+\sigma_3-p+1.$$

Here $A(L_{12},L_{23},L_{13})$ are arbitrary constants depending on the intermediate momentums and $C^{L_i,L_{ik},L_{ij}}$ are the Wigner coefficients for the group SO(p-1).

TOPOLOGICAL AND ALGEBRAIC STRUCTURE OF LINEAR PROBLEMS ASSOCIATED WITH COMPLETELY INTEGRABLE SYSTEMS

D.V. Chudnovsky*)

Dept. of Mathematics
Columbia University
New York, N.Y. 10027, USA

---

*) This work was supported by the U.S. Air Force under Grant AFOSR-81-0190, and by the John Simon Guggenheim Memorial Foundation.

CHAPTER I

BÄCKLUND TRANSFORMATIONS

"Names are not always what they seem. The common Welsh name Bzjxxllwep is pronounced Jackson"

M.Twain. Pudd'nhead Wilson's New Calendar.

Bäcklund transformations have long ago established themselves as one of the most important tools in the investigation of completely integrable systems of p.d.e. Various geometric and analytic interpretations of Bäcklund transformations (BTs) as differential correspondences were examined in recent years especially in connection with isospectral deformation equations[1,2]. D.V. and G.V. Chudnovsky in a series of papers starting from 1979, inspired by striking analogy with the Padé approximation theory, were developing the concept of BT as an isomonodromy deformation of solutions of the Riemann boundary value problem consisting in creation of apparent singularities and addition of integers to local multiplicities[3,4,5,6,7,9]. In this chapter we try to summarize this approach together with topological interpretation of BT based on category theory. We also touch upon BTs realized as Darboux transformations applied to three dimensional completely integrable systems; subject to be elaborated upon in Chapter II.

I.1. BTs for solutions of the Riemann boundary value problem

The definition of BT can be roughly formulated as a linear transformation with rational function coefficients acting on multivalued matrix solution $\phi(\lambda)$ of the Riemann boundary value problem in the λ-plane[10,5-7*)]. From the point of view of nonlinear completely integrable systems this means that BTs are directly associated with initial linear spectral problem. This point of view on BT can be traced in the classical literature to Lappo-Danilevski[12], Riemann[13] (contiguous relations),... etc. and was studied as "addition of

---

*) For many important cases of "reductions", when additional symmetries are imposed on $\phi(\lambda)$ the λ-plane is substituted by an arbitrary Riemann surface Γ. This situation can be, however, restated for the λ-plane, increasing the rank of vector bundle defined by $\phi(\lambda)$, considering Γ as a covering of $\mathbb{C}$ (cf. 11).

integers" in[3,4,6]. Zakharov-Mikhailov[14] and Zakharov-Shabat[35] studied essentially the same concept under the name of "dressing of the vacuum" or through the solution of the Riemann boundary value problem with zeroes. In a series of papers of Miwa and Jimbo[15,16] an equivalent concept was described as Schlesinger transformations*).

The definition of BT from[5,6,9,17] identify the Bäcklund transformation of the initial n x n matrix $\phi(\lambda)$ as its linear transformation

$$\phi'(\lambda) = L(\lambda)\phi(\lambda) \tag{1.1}$$

For $\phi(\lambda)$ satisfying the Riemann boundary value problem and $\Gamma = \mathbb{CP}^1$, BT (1.1) can be invariantly described in terms of regular exponential matrices W_a of $\phi(\lambda)$ at singular points $\lambda=a$.

At a singular point $\lambda=a$ $\phi(\lambda)$ has the following local expansion

$$\phi(\lambda) = \tilde{\phi}_a(\lambda).(\lambda-a)^{W_a}.H_a(\lambda) \tag{1.2}$$

where $\tilde{\phi}(\lambda)$ is holomorphic and invertible at $\lambda=a$, W_a is called a regular exponential matrix, and $H_a(\lambda)$ corresponds to irregular singular behaviour at $\lambda=a$ and is an entire function in inverse powers of the local parameter $\lambda-a$.

According to[5,6,7] the space of BT data is determined by the isomonodromy deformation conditions

$$\exp(2\pi\sqrt{-1}W_a) = \text{const}$$

for all a (being a regular or a singular point of $\phi(\lambda)$). Following[6,7,12] the condition of isomonodromy can be expressed as follows. Let W_a be reduced to its Jordan normal form

$$W_a = S_a^{-1} . (J_{\rho_1},\ldots,J_{\rho_\ell}).S_a \tag{1.3}$$

where J_{ρ_i} is a Jordan block of the size k_i corresponding to the eigenvalue ρ_i, $k_1 + \ldots + k_\ell = n$. While a Jordan normal form is unique

*) For Painlevé equations (manifestly Painlevé VI), BTs were discovered by R. Garnier, who remarked that they are simple consequences of contiguous relations between auxiliary Fuchsian l.o.d.e.

the choice of S_a is not and it depends on an arbitrary element of centralizer of W_a. Hence the space of all possible BT data of $\phi(\lambda)$ is described as a space of all possible regular exponential matrices W'_a such that

$$W'_a = S_a^{-1} \cdot (J_{\rho_1} + m_1 I_{k_1}, \ldots, J_{\rho_\ell} + m_\ell I_{k_\ell}) \, S_a,$$

where (i) S_a are arbitrary matrices satisfying (1.3), with the proper normalization at singularities $\lambda=a$ in (1.2) (continuous parameters); and (ii) $m_1, \ldots, m_\ell$ are arbitrary integers and I_{k_i} are $k_i \times k_i$ unit matrices (discrete parameters).

The case of reductions was examined in[17,18], where infinite dimensional "Lie-Mikhailov" algebras were introduced to describe the space of BT data, preserving additional symmetries.

In applications to completely integrable systems, $\phi(\lambda)$ is a function of parameters x, t, y, ... and satisfies linear o.d.e. with coefficients being rational in λ. An important remark here is the following one, showing the invariance of the structure of this l.o.d.e. under the action of BT:

<u>Lemma 1.1.</u> Let $\phi(\lambda)$ satisfy l.o.d.e.

$$\partial\phi(\lambda)/\partial x = U(\lambda)\phi(\lambda),$$

and let $\phi'(\lambda)$ be a BT (1.1) of $\phi(\lambda)$ for which BT data are x-independent. Then $\phi'(\lambda)$ satisfies l.o.d.e.

$$\partial\phi'(\lambda)/\partial x = U'(\lambda)\phi'(\lambda).$$

Here $U'(\lambda)$ is regular at those points in the λ-plane where $U(\lambda)$ is regular.

It is convenient to reformulate BT data using the language of inverse scattering. We define the scattering matrix for solutions of l.o.d.e. with coefficients depending on λ:

$$\partial\phi(\lambda,x)/\partial x = U(\lambda,x)\phi(\lambda,x) \tag{1.4}$$

where $U(\lambda,x)$ is assumed to be rational in λ:

$$U = \sum_{j=1}^{m} \sum_{r=1}^{q_j} U_{i,r}(\lambda-a_i)^{-r} + \sum_{r=0}^{q_\infty} U_{\infty,r}\lambda^{-r}.$$

The main assumption here is simple behavior of $U(\lambda,x)$ at x-infinity: $U(\lambda,x)\to U_\infty(\lambda)$ as $|x| \to \infty$, where $U_\infty(\lambda)$ is x-independent. Hence we define the scattering matrix $S(\lambda)$ of linear problem (1.4) as

$$S(\lambda) = \phi_-(x,\lambda)^{-1}.\phi_+(x,\lambda).$$

Here, ϕ_+ and ϕ_- denote two fundamental solutions of (1.4) for which

$$\phi_1(x,\lambda) \to e^{U_\infty(\lambda)x} \quad \text{as } x \to \pm\infty$$

The BT of a (solution of) linear differential operator $L_\lambda=\partial/\partial x-$

$U(\lambda,x)$ is a new linear differential operator $L'_\lambda = \partial/\partial x - U'(\lambda,x)$, given by

$$L'_\lambda = L(\lambda).L_\lambda.L(\lambda)^{-1} \tag{1.5}$$

(1.5) is equivalent to the BT expression (1.1) with $\phi(\lambda)$ and $\phi'(\lambda)$ being fundamental solutions of, respectively, $L_\lambda\phi=0$, $L'_\lambda\phi'=0$. This definition of BT allows us to reexpress BT data. The scattering matrices $S'(\lambda)$ of L'_λ and $S(\lambda)$ of L_λ are related by a linear fractional transformation

$$S'(\lambda) = B(\lambda)^{-1}.S(\lambda).B(\lambda)$$

for an element $B(\lambda)$ from $gl(n) \otimes \mathbb{C}(\lambda)$. In general, for symmetries added $B(\lambda)$ belongs to infinite-dimensional "Lie-Mikhailov" groups[7].

I.2. Some categorial considerations

The group of BTs turns out to be a very complicated infinite dimensional object (even within a given class of reduction). In order to describe it, it is useful to use the analogy with algebraic topology and to accept the definitions from category theory[19]. This language turns out to be particularly useful for studies of BTs applied to the quantum versions of completely integrable equations. In this quantum setting, the BT formula (1.1) is interpreted as a quantum transfer matrix equation with elements of the matrix $L(\lambda)$

being elements of a certain associative algebra B, i.e. $L(\lambda)$ is considered as an element of an infinite dimensional algebra $B \otimes \mathbb{C}(\lambda)$. In order to insure that the successive applications of BTs (1.1) generate completely integrable systems, one imposes on the structure of $L(\lambda)$ certain restrictions that in fact determine the symplectic structure of the lattice systems generated by BT (1.1). The restrictions on $L(\lambda)$, determining commutation relations between its elements, takes the form of the so called Baxter lemma[18,20]:

$$R(\lambda,\mu)(L(\lambda) \otimes L(\mu)) = (L(\mu) \otimes L(\lambda))R(\lambda,\mu) \tag{2.1}$$

for a certain (scalar) $n^2 \times n^2$ matrix $R(\lambda,\mu)$. Similar relations can be written between different transfer matrices $L(\lambda)$. These relations can be properly understood only in the context of monoidal symmetric categories introduced by MacLane[19]. These categories $C = \langle C, \otimes, \alpha, \gamma \rangle$ are characterized by the existence of a bifunctor $\otimes: C \times C \to C$ and two natural isomorphisms α and β. Explicitly

$$\alpha = \alpha_{x,y,z} : x \otimes (y \otimes z) \cong (x \otimes y) \otimes z$$

and

$$\gamma_{x,y} : x \otimes y \cong y \otimes x$$

are natural for all $x, y, z \in C$. Isomorphisms α and γ express "associativity and commutativity up to isomorphism". Axioms of monoidal symmetric category consist of diagrams describing the compatibility of α and γ called hexagon and pentagon[19, Chapter 7]. This general definition of monoidal symmetric category is applied in the case when C is a category of vector spaces and the bifunctor $\otimes: C \times C \to C$ is a natural operation of tensor product of vector spaces. This category C is strictly associative, which means that the isomorphism $\alpha = \alpha_{x,y,z}$: $x \otimes (y \otimes z) \cong (x \otimes y) \otimes z$ is fixed as an identity isomorphism for arbitrary vector spaces $x, y, z \in C$.

The isomorphism $\gamma = \gamma_{x,y} : x \otimes y \cong y \otimes x$ is then the operation that determines operation of Bäcklund transformation and "S-matrices" $R(\lambda,\mu)$. The functoriality axiom itself, without additional conditions of monoidal symmetry, turns out to be equivalent to the generalized Baxter lemma (2.1). Conditions of monoidal symmetry is expressed then as an "axiom of triangle" (L. Breen):

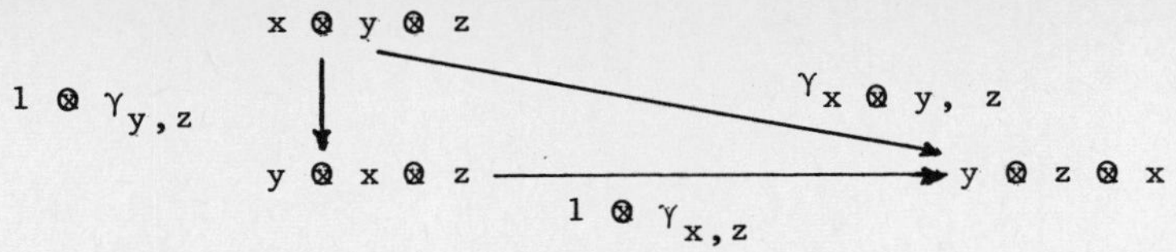

$$\begin{array}{ccc} x \otimes y \otimes z & & \\ {\scriptstyle 1 \otimes \gamma_{y,z}} \downarrow & \searrow {\scriptstyle \gamma_{x \otimes y,\, z}} & \\ y \otimes x \otimes z & \xrightarrow[\ 1 \otimes \gamma_{x,z}\ ]{} & y \otimes z \otimes x \end{array}$$

This axiom together with functoriality, read in terms of S-matrix $R(\lambda,\mu)$ is equivalent to the factorizations axioms of S-matrices introduced by Zamolodchikov and Baxter[20].

In this approach, preserving the functoriality axiom, one can study categories that do not satisfy the triangle axiom, but satisfy higher analogs of commutativity (symmetry) borrowed from algebraic topology. The axiom next in complexity to the triangle one is equivalent to Zamolodchikov tetrahedron axiom[21], which is used to quantize three dimensional systems. Similarly one can consider non factorized S-matrices leading to completely integrable systems.

I.3. Darboux transformations as Bäcklund transformations

Historically first BTs were presented by Darboux[22], who studied the following problem: how to construct Schrödinger operator $\phi''+u(x)\phi=\lambda\phi$ all of whose eigenfunctions ϕ are explicitly known. He devised an iterative procedure for construction of such potentials starting from a given one (e.g. zero one). Later Darboux extended this construction to other linear problems, in particular to the linear problem corresponding to the sin-Gordon equation (thus giving BT for sin-Gordon). Later Darboux transformation was rediscovered by people doing spectral theory (by Krein and Crum[23]) and the scattering interpretation of the Darboux transformation was given Flaschka and McLaughlin[1]. Later in[24] we extended Darboux transformations to arbitrary non-stationary linear differential operators $Ly=\frac{\partial}{\partial t}y$, $L=(\frac{\partial}{\partial x})^n+u_{n-2}(\frac{\partial}{\partial x})^{n-2}+\ldots$ and obtained from this BTs and multisoliton formulas for three dimensional completely integrable systems of Zakharov-Shabat type. Darboux transformations (DTs) are particular cases of BTs and can be taken as a basis for "elementary" BTs. Such a choice of the basis for "elementary" BTs is particularly useful when we are examining universal relations between BTs. Their usefulness lies in a simple Wronskian type structure of the iterations of BTs.

Studying DT as a means of generation of BTs for completely integrable systems we want to find transformations of linear p.d.e.s. In the scalar case these linear problems take the form

$$L_n\phi = \frac{\partial}{\partial t_n}\phi,$$

$$L_n = \sum_{i=0}^{n} u_i^{(n)} \left(\frac{\partial}{\partial x}\right)^i, \quad u_n^{(n)} = 1, \quad u_{n-1}^{(n)} = 0 \tag{3.1}$$

$n=1,2,\ldots$ (and $t_1 = x$). The essence of the iterative applications of DTs lies in the following.

<u>Proposition 3.1.</u> Let $\phi_1,\ldots,\phi_m$ be arbitrary solutions of the system (3.1) and let ψ be its general solution. Then a new function

$$\Psi = \frac{W(\phi_1,\ldots,\phi_m,\psi)}{W(\phi_1,\ldots,\phi_m)} \tag{3.2}$$

is a general solution (eigenfunction) of a new system of linear problems of type (3.1):

$$\hat{L}_n\Psi = \frac{\partial}{\partial t_n}\Psi,$$
$$\hat{L}_n = \left(\frac{\partial}{\partial x}\right)^n + U_{n-2}^{(n)} \left(\frac{\partial}{\partial x}\right)^{n-2},\ldots \ : n=1,2,\ldots \ . \tag{3.1'}$$

The coefficients of operators $\hat{L}_n$ are determined explicitly in terms of coefficients of L_n and Wronskian in ϕ_i. Here W denotes a Wronskian:

$$W(f_1,\ldots,f_k) = \det\left(\frac{\partial^{i-1} f_j}{\partial x^{i-1}}\right) \ i.j=1,\ldots,k^{*)}.$$

One has explicit formulas for all coefficients of new $\hat{L}_n$. However, only one of these coefficients is really an independent function since all the linear problems (3.1) are consistent: $\left[\frac{\partial}{\partial t_n} - L_n, \frac{\partial}{\partial t_k} - L_k\right] = 0$ for $n,k=1,2,\ldots$. Hence all "pseudo-

---

*) In view of a particular form of system (3.1) Wronskians from their solutions admit a more interesting multidimensional form of generalized Wronskians:

$W(\phi_1,\ldots,\phi_m) = \det\left(\frac{\partial}{\partial t_{i-1}}\phi_j\right)_{i,j=1}^{m}, \ \frac{\partial}{\partial t_0}\phi_j = \phi_j.$

potentials" $u_i^{(n)}$ can be determined via differential equations from the first nontrivial coefficient $u=\frac{1}{2}u_0^{(2)}$ $(L_2 \overset{def}{=\!=} \frac{\partial^2}{\partial x^2}+2u)$, e.g. $u_{n-2}^{(n)}=nu$, etc. The transformation formula from u to U in (3.1)' under the action of DT (3.2) is the following

$$U = u + \frac{d^2}{dx^2} \log W(\phi_1,\ldots,\phi_m). \tag{3.3}$$

This formula takes even more familiar form (of Fredholm determinant or logarithmic derivative of the θ-function) if one introduces, following Wahlquist[1] the potential σ such that $u=\frac{d^2}{dx^2}\log\sigma$. The transformation formula from the potential σ to a new potential σ' under the action of DT (3.2) now takes the form

$$\sigma' = \sigma\cdot W(\phi_1,\ldots,\phi_m). \tag{3.4}$$

Remark 3.2. Let ϕ_i now be normalized eigenfunctions of (3.1) corresponding to spectral parameters λ_i with a general eigenfunction $\psi=\psi(\lambda)$ of (3.1) having the following expansion at $\lambda\to\infty$: $\psi(\lambda) = \{\sum_{j=0}^{\infty} \psi_j\lambda^{-j}\}\times\exp\{\sum_{i=1}^{\infty}\lambda^i t_i\}$, $\psi_0=1$[24]; $\psi_i=\psi(\lambda_i)$: $i=1,\ldots,m$.

Then the formula for normalized potential in (3.4) takes the following normalized form (cf. Wahlquist[1]):

$$\sigma' = \sigma\cdot\frac{W(\phi_1,\ldots,\phi_m)}{W(\lambda_1,\ldots,\lambda_m)}, \tag{3.4'}$$

where $W(\lambda_1,\ldots,\lambda_m)$ is the Wandermond determinant in $\lambda_1,\ldots,\lambda_m$.

We show now why DTs can be taken as a basis of "elementary" BTs. For this we want to remind a general formula for another basis of "elementary" BTs from[7] corresponding to the addition of one pole at $\lambda=\mu_1$ and one zero at $\lambda=\mu_2$ to $\Phi(\lambda)$. If $\mu_1\neq\mu_2$ then the BT $\Phi'(\lambda) = \mathcal{B}_{\mu_1}^{\mu_2}\Phi(\lambda)$ has the form

$$\Phi'(\lambda) = \{I + \frac{\mu_1-\mu_2}{\lambda-\mu_1}P\}\Phi(\lambda) \tag{3.5}$$

with $P = \phi_2^t\phi_1/\phi_1\phi_2^t$ and two eigenfunctions $\phi_2^t=\Phi(\mu_2).C_2^t$, $\phi_1=C_1.\Phi(\mu_1)^{-1}$ (with arbitrary vectors C_1, C_2). However the "elementary" BT $\mathcal{B}_{\mu_1}^{\mu_2}$ can be decomposed into the following action of BTs $\mathcal{B}_{\mu_1}^{\mu_2} = \mathcal{B}_{\infty}^{\mu_2} \circ (\mathcal{B}_{\infty}^{\mu_1})^{-1}$.

Here $\mathcal{B}_{\infty}^{\mu}$ is a BT adding a zero at $\lambda=\mu$ and changing exponent at infinity. This BT is described in[6], in fact, coincides with elementary DT (3.2) for m=1, $\lambda_1=\mu$. Hence one can take for basis of BTs Darboux transformations.

CHAPTER II

COMPLETELY INTEGRABLE NONLINEAR DIFFERENCE AND DIFFERENTIAL EQUATIONS GENERATED BY BTs

"It is easy to find fault, if one has that disposition. There was once a man who, not being able to find any other fault with his coal, complained that there were too many prehistoric toads in it".

M.Twain. Pudd'nhead Wilson's Calendar.

II.1. General matrix two dimensional isospectral deformation equations and three dimensional systems

In this section we briefly review the generation of two dimensional isospectral deformation equations from an arbitrary linear differential operator with matrix coefficients. All known two dimensional completely integrable systems and hierarchies are particular reductions of this class of equations[25,26]. This review is presented parallel to the examination of three dimensional scalar completely integrable systems of Kadomtzev-Petviashvili (KP) type. Namely, we show how solutions of all matrix two dimensional systems are naturally imbedded into solutions of scalar three dimensional systems of KP type. This is important in order to incorporate later all matrix two dimensional isospectral deformation systems into a single universal "difference" relation. We also briefly examine pole interpretation of two and three dimensional completely integrable systems.

In the description of two dimensional isospectral deformation equations we follow Newell[25] and Dubrovin[27], see also[28]. The size of matrices under consideration is chosen to be n and for the space $\mathbb{A}$ of diagonal matrices, variables are labeled by elements $A \in \mathbb{A}$, and we denote $A_k = (\delta_{ik}\delta_{jk})_{i,j=1}^{n} \in \mathbb{A}$: $k = 1,\ldots,n$. The commuting family of matrix linear differential spectral problems determining isospectral deformation equations has the following form

$$L_A(\lambda) \overset{\text{def}}{=\!=} \frac{\partial}{\partial x_A} + [\,A,V\,] - \lambda\cdot A \qquad (1.1)$$

where $\frac{\partial}{\partial x_A} = \sum_{i=1}^{n} a_i \frac{\partial}{\partial x_{A_i}} \equiv \sum a_i \partial/\partial x_i, A = (a_i \delta_{ij}) \in \mathcal{A}$, $[\cdot,\cdot]$ is a commutator, λ is the spectral parameter and $V = (v_{ij})_{i,j=1}^{n}$ is the matrix of "pseudopotentials", where one assumes that diagonal elements of V are zeroes.

Now an arbitrary two dimensional matrix isospectral deformation equation associated with (1.1) can be represented as a consistency condition of two linear problems in variables x_A and t_B for $A,B \in \mathcal{A}$:

$$L_A(\lambda) \cdot \phi(\lambda) = 0$$
$$L_B^{(2)}(\lambda) \cdot \phi(\lambda) = 0 \qquad (1.2)$$

where $L_B^{(2)}(\lambda) = \frac{\partial}{\partial t_B} + B\cdot\lambda^N + U_{N-1}\cdot\lambda^{N-1} + \ldots + U_0$ for $N \geq 1$. Coefficients of $L_B^{(2)}(\lambda)$ can be determined using the following simple rule: we consider a solution of the following matrix differential equations $\frac{d\eta}{dx_A} = [\eta,[A,V]-\lambda A]$. Then for every $B \in \mathcal{A}$ these equations have a unique solution $\eta_B = B + \Sigma_{i=1}^{\infty} \eta_{i,B}\cdot\lambda^{-i}$ such that $\eta_B \equiv B$ if V=0. Then the operator $L_B^{(2)}(\lambda)$ has the form

$$L_B^{(2)}(\lambda) = \frac{\partial}{\partial t_{N,B}} + B\lambda^N + \eta_{1,B}\lambda^{N-1} + \ldots + \eta_{N,B} \quad , \qquad (1.3)$$

(or is a linear combination of operators of this form). Equations (1.2) can be represented as a local system of nonlinear p.d.e. on "pseudo-potentials" v_{ij}:

$$[L_A(\lambda), L_B^{(2)}(\lambda)] = 0. \qquad (1.2')$$

<u>Remark 1.1.</u> The locality of equations (1.2') follows from the local expressions of elements of $\eta_{i,B}$ in terms of v_{ij} and their derivatives in $\partial/\partial x_{A_k} \equiv \partial/\partial x_k$. One should bear in mind a remark made in[9] that since all linear problems of the form (1.2) are commuting, "pseudo-potentials" v_{ij} satisfy systems of commuting nonlinear differential equations in all variables $\partial/\partial x_A$, $\partial/\partial t_{N,B}$ (see (1.3)). Also other classes of two dimensional completely integrable systems arise from spectral problem (1.1) if one considers commutativity condition of the form (1.2) with $L_B^{(2)}$ substituted by a linear operator with arbitrary singularities in the λ plane. Again v_{ij} can be considered as satisfying overdetermined systems (1.2') of all possible commuta-

tivity conditions in a given class of reductions.

Apparently two dimensional systems of the form (1.2) can be imbedded in three dimensional systems of KP-type. As in Chapter I, §3 the KP-systems are represented as commutativity conditions for scalar linear differential operators

$$\left[\frac{\partial}{\partial t_n} - L_n, \frac{\partial}{\partial t_m} - L_m\right] = 0: n,m = 1,2,... \tag{1.4}$$

for $L_n = (\frac{\partial}{\partial x})^n + u_{n-2}^{(n)}(\frac{\partial}{\partial x})^{n-2} + \ldots$.

Remark 1.2. Alternatively L_n can be described using pseudodifferential operator $L_0 = \frac{\partial}{\partial x} + v_{-1}(\frac{\partial}{\partial x})^{-1} + \ldots v_{-n}(\frac{\partial}{\partial x})^{-n} + \ldots$ as $L_n = [L_0^n]_+$ (the differential part of L_0^n).

The main result on the relationship between two dimensional matrix spectral problems (1.1), (1.2) and scalar KP equation is the following.

Theorem 1.3. Let $L_A(\lambda)$ and $L_B^{(2)}(\lambda)$ be commuting linear matrix differential operators from (1.1), (1.2) and $\bar{\phi}$ be their common eigenfunction, corresponding to the solution of equations of (1.2'):

$$L_A(\lambda)\bar{\phi} = 0, \; L_B^{(2)}(\lambda)\bar{\phi} = 0 \text{ for all } A,B \in \mathbb{A}.$$

Then the i-th component of ϕ_i of $\bar{\phi}$ satisfies scalar linear differential equations

$$L_m^{<i>}\phi_i = \frac{\partial}{\partial t_{m,A_i}}\phi_i \; : \; m = 1,2,3,...$$

with $L_m^{<i>} = \Sigma_{\alpha=0}^m a_{\alpha,m}(\partial/\partial x_{A_i})^\alpha$, $a_{m,m} = 1, a_{m-1,m} = 0$ with coefficients $a_{\alpha,m}$ expressed as polynomials in $v_{k,\ell}$ and their derivatives in $\partial/\partial x_{A_i}$.

Example 1.4. For n=2 systems (1.2') with $A=B=\begin{pmatrix}1 & 0\\ 0 & -1\end{pmatrix}$ and n=2,3 are equivalent to the coupled nonlinear Schrödinger and the coupled modified KdV equations:

$$\frac{\partial}{\partial t_2} p = p_{xx} + 2pqp, \quad \frac{\partial}{\partial t_2} q = -q_{xx} - 2qpq;$$

$$\frac{\partial}{\partial t_3} p = p_{xxx} + 6p_x qp, \quad \frac{\partial}{\partial t_3} q = q_{xxx} + 6q_x pq$$

$\partial/\partial x = \partial/\partial x_1$. It follows from Theorem 1.3 (or can be checked directly) that pq satisfies KP equations. Here u=pq is the potential in the sense of Ch. I, §3 (so that $L_2 = \frac{d^2}{dx^2} + 2pq$).

In particular, the imbedding of matrix two-dimensional completely integrable systems into a scalar three-dimensional system of KP type, allows us to determine the evolution of equation for poles of meromorphic solutions of two-dimensional completely integrable systems. For systems of KP type, if no reductions are imposed, the motion of poles of meromorphic functions $u_i^{(n)}$ is described by Hamiltonian flows commuting with the one describing one-dimensional particles interacting with the potential Gx^{-2}, $G = -4$[29,30,31]. Here as above, one can describe equations of evolution in terms of poles of a single function $u = u(x,t_2,t_3,...)$ such that $u_{n-2}^{(m)} = mu$, $m = 2,3,$ ... . We assume that for t_i close to 0, u is meromorphic as a function of x with $a_i = a_i(t_2,t_3,...)$ being the poles in the complex x-plane. Then the evolution of a_i in the t_2 direction is governed by the following completely integrable many-particle Hamiltonian

$$H = \tfrac{1}{2}\Sigma\, b_i^2 - 4 \sum_{i<j} (a_i - a_j)^{-2},$$

$b_i = \dot{a}_i$. This Hamiltonian is known to have an infinite family of commuting Hamiltonians J_m, $m = 2,3,...$, $J_2 = H$ such that J_m is of degree m in b_i and such that $J_m = \frac{1}{m}\Sigma_i\, b_i^m - 4 \sum_{i<j} (b_i^{m-2} + ... + b_j^{m-2}) x (a_i - a_j)^{-2} + \ldots$.*)
In these notations the motion of poles a_i for the solution of the KP hierarchy (1.4) in the t_m-direction is governed by the Hamiltonian J_m. Hence for an arbitrary two dimensional matrix completely integrable system (1.2) with $A=B=A_\alpha$, the motion of poles in the x -plane of meromorphic solutions is governed by the Hamiltonian J_N in the $t_{N,B}$ direction.

Such a simple description of the evolution of poles of mero-

*) J_m is uniquely determined if one imposes on J_m the condition of homogeneity of "degree" $-m$ (b_i is of "degree" -1 as well as $(a_i - a_j)^{-1}$).

morphic solutions of completely integrable systems in dimensions 2 and 3 (typical "Painlevé" property), is completely destroyed if one imposes constraints or reductions on linear problems. The first such instance occurs in the reduced KP hierarchy (1.4), where n and m are odd integers, and, additionally, $u_0^{(n)}=0$. As it was shown in our early paper, already for n=3, m=5 the motion of poles gives rise to new completely integrable Hamiltonians many-particle systems. Other Hamiltonian many-particle systems arise from different reductions.

II.2. The reduction and decomposition of nonlinear completely integrable systems of differential equations into systems of difference equations

Completely integrable systems in dimensions 2 and 3 of the type described in II.1., can be represented instead as a common action of 2 or 3 commuting Hamiltonian systems (depending on an infinite number of degrees of freedom). Moreover, general completely integrable systems of differential equations can be substituted by equivalent systems of difference-differential or difference equations. One such substitution was briefly discussed in II.1., where a pole interpretation served as a decomposition method. Another decomposition method was studied recently, in particular in[28,34], where the moment interpretation of linear spectral problems was used to decompose two or three dimensional completely integrable systems into a common action of 2 or 3 quadratic Hamiltonians with quadratic constraints. For example, the reduction of equations of KP type (1.4) to the system of two dimensional equations is achieved using the expression of "pseudopotentials" in terms of squared eigenfunctions of associate linear problems. Namely $u = \int\psi(\lambda)\psi^*(\lambda)\, d\sigma_\lambda$, and similar expressions for all coefficients of operators $L_n = [(L_0)^n]_+$ follow from the following representation of pseudodifferential operator L_0 as $L_0 = \partial+\Sigma_{j=-1}^{\infty} v_j(\frac{\partial}{\partial x})^{-j}$, for $v_{-j} = \int\Psi(\lambda)(\frac{\partial}{\partial x})^{j-1} x\ \Psi^*(\lambda)\, d\sigma(\lambda)$ for j=1,2,... and eigenfunctions $\Psi(\lambda)$ of $L_0\Psi(\lambda) = \lambda\Psi(\lambda)$, $L_o^*\Psi^*(\lambda) = \lambda\Psi^*(\lambda)$.

Continuing this process, we decompose two-dimensional isospectral deformation equations into the common action of 2 moment - like Hamiltonians, and three-dimensional systems of KP type into the common action of 3 Hamiltonians through the representation $u = tr\{\int\phi(\lambda)\phi^*(\lambda)\, d\Sigma_\lambda\}$, see[28].

We found recently a much more explicit and efficient way to

decompose multidimensional completely integrable systems into a simpler lattice Hamiltonian systems. This approach is based on the reduction of isospectral deformation equations to lattice (difference -differential and difference-difference) systems using a successive application of BTs[5,6,7,17]. In fact, this approach turns out to be more geometrically motivated because the nonlinear completely integrable systems of p.d.e.s appear as a result of infinitesimal action of BTs. Hence the two- or three- (four-) dimensional completely integrable systems of p.d.e.s are limits of the completely integrable lattice models when the spacings of the lattice (in different space-time directions) tend to zero. Completely integrable lattice models themselves are written down as relations between BTs considered as elements of an appropriate infinite-dimensional Lie group of BTs. This procedure was elaborated upon in[5,6,7,17], and we present a few examples of lattice systems generated by the application of BTs in the two-dimensional case.

Successive applications of different BTs and their commutation laws give rise to difference (lattice) equations expressed as algebraic equations on elements of L(λ). We denote BT $\Phi'(\lambda)$ of $\Phi(\lambda)$ corresponding to BT data $\mathcal{B}$ as

$$B_{\mathcal{B}}(\Phi(\lambda)) \overset{\mathrm{def}}{=\!=} \Phi'(\lambda) = L(\lambda)\phi(\lambda), \tag{2.1}$$

and, $L(\lambda)$ is denoted as $L_{\mathcal{B}}(\lambda)$. The law of composition of two BTs can be expressed as follows

$$B_{\mathcal{B}_1\oplus\mathcal{B}_2}(\Phi(\lambda)) = B_{\mathcal{B}_2}(B_{\mathcal{B}_1}(\Phi(\lambda))).$$

or (2.2)

$$L_{\mathcal{B}_1\oplus\mathcal{B}_2}(\lambda) = L_{\mathcal{B}_2}(\lambda)L_{\mathcal{B}_1}(\lambda).$$

In order to understand what lattice models arise from conditions of commutativity of BTs, we consider first two-dimensional difference equations. In this case one takes an abelian subgroup of a group of BTs, generated by two independent BTs. Then the space of BT data under consideration is a free abelian group $A = \{n\mathcal{B}_1 \oplus m\mathcal{B}_2\}n,m\ \mathbb{Z}$ with two generators $\mathcal{B}_1$ and $\mathcal{B}_2$.

Starting from an arbitrary $\Phi(\lambda)$ we consider its BT corresponding to an arbitrary element of A:

$$\Phi_{n,m}(\lambda) \stackrel{\text{def}}{=\!=} B_{n\mathcal{B}_1 \oplus m\mathcal{B}_2}(\Phi(\lambda)).$$

$\Phi_{n,m}$ can be inductively defined in terms of successive one - step applications of BTs:

$$\Phi_{n+1,m}(\lambda) = L^1_{n,m}(\lambda)\Phi_{n,m}(\lambda) \quad [=B_{\mathcal{B}1}(\Phi_{n,m}(\lambda))];$$
$$\Phi_{n,m+1}(\lambda) = L^2_{n,m}(\lambda)\Phi_{n,m}(\lambda) \quad [=B_{\mathcal{B}2}(\Phi_{n,m}(\lambda))]. \tag{2.3}$$

The commutativity conditions between BTs can be expressed in the form (2.2). In our case, since the BTs are commuting, the only equation of the form (2.2) to be checked is $B_{\mathcal{B}_1\oplus\mathcal{B}_2} = B_{\mathcal{B}_2\oplus\mathcal{B}_1}$. In notations (2.3) this operator identity is equivalent to the following systems of quadratic difference equations for coefficients of $L^1_{n,m}(\lambda)$ and $L^2_{n,m}(\lambda)$:

$$L^1_{n,m+1}(\lambda)L^2_{n,m}(\lambda) = L^2_{n+1,m}(\lambda)L^1_{n,m}(\lambda) \tag{2.4}$$

Eq. (2.4) can be compared with Hirota's and Levi's method of introduction of difference-difference[32,33] analogs of two dimensional isospectral deformation equations. In fact, eq. (2.4) in limit cases turn into the Lax representation of two-dimensional isospectral deformation equations. First of all we consider the continuous limit in the variable m.

Substituting $x = m\varepsilon$ as $\varepsilon\to 0$ into (2.4) one gets a familiar operator identity equivalent to a system of difference-differential equations associated with the inverse scattering method:

$$A_{n+1}(\lambda)L_n(\lambda) - L_n(\lambda)A_n(\lambda) = \partial L_n(\lambda)/\partial x. \tag{2.5}$$

Further, making a continuous limiting process in the variable n, operation equation (2.5) turns into the Lax-Zakharov-Shabat pair representation[14,35].

$$\frac{\partial}{\partial x}L(\lambda) - \frac{\partial}{\partial t}A(\lambda) = [A(\lambda), L(\lambda)]. \tag{2.6}$$

The linear differential spectral problem appearing here is naturally generated by an infinitesimal application of the BT, and corresponds to a one-parametric subgroup of BTs. Different classes of equations (2.5) or (2.6) arise if one applies BTs corresponding to (2.3) with different BT data $\mathcal{B}_i$. The difference-differential (lattice) systems appear in the scheme (2.3)-(2.5) using the application of a sequence of BTs with BT data $\mathcal{B}_0, \mathcal{B}_1, \ldots$ to a given $\phi(\lambda) = \phi_0(\lambda)$:

$$\phi_{N+1}(\lambda) = L_N(\lambda)\phi_N(\lambda),$$
$$L_N(\lambda) = L_{\mathcal{B}_N}(\lambda), \ \phi_{N+1}(\lambda) = B_{\mathcal{B}_N}(\phi_N(\lambda)). \tag{2.7}$$

The x- or t-dependence of $\phi_N(\lambda)$ is supplied by the l.o.d.e. of the form (1.4), generated by the infinitesimal action of the BT. One should bear in mind that according to Lemma 1.1, $\phi_N(\lambda)$ satisfies l.o.d.e. of the same form as $\phi_0(\lambda)$, if BT data are x-independent.

For example, one can consider systems (2.7) where BTs $B_{\mathcal{B}_{\mu i}}$ are elementary BTs described in (3.5), Ch. I.3. This way we derive generalizations of XXX-models of statistical mechanics (in the multi-spin case and for an arbitrary representation of gl(n)). For n=2 one of the Hamiltonians arising from (2.6) (and belonging to an infinite sequence of commuting Hamiltonians) is[7,17,18]

$$\Sigma_{i=1}^{N} \{1 + \cosh(q_i - q_{i+1})\}p_i p_{i+1},$$

which is a quantum lattice XXX model with the Hamiltonian H''_N, written in Darboux coordinates. Here $H''_N = \Sigma_{i=1}^{N}\{\sigma_i^1\sigma_{i+1}^1 + \sigma_i^2\sigma_{i+1}^2 + \sigma_i^3\sigma_{i+1}^3\}$ and $[\sigma_\ell^i, \sigma_m^j] = \delta_{\ell m}\sigma_\ell^k$ for a permutation (i,j,k) of (1,2,3) and σ_ℓ^i corresponds to elements of U(sl(2)).

Similarly, a true lattice nonlinear Schrödinger model is derived in[5,6,17] by applying elementary BTs with data $\mathcal{B}_\mu^\infty$ (see Ch. I.3.) to solutions of the Dirac equation. In the most general case, the system of difference-differential equations (2.5) is the following one (A_N is of size $n_1 \times n_2$, B_N is of size $n_2 \times n_1$, $n_1 + n_2 = n$):

$$A_{N,x} = A_{N+1} + \mu_N A_N - A_N B_N A_N,$$

$$B_{N,x} = -B_{N-1} - \mu_N B_N + B_N\, A_N B_N. \tag{2.8}$$

System (2.8) is equivalent to the lattice version of the nonlinear Schrödinger equation studied in[6]. In particular, (2.8) includes the following quantum Hamiltonians ($p_n = A_N, q_N = B_N$):

$$H = \sum_N (\tfrac{1}{2} q_N^2 p_N^2 + p_N q_N - \mu_N p_N q_N - p_N q_{N-1}).$$

Such an approach is especially important for quantum two - dimensional models because they allow a rigorous treatment of the "exactly solvable" quantum field theory models. For example, the application of the Baxter Lemma (2.1), Ch. I.2. is easily proved in this case, together with the proof of the usual algebraic version of the Bethe Ansatz easy to prove for completely integrable lattice models. This way we avoid the familiar problems of the "quantum inverse scattering method" where the lattice approximation, which is "approximately" completely integrable, is introduced. In our case, the lattice models are equivalent and approximate the original continuous systems, with the symplectic structure in the classical case or the S-matrix in the quantum case to be preserved.

II.3. Relations between BTs as a universal form of completely integrable equations

In this section we present algebraic relations between elements of transfer matrices defining successive iteration of BTs. These relations between transfer matrices are completely integrable lattice models. Two-dimensional lattice models were introduced this way in (2.4), and in II.2. we showed how various two-dimensional difference-differential and differential-differential completely integrable systems arise this way. It is remarkable that the system (2.4) is a corollary of a simple action of two commuting BTs in (2.3). One can describe three-dimensional systems in the same way using iterative application of three distinct BTs.

Relations between iterative applications of BTs seem to be concentrated around generalizations of the famous Bianchi theorem, see[1], expressing the composition of BTs $B_{k_1} \circ B_{k_2}$ of the sin-Gordon equation algebraically in terms of B_{k_1} and B_{k_2}. There seems to be a general belief that such statements are universal for (say, two - dimensional) completely integrable systems. However, such a statement is a simplification of the complex structure of BTs acting as cannonical transformations on completely integrable systems. First of all, BTs form a highly non-commutative group even for the simplest equations. The reason for the apparent "commutativity" of BTs is that in the composition of BTs, continuous parameters are often disregarded so that BT seems to be reduced to an addition of apparent singularities and addition of integers (commutative operations). For matrix linear problems, the non-commutativity of BTs is better known because of its relationship with non-Abelian gauge transformations.

The lattice (difference-difference) equations (2.4) generated by the composition formula (2.2) produce in different limits various completely integrable one, two and three dimensional systems.

Since BTs form an infinite dimensional group, the method of II.2. produces a large class of nonlinear lattice equations obtained by successive applications of different BTs. The existence of this large class of lattice equations, which can be classified using topological methods of I.2., is responsible for infinitely many commuting dynamical flows in the continuous limits. Among lattice equations are "universal" ones equivalent to the commutation relations between elementary BTs.

Typically one can take as elementary BTs those that create a single apparent singularity (if the positions of singularities are distinct, corresponding elementary BTs commute).

We present an example of such a "universal" equation in the case of BTs applied to solution of the Dirac equation $\{\frac{\partial}{\partial x} + U_0(\lambda)\}\phi(\lambda) = 0$ for n=2 and $U_0(\lambda) = \begin{pmatrix} \lambda & p \\ q & -\lambda \end{pmatrix}$ with two "pseudopotentials" p and q. We apply the "generator" BTs with data $\mathfrak{B}_\mu^\infty$ (see I.3. and II.2.).

To stress the two-dimensionality of corresponding lattice equations we use two sequences of parameters $\vec{\mu}=(\mu_1,\ldots,\mu_N)$, $\vec{\eta}=(\eta_1,\ldots,\eta_M)$ such that BTs $B_{\mathcal{B}\mu_i}$ and $(B_{\mathcal{B}\eta_j})^{-1}$ are applied: $i=1,\ldots,N$; $j=1,\ldots,M$. If all μ_i and η_j are distinct (though the limiting case can be easily analyzed), the new potential can be represented as
$U_{\vec{\mu};\vec{\eta}} = B_{\mathcal{B}\mu_1} \ldots B_{\mathcal{B}\mu_N} \cdot B^{-1}_{\mathcal{B}\eta_1} \ldots B^{-1}_{\mathcal{B}\eta_M} \circ U_0 =$

$= \begin{pmatrix} \lambda & p_{\bar{\mu};\bar{\eta}} \\ q_{\bar{\mu};\bar{\eta}} & -\lambda \end{pmatrix}$ with "pseudopotentials" $p_{\bar{\mu};\bar{\eta}}$ and $q_{\bar{\mu};\bar{\eta}}$.

The difference equation satisfied by $p_{\bar{\mu};\bar{\eta}}$ and $q_{\bar{\mu};\bar{\eta}}$ takes the following form[36]:

$$p_{(\cdot,\mu);(\cdot,\eta)} = p_{(\cdot);(\cdot)} + \frac{4(\eta-\mu)p_{(\cdot);(\cdot,\eta)}}{p_{(\cdot);(\cdot,\eta)}q_{(\cdot,\mu);(\cdot)}+4} ; \qquad (3.1)$$

$$q_{(\cdot,\mu);(\cdot,\eta)} = q_{(\cdot);(\cdot)} + \frac{4(\mu-\eta)p_{(\cdot,\mu);(\cdot)}}{p_{(\cdot);(\cdot,\eta)}q_{(\cdot,\mu);(\cdot)}+4}$$

In proper limits these equations turn into lattice and continuous NLS equations. At the same time, equations (3.1) give rise to similar formulas on the composition of BTs known for KdV, mKdV, Sin-Gordon, and other similar equations[1]. E.g. for the KdV equation we have Wahlquist and Estabrook's composition formula[1]:

$$w_{\mu;\eta} = w + \frac{\mu^2-\mu^2}{w_{\mu;\cdot}-w_{\cdot;\eta}}$$

where $u=w_x$ satisfies KdV. Similarly for solutions w of mKdV and sin-Gordon equations we have another superposition formula

$$\tan\left(\frac{w_{\mu;\eta}-w}{2}\right) = \frac{\mu+\eta}{\mu-\eta}\tan\left(\frac{w_{\mu;\cdot}-w_{\cdot;\eta}}{2}\right).$$

Though these and similar formulas are considered purely algebraic, they are not very useful in the determination of proper continuous limits because infinitesimally (for a large μ or η) w_μ is expressed in inverse powers of μ through algebraico-differential polynomials in w. It is important to use the appropriate variables in which the infinitesimal action of BTs is linear (and is represented by linear differential operators). According to II.2. such variables are not the pseudopotentials but rather eigenfunctions ϕ (or their expansions in λ-plane), since BTs are, by definition, acting linearly on ϕs. Hence new variables can be easily found if it is remembered that e.g., BTs of pseudopotentials are expressed through solutions of **Riccati** equations--ratios of components of eigenfunctions. For example for coupled NLS equation with pseudopotentials p and q, new variables are $p = \frac{f_1}{f}$, $q = \frac{f_2}{f}$ with $pq = d^2/dx^2 \log f$, so that the auxiliary variable f (or f_1, f_2) is, in fact, Hirota's variable (see[1]) in his direct method of studies of nonlinear equations. These variables are studied in[15,16] as "τ-functions".

In these new variables, algebraic relations between iterations of BTs take universal form though there are no more simple relations between $B_1 \circ B_2$ and B_1, B_2. For example, here is the difference equation satisfied by iteration of BTs to pseudopotentials, $p_{\vec{\mu}} q_{\vec{\mu}} = \frac{d^2}{dx^2} \log f_{\vec{\mu}}$ in new variables $f_{\vec{\mu}}$:

$$(\eta-\rho) f_{(\vec{\mu},\lambda)} f_{(\vec{\mu},\eta,\rho)} + (\rho-\lambda) f_{(\vec{\mu},\eta)} f_{(\vec{\mu},\rho,\lambda)} + (\lambda-\eta) f_{(\vec{\mu},\rho)} f_{(\vec{\mu},\lambda,\eta)} = 0$$

for (distinct) μ, λ, η, ρ. (3.2)

The origin of this equation will become clearly understood only by looking to the three-dimensional equations of KP type and their BTs expressed as Darboux transformations in I.3. First of all, one should remember, that according to the results of II.1. all matrix isospectral deformation equations can be imbedded into scalar three-dimensional equations of KP type, and their pseudopotential can be expressed in terms of a single function u and its potential σ, $u = \frac{d^2}{dx^2} \log \sigma$ (and derivatives of σ). On the other hand, it is easy to determine algebraic relations between BTs of the potential σ using the Wronskian formula (3.4), Ch. I.3. In this formula, BT data are determined by a sequence of eigenfunctions $\bar{\phi} = (\phi_1, \ldots, \phi_m)$ of (3.1), $\sigma_{\bar{\phi}} = \sigma . W(\phi_1, \ldots, \phi_m)$. One can then use algebraic considerations from I.2. to find the basis of identities satisfied by BTs $\sigma_{\bar{\phi}}$. Such basic identities are of the form (3.2):

$$\sigma_{(\bar{\phi},\phi_1)} \sigma_{(\bar{\phi},\phi_2,\phi_3)} + \sigma_{(\bar{\phi},\phi_2)} \sigma_{(\bar{\phi},\phi_3,\phi_1)} + \sigma_{(\bar{\phi},\phi_3)} \sigma_{(\bar{\phi},\phi_1,\phi_2)} = 0 \qquad (3.3)$$

We represent this equation in an abstract form, denoting by

B_1, B_2, B_3 three distinct elementary BTs (corresponding, say, to the addition of different apparent singularities):

$$\sigma_{B_1} \cdot \sigma_{B_2 \circ B_3} + \sigma_{B_2} \cdot \sigma_{B_3 \circ B_1} + \sigma_{B_3} \cdot \sigma_{B_1 \circ B_2} = 0. \tag{3.4}$$

Apparently the equation (3.4) is a truly universal one, giving in different continuous limits all matrix two-dimensional isospectral deformation equations from II.2. and three-dimensional equations of KP type (without discrimination for reductions among them). Moreover the equation (3.4), being quadratic in σ, can be used for Hirota's bilinear representation. The form (3.3) of the equation on BTs can be rewritten in a more conventional form if one uses the normalization (3.4'), I.3. of the Wronskian formula, corresponding to normalized eigenfunctions $\phi_i = \phi(\lambda_i)$ with eigenvalues λ_i. To stress the three-dimensionality of this equation one treats the sequence $\vec{\mu}$ of spectral data as the combination of three different sequences $\vec{\mu} = (\lambda_1,\ldots,\lambda_k;$ $\eta_1,\ldots,\eta_\ell;\ \rho_1,\ldots,\rho_m)$. Then equation (3.3) takes the form of the generalization of a discrete analog of a generalized Toda equation proposed recently by Hirota[37]:

$$(\eta_{\ell+1}-\rho_{m+1})\sigma_{(\bar{\mu},\lambda_{k+1})}\sigma_{(\bar{\mu},\eta_{\ell+1},\rho_{m+1})} + (\rho_{m+1}-\lambda_{k+1})\sigma_{(\bar{\mu},\eta_{\ell+1})}\sigma_{(\bar{\mu},\lambda_{k+1},\rho_{m+1})}$$
$$+ (\lambda_{k+1}-\eta_{\ell+1})\sigma_{(\bar{\mu},\rho_{m+1})}\sigma_{(\bar{\mu},\lambda_{k+1},\eta_{\ell+1})} = 0. \tag{3.3'}$$

Hirota's equation itself follows from (3.3), if BTs are applied to the same apparent singularities $\lambda_1 = \ldots = \lambda_k = \lambda$, $\eta_1 = \ldots = \eta_\ell = \eta$, $\rho_1 = \ldots = \rho_m = \rho$ and the application of BTs is interpreted additively $\sigma(k,\ell,m) \stackrel{\text{def}}{=\!=} \sigma_{(k\cdot\lambda,\ell\cdot\eta,m\cdot\rho)}$. We get

$$(\eta-\rho)\sigma(k+1,\ell,m)\sigma(k,\ell+1,m+1) + (\rho-\lambda)\sigma(k,\ell+1,m)\sigma(k+1,\ell,m+1)$$

$$+ (\lambda-\eta)\sigma(k,\ell,m+1)\sigma(k+1,\ell+1,m) = 0.$$

This equation was nicely written by Hirota in the bilinear form

$$\{\exp(D_1) + \exp(D_2) + \exp(D_3)\}f\cdot f = 0, \tag{3.5}$$

where D_1, D_2, D_3 are linear differential operators or their linear combinations with a constant. In[37] it is shown that the KP equation, two-dimensional Toda lattice, sin-Gordon equation, mKdV, KdV and Benjamin-Ono equations are particular limit cases of the equation (3.5). The equation (3.5) is however, a simplified version of a more abstract universal equation (3.4). From equation (3.4), taking an infinitesimal action of elementary BTs, one can recover all two- and three-dimensional systems of equations studied in Chapters I - II. The infinitesimal action of BTs in terms of variables σ is represented by a linear differential operator or by linear transformations (discrete, or periodic case). E.g. the infinitesimal action of BT for the KP system Ch. I.3., Ch. II.1. with the eigenvalue λ as $\lambda \to \infty$ can be represented as

$$\sigma_{B_\lambda} = \exp \{\Sigma_{i=1}^{\infty} \lambda^{-i} \frac{\partial}{\partial t_i}\}\sigma$$

as $\lambda \to \infty$. Substituting this expansion in (3.4) for different λ_1, λ_2, λ_3, one recovers various Hirota's bilinear representations for equations of the KP hierarchy. Two-dimensional equations of II.1. are recovered from (3.4) if two of the BTs B_1 and B_2 are realized through their infinitesimal representation as differential operators and B_3 acting as difference operator.

Universal relations (3.4) can be used to generate other similar relations between successive applications of four or more BTs. For example, one can deduce the following identity

$$\sigma_{B_1 o B_2} \cdot \sigma_{B_3 o B_4} + \sigma_{B_1 o B_4} \cdot \sigma_{B_2 o B_3} - \sigma_{B_1 o B_3} \cdot \sigma_{B_2 o B_4} = 0, \qquad (3.6)$$

etc. These and other equations can be also used in algebraic geometry to characterize Jacobians of algebraic curves among Abelian varieties because σ can be realized as the Riemann θ-function for solutions of KP equations associated with an arbitrary algebraic curve Γ. The BT $\sigma \to \sigma_{B_\lambda}$ corresponds to the addition in Jacobian of a point corresponding to the point on Γ.

Finally we want to remark that other three-dimensional completely integrable systems can be realized as lattice equations of the form (3.4). This include reductions of the KP hierarchy and other three-dimensional systems like two-dimensional NLS and also monopole equations for different gauge groups. Cf. in this respect Prasad formula for iterative application of AW Ansatz. Similarly one can obtain a lattice reformulation of non-linear σ-models and self-dual Yang-Mills models (written in Yang gauge), together with various 2 and 4-dimensional models arising from flows commuting with these models.

E.g. the stationary solutions of self-dual Yang-Mills equations: $(g^{-1}g_y)_{\bar{y}} + (g^{-1}g_z)_{\bar{z}} = 0$ are directly derivable from the equations (3.4) with various B_is.

REFERENCES

1- R.Miura, ed., Bäcklund transformations, Lecture Notes Math., v.515, Springer, 1976.

2- R.Hermann, Geometric theory of non-linear differential equations, Backlund transformations and solitons, Part B. Math. Sci. Press, 1977.

3- D.V.Chudnovsky, G.V.Chudnovsky, Lett. Math. Phys. 4(1980), 373.

4- G.V.Chudnovsky, Lecture Notes Phys. v.120, Springer, 1980, 103; Lecture Notes Phys. v.126, 1980, 136.

5- D.V.Chudnovsky, G.V.Chudnovsky, Phys. Lett., 82A(1981), 271.

6- D.V.Chudnovsky, G.V.Chudnovsky, Phys. Lett., 87A (1982), 325.

7- D.V.Chudnovsky, G.V.Chudnovsky, Phys. Lett., 89A(1982), 117.

8- D.V.Chudnovsky, Cargèse lectures in bifurcation phenomena in mathematical physics and related topics, D.Reidel, Boston, U.S.A., 1980, 385.

9- D.V.Chudnovsky, G.V.Chudnovsky, J. Math. Phys. 22(1981), 2518.

10- D.V.Chudnovsky, G.V.Chudnovsky, ed., The Riemann problem, Complete Integrability and Arithmetic Applications, Lect. Notes, Math., v.925, Springer, 1982.

11- J.Plemelj, Problems in the sense of Riemann and Klein, John Wiley, 1969.

12- J.A.Lappo-Danilevsky, Mémoires sur la théorie des systèmes des equation differentielles linéaires, 1930 (Chelsea reprint 1953).

13- B.Riemann, Oeuvres mathématiques, Blanchard, Paris, 1968.

14- V.E.Zakharov, A.V.Mikhailov, Zh. Eksp. Teor. Fiz 74 (1978), 1953.

15- M.Jimbo, T.Miwa, RIMS Preprints 319, 327, 346 Kyoto, Japan 1980.

16- E.Date, M.Jimbo, M.Kashiwara, T.Miwa, RIMS Preprints 357, 358, 359, 360, 361, 362, Kyoto, Japan 1981-1982.

17- D.Chudnovsky, in Festschrift for F.Gürsey (to appear).

18- D.V.Chudnovsky, G.V.Chudnovsky, Phys. Rev. Lett. 47 (1981), 1093.

19- S.MacLane, Categories for the working mathematicians, Springer, 1971.

20- A.B.Zamolodchikov, Comm. Math. Phys. 69(1979), 165.

21- A.B.Zamolodchikov, Comm. Math. Phys. 79(1981), 489.

22- G.Darboux, C.R. Acad. Sci. Paris 94(1882), 1465; Théorie des Surfaces, v.II, p.210, 1890.

23- M.M.Crum, Quart. J. Math. 2(1955), 121.

24- D.V.Chudnovsky, Proc. Natl. Acad. Sci. 75(1978), 4082; D.V. Chudnovsky, G.V.Chudnovsky, Lett. Nuovo Cimento, 25(1979), 263.

25- A.C.Newell, Proc. Royal. Soc. London, A365(1979), 283.

26- F.Calogero, A. Degasperis, Nuovo Cimento 32B(1976), 201; 39B(1977), 1.

27- B.A.Dubrovin, Funct. Anal. Appl. 11(1977), No.4.

28- D.V.Chudnovsky, Lecture Notes Phys. v.126, Springer, 1980, 352.

29- D.V.Chudnovsky, G.V.Chudnovsky, Nuovo Cimento, 40B(1977), 339.

30- H.Airault, H.McKean, J.Moser., Comm. Pure Appl.Math. 55(1977), No 4.

31- D.V.Chudnovsky, J.Math. Phys. 20(1979), 2416.

32- R.Hirota, J.Phys. Soc. Japan, 43(1977), 2079.

33- D.Levi, J. Phys. A 14(1981), 1083; D.Levi, R.Benguria, Proc. Natl. Acad. Sci. 77(1980), 5025.

34- E.Trubowitz, P.Dieft, F.Lund, Com. Pure Appl. Math. 74(1980), 141.

35- V.E.Zakharov, A.B.Shabat, Funct. Anal. Appl. 13(1979), 13.

36- B.G.Konopelchenko, Phys. Lett 87A(1982), 445-448.

37- R.Hirota, J.Phys. Soc. Japan, 50(1981), 3785.

INFINITE DIMENSIONAL SYMMETRY ALGEBRAS IN INTEGRABLE SYSTEMS

Harald Eichenherr
CERN
CH-1211 Geneva 23
SWITZERLAND

ABSTRACT

An infinite dimensional symmetry algebra of the Heisenberg spin chain is described and some of its properties are discussed.

It is shown that the principle of gauge equivalence of Lax pairs leads to the existence of such symmetry algebras even in models which do not have a global non-Abelian symmetry. This is explained for the examples of the non-linear Schrödinger equation and the complex sine-Gordon equation. In the latter case one also uses the fact that symmetry algebra and conformal transformations commute.

1. Introduction

It has been known for several years that the stationary, axially symmetric Einstein equations possess an infinite dimensional algebra of infinitesimal symmetry transformations[1] which is isomorphic to the $\mathbb{Z}$ graded Lie algebra $s\ell(2,\mathbb{R})\otimes\mathbb{R}[\lambda,\lambda^{-1}]$ of $s\ell(2,\mathbb{R})$ valued Laurent polynomials in the undeterminate λ. These symmetries have been used to produce new solutions of the gravitational field equations. Since these under the above mentioned restrictions coincide with the equations of motion of the two-dimensional non-linear σ model on the noncompact symmetric coset space $SL(2,\mathbb{R})/SO(2)$, one can ask whether such symmetry algebras exist also for other representatives of the class of generalized σ models which have been studied recently.

For the chiral models (i.e. non-linear σ models where the field takes values in a Lie group G with Lie algebra g), this question has been answered in the positive sense by various authors[2]. L. Dolan has shown that the infinitesimal symmetries related to the usual non-local charges generate a Lie algebra $g\otimes\mathbb{R}[\lambda^{-1}]$. K. Ueno used the Riemann-Hilbert formalism which had been developped by Zakharov and Mikhailov[3] to compute solutions of σ models. In Ueno's approach, a representation of the Lie algebra $g\otimes\mathbb{R}[\lambda,\lambda^{-1}]$ acts on the manifold of "wave functions" of the Lax pair, the "negative" subalgebra $g\otimes\mathbb{R}[\lambda^{-1}]$ coinciding with Dolan's transformations, the "positive" subalgebra $g\otimes\mathbb{R}[\lambda]$ leaving the chiral field invariant (but not the wave functions).

In this talk, I shall discuss the symmetry algebra of the classical two-dimensional Heisenberg spin chain. It turns out that this algebra does not correspond to a Poisson bracket algebra of conserved charges, because the symmetries in general change the energy, although they commute with time displacements of the spin field.

Further, I shall show that non-Abelian symmetry algebras exist also for models which do not have a global non-Abelian symmetry, provided they are gauge equivalent (in the sense of Zakharov and Takhtajan[4]) to a model in which such a symmetry algebra is already known. This applies, for example, to the non-linear Schrödinger equation and the complex sine-Gordon equation.

2. Symmetry algebra of the Heisenberg model

The Heisenberg model involves a field

$$S(t,x) = \sum_{i=1}^{3} S_i(t,x)\sigma_i \quad \text{with } S^2 = \mathbf{1}$$

where σ_i $i=1,2,3$ denote the Pauli matrices. The Hamiltonian

$$H = \frac{1}{4}\int_{-\infty}^{+\infty} dx\ \mathrm{tr}(\partial_x S)^2$$

and the Poisson brackets

$$\{S_i(t,x),S_j(t,y)\} = \varepsilon_{ijk}S_k(t,x)\delta(x-y)$$

generate the equation of motion

$$\partial_t S = \frac{1}{2i}\partial_x[S,\partial_x S] \tag{1}$$

which is the integrability condition of the Lax pair[5)]

$$\partial_x R^{(\lambda)} = -i\lambda S R^{(\lambda)} \tag{2a}$$

$$\partial_t R^{(\lambda)} = (-\tfrac{\lambda}{2}[S,\partial_x S] + 2i\lambda^2 S)R^{(\lambda)} \quad . \tag{2b}$$

The "wave function" $R^{(\lambda)}(t,x)$ takes values in SU(2) (SL(2,$\mathbb{C}$)) for real (complex) values of the spectral parameter λ. We consider the Lie algebra

$$\mathcal{L} = su(2)\otimes\mathbb{R}[\lambda,\lambda^{-1}];$$

its elements can be represented as su(2) valued Laurent polynomials

$$X \in \mathcal{L}: \qquad X(\lambda) = \sum_{-\infty}^{+\infty} \lambda^{-j}X^{(j)} \quad , \quad X^{(j)} \in su(2) \tag{3}$$

$$[X,Y](\lambda) = \sum_{-\infty}^{+\infty} \lambda^{-(j_1+j_2)}[X^{(j_1)},Y^{(j_2)}] \quad .$$

$\mathcal{L}$ is a real form of the loop algebra corresponding to the Kac-Moody algebra A_1^1 . The representation of $\mathcal{L}$ to be defined acts on the space of wave functions $R^{(\lambda)}$ solving (2a/b) for some smooth field S, such that $R^{(o)}(t,x) = \mathbf{1}$. $R^{(\lambda)}$ is assumed to be analytic with respect to λ in a certain neighbourhood of o, and all the following statements refer to this domain of analyticity. The representation is defined by[6)]

$$X \in \mathcal{L}: R^{(\lambda)} \rightarrow (\mathbf{1}+\delta M_X)R^{(\lambda)} \qquad (\delta = \text{infinitesimal parameter})$$

where

$$M_X R^{(\lambda)} = -\frac{1}{2\pi i}\int_{C_{o\lambda}} \frac{d\mu}{\mu}\,\frac{\lambda}{\lambda-\mu}\,V_X(\mu)R^{(\lambda)} \tag{4}$$

$$V_X(\mu) = R^{(\mu)}X(\mu)R^{(\mu)^{-1}} \tag{5}$$

$$M_X S = -\frac{1}{2\pi i} \int_{C_o} \frac{d\mu}{\mu} [V_X(\mu), S] \tag{6}$$

$$M_X[S,\partial_x S] = -\frac{1}{2\pi i} \int_{C_o} \frac{d\mu}{\mu} [V_X(\mu), [S,\partial_x S] - 4i\mu S] \tag{7}$$

In all these formulae, the contour C_o ($C_{o\lambda}$) surrounds the point o (o,λ) once counterclockwise and is contained in the analyticity domain which is not changed by M_X. A straightforward computation shows that $(\mathbb{1}+\delta M_X)R^{(\lambda)}$ solves (2a/b) to first order in δ with S and $[S,\partial_x S]$ replaced by $(\mathbb{1}+\delta M_X)S$ and $(\mathbb{1}+\delta M_X)[S,\partial_x S]$. Because of $M_X R^{(o)}=0$ and $M_X R^{(\lambda)} R^{(\lambda)-1} \in su(2)$ for real λ , the function space defined above is, to first order in δ , mapped into itself by $(\mathbb{1}+\delta M_X)$. Appropriately deforming the integration contours, one can prove that $X \to M_X$ is in fact a representation of ℓ, i.e.

$$[M_X, M_Y] = M_{[X,Y]} \quad .$$

In terms of a basis $X_a^n = \lambda^{-n} T_a$, $T_a = \frac{\sigma_a}{2i}$, we have

$$[M_a^n, M_b^m] = \varepsilon_{abc} M_c^{(n+m)} \quad .$$

Because (4)-(7) is an infinitesimal symmetry of the Lax pair (2a/b), the transformation (6) is an infinitesimal symmetry of the equation of motion (1), the latter being the integrability condition of the former.

To evaluate the above integral formulae, we insert into (4)-(6) the expansion (cf. Ref.7)

$$R^{(\lambda)} T_a R^{(\lambda)-1} = \sum_{n=o}^{\infty} \lambda^n V_a^n$$

which is valid in the analyticity domain of $R^{(\lambda)}$. The V_a^n fulfill

$$V_a^o = T_a$$

$$\partial_x V_a^n = -i[S, V_a^{n-1}]$$

$$\partial_t V_a^n = [-\tfrac{1}{2}[S,\partial_x S], V_a^{n-1}] + 2i[S, V_a^{n-2}] \quad .$$

Then we find ($\sum_o^n = o$ for $n < o$)

$$M_a^n R^{(\lambda)} = R^{(\lambda)} \lambda^{-n} T_a - \sum_{m=o}^{n} V_a^m \lambda^{m-n} R^{(\lambda)} \tag{8}$$

$$M_a^n S = \begin{cases} [S, V_a^n] & \text{for } n \geq o \\ o & \text{for } n < o \end{cases} \quad . \tag{9}$$

From (8) we see that the M_a^n for $n<o$ simply multiply $R^{(\lambda)}$ from the right with the defining tensor product representation (3) of $\mathcal{L}$, thus leaving S invariant. In other words, the M_a^n ($n<o$) change the infinitely many integration constants in $R^{(\lambda)}$, nothing else.

In spite of the fact that $S \to (\mathbf{1}+\delta M_a^n)S$ is a symmetry of the equation of motion, in general the energy is changed under this transformation. We compute

$$\begin{aligned} M_a^n H &= \frac{1}{2} \int dx \; \mathrm{tr}(\partial_x S \; M_a^n \partial_x S) \\ &= \frac{1}{2} \int dx \; \mathrm{tr}(\partial_x S \; \partial_x [S, V_a^n]) \\ &= \frac{1}{2} \int dx \; \mathrm{tr}(\partial_x S \; [S, -i[S, V_a^{n-1}]]) \\ &= \int dx \; \mathrm{tr}(-i\partial_x S \; V_a^{n-1}) \qquad \text{using } S^2 = \mathbf{1} \\ &= \mathrm{tr}(-iSV_a^{n-1})\big|_{-\infty}^{+\infty} \end{aligned} \tag{10}$$

which in general is nonzero. Hence, for the Heisenberg model there cannot exist a Poisson bracket algebra of conserved charges Q_a^n generating the symmetries by

$$M_a^n S = \{Q_a^n, S\} \quad .$$

Equation (10) shows that the asymptotic behaviour of the spin field plays an important rôle. For the rest of this chapter, let us look at the action of the symmetry algebra on fields behaving like

$$S(t,x) \to \sigma_3 \quad \text{as} \quad |x| \to \infty \quad . \tag{11}$$

This is the usual situation if one employs inverse scattering techniques. Of course, the transformed solutions in general will not obey condition (11), since it is spoilt already by simple rotations $S \to [S,T_a]$ effected by the M_a^o . In fact, it is obscure what is the relation between the inverse scattering method (which is very sensitive against the boundary conditions) and the symmetry algebra (which drastically changes the boundary conditions), both being based on the Lax representation.

Let now $R^{(\lambda)}$ coincide with one of the two Jost solutions of the scattering problem (2a) :

$$\begin{aligned} F^{(\lambda)}(x) &\to \exp(-i\lambda\sigma_3 x) \qquad && \text{as } x \to +\infty \\ G^{(\lambda)}(x) &\to \exp(-i\lambda\sigma_3 x) \qquad && \text{as } x \to -\infty \end{aligned} \tag{12}$$

$$F^{(\lambda)}(x) = G^{(\lambda)}(x)\; T(\lambda)$$

where $$T(\lambda) = \begin{pmatrix} a(\lambda) & -\overline{b}(\lambda) \\ b(\lambda) & \overline{a}(\lambda) \end{pmatrix} \tag{13}$$

is the monodromy (transition) matrix. The Jost functions $a(\lambda)$ and $b(\lambda)$ provide action and angle variables for the spin fields obeying (11). Choosing $R^{(\lambda)} = G^{(\lambda)}$ and inserting (12),(13) into (1o), we find

$$M_3^n H = \frac{1}{n!}\frac{d^n}{d\lambda^n}\, 4\lambda(|a(\lambda)|^2-1)\Big|_{\lambda=o} \qquad = \text{fct.(action variables)}$$

$$\left.\begin{aligned} M_1^n H &= \frac{1}{n!}\frac{d^n}{d\lambda^n}\, 4\lambda \mathrm{Re}\; a(\lambda)\overline{b}(\lambda)\Big|_{\lambda=o} \\ M_2^n H &= -\frac{1}{n!}\frac{d^n}{d\lambda^n}\, 4\lambda \mathrm{Im}\; a(\lambda)\overline{b}(\lambda)\Big|_{\lambda=o} \end{aligned}\right\} \quad = \text{fct.(action+angle var.)}$$

From these formulae we see that the discrete part of the

spectrum of the scattering problem (2a) (i.e. the multi-soliton solutions) is transformed into itself by the Cartan subalgebra of $\mathcal{t}$ spanned by the M_3^n. For multisolitons we have

$$b(\lambda) = o \ , \ |a(\lambda)| = 1 \quad \text{for real } \lambda,$$

hence $M_X H = o$ for all $X \in \mathcal{t}$. Further, $[T(\lambda),\sigma_3] = o$, implying that the boundary condition (11) is stable under M_3^n, moreover,

$$M_3^n \, T(\lambda) = M_3^n \, (G^{(\lambda)^{-1}} F^{(\lambda)}) = o$$

for real λ. An N - soliton configuration is characterized by two sets of parameters :

i) $\zeta_j \in \mathbb{C}$, $\text{Im } \zeta_j < o$, $j=1...N$.

The ζ_j are the zeroes of the analytically continued Jost function $a(\lambda)$ in the lower half plane; they determine the action variables.

$$a(\lambda) = \det(F_1^{(\lambda)}, G_2^{(\lambda)}) \quad , \ a(\zeta_j) = o$$

where $F_1^{(\lambda)}$ $(G_2^{(\lambda)})$ denotes the first (second) column vector of the Jost solution $F^{(\lambda)}$ $(G^{(\lambda)})$.

ii) $m_j \in \mathbb{C}$, $j=1...N$.

The m_j are given in terms of the eigenfunctions of the scattering operator corresponding to $\lambda = \zeta_j$ by

$$F_1^{(\zeta_j)} = b_j \ G_2^{(\zeta_j)} \quad , \quad m_j = \frac{b_j}{i\partial_\lambda a(\zeta_j)} \quad .$$

They determine the angle variables.

By inspection of the transformation formulae it turns out that

$$M_3^n \, \zeta_j = o \quad , \quad M_3^n \, m_j = 2\zeta_j^{-n} \quad , \ n \geq o \ .$$

Thus the Cartan subalgebra leaves the action variables of a

multisoliton configuration invariant whereas the angle variables can be shifted arbitrarily.

3. Symmetry algebras and gauge equivalence of Lax pairs

It is well known that the Lax pairs of various couples of two-dimensional models are related by gauge transformations : The integrability condition of a Lax pair expresses the flatness of a certain connection, and this flatness is preserved by gauge transformations which effect a transition to new field variables. Usually, one of both systems has a non-Abelian global symmetry group G. Defining an appropriate G-valued gauge transformation leads to a new Lax pair with an equation of motion for the new variables which are invariants with respect to G. Thus, the second system has no longer a non-Abelian inner symmetry. Nevertheless, if one knows a Kac-Moody symmetry algebra for the first system and if one can find its action on the gauge transformation, one can determine its action on the second system. As examples for this relation, I shall discuss the non-linear Schrödinger equation (NLS) which is gauge equivalent to the Heisenberg model, and the complex sine-Gordon equation which is gauge-equivalent to the SU(2) chiral model. Actually, in the latter case a breaking of conformal invariance is involved which requires to look at the relation of Kac-Moody algebra and conformal symmetry.

3.1. Heisenberg model and non-linear Schrödinger equation

The gauge equivalence of the Heisenberg model and the NLS is mediated by an SU(2) valued field $\Gamma(t,x)$ such that[4)]

$$S = \Gamma\sigma_3\Gamma^{-1} \quad \text{and} \quad \mathrm{diag}(\Gamma^{-1}\partial_x\Gamma) = o \quad . \tag{14}$$

Writing $\Gamma^{-1}R^{(\lambda)} = \Phi^{(\lambda)}$,

$$\Gamma^{-1}\partial_x\Gamma = \begin{pmatrix} & -\overline{\Psi} \\ \Psi & \end{pmatrix} \quad , \quad \Gamma^{-1}\partial_t\Gamma = i\begin{pmatrix} |\Psi|^2 & \partial_x\overline{\Psi} \\ \partial_x\Psi & -|\Psi|^2 \end{pmatrix} \tag{15}$$

we get from (2a/b)

$$\partial_x \Phi^{(\lambda)} = -(\Gamma^{-1}\partial_x\Gamma + i\lambda\sigma_3)\Phi^{(\lambda)}$$

$$\partial_t \Phi^{(\lambda)} = (-\Gamma^{-1}\partial_t\Gamma + 2\lambda\Gamma^{-1}\partial_x\Gamma + 2i\lambda^2\sigma_3)\Phi^{(\lambda)}$$

which is the Lax pair for the NLS

$$i\partial_t\Psi + \partial_x^2\Psi + 2|\Psi|^2\Psi = 0 \quad .$$

Defining now

$$M_X\Gamma = -\frac{1}{2\pi i}\int_{C_o} \frac{d\mu}{\mu}\,\Gamma\, W_X(\mu) \quad , \tag{16}$$

$$W_X(\mu) = \Phi^{(\mu)}X(\mu)\Phi^{(\mu)-1} = \Gamma^{-1}V_X(\mu)\Gamma \quad ,$$

one easily verifies that (16) and (17) in fact give the transformation (6) for the field S. On the other hand, from (15) we get the formulae

$$X \in \mathcal{L}: \quad M_X\Phi^{(\lambda)} = \Phi^{(\lambda)}X(\lambda) - \frac{1}{2\pi i}\int_{C_o} \frac{d\mu}{\lambda-\mu}\,W_X(\mu)\Phi^{(\lambda)}$$

(λ lies outside the contour C_o)

$$M_X(\Gamma^{-1}\partial_x\Gamma) = -\frac{1}{2\pi i}\int_{C_o} d\mu\,[W_X(\mu), i\sigma_3] \tag{17}$$

$$M_X(\Gamma^{-1}\partial_t\Gamma) = -\frac{1}{2\pi i}\int_{C_o} d\mu\,[2\Gamma^{-1}\partial_x\Gamma + 2i\mu\sigma_3, W_X(\mu)] \quad .$$

From (17) we see that the gauge condition $\mathrm{diag}(\Gamma^{-1}\partial_x\Gamma) = 0$ is left invariant by M_X. Projecting out the transformation of Ψ itself, we find

$$M_X\,\Psi = -\frac{1}{2\pi i}\int_{C_o} d\mu\,\mathrm{tr}\Big((i\sigma_1-\sigma_2)W_X(\mu)\Big) \quad ,$$

in particular,

$$M_a^o\,\Psi = 0 \; , \quad M_a^1\,\Psi = -\,\mathrm{tr}\Big((i\sigma_1-\sigma_2)\Phi^{(o)}T_a\Phi^{(o)-1}\Big) .$$

So the SU(2) rotations of course act trivially on Ψ;

however, already M_a^1 is a very complicated operator involving $\Phi^{(o)}$, i.e. a path ordered exponential of matrices which contain Ψ and its derivatives.

3.2 SU(2) chiral model and complex sine-Gordon equation

The Lagrangian density and the field equation for the SU(2) chiral field g(t,x) are

$$\mathcal{L} = -\frac{1}{4}\,\mathrm{tr}(g^{-1}\partial_\mu g)^2 \quad \text{and} \quad \partial_\mu\partial^\mu g - \partial_\mu g\, g^{-1}\partial^\mu g = o \quad .$$

In terms of the currents $j_\mu = g^{-1}\partial_\mu g$, and employing light cone coordinates $\xi = \frac{1}{2}(t+x)$, $\eta = \frac{1}{2}(t-x)$, the Lax pair is[8)3)]

$$\partial_\xi R^{(\lambda)} = \frac{\lambda}{1-\lambda}\, j_\xi\, R^{(\lambda)} \quad , \quad \partial_\eta R^{(\lambda)} = -\frac{\lambda}{1+\lambda}\, j_\eta\, R^{(\lambda)} \quad . \qquad \text{(18a/b)}$$

From the work of Ueno[2)], we know the transformation formulae

$$X \in \mathfrak{k}: \quad M_X R^{(\lambda)} = -\frac{1}{2\pi i}\int_{C_{o\lambda}} \frac{d\mu}{\mu}\,\frac{\lambda}{\lambda-\mu}\, V_X(\mu)\, R^{(\lambda)} \qquad (19)$$

$$V_X(\mu) = R^{(\mu)} X(\mu) R^{(\mu)-1}$$

$$M_X g = \frac{1}{2\pi i}\int_{C_o} \frac{d\mu}{\mu}\, g\, V_X(\mu) \qquad (2o)$$

$$M_X j_\xi = -\frac{1}{2\pi i}\int_{C_o} \frac{d\mu}{\mu(1-\mu)}\, [V_X(\mu), j_\xi] \qquad (21)$$

$$M_X j_\eta = -\frac{1}{2\pi i}\int_{C_o} \frac{d\mu}{\mu(1+\mu)}\, [V_X(\mu), j_\eta] \quad . \qquad (22)$$

In these formulae, the contours do not surround the points $\lambda=\pm 1$.

Let now the gauge transformation $\Gamma(\xi,\eta) \in SU(2)$ be such that

$$j_\xi = -i\Gamma\sigma_3\Gamma^{-1}, \quad \mathrm{diag}(\Gamma^{-1}\partial_\xi\Gamma) = \frac{i}{2}\sigma_3 \quad , \quad \partial_\eta\Gamma\Gamma^{-1} + \frac{1}{2}j_\eta = o \quad . \qquad (23)$$

Writing $\quad \Gamma^{-1}R^{(\lambda)} = \Phi^{(\lambda)} \quad ,$

$$\Gamma^{-1}\partial_\xi\Gamma = \frac{i}{2}\begin{pmatrix} 1, & (\partial_\xi\alpha - i\partial_\xi\beta \, tg\frac{\alpha}{2})\exp(-i\omega) \\ (\partial_\xi\alpha + i\partial_\xi\beta \, tg\frac{\alpha}{2})\exp(i\omega), & -1 \end{pmatrix}$$

$$\Gamma^{-1}j_\eta\Gamma = -i\begin{pmatrix} \cos\alpha, & i\sin\alpha\exp(-i\omega) \\ -i\sin\alpha\exp(i\omega), & -\cos\alpha \end{pmatrix} \qquad (24)$$

we get from (18a/b)

$$\partial_\xi\Phi^{(\lambda)} = -\left(\frac{i\lambda}{1-\lambda}\sigma_3 + \Gamma^{-1}\partial_\xi\Gamma\right)\Phi^{(\lambda)}$$

$$\partial_\eta\Phi^{(\lambda)} = \frac{1}{2}\frac{1-\lambda}{1+\lambda}\Gamma^{-1}j_\eta\Gamma\,\Phi^{(\lambda)}$$

which is the Lax pair for the so-called complex sine-Gordon equation[8)]

$$\partial_\xi\partial_\eta\alpha - \frac{tg^2\frac{\alpha}{2}}{\sin\alpha}\partial_\xi\beta\partial_\eta\beta + \sin\alpha = 0$$

$$\partial_\xi\partial_\eta\beta + \frac{1}{\sin\alpha}(\partial_\xi\beta\partial_\eta\alpha + \partial_\eta\beta\partial_\xi\alpha) = 0 \quad .$$

ω is related to α and β by

$$\partial_\xi\omega = \partial_\xi\beta\frac{\cos\alpha}{2\cos^2\frac{\alpha}{2}}$$

$$\partial_\eta\omega = \partial_\eta\beta\frac{1}{2\cos^2\frac{\alpha}{2}} \quad .$$

From (23) and (24) we see that it has been assumed that

$$-\frac{1}{2}\mathrm{tr}\, j_\xi^2 = -\frac{1}{2}\mathrm{tr}\, j_\eta^2 = 1 \quad . \qquad (25)$$

For nonvanishing j_ξ, j_η this can always be achieved by appropriate (local) conformal transformations[8)]. However, the breaking of conformal symmetry implied by the requirements (25) does not prevent us from carrying over the symmetry algebra (19)-(22) from the conformal invariant chiral model to the complex sine-Gordon system, because conformal and Kac-Moody transformations of the chiral field commute.

In fact, the infinitesimal conformal transformations on

the coordinates ξ and η are generated by the conserved charges

$$Q_\xi(F) = -\frac{1}{2} \int dx\ F(\xi)\ \mathrm{tr}\ j_\xi^2$$

$$Q_\eta(G) = -\frac{1}{2} \int dx\ G(\eta)\ \mathrm{tr}\ j_\eta^2$$

where the functions F and G depend only on ξ and η , respectively. We have

$$\{Q_\xi(F),g\} = F(\xi)\partial_\xi g$$

$$\{Q_\eta(G),g\} = G(\eta)\partial_\eta g$$

and the $Q_\xi(F)$ and $Q_\eta(G)$ span the two commuting infinite dimensional conformal Lie algebras of two-dimensional space-time[9] :

$$\{Q_\xi(F_1),Q_\xi(F_2)\} = Q_\xi(F_1\partial_\xi F_2 - \partial_\xi F_1 F_2)$$

$$\{Q_\eta(G_1),Q_\eta(G_2)\} = Q_\eta(G_1\partial_\eta G_2 - \partial_\eta G_1 G_2)$$

$$\{Q_\xi(F),Q_\eta(G)\} = 0 \quad .$$

Choosing powers of ξ and η as basis functions, we recover the classical Virasoro algebra. Now the commutator of M_X and conformal transformations is

$$\begin{aligned}[M_X,F(\xi)\partial_\xi]g &= M_X\{Q_\xi(F),g\} - \{Q_\xi(F),M_X g\} \\ &= \{M_X Q_\xi(F),g\} \\ &= -\int dx\ F(\xi)\ \{\mathrm{tr}(j_\xi M_X j_\xi),g\} \\ &= 0\end{aligned}$$

upon inserting (21).

Proceeding as for the NLS, we can now derive the transformations of the complex sine-Gordon fields which read

$$X \in \mathcal{L}: \quad M_X \Phi^{(\lambda)} = \Phi^{(\lambda)}X(\lambda) - \frac{1}{2\pi i}\int_{C_o} \frac{d\mu}{\lambda-\mu}\,\frac{1-\lambda}{1-\mu}\,W_X(\mu)\ \Phi^{(\lambda)}$$

$$W_X(\mu) = \Phi^{(\mu)} X(\mu) \Phi^{(\mu)^{-1}}$$

$$M_X(\Gamma^{-1} j_\eta \Gamma) = \frac{1}{2\pi i} \int_{C_o} \frac{d\mu}{1-\mu^2}\, 2[W_X(\mu), \Gamma^{-1} j_\eta \Gamma]$$

$$M_X(\Gamma^{-1} \partial_\xi \Gamma) = \frac{1}{2\pi i} \int_{C_o} \frac{d\mu}{(1-\mu)^2}\, [i\sigma_3, W_X(\mu)] \quad .$$

Concerning M_a^o and M_a^1, the situation is analogous to the NLS case. The symmetry algebra does not restrict to a symmetry algebra of the sine-Gordon equation itself: application of M_X to $\alpha \neq o$, $\beta = o$ yields $\alpha \neq o$, $\beta \neq o$. Equivalently, the symmetry algebra (19)-(22) does not restrict to a symmetry algebra of the $\mathbb{C}P^1$ or S^2 non-linear σ model which is gauge equivalent to the sine-Gordon equation itself: The $\mathbb{C}P^1$ model corresponds to the SU(2) chiral model together with the constraint[3] $g^2 = \mathbb{1}$ which is not compatible with the action of M_X.

4. Final remarks

We know how to construct Kac-Moody symmetry algebras for several two-dimensional models as well as for the self-dual sector of the four-dimensional Yang-Mills theory[10]. The Lie algebras discussed here are built up from symmetries which exceed the Galilei or Poincaré space-time symmetries by far. In this respect, similar situations are known from classical and quantum mechanics: examples are the O(4) symmetry of the Kepler/Coulomb problem or the SU(N) symmetry of the N-dimensional isotropic harmonic oscillator. A mass point moving on the manifold of a group G has G×G as symmetry group (replacing the translation group of flat space). Common to these examples is the fact that the energy can be expressed in terms of the (quadratic) Casimir operators of the respective groups. This indicates that the symmetry groups are "complete" in the sense that they determine the energy spectrum. In the case of infinitely many degrees of freedom, one can expect the Casimir operators of the Kac-Moody algebras to play a distinguished rôle. Formally, the expressions

$$C_p = \sum_{a_1 \ldots a_p = 1}^{3} \sum_{\substack{n_1 \ldots n_p = -\infty \\ \Sigma n_j = o}}^{+\infty} tr(T_{a_1} \ldots T_{a_p}) \, M_{a_1}^{n_1} \ldots M_{a_p}^{n_p} \qquad (26)$$

are Casimir operators of $\mathcal{L}$: $[C_p, M_a^n] = o$ for all p, a, n, as can be shown by an argument used by Okubo for finite dimensional groups[11]. I say formally, because the sums in (26) may well diverge. However, at present it is unclear what might be the significance of the C_p in the context of the various two-dimensional models, and we are lacking an argument for the completeness of the symmetry algebras discussed here.

Finally, we note that both the non-Abelian Kac-Moody symmetries and the inverse scattering formalism leading to infinitely many commuting integrals of the motion (and thus to an Abelian group) are based on the same structure, namely on the Lax representation. Which is the relationship of non-Abelian Kac-Moody symmetry algebras and the notion of complete integrability requiring the existence of Abelian symmetry algebras?

References

1) W. Kinnersley, J. Math. Phys. 18 (1977) 1529
W. Kinnersley and D.M. Chitre, J. Math. Phys. 18 (1977) 1538
I. Hauser and F.J. Ernst, Phys. Rev. D2o (1979) 362
2) L. Dolan, Phys. Rev. Lett. 47 (1981) 1371
B. Julia, Paris preprint LPTENS 81/14
K. Ueno, Kyoto University preprint RIMS-374 (1981)
K. Ueno and Y. Nakamura, Kyoto university preprint RIMS-376 (1981)
See also the footnote in: Th. Curtright and C. Zachos, Phys Rev. D24 (1981) 2661
3) V.E. Zakharov and A.V. Mikhailov, Sov. Phys. JETP 47 (1978) 1o17

4) V.E. Zakharov and L.A. Takhtajan, Theor. Math. Phys. 38 (1979) 17
5) L.A. Takhtajan, Phys. Lett. 64A (1977) 235
6) H. Eichenherr, CERN TH-3299 (1982), to appear in Phys. Lett. B
7) C. Devchand and D.B. Fairlie, Nucl. Phys. B194 (1982) 232
8) K. Pohlmeyer, Commun. Math. Phys. 46 (1976) 2o7
9) M. Ademollo, A. d'Adda, R. d'Auria, E. Napolitano, S. Sciuto, P. di Vecchia, F. Gliozzi, R. Musto and F. Nicodemi, Nuovo Cimento 21A (1974) 77
1o) L.L. Chau, M.L. Ge and Y.S. Wu, Phys. Rev. D25 (1982) 1o86
K. Ueno and Y. Nakamura, Phys. Lett. 1o9B (1982) 273
L. Dolan, Rockefeller preprint RU82/B2o (1982)
11) S. Okubo, J. Math. Phys. 18 (1977) 2382

SCATTERING AND TRANSFER IN SOME GROUP THEORETICAL POTENTIALS

F. GÜRSEY

J. W. GIBBS LABORATORY
YALE UNIVERSITY
NEW HAVEN, CONNECTICUT 06511
U.S.A.

I. INTRODUCTION

In this talk I am going to report on some recent application of group theory to quantum mechanical bound states and scattering problems and to the derivation of the transfer matrix in certain periodic potentials. This work was done in collaboration with Y. Alhassid and F. Iachello[1].

The two best known examples of the application of group theory to quantum mechanics are given by the H-atom and the harmonic oscillator. In the former case the group that describes the degeneracy of the bound state levels is 0(4), while it is SU(3) in the latter case. For the H-atom the Hamiltonian which commutes with the generators of 0(4) is itself a generator of the non-compact group 0(4,2). The harmonic oscillator Hamiltonian is a generator of SU(3,1) and commutes with SU(3). Group theory can also be applied to the scattering problem in the case of the H-atom. In this case the energy changes sign and the group 0(4) of the bound states turns into the non-compact group 0(3,1) for the continuum states that describes scattering[2].

In most practical cases that one encounters in nuclear, molecular physics, the number of bound states is finite, unlike the H-atom case. A good approximation to a potential giving a finite number of bound states is the square-well potential. Rounding off the square-well does not change this finiteness property. Among the rounded off square-well potentials there is one kind, namely, the Pöschl-Teller potential[3] $V(x)=\alpha[\cosh x]^{-2}$ which is associated with the finite number of bound states and is also derivable from the representation theory of the group SU(2). Scattering off such a potential can also be described group theoretically provided SU(2) is replaced by the non-compact group SU(1,1). it follows that for both the bound states and scattering in a Pöschl-Teller potential the wave functions can be directly obtained from the unitary representation functions for SU(2) and SU(1,1). The situation is much simpler than the H-atom case and it is surprising that group theoretical methods have not been used up to now for this scattering problem in a well with finite number of bound states. Results we have obtained for the Pöschl-Teller problem will be summarized in this talk.

There is a related periodic potential, namely, the Scarf potential[4)] $V(x)=\beta[\sin x]^{-2}$ that is also related to groups SU(2) and SU(1,1). It arises as a sophisticated version of a Kronig-Penney[5)] type potential for a one-dimensional crystal leading to a complete solution for conduction bands which are finite in number and an explicit calculation of the effective mass of the electron in such a solid. The two problems, one with application in nuclear physics and the other in solid-state physics, are in fact related. They are both special cases of the Lamé equation which is a Schrödinger equation with potential $\wp(x)$ (the Weierstrass elliptic function) which interpolates between the Pöschl-Teller potential and the Scarf potential. $\wp(x)$ has two periods, one real, one complex. Special degenerate cases give the $(\sin x)^{-2}$ with one real period and $(\cosh x)^{-2}$ with one imaginary period. When both periods are infinite the Weierstrass function reduces to x^{-2} (the a-periodic case), which is the potential studied by Calogero[6)]. Calogero has also shown that the quantum mechanical many-body problem with two-body potential $V(x_i-x_j)$ between particles located at x_i and x_j with V(x) being the Weierstrass function or one of its degenerate cases is a completely integrable many-body system. The classical limit is also completely integrable. It can also be shown that the Lamé equation is also related to the groups SU(2) and SU(1,1) provided that the group element is parametrized by means of conical coordinates. Thus, in principle, scattering and bound state problems for all Calogero potentials can be obtained from unitary representations of SU(2) and SU(1,1) in appropriate coordinate systems. In the case of a periodic potential, there is no scattering. Instead, the transfer matrix is derived group theoretically. The location of bound states is replaced in this case by the location of a finite number of conduction bands. The Calogero potentials also arise in other problems. For instance, the mean-field in the Hartree-Fock approximation to the n-body problem with $\delta(x)$ potential is described by Pöschl-Teller potential. The non-relativistic limit of the S-matrix associated with the sine-Gordon field theory in two dimensions coincides with scattering in a Pöschl-Teller potential. The Morse potential[7)] that arises in atomic physics can also be connected with the same groups, SU(2) and SU(1,1). The group theoretical approach provides a unified approach for these solvable problems in nuclear, atomic, molecular and solid-state physics, unveiling their underlying unity of structure. We hope that the detailed study of the unitary representations of higher compact and non-compact groups will have applications to the solution of three-dimensional bound state scattering and transfer problems in physics.

II. RELATION OF SU(2) TO THE PÖSCHL-TELLER, SCARF AND MORSE POTENTIALS

We start with the generators J_i (i=1,2,3) of SU(2) and diagonalize the Casimir operator $\vec{J}\cdot\vec{J}$ and J_3 with eigenstates $|j,m\rangle$.

$$\vec{J}\cdot\vec{J}|j,m\rangle = j(j+1)|j,m\rangle \quad ,$$
$$J_3|j,m\rangle = m|j,m\rangle \quad . \tag{2.1}$$

SU(2) acts on the sphere S^2 with line element

$$ds^2 = dr^2 + r^2(d\theta^2 + \sin^2\theta d\phi^2) \tag{2.2}$$

in polar coordinates r,θ,ϕ. J_3 and the Casimir operator are then realized by the following Casimir operators

$$J_3 = -i\partial_\phi$$
$$\vec{J}\cdot\vec{J} = -\frac{1}{\sin^2\theta}\partial_\phi^2 - \frac{1}{\sin\theta}\partial_\theta\sin\theta\partial_\theta \tag{2.3}$$

acting on eigenfunctions

$$\chi_j^m(\theta,\phi) = U_j^m(\theta)e^{im\phi} \tag{2.4}$$

where

$$U_j^m(\theta) = d_j^{mm'}(\theta)\Big|_{m'=0} = P_j^m(\cos\theta) \tag{2.5}$$

The functions,

$$\Phi_j^m = \sqrt{\sin\theta}\, U_j^m = \sqrt{\sin\theta}\, P_j^m(\cos\theta) \tag{2.6}$$

satisfy the Schrödinger equation

$$\left(-\frac{d^2}{d\theta^2} + \frac{m^2-1/4}{\sin^2\theta}\right)\Phi_j^m = (j+\tfrac{1}{2})^2\,\Phi_j^m \tag{2.7}$$

in the form $H\psi=E\psi$ with the Scarf potential and positive energy eigenvalue, leading to the group theoretical solution of the solid-state model proposed by Scarf.

Now choose a new variable x related to θ by

$$\cos\theta = \tanh x \tag{2.8}$$

then the equation $\vec{J}\cdot\vec{J}\chi = j(j+1)\chi$ takes the form

$$\left(-\frac{d^2}{dx^2} - \frac{j(j+1)}{\cosh^2 x}\right)U_j^m(x) = -m^2U_j^m(x) \tag{2.9}$$

which is a Schrödinger equation with Pöschl-Teller potential

$$V(x) = -\frac{j(j+1)}{\cosh^2 x} \tag{2.10}$$

and energy eigenvalues $E=-m^2<0$, describing bound state solutions in such a potential. From group theory we also have $|m|\leq j$ showing that there is a finite number of bound states.

The normalized bound state wave functions are given by

$$U_j^m(x) = \sqrt{\frac{(j-m)!}{(j+m)!}\,(j+\tfrac{1}{2})}\,P_j^m(\tanh x) \quad . \tag{2.11}$$

Another form which is important for analytic continuation is

$$P_j^m(z) = \frac{1}{\Gamma(1-m)}\left(\frac{z+1}{z-1}\right)^{m/2} F\left(-j,j+1,1-m,\frac{1-z}{2}\right) \tag{2.12}$$

where $z=\cos\theta=\tanh x$.

The Morse potential can also be derived in the following way. Let us start with the oscillator representation of SU(2).

$$\begin{aligned} J_+ &= J_1 + iJ_2 = a^\dagger b \quad , \\ J_- &= J_1 - iJ_2 = b^\dagger a \quad , \\ J_3 &= \frac{a^\dagger a - b^\dagger a}{2} \quad , \end{aligned} \tag{2.13}$$

then $\vec{J}\cdot\vec{J}=J(J+1)$ where the operator $2J$ is the number operator $N=a^\dagger a+b^\dagger b$. Here a and b are boson operators with the differential operator realization

$$\begin{aligned} a &= x + \partial_x/\sqrt{2} \quad , \quad a^\dagger = x - \partial_x/\sqrt{2} \quad , \\ b &= y + \partial_y/\sqrt{2} \quad , \quad b^\dagger = y - \partial_y/\sqrt{2} \quad , \end{aligned} \tag{2.14}$$

then we find

$$J_2 = -\frac{i}{2}(x\partial_y - y\partial_x) \quad . \tag{2.15}$$

Switching to the plane-polar coordinates r and ϕ given by $x=r\cos\phi$, $y=r\sin\phi$ we obtain,

$$\begin{aligned} N = 2J &= \frac{1}{2}\left(-\frac{1}{r}\partial_r(r\partial_r) - \frac{1}{r^2}\partial_\phi^2 + r^2\right) - 1 \quad , \\ J_2 &= \frac{-i}{2}\partial_\phi \quad , \end{aligned} \tag{2.16}$$

with eigen-equations,

$$J_2\psi = m\psi \quad , \quad J\psi = j\psi \quad . \tag{2.17}$$

Let $r^2=(2j+1)e^{-\xi}$ and $\psi = R_j^m(\xi)e^{2im\phi}$. The equation for R_j^m takes the form

$$\left[-\frac{\partial^2}{\partial\xi^2} + (j+\tfrac{1}{2})^2(e^{-2\xi} - 2e^{-\xi})\right] R_j^m = -m^2 R_j^m \tag{2.18}$$

which is a Schrödinger equation with the Morse potential

$$V(\xi) = (j+\tfrac{1}{2})^2(e^{-2\xi} - 2e^{-\xi}) \tag{2.19}$$

and negative energy eigenvalues $E=-m^2(|m|\leq j)$ leading to a finite number of bound states.

III. RELATION OF SU(1,1) TO SCATTERING IN THE PÖSCHL-TELLER POTENTIAL

Instead of $\vec{J} = -i\vec{r}\times\vec{\nabla}$ for SU(2), we start from the hermitian representation

$$H_x = i(y\partial_z + z\partial_y) \quad , \quad H_y = -i(z\partial_x - x\partial_z) \quad , \quad H_z = -i(x\partial_y + y\partial_x) \tag{3.1}$$

for the SU(1,1) Lie algebra. Indeed

$$[H_x,H_y] = iH_z \quad , \quad [H_y,H_z] = iH_x \quad , \quad [H_z,H_x] = -iH_y \tag{3.2}$$

The H_i do not depend on r where $r^2=x^2-y^2+z^2$ (hyperboloid). Let

$$\begin{aligned} x &= r\sin\theta\cosh\phi \quad , \\ y &= r\sin\theta\sinh\phi \quad , \\ z &= r\cos\theta \quad , \end{aligned} \tag{3.3}$$

with this parametrization of the hyperboloid, we have

$$\begin{aligned} &H_z = -i\partial_\phi \\ &C = -H_x^2 + H_y^2 - H_z^2 \quad , \\ &C\psi_j^k = j(j+1)\psi_j^k \quad , \quad H_z\psi_j^k = k\psi_j^k \quad , \end{aligned} \tag{3.4}$$

where k is now a continuous variable. Note that the parameter ϕ is not the argument of a periodic function, hence is not an angle. Eq. (3.4) separates with

$$\psi_j^k = U_j^k(\cos\theta)e^{ik\phi} \quad . \tag{3.5}$$

Putting as before $z=\cos\theta$, $U_j^k(z)$ satisfies

$$\left|\frac{d}{dz}(1-z^2)\frac{d}{dz} + \frac{k^2}{1-z^2} + j(j+1)\right| U_j^k(z) = 0 \quad . \tag{3.6}$$

Transforming $z=\cos\theta$ to $\cos\theta=\tanh x$ as before, we get

$$\left| -\frac{d^2}{dx^2} - \frac{j(j+1)}{\cosh^2 x}\right| U_j^k = k^2 U_j^k \tag{3.7}$$

which is the Schrödinger Hamiltonian for the Pöschl-Teller potential but with positive eigenvalues $E=k^2>0$, describing scattering states instead of bound states in the same potential. The discrete quantum number m has been analytically continued into the continuous eigenvalue ik. The continuum wave functions U^k are now proportional to $P_j^{(ik)}(\tanh x)^j$ or

$$U_j^k = \frac{e^{-\frac{\pi k}{2}}}{(1-ik)} e^{ikx} F_{(-j,j+1,1-ik,\frac{1}{1+e^{2x}})} \tag{3.8}$$

For $x\to\infty$ $(z\to 0)$ we have $F\to 1$ and $U_j^k \sim e^{ikx}/\Gamma(1-ik)$, a plane wave function. In the limit $x\to-\infty$ $(z\to 1)$ F becomes a rational expression in Γ functions and U_j^k consists of two terms, one with a factor e^{ikx}, the other with e^{-ikx}. Hence for $x\to-\infty$ we have a superposition of an incoming and reflected wave while for $x\to\infty$ we have a single transmitted wave. The above group theoretical expression leads to the following values for R (reflection coefficient) and T (transmission coefficient) in the potential

$$V(x) = -\frac{j(j+1)}{\cosh^2 \frac{x}{a}} \tag{3.9}$$

(a=width of the potential)

$$R = \frac{\Gamma(ika)\Gamma(-ika-j)\Gamma(-ika+j+1)}{\Gamma(-ika)\Gamma(-j)\Gamma(j+1)} , \tag{3.10}$$

$$T = \frac{\Gamma(1-ika+j)\Gamma(-ika-j)}{\Gamma(1-ika)\Gamma(-ika)} . \tag{3.11}$$

Scattering in the Morse potential can be treated in the same way.

IV. HARMONIC ANALYSIS WITH SU(1,1)

If we had started with Euler-like angles for SU(1,1) : α,β,γ with β associated with pseudo-rotations then

$$D_j^{m,m'}(\alpha,i\beta,\gamma) = e^{im\alpha} d_j^{mm'}(i\beta) e^{im'\gamma} \tag{4.1}$$

For functions

$$\psi_j^{mm'}(\beta) = \sqrt{\sinh\beta}\; d_j^{mm'}(i\beta) \tag{4.2}$$

$\langle\psi|\Phi\rangle = \int_0^\infty \psi^*\Phi d\beta$ is invariant.

$$H\psi_j^{mm'}(x) = -(2j+1)^2 \psi_j^{mm'}(x) \tag{4.3}$$

with $\beta = x/a$ and

$$H = -\frac{d^2}{dx^2} - \frac{(m+m')^2 + \frac{1}{4}}{\cosh^2 \frac{x}{a}} + \frac{(m-m')^2 - \frac{1}{4}}{\sinh^2 \frac{x}{a}} \quad . \tag{4.4}$$

This is a Schrödinger equation with a modified Pöschl-Teller potential and an energy eigenvalue, the sign of which depends on the unitary representations of SU(1,1). There are three classes of unitary representations of SU(1,1), namely the analytic or discrete representations, the principal series representations and the complementary series representations. For our purpose we shall only consider the first two classes. For the discrete series 2j is a positive integer, m and m' can take values j, j+1, j+2, ... When m and m' are fixed, only a finite number of j values are possible and the energy is negative. Hence this case corresponds to bound states in the modified Pöschl-Teller potential. For the principal series $j^* = -j-1$, j is complex with the value $j = -\frac{1}{2} + ik$. The energy becomes $E = k^2 > 0$, describing scattering solutions for the above Hamiltonian. Thus we achieve a unified description of the bound states and scattering problems in the modified Pöschl-Teller potential. The two kinds of solutions can be brought together by means of a fundamental theorem in the harmonic analysis with respect to SU(1,1). This is the famous completeness theorem which states that any square integrable function F(x) for $0 < x < \infty$ can be expanded in terms of the principal series and a finite number of analytic representations of SU(1,1). (Note that the complementary series does not enter in this expansion). Thus we can write

$$F(x) = \frac{1}{2\pi^2}\int_0^\infty C(k)\psi^{mm'}_{-\frac{1}{2}+ik}(x)k\,\tanh(k\pi)dk + \frac{1}{2\pi^2}\sum_{\ell=1}^{M}(\ell-\tfrac{1}{2})C_\ell\psi^{mm'}_\ell(x) \tag{4.5}$$

$$M = \begin{cases}\min(|m|,|m'|) & \text{for } m,m'>0\\ 0 & \text{for } m,m'<0\end{cases}$$

where

$$C(k) = \int_0^\infty \overline{\psi^{mm'}_{-\frac{1}{2}+ik}(x)F(x)dx} \tag{4.6}$$

has the physical interpretation of the scattering component of the wave function F(x) and

$$C_\ell = \int_0^\infty \overline{\psi^{mm'}_\ell(x)F(x)dx} \tag{4.7}$$

can be interpreted as the bound state component of the same physical state described by F(x).

The SU(1,1) harmonic analysis generalizes the Fourier expansion connected with U(1). The Completeness Theorem states that for a physical wave function (square integrable function) scattering states (described by the principal series) and bound states (described by the discrete series) form a complete set. Note that in this approach both bound states and scattering states become related to unitary representations of SU(1,1).

V. A REVIEW OF RELATIONS BETWEEN SCATTERING AND TRANSFER MATRICES

In order to see the difference of treatment in the case of a periodic potential and a square-well potential, let us first consider the case of a potential V(x) which vanishes at $x=\pm\infty$. In this case the wave functions with positive energy behave like plane waves asymptotically and we can introduce a scattering matrix in the following way. Call $\psi_L(x)$ the asymptotic form of the wave function at $x=-\infty$. $\psi_L(x)$ is taken to be the superposition of an incoming and an outgoing plane wave of form

$$\psi_L(x) = A_0 e^{ikx} + B_0 e^{-ikx} \tag{5.1}$$

Similarly the right asymptotic wave function at $x=+\infty$ is written as

$$\psi_R(x) = A_1 e^{ikx} + B_1 e^{-ikx} \tag{5.2}$$

Thus A_0 and B_1 are associated with incoming waves while A_1 and B_0 correspond to outgoing waves. For a real potential the conservation of flux gives

$$|A_0|^2 + |B_1|^2 = |A_1|^2 + |B_0|^2 \quad . \tag{5.3}$$

Hence, if

$$\begin{pmatrix} B_0 \\ A_1 \end{pmatrix} = \begin{pmatrix} S_{11} & S_{12} \\ S_{21} & S_{22} \end{pmatrix} \begin{pmatrix} A_0 \\ B_1 \end{pmatrix} \tag{5.4}$$

or

$$\psi_{out} = S\psi_{in} \tag{5.5}$$

the scattering operator S is a unitary U(2) matrix ($SS^\dagger=1$). Furthermore, S can be chosen to be real and symmetric if V is real from time reversal invariance. Then S_{11} is the reflection coefficient to the left and S_{22} is the reflection coefficient to the right. $S_{12}=T$ is the transmission coefficient.

In the case of a Kronig-Penney type periodic potential, no asymptotic states exist. Hence, we cannot introduce a scattering matrix. However, we can still talk about a transformation of the wave functions on the left into the wave functions on the right. The left and right functions are associated respectively with the coefficient pairs (A_1,B_1) and (A_0,B_0). In this case we write

$$\psi_R = M\psi_L \quad \text{or} \quad \begin{pmatrix} A_1 \\ B_1 \end{pmatrix} = \begin{pmatrix} M_{11} & M_{12} \\ M_{21} & M_{22} \end{pmatrix} \begin{pmatrix} A_0 \\ B_0 \end{pmatrix} \tag{5.6}$$

Since $|A_0|^2 - |B_0|^2 = |A_1|^2 - |B_1|^2$ it follows that M is an SU(1,1) matrix of the form

$$M = \begin{pmatrix} \alpha & \beta \\ \beta^* & \alpha^* \end{pmatrix} \quad , \qquad \det M = 1 \quad . \tag{5.7}$$

M is called the transfer matrix. It can be introduced as a matrix which transforms the wave functions in one cell to those into a neighboring cel for a periodic potential. Note that the transfer matrix can also be defined when the S-matrix exists in the non-periodic case through Eq.(5.6 Then the relation between M and S is given by

$$M = \begin{pmatrix} \dfrac{1}{S_{12}{}^*} & -\left(\dfrac{S_{11}}{S_{12}}\right)^* \\ -\left(\dfrac{S_{11}}{S_{12}}\right) & \dfrac{1}{S_{12}} \end{pmatrix} \tag{5.8}$$

For example, in the Pöschl-Teller scattering, we find

$$\alpha = \frac{\Gamma(ik)\Gamma(1+ik)}{\Gamma(1+ik+j)\Gamma(ik-j)} \quad ,$$
$$\beta = \frac{\Gamma(ik)\Gamma(1-ik)}{\Gamma(-j)\Gamma(j+1)} \quad . \tag{5.9}$$

The S-matrix elements (R and T) can be calculated from Eqs. (5.7) and (5.8).

In the periodic case similar expressions exist for M with the Scarf potential $(\sin x)^{-2}$. Instead of asymptotic states, one takes two independent solutions ψ_1 and ψ_2 in the range $0<x\leq a$ where a is the period of the potential. In the nth cell we can write

$$\psi = A_n \psi_1(x-na) + B_n \psi_2(x-na) \quad . \tag{5.10}$$

The transfer matrix is defined by

$$\begin{pmatrix} A_n \\ B_n \end{pmatrix} = M \begin{pmatrix} A_{n-1} \\ B_{n-1} \end{pmatrix} \tag{5.11}$$

Matching the solutions at the boundary x=na, one finds

$$M = \begin{pmatrix} \psi_1(0) & \psi_2(0) \\ \psi_1'(0) & \psi_2'(0) \end{pmatrix}^{-1} \begin{pmatrix} \psi_1(a) & \psi_2(a) \\ \psi_1'(a) & \psi_2'(a) \end{pmatrix} \tag{5.12}$$

M is unimodular. For a real potential one can choose $\psi_2=\psi_1^*$ then, M is an SU(1,1) matrix. Note that if V is complex (optical potential), M is an SL(2,c) matrix.

When the periodic potential is obtained from the group SU(2) that leaves invariant the sphere parametrized by r,θ,ϕ we obtain a Schrödinger equation in θ. The wave functions are obtained from the representation functions $d_j^{mm'}(\theta)$. O(3) functions are single valued on the sphere, hence the corresponding wave functions of the Schrödinger equation with a periodic potential are periodic with respect to $\theta\to\theta+2n\pi$. Now the eigenvalues of M given by Eq. (5.12) are $e^{\pm i\kappa a}$ where κ is the crystal momentum. The eigenvectors are the Bloch waves

$$\psi_{n,\kappa}(x) = e^{i\kappa x}u_{n,\kappa}(x) \tag{5.13}$$

with

$$u_{n,\kappa}(x+a) = u_{n,\kappa}(x) \tag{5.14}$$

where $x=a\theta/\pi$ and a is the lattice constant. The periodicity of ψ in θ gives

$$\psi_{n,\kappa}(x+2a) = e^{2i\kappa a}\psi_{n,\kappa}(x) = \psi(x) \tag{5.15}$$

Thus we must have

$$\kappa a = \pi m \quad . \tag{5.16}$$

When κ is limited to the Brillouin zone we have solutions for $\kappa=0$ and at the edges of the zone where $\kappa=\pm\pi/a$. Because $V(x)=V(-x)$, $u_{n,-\kappa}(x)= u_{n,\kappa}(-x)$ and parity is a good quantum number. For $m\neq0$, there is an odd solution and an even solution with lower energy. For m=0 only one solution exists. The even solution is at the top of the energy band while the odd one is at the bottom of the next band, the energy difference between the two being the energy gap. The number m is just the eigenvalue of J_z and classifies the Brillouin zones or bands. For single valued representations j is an integer and gives the number of bands. There are 2j+1 states classified by m.

When E is in a conduction band the particle travels with an effective mass m* which can be calculated. At the edge of the band m* becomes infinite and the forbidden band starts.

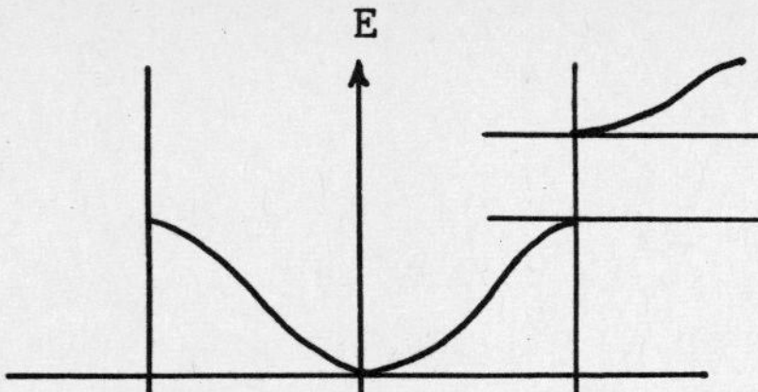

VI. INTERPOLATION BETWEEN SCATTERING IN A NON-PERIODIC POTENTIAL AND TRANSFER IN A PERIODIC POTENTIAL: THE LAMÉ EQUATION

A sphere S^2 with radius r can be parametrized in terms of conical coordinates instead of the usual polar coordinates that appear in Eq. (2.2), by means of the formulae

$$\begin{aligned} x &= \frac{r}{k'}\, \mathrm{dn}\theta\ \mathrm{dn}\phi \quad , \\ y &= ir\,\frac{k}{k'}\, \mathrm{cn}\theta\ \mathrm{cn}\phi \quad , \\ z &= rk\ \mathrm{sn}\theta\ \mathrm{sn}\phi \quad , \end{aligned} \tag{6.1}$$

where $\mathrm{cn}\theta$, $\mathrm{sn}\theta$ and $\mathrm{dn}\theta$ are Jacobi elliptic functions with modulus k and co-modulus k' such that $k^2+k'^2=1$, $x^2+y^2+z^2=r^2$. 4K and 2iK' are respectively the real and imaginary periods of $\mathrm{sn}\theta$. θ is real in the interval $-2K<\theta<2K$. ϕ is complex with $\mathrm{Re}\phi=K$, $0<\mathrm{Im}\phi<2K'$. Then x, y and z are real.

The generators of SU(2) (angular momentum operators) do not depend on r. The Casimir operator is given by[8)]

$$\vec{J}\cdot\vec{J} = \frac{1}{k^2}\,\frac{1}{\mathrm{sn}^2\theta-\mathrm{sn}^2\phi}\,(\partial_\theta^2-\partial_\phi^2) \quad . \tag{6.2}$$

Assuming a quadratic Hamiltonian of the form

$$H = J_x^2 + k^2 J_y^2 \quad , \tag{6.3}$$

one finds

$$H = \frac{1}{\mathrm{sn}^2\theta-\mathrm{sn}^2\phi}\,(\mathrm{sn}^2\phi\,\partial_\theta^2 - \mathrm{sn}^2\theta\,\partial_\phi^2) \quad . \tag{6.4}$$

Let Ψ be the eigenfunction of these commuting operators. We have

$$\vec{J}\cdot\vec{J}\Psi = j(j+1)\Psi \quad , \quad H\Psi = E\Psi \quad . \tag{6.5}$$

Putting $\Psi = \Theta(\theta)\Phi(\phi)$, Θ satisfies the Lamé equation

$$\left[- \frac{d^2}{d\theta^2} + j(j+1)k^2 sn^2\theta\right]\Theta = E\Theta \quad . \tag{6.6}$$

Φ satisfies the same equation with θ replaced by ϕ. This is a Schrödinger equation for the potential $V(x)=k^2 j(j+1)sn^2(x,k)$ if we take $\theta=x$. $V(x)$ is periodic with real period 2K. The edges of the Brillouin zones are given by the conditions $\kappa= \frac{\pi}{2K}\, m$ ($m=0, \pm1, \ldots,\pm j$) where κ is a crystal momentum. The number of bands is j that characterizes the irreducible representation of O(3) (single valued representations of SU(2)). The number of states is 2j+1.

In the limit $k\to1$, the real period disappears, V(x) becomes non-periodic (with only an imaginary period) and we obtain

$$V(x) = - \frac{j(j+1)}{\cosh^2 x} + j(j+1) \tag{6.7}$$

which is a Pöschl-Teller potential. In that limit $H=\vec{J}^2-J_z^2$ with eigenvalues $E_m=j(j+1)-m^2$ where m is eigenvalue of J_z. In this limit the Schrödinger equation admits scattering solutions.

In the limit $k\to0$, after a proper normalization, $\mathrm{sn}x\to\sin x$ giving a Kronig-Penney model with a periodic potential. In this case a transfer matrix can be defined. Finally, if x is replaced by ix and k by k', we have $\mathrm{sn}x \to i\frac{1}{\mathrm{sn}x}$. In the limit $k'\to0$ $\mathrm{sn}^2x \to \frac{1}{\sin^2 x}$ giving the Scarf model which is again periodic and admits a transfer matrix.

Thus the Lamé equation interpolates smoothly between the Pösch-Teller, Kronig-Penney and Scarf Schrödinger equations and contains as special cases periodic and non-periodic potentials that are related to each other through the groups SU(2) and SU(1,1). Bound states in the non-periodic case correspond to conduction bands and scattering in the former case corresponds to transfer in the latter case. When $k\to0$, the edges of the forbidden band shrink and the conduction bands collapse to scattering states with the energy relation $E = \frac{1}{2}\kappa^2$.

VII. CONCLUDING REMARKS

We have carried out a similar analysis for the groups $O(4)\sim SU(2)\times SU(2)$, $O(2,2)\sim SU(1,1)\times SU(1,1)$ and $SO(3,2)\sim Sp(4)$ and obtained a classification of most known solvable cases of the Schrödinger equation with periodic and non-periodic potentials[9].

The transfer matrix in the Lamé case has not yet been worked out explicitly. In this case the matrix elements have to involve the elliptic generalization of the Euler Gamma function $\Gamma(\kappa)$ which must depend on the modulus k or $q=e^{i\pi\tau}$ where τ is a ratio of the two periods of the elliptic function. The transfer matrix must then be expressed in terms of a function $\Gamma_q(\kappa)$ where κ is a crystal momentum. Such a function was introduced by F.H. Jackson[10] and called the q-Gamma function.

It is known that scattering in the Pöschl-Teller potential is related to the transfer matrix of the 6-Vertex model in statistical mechanics. The generalization of the 6-Vertex model is the 8-Vertex Baxter[11] model whose transfer matrix is in terms of elliptic functions. The free energy in this model first calculated by Baxter can be re-expressed in terms of q-Gamma functions[12]. As shown by Zamolodchikov and Shankar[13] there is a connection between the free energy in the statistical mechanical vertex model and the S-matrix of a corresponding, completely integrable two-dimensional relativistic quantum field theoretical model which in turn, in the non-relativistic approximation, becomes connected with the scattering functions of a Schrödinger theory with solvable potential. For instance, the 6-Vertex model is related to the sine-Gordon field theory which, in the non-relativistic limit, is connected with scattering in a Pöschl-Teller potential. This leads us to conjecture that the quantum field theory associated with the Baxter model will have an S-matrix or transfer matrix expressible in terms of q-Gamma functions and will admit a non-relativistic limit in which the transfer matrix coincides with that obtained from the Lamé equation, again by means of $\Gamma_q(\kappa)$.

We also note that Lamé potential being a Calogero potential, the n-body scattering and n-body transfer matrix can be expressed by means of two-body S or M matrices.

When the one-dimensional variable in the examples we have studied is interpreted as the radial variable in three dimensions, the same group theoretical approach can also be used in three-dimensional scattering and transfer problems at least as a useful approximation.

In principle, group theory can be applied to generate solvable potentials in one-dimensional relativistic quantum mechanics using the Dirac equation instead of the Schrödinger equation. It is not known if n-body solvable problems exist in this case. If they do (as in the case of the well-known $\delta(x)$ potential) they should be directly related to the solutions of a solvable quantum field theory model such as the sine-Gordon theory. It would also be interesting to explore how far these group theoretical methods can be extended to supersymmetric models related to scattering of bosons and fermions in solvable potentials.

I would like to conclude with some remarks concerning the possible relevance of our approach to some problems of Quantum Field Theory in four dimensions. If the theory is renormalizable, a cutoff Λ can be introduced (corresponding to a length $L=\Lambda^{-1}$) by quantizing the fields in a cell with periodic boundary conditions. Then the independence of the short distance behavior from L leads to renormalization group equations. Imposing periodic boundary conditions is equivalent to putting the fields in an external periodic potential in analogy with the case of an electron in a crystal. The characteristic distance L in Quantum Chromodynamics (QCD) is known to be of the order of the pion Compton wave length, or the confinement distance for quarks (bag size). Just as in the solid state situation, there is no S-matrix for quarks and gluons. However, using periodic boundary conditions one can define a transfer matrix relating quantum numbers of colored particles from one cell to the next and from its properties infer the existence of conduction bands and forbidden bands, the latter being associated with the confinement of quarks in one cell. The effective mass m* of the quark should diverge at the edge of the bag like that of the localized electron. In general, in a non-abelian gauge field theory, because of confinement there will be no asymptotic states and hence no meaningful S-matrix. Hence, solid-state physics concepts like transfer matrix, effective mass, forbidden bands, etc., should be more appropriate in the analysis of gauge field theories. If such theories or their self-dual sectors are in some sense completely integrable and associated with some infinite group structure (like a Kac-Moody algebra), the properties of the transfer matrix and the location of forbidden bands could be, as in solid state physics, related to the representations of the associated group.

Research supported in part by U.S. Department of Energy under Contract No. DE-AC02-76ER03075.

REFERENCES

1. Y. Alhassid, F. Gürsey and F. Iachello, Potential Scattering, Transfer Matrix and Group Theory, Phys. Rev. Letters (1983) (to be published); Y. Alhassid, F. Gürsey and F. Iachello, Group Theory Approach to Scattering, Ann. of Phys., NY, in press (1983); Y. Alhassid, F. Gürsey and F. Iachello, Group Theory Approach to Scattering and Transfer II, to be published in Ann. of Phys. NY (1983).
2. A. O. Barut and H. Kleinert, Phys. Rev. 157 (1967), 1180; 160 (1967) 1149; A. Barut and E. C. Philips, Commun. Math. Phys. 8 (1969), 52.
3. G. Pöschl and E. Teller, Z. Physik 83 (1933), 143.
4. F. L. Scarf, Phys. Rev. 112, (1958), 1137.

5. R. de L. Kronig and W.G. Penney, Proc. Roy. Soc. (London) A130, (1931), 499.
6. F. Calogero, Lett. Nuovo Cimento, 13 (1975) 411.
7. P.M. Morse, Phys. Rev. 34 (1929) 57.
8. J. Patera and P. Winternitz, J. Math. Phys. 14, 1130 (1973).
9. For separable potentials see: L. Infeld and T.E. Hull, Rev. of Mod. Phys. 23 (1951) 21.
For group theoretical treatment of potentials see: A.M. Perelomov, Sov. J. Part. Nucl. Phys. 10 (1979) 336.
J. Marsden and A. Weinstein, Rep. Math. Phys. 5 (1974) 121.
J.M. Gipson, Phys. Rev. Lett. 48 (1982) 1511 and references therein.
10. F. H. Jackson, Quart. Jour. Pure and Appl. Math. 41 (1910) 193.
11. J.R. Baxter, Ann. of Phys. 70 (1972) 193.
12. Private communication by G. Chudnovsky.
13. A.B. Zamolodchikov and Al.B. Zamolodchikov, Ann. of Phys. 120 (1979) 153;
R. Shankar, Phys. Rev. Letters 47 (1981) 1177.

REDUCTION OF SUPERSYMMETRIC σ-MODELS ON GRADED MANIFOLDS

Stefano Sciuto

Istituto di Fisica Teorica dell'Università di Torino
Sezione di Torino dell'Istituto Nazionale di Fisica Nucleare

I-10125 TORINO - ITALY

ABSTRACT

A geometrical treatment of supersymmetric σ-models on ordinary and graded manifolds is given.
It is shown that the SuSy σ-models on graded manifolds can be reduced to give generalizations of the SuSy sine-Gordon equation, endowed with an associate linear set. In particular the SuSy generalization of the Complex sine-Gordon model is briefly discussed.

1. Introduction

It is well known that two-dimensional σ-models on symmetric spaces G/H [1] can be formulated in terms of integrability conditions

$$R := d\omega - \omega\wedge\omega = 0 \tag{1.1}$$

of an associate linear set

$$dV = \omega V . \tag{1.2}$$

In eqs. (1.1), (1.2) V is a column of 0-forms on Minkowski space-time and $\omega = \omega(\lambda)$ is a 1-form matrix, valued on the Lie algebra $\mathbb{G}$ and depending on a parameter λ ; finally d is the exterior derivative.
As the quotient space G/H between the group G and its subgroup H is a symmetric space, the Lie algebra $\mathbb{G}$ can be decomposed in the following way:

$$\mathbb{G} = \mathbb{H} + \mathbb{K} , \tag{1.3}$$

with the commutation relations

$$[\mathbb{H},\mathbb{H}]\subset\mathbb{H},\quad [\mathbb{H},\mathbb{K}]\subset\mathbb{K},\quad [\mathbb{K},\mathbb{K}]\subset\mathbb{H}. \tag{1.4}$$

Then the projections of eq. (1.1) on the subalgebra $\mathbb{H}$ and on the complement $\mathbb{K}$ respectively read:

$$d\omega_H - \omega_H\wedge\omega_H - \omega_K\wedge\omega_K = 0 \tag{1.5a}$$

$$d\omega_K - \omega_H\wedge\omega_K - \omega_K\wedge\omega_H = 0\ , \tag{1.5b}$$

where ω_H and ω_K are the vertical and horizontal projections of the connection ω :

$$\omega = \omega_H + \omega_K\ ,\qquad \omega_H\in\mathbb{H}\ ,\quad \omega_K\in\mathbb{K}\ . \tag{1.6}$$

The λ dependence of eqs. (1.1) and (1.2) is introduced via the duality transformation [2, 3, 4]:

$$\begin{cases} \omega_H(1) \longrightarrow \omega_H(\lambda) = \omega_H(1) \\ \omega_K(1) = dx^+\omega_{K+} + dx^-\omega_{K-} \longrightarrow \omega_K(\lambda) = \dfrac{dx^+}{\lambda}\omega_{K+} + \lambda dx^-\omega_{K-}\ , \end{cases} \tag{1.7}$$

where the differentials of the light cone variables $x^\pm = \frac{1}{2}(x^1 \pm x^0)$ have been used as a basis for the 1-forms ω_H, ω_K. The requirement that the integrability conditions (1.1) hold for any value of the parameter λ amounts to add to eqs. (1.5a, b) a new equation:

$$d{}^*\omega_K - \omega_H\wedge{}^*\omega_K - {}^*\omega_K\wedge\omega_H = 0\ , \tag{1.5c}$$

where the star means the Hodge duality operation:

$${}^*\omega_K = dx_\mu\,\varepsilon^{\mu\nu}\,\omega_{K\nu} = dx^+\omega_{K+} - dx^-\omega_{K-}\ . \tag{1.8}$$

As eqs. (1.5a, b) only mean that ω is a pure gauge connection:

$$\omega = -g^{-1}dg\ ,\qquad \forall\, g = g(x)\in G, \tag{1.9}$$

one can take $g(x)$ as a new dynamical variable and interpretate eqs. (1.5c) as the true equations of motion of the σ-model. It is easy to see that eqs. (1.5c) can be obtained by varying $g(x)$ in the action:

$$S=\int d^2x \, Tr\left(g^{-1}\mathcal{D}g \cdot g^{-1*}\mathcal{D}g\right) , \qquad (1.10)$$

where $\mathcal{D}$ means the covariant exterior derivative under the gauge group H ; then eq. (1.9) can be seen as the definition of the 1-form matrix $\omega(\lambda)$ of the associate linear set (1.2).
In this talk we will show that the formulation of the σ-models sketched above(*), which is particularly suitable to find the conserved charges and to discuss the "reduction" problem, can be almost literally translated in superfield language to build supersymmetric σ-models (SuSy σ) on ordinary or graded manifolds. After a discussion of peculiar features of exterior calculus on superspace, in Sect. 2 we will present the general scheme to build SuSy σ on ordinary and graded manifolds.
In Sect. 3 we will discuss the reduction procedure that relates SuSy σ on graded manifolds with the SuSy sine-Gordon equation and its generalizations.
Finally in Sect. 4 the reduction of the SuSy σ on $SU(2|1)/U(2)$ is exhibited, leading to the SuSy generalization of the Pohlmeyer, Lund-Regge, Getmanov model [2, 6, 7] .

2. Exterior calculus in superspace and SuSy σ-Models.

In order to supersymmetrize the procedure outlined in Section 1, the first step is to extend exterior calculus to superspace.
Of course eqs. (1.1), (1.2) can be written on any manifold irrespective of its dimensionality and of its grading. However two-dimensional space time has the interesting peculiarity that eq.(1.1) reduces to a single matrix equation, as any 2-form is proportional to $dx^0 \wedge dx^1$. Superspace, instead, is parametrized by 2 ordinary variables $(x^\mu, \mu=0,1)$ and 2 Grassmann variables $(\theta_\alpha, \alpha=1,2)$; then the 2-form equation (1.1) has components along the 8 directions (**):

(*) For sake of brevity we do not explicitly discuss the principal (or chiral field) σ-models [4,5] , which can be seen as particular cases of the σ-models on symmetric spaces.

(**) The brackets $\{\ \}$ and $[\ ]$ mean symmetrization and antisymmetrization respectively, according to the (anti) commutation properties of the wedge product among the differentials $d\theta^\alpha$ and dx^μ.

$$d\theta^{\{\alpha} \wedge d\theta^{\beta\}}$$

$$d\theta^{\alpha} \wedge dx^{\mu} \qquad (2.1)$$

$$dx^{[\mu} \wedge dx^{\nu]} \quad .$$

Actually it happens that only 3 out of the 8 equations contained in eq. (1.1) are independent; even more, a single equation gives the most relevant information [8] .

To prove this property it is convenient to use the contravariant basis (*):

$$E^{\alpha} := d\theta^{\alpha}$$

$$e^{\mu} := dx^{\mu} - i\, d\theta \gamma^{\mu} \theta \qquad (2.2)$$

of left invariant forms under rigid supersymmetry transformations

$$\theta^{\alpha} \longrightarrow \theta^{\alpha} + \varepsilon^{\alpha}$$

$$x^{\mu} \longrightarrow x^{\mu} - i\, \varepsilon \gamma^{\mu} \theta \quad ; \qquad (2.3)$$

let us remark that the basis (2.2) is dual to the basis

$$D_{\alpha} := \frac{\partial}{\partial \theta^{\alpha}} + i (\gamma^{\mu} \theta)_{\alpha} \frac{\partial}{\partial x^{\mu}}$$

$$\partial_{\mu} := \frac{\partial}{\partial x^{\mu}} \qquad (2.4)$$

of tangent vectors (or supersymmetric covariant derivatives). By inserting the 2-form

$$R = E^{\alpha} \wedge E^{\beta} R_{\alpha\beta} + E^{\alpha} \wedge e^{\mu} R_{\alpha\mu} + e^{\mu} \wedge e^{\nu} R_{\mu\nu} \qquad (2.5)$$

into the Bianchi identity

$$dR = \omega \wedge R - R \wedge \omega \qquad (2.6)$$

and by using the Maurer Cartan equation for the graded translation

---

(*) Spinorial indices are raised by the antisymmetric Ricci tensor $\theta^{\alpha} = \theta_{\beta} \varepsilon^{\beta\alpha}$ and saturated according to the convention $\psi\chi = \psi^{\alpha}\chi_{\alpha}$; the γ matrices are chosen to be real.

group

$$de^{\mu} = i\, E\gamma^{\mu} E \tag{2.7}$$

(which directly follows from the definition (2.2)) one easily gets [9] that the equation

$$R_{\alpha\beta} = 0 \tag{2.8}$$

implies the equations

$$R_{\alpha\mu} = 0 \tag{2.9a}$$

$$R_{\mu\nu} = 0 \; . \tag{2.9b}$$

In other words, in the basis (2.2), the necessary and sufficient condition that the 2-form R vanish is the vanishing of its components along the directions $d\theta^{\{\alpha} \wedge d\theta^{\beta\}}$.
It is useful to write Eq. (2.8) in terms of the components $\Omega_\alpha , \omega_\mu$ of the 1-form ω :

$$\omega = E^{\alpha}\Omega_{\alpha} + e^{\mu}\omega_{\mu} \quad ; \tag{2.10}$$

to this aim it is convenient to use the notations of ref. [8] and introduce the <u>truncated 1-form</u> (t1-form)

$$\hat{\Omega} := E^{\alpha}\Omega_{\alpha} \equiv d\theta^{\alpha}\Omega_{\alpha} \tag{2.11}$$

and the <u>truncated exterior derivative</u>

$$\hat{D} := E^{\alpha} D_{\alpha} \equiv d\theta^{\alpha} D_{\alpha} \; . \tag{2.12}$$

In this language, the definitions (1.1), (2.5), (2.10) and Eq.(2.7) give:

$$E^{\alpha}\wedge E^{\beta} R_{\alpha\beta} = i\, E\gamma^{\mu}E\, \omega_{\mu} + \hat{D}\hat{\Omega} - \hat{\Omega}\wedge\hat{\Omega} \; . \tag{2.13}$$

Moreover, by inserting the definition

$$\hat{\mathcal{S}} := \hat{D}\hat{\Omega} - \hat{\Omega}\wedge\hat{\Omega} = E^{\alpha}\wedge E^{\beta}\mathcal{S}_{\alpha\beta} \tag{2.14}$$

in Eq. (2.13), Eq. (2.8) becomes:

$$\mathcal{S}_{12} = 0 \; , \tag{2.15}$$

$$\begin{cases} \omega_+ = -i\rho_{11} \\ \omega_- = i\rho_{22} \end{cases} \tag{2.16}$$

Then the most convenient way to discuss the linear set (1.2) in the supersymmetric case is to start from the truncated linear set

$$\hat{D} V = \hat{\Omega} V , \tag{2.17}$$

whose integrability condition is the single matrix equation (2.15); following ref. [8] it can be written in the form:

$$\hat{D}\hat{\Omega} - \hat{\Omega}\wedge\hat{\Omega} \doteq 0 , \tag{2.15'}$$

where the symbol $\doteq 0$ means that the coefficient of $d\theta^1 \wedge d\theta^2$ in the left hand side must vanish. The Eq. (2.15) does not involve the coefficients of $d\theta^1 \wedge d\theta^1$ and of $d\theta^2 \wedge d\theta^2$; however the calculation of these coefficients is very useful, because Eqs. (2.16) can be used to define ω_μ in terms of Ω_α .
Then the truncated linear set (2.17) can be completed by adding the equations

$$\partial_\mu V = \omega_\mu V \tag{2.18}$$

with the usual consistency conditions:

$$\partial_\mu \omega_\nu - \partial_\nu \omega_\mu - [\omega_\mu , \omega_\nu] = 0 \tag{2.19}$$

Let us remark again that Eq. (2.19), which coincides with Eq. (2.9b), is not a new one but is implied by Eq. (2.15) on Ω_α and by the definition (2.16) of ω_μ ; however eq. (2.19) can be very useful, as its θ independent part gives the equations of motion of our models [10, 12] in terms of ordinary fields, while eq. (2.15) gives the same equations in terms of superfields.
Following the procedure outlined in Section 1 we have now to specify the manifold in which the 1-form ω (or the t1-form $\hat{\Omega}$) takes values.
Two main possibilities can be considered:

A) $\omega, \hat{\Omega} \in \mathbb{G}$ where $\mathbb{G}$ is an ordinary Lie algebra, decomposable according to eqs. (1.3), (1.4).

B) $\omega, \hat{\Omega} \in \mathbb{G}$ where $\mathbb{G}$ is a graded Lie algebra; eqs.(1.3), (1.4) are then replaced by the analogous ones:

$$\mathbb{G} = \mathbb{H} + \mathbb{K} , \tag{2.20}$$

where $\mathbb{H}$ is the maximal even subalgebra of $\mathbb{G}$ and $\mathbb{K}$ is the odd sector of $\mathbb{G}$;

$$[\mathbb{H},\mathbb{H}]\subset\mathbb{H}\,,\ [\mathbb{H},\mathbb{K}]\subset\mathbb{K}\,,\ \{\mathbb{K},\mathbb{K}\}\subset\mathbb{H}\,. \tag{2.21}$$

Because of the strict analogy between eqs. (2.21) and eqs. (1.4), which respectively define a graded Lie algebra and a symmetric space G/H , it is possible to give a unified treatment of SuSy σ on the two kinds of manifolds [11] .
Once that the t1-form $\hat{\Omega}$ is projected on $\mathbb{H}$ and $\mathbb{K}$:

$$\hat{\Omega}=\hat{A}+\hat{K}\,,\qquad \hat{A}\in\mathbb{H}\,,\quad \hat{K}\in\mathbb{K}\,, \tag{2.22}$$

eq. (2.15') reads:

$$\hat{D}\hat{A}-\hat{A}\wedge\hat{A}-\hat{K}\wedge\hat{K}\doteq 0 \tag{2.23a}$$

$$\hat{D}\hat{K}-\hat{A}\wedge\hat{K}-\hat{K}\wedge\hat{A}\doteq 0\,. \tag{2.23b}$$

It is easy to see that under the duality transformation defined by:

$$\begin{cases} \hat{A}(1) \longrightarrow \hat{A}(\lambda)=\hat{A}(1) \\ \hat{K}(1)=d\theta^1K_1+d\theta^2K_2 \longrightarrow \hat{K}(\lambda)=\dfrac{d\theta^1}{\sqrt{\lambda}}K_1+\sqrt{\lambda}\,d\theta^2K_2\,, \end{cases} \tag{2.24}$$

eq. (2.23a) does not change; instead eq. (2.23b) becomes the sum of two terms proportional respectively to $\lambda^{\pm 1/2}$.
Thus the λ independence requirement amounts to add to eqs. (2.23) a new equation:

$$\hat{D}\underset{*}{\hat{K}}-\hat{A}\wedge\underset{*}{\hat{K}}-\underset{*}{\hat{K}}\wedge\hat{A}\doteq 0\,, \tag{2.23c}$$

where star operation on truncated 1-forms is defined by:

$$\underset{*}{\hat{K}}:=d\theta\gamma^5K=d\theta^1K_1-d\theta^2K_2\,. \tag{2.25}$$

Eqs. (2.23 a, b) guarantee that the curvature R vanishes (i.e. that $\hat{\Omega}(\lambda)=-g^{-1}\hat{D}g$); instead eq. (2.23c) is the σ- model equation of motion for the (graded) group element $g(x,\theta)$.

It is worthwhile to note that the use of truncated forms makes the treatment of this Section very close to that of Sect. 1, where the purely bosonic case was discussed; instead if the complete 1-form ω were used the duality transformation (2.25) and the equations of motion (2.23c) would be much more involved.
As in the purely bosonic case, the equations of motion (2.23c) can be obtained from an action written in terms of differential forms on superspace; to this aim the generalization of the Hodge duality operation proposed by Berezin [13] must be used; however we have no time to discuss this point here and send the interested reader to ref. [11] .
As a final remark of this Section let us note that two-dimensional SuSy σ on symmetric spaces (case A) have been largely studied [14, 15, 5] both at the classical and at the quantum level, getting inspiring suggestions about four dimensional theories of physical relevance; instead SuSy σ on graded manifolds (case B) seem less interesting as their equations of motion treat on the same foot bosonic and fermionic degrees of freedom (matrix elements of the graded group element $g(x,\theta)$) in contrast with the spin-statistic theorem.
However with respect to the "reduction" problem, that we will discuss in Sect. 3, the SuSy σ on ordinary and graded manifolds exchange their rôles; the SuSy σ on graded manifolds can be easily reduced to give perfectly sensible models, like the SuSy sine-Gordon equation [14] and its generalizations [10] ; on the contrary no analogous of the reduction procedure is known for SuSy σ on ordinary manifolds.

3. The reduction of the SuSy σ-models.

Apart the grading of the conformal group, the SuSy σ on ordinary or graded manifolds built in Section 2 share with their purely bosonic counterparts (Section 1) the same invariance properties: rigid invariance under the (graded) group G and local invariance under the subgroup H, according to the transformations:

$$g \to g_0 g h \quad , \qquad \forall g_0 \in G \;, \; \forall h = h(x,\theta) \in H. \qquad (3.1)$$

Roughly, the reduction procedure consists in replacing the (graded) group element $g(x,\theta)$ by new dynamical variables invariant under the transformations (3.1).
The first and very well known example is due to Pohlmeyer [2] and relates (at the classical level) the O(3) σ-model (more precisely the O(3)/O(2) σ-model) with the sine-Gordon equation.
In this Section we will show that an analogous procedure can be ap

plied to reduce SuSy σ on graded manifolds.
Firstly we note that the t1-form matrix $\hat{\Omega}(\lambda)=-g^{-1}\hat{D}g \in \mathbb{G}$ is already invariant under rigid G transformations $g \to g_0 g$; instead it transforms inhomogeneously under local H transformations $g \to gh$.
Secondly we observe that the requirement that the Maurer Cartan formula make sense, i.e. $\hat{\Omega} \wedge \hat{\Omega} \in \mathbb{G}$, determines the bosonic or fermionic character of the projections $\hat{A}$ and $\hat{K}$ of $\hat{\Omega}$; namely, in both cases A and B of Section 2 the projection $\hat{A}$ on the subalgebra $\mathbb{H}$ must be a matrix of bosonic t1-forms (and then the components A_1, A_2 are fermionic superfields); instead the projection $\hat{K}$ on $\mathbb{K}$ must be a matrix of bosonic or fermionic t1-forms if one deals with ordinary (case A) or graded (case B) manifold SuSy σ respectively.
Then in the case B (and only in this case) the components K_1, K_2 of the t1-form $\hat{K}$ are bosonic superfields.
In this case the reduction procedure can be performed by the following steps:

i) fix a suitable gauge in order to eliminate the gauge freedom under the local H transformations (*);
ii) parametrize the matrix elements of K_1 and K_2 by Lorentz scalar bosonic superfields (**);
iii) solve eqs. (2.23b,c) to get A_1 and A_2 in terms of the matrix elements of K_1 and K_2;
iv) substitute in eqs. (2.23a); they turn out to be a set of second order equations of motion for the scalar bosonic superfields introduced at the step ii), and then agree with the spin statistic theorem.

The reduced models obtained by the recipe (i-iv) are relativistically invariant; in fact with the choices ii) and iii) (K_1 and K_2 Lorentz scalars and A_1 (A_2) linear homogeneous in the first spinorial derivatives D_1 (D_2) of scalar superfields) the duality trans

---

(*) A thorough discussion of the "reduction gauge" in the purely bosonic case can be found in refs. [1, 16].

(**) For rank 1 graded manifolds (like $SU(n|1)/U(n)$) the graded conformal invariance can be suitably broken to make these superfields actually Lorentz invariant; for higher rank manifolds the Lorentz invariance requirement can be fulfilled by selecting a subclass of solutions of the σ-models.

formation (2.24) becomes the Lorentz boost:

$$x^+ \rightarrow \frac{x^+}{\lambda}, \quad x^- \rightarrow \lambda x^-, \quad \theta^1 \rightarrow \frac{\theta^1}{\sqrt{\lambda}}, \quad \theta^2 \rightarrow \sqrt{\lambda}\,\theta^2.$$

The SuSy models obtained in this way generalize the SuSy sine-Gordon equation; actually to get exactly the SuSy sine-Gordon model (and the other SuSy models whose purely bosonic limits are reduced σ-models on symmetric spaces H/H_1 with non trivial H_1 [17]) one needs a finer (Z_4) grading of the graded Lie algebra $\mathcal{G}$ (see refs. [10,12,18]); the general scheme however remains the one outlined above.
The last important remark of this Section is that the reduction procedure does not work for SuSy σ on ordinary manifolds (case A of Section 2); in fact in such a case the matrix elements of K_1 and K_2 are fermionic superfields and eqs. (2.23b) cannot be solved to express A_1, A_2 in terms of K_1 and K_2.

4. An example.

The simplest example of SuSy σ on a graded manifold G/H corresponds to the choice $G = SU(2|1)$, $H = U(2)$. Its action [11,18] is very similar to the action [15] of the CP_2 $(\sim SU(3)/U(2))$ SuSy σ :

$$S = \frac{i}{2}\int d^2x\, \bar{d}^2\theta \left[(\nabla^\alpha Z)^+ \nabla_\alpha Z + \lambda (Z^+ Z - 1) \right], \tag{4.1}$$

where ∇_α is the spinorial derivative covariant under $U(2)$ transformations:

$$\nabla_\alpha Z = (D_\alpha + D_\alpha Z^+ \cdot Z) Z \tag{4.2}$$

and λ is a Lagrange multiplier (*). However the fundamental object $Z(x,\theta) \in G/H$ is now a complex vector whose components are one fermionic and two bosonic superfields.
The reduction of this model follows the scheme of Section 3. In particular, by suitably breaking the graded conformal invariance one gets from the equations

$$D_2 \left[(D_1 Z)^+ D_1 Z \right] = D_1 \left[(D_2 Z)^+ D_2 Z \right] = 0, \tag{4.3}$$

(*) We use the Berezin [13] notation $\bar{d}\theta$ for the integration on Grassmannian variables in order to avoid any confusion with the differentials $d\theta$ introduced in Section 2.

which follows from the equations of motion, the stronger ones (*)

$$(D_1 Z)^+ D_1 Z = (D_2 Z)^+ D_2 Z = m . \qquad (4.4)$$

The reduced model [12] is described in terms of two real scalar superfields $\Phi(x,\theta)$ and $H(x,\theta)$ defined by

$$(D_1 Z)^+ D_2 Z = i m \cos\Phi \, e^{iH} . \qquad (4.5)$$

Its equations of motion are:

$$\begin{cases} \frac{i}{2} D^\alpha D_\alpha \Phi + \frac{i}{2} \frac{\cos\Phi}{\sin^3\Phi} D^\alpha H D_\alpha H + m \cos\Phi \cos H = 0 \\ \frac{i}{2} D^\alpha D_\alpha H - \frac{2i}{\sin 2\Phi} D^\alpha \Phi D_\alpha H + m \frac{\sin^2\Phi}{\cos\Phi} \sin H = 0 \end{cases} ; \qquad (4.6)$$

the t1-form matrix $\hat{\Omega}$ of the associate linear set (2.17) is rather simple and can be found in ref. [12] .

This reduced model, which reproduces the SuSy sine-Gordon model for H=0, is rather interesting for several reasons.

i) An action exists whose variation reproduces eqs. (4.6):

$$S = \int d^2x \, d^2\theta \left\{ \frac{1}{4} D^\alpha \Phi D_\alpha \Phi + \frac{1}{4} \cot^2\Phi \, D^\alpha H D_\alpha H - i m \cos\Phi \cos H \right\} ; \qquad (4.7)$$

this fact is not trivial because, as in the purely bosonic case, no general recipe is known to write actions for the reduced models.

By expanding the superfields Φ and H in powers of θ

$$\begin{cases} \Phi(x,\theta) = \varphi(x) + i\,\theta^\alpha \psi_\alpha(x) + \frac{i}{2} \theta^\alpha \theta_\alpha F(x) \\ H(x,\theta) = h(x) + i\,\theta^\alpha \chi_\alpha(x) + \frac{i}{2} \theta^\alpha \theta_\alpha G(x) \end{cases} , \qquad (4.8)$$

the action (4.7) can be written in terms of ordinary fields:

(*) Eq. (4.4) shows very neatly why the O(3) SuSy σ and the SuSy sine-Gordon model are <u>not</u> classically equivalent, differently from their purely bosonic sectors: if $Z(x,\theta) \in S_2 \sim O(3)/O(2)$, its components are real bosonic superfields and the left hand sides of eq. (4.4) identically vanish.

$$S=\int d^2x\left\{-\frac{1}{2}\left(\partial^\mu\varphi\partial_\mu\varphi+\cot^2\varphi\,\partial^\mu h\,\partial_\mu h+m^2\sin^2\varphi\right.\right.$$

$$\left.-i\bar\eta\partial\!\!\!/\eta-\frac{1}{\sin^2\varphi}\left[(\bar\eta\eta)^2+\bar\eta\,\partial\!\!\!/ h\gamma^5\eta\right]-im\cos\varphi\,\bar\eta(\cos h-i\gamma^5\sin h)\eta\right\}, \tag{4.9}$$

where the auxiliary fields F and G have already been eliminated and, besides the real scalar fields φ and h, the complex spinor

$$\eta=\frac{1}{\sqrt{2}}\left(\psi+i\cot\varphi\,\gamma^5\chi\right) \tag{4.10}$$

has been introduced.

ii) The purely bosonic limit $(\eta=0)$ of the action (4.9) describes the Complex sine-Gordon model built by Pohlmeyer [2], Lund Regge [6] and Getmanov [7] by reduction of the $O(4)$ σ-model.

iii) The associate linear set (2.17) allows to write an infinite set of local conservation laws [12], which generalize those already known [19] for the SuSy sine-Gordon model.

iv) By construction, the action (4.7), (4.9) is invariant under N=1 supersymmetry; actually it turns out to be invariant under the extended (N=2) supersymmetry group generated by a complex spinorial charge Q_α and a $U(1)$ generator T, with the (anti)commutation relations [20]:

$$\left[T,Q_\alpha\right]=-\frac{i}{2}Q_\alpha\;,\qquad\left[T,\bar Q_\alpha\right]=\frac{i}{2}\bar Q_\alpha\;,$$
$$\left\{Q_\alpha,\bar Q^\beta\right\}=2(\gamma^\mu P_\mu)_\alpha{}^\beta-2i\,K_\alpha{}^\beta \tag{4.11}$$

where, besides the momentum P_μ, the central charge

$$K=m\cos\varphi\left(\gamma^5\cos h+i\sin h\right)\Big|_{x^1=-\infty}^{x^1=+\infty} \tag{4.12}$$

appears, which generalizes the analogous one found by Witten and Olive [21] in the SuSy sine-Gordon model.

v) The invariance under N=2 supersymmetry can be made manifest by complexification of the Grassmann coordinates θ^α of the superspace. The action obtained in this way [20] could also be obtained (for $m=0$) by applying the Zumino recipe [22] to the Pohlmeyer, Lund, Regge, Getmanov model, treated as a nonlinear σ-model on a very simple Kähler manifold. The procedure of refs. [8, 10] can be extended to build N=2 SuSy models, endowed of an associate linear set, on the graded manifolds $SU(n|1)/U(n)$ for any $n\geq 2$ [23]; in this framework one can show that for n=2 the model discussed in this Section is essentially unique.

REFERENCES

[1] For a very good review see for instance: H. Eichenherr, Tvärminne Lectures, CERN preprint TH-3073 (1981).

[2] K. Pohlmeyer, Commun. Math. Phys. 46 (1976) 207.

[3] H. Eichenherr and M. Forger, Nucl. Phys. B155 (1979) 381.

[4] V.E. Zakharov and A.V. Mikhailov, JETP Lett. 27 (1978) 42; Sov. Phys. JETP 47 (1978) 1017.

[5] A.V. Mikhailov, JETP Lett. 28 (1978) 512.

[6] F. Lund and T. Regge, Phys. Rev. 14 (1976) 1524.

[7] B.S. Getmanov, JETP Lett. 25 (1977) 119.

[8] S. Sciuto, Phys. Lett. 90B (1979) 75.

[9] M.F. Sohnius, Nucl. Phys. B136 (1978) 461.

[10] R. D'Auria and S. Sciuto, Nucl. Phys. B171 (1980) 189.

[11] E. Napolitano and S. Sciuto, Nuovo Cimento 64A (1981) 406.

[12] E. Napolitano and S. Sciuto, Commun. Math. Phys. 84 (1982)171.

[13] F.A. Berezin, Sov. J. Nucl. Phys. 30 (1979) 605.

[14] P. Di Vecchia and S. Ferrara, Nucl. Phys. B130 (1977) 93; E. Witten, Phys. Rev. D16 (1977) 299.

[15] O. Alvarez, Phys. Rev. D17 (1978) 1123; R. Shankar and E. Witten, Phys. Rev. D17 (1978) 2126; A. D'Adda, P. Di Vecchia and M. Luscher, Nucl. Phys. B152 (1979) 125; A.V. Mikhailov and A.M. Perelomov, JETP Lett. 29 (1979) 445; J.F. Schonfeld, Nucl. Phys. B169 (1980) 49 and references contained therein.

[16] H. Eichenherr and M. Forger, Commun. Math. Phys. 82 (1981)227.

[17] R. D'Auria, T. Regge and S. Sciuto, Phys. Lett. 89B (1980)363; Nucl. Phys. B171 (1980) 167.

[18] E. Napolitano and S. Sciuto, Phys. Lett. 104B (1981) 285.

[19] S. Ferrara, L. Girardello and S. Sciuto, Phys. Lett. 76B (1978) 303; L. Girardello and S. Sciuto, Phys. Lett. 77B (1978) 267; M. Chaichian and P.P. Kulish, Phys. Lett. 78B (1978) 413.

[20] E. Napolitano and S. Sciuto, Phys. Lett. 113B (1982) 43.

[21] E. Witten and D. Olive, Phys. Lett. 78B (1978) 97.

[22] B. Zumino, Phys. Lett. 87B (1979) 203

[23] L. Grabbi and S. Sciuto, in preparation.

TWO DIMENSIONAL σ-MODELS AND HARMONIC MAPS FROM S^2 TO S^{2n}

J.L. Verdier
Ecole Normale Supérieure
PARIS FRANCE

I. THE PROBLEM

Let G be a compact group, (V,g) be a compact Riemannian manifold and $B \to V$ a principal G-bundle. Let Γ be a G-invariant connection on B, $R(\Gamma)$ its curvature tensor. Set

$$E(\Gamma) = \int_V \| R(\Gamma) \|^2 \, dv .$$

It is the energy of Γ. The Yang-Mills problem is to study the space of Γ's such that $E(\Gamma)$ is extremum or minimum.

The most famous example is the case $G = SU(n)$, $V = S^4$ (case of *Instantons*, results of Atiyah Drinfeld Hitchin Manin). Another example is the case $G = SU(n)$, V = 2-dimension Riemannian Surface. The solutions are in 1-1 correspondance with the *stable vector bundles over* V, of rank n.

The Penrose transform allows to reformulate the problem of finding instantons : there is a classifying space X and the Instantons are in 1-1 correspondance with maps $f : S^4 \to X$, such that some functional $E(f)$ is extremal.

In this lecture, we will consider a simpler problem of the same type. Given two Riemannian varieties X and Y and a map $f : X \to Y$, define the energy of f as being

$$E(f) = \int_X \| T_x f \|^2 \, d_x v .$$

The *harmonic maps* are extremal maps for this functional, and we will study the space of those maps.

In the 2-dimensional case the simplest surface is $S^2 = \mathbb{C}P^1$, and as classifying spaces we will consider the spheres S^{2n}. I wish to thank R. STORA for having pointed out to me this problem and for stimulating discussions.

II. THE PENROSE TRANSFORMATION AND CALABI LIFTING

Let V an even dimensional oriented variety with a metric g. Let P(V) the variety of couples (x, I_x) where $x \in V$ and $I_x : T_x \to T_x$ is a direct orthogonal transformation of the tangent space T_x such that $I_x^2 = -1$. In other word P(V) consists of the complex structures on the fibers of the tangent bundle of V , compatible with the metric. It is called the *Penrose transform of* V [D].

When $V = S^{2n}$, P(V) is the *quadratic Grassman variety* $QG(\mathbb{C}^{2n+1})$ i.e. the variety of n dimensional subspaces of $\mathbb{C}^{n+1}$, isotropic for the standard quadratic form. It carries a $\mathbb{C}$-analytic structure. We have a natural projection $P(S^{2n}) \xrightarrow{\pi} S^{2n}$ and since $P(S^{2n})$ is equipped naturally with a metric we can consider the subbundle S of the tangent bundle $T_{P(S^{2n})}$ which is orthogonal to the fibers of π . Notice that π is not a $\mathbb{C}$-analytic map, and taking the orthogonal is not a $\mathbb{C}$-analytic operation. However it is a *fact* that S is *an analytic subbundle of* $T_{P(S^{2n})}$.

When $V = S^4$ then P(V) is $\mathbb{C}P^3$ and the Penrose transform is the classical one. The subbundle S in this case can be described as follow :if $\mathbb{C}P^3 = \mathrm{Proj}(\mathbb{C}^4)$, there exists on $\mathbb{C}^4$ a non degenerate alternate bilinear form B such that for any $x \in \mathbb{C}P^3$, S_x is the two-dimensional subspace of T_x corresponding to the orthogonal of x for B (nul correlation bundle).

The Calabi transform is the following. Let $V = S^4$ and $f : S^2 \to S^4$ an immersion (to simplify). Let $x \in S^2$. Then $df(T_x)$ is a 2-dimensional subspace of $T_{f(x)}$. There exists a unique, direct and orthogonal complex structure I_x on $T_{f(x)}$ such that $df(T_x)$ is a complex line. Hence we get a point $(f(x), I_x)$ of $P(S^4)$. Therefore we have defined a lifting

$\tilde{f} : S^2 \to P(S^4)$ such that the following diagram commutes

(*)

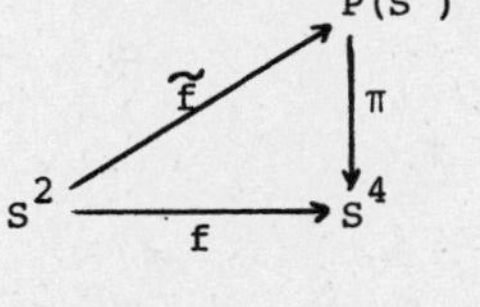

The map $\tilde{f}$ is called the Calabi lifting of f . When $V = S^{2n}$ a similar

procedure involving more derivation of f allows to define the Calabi lifting, when the map f is harmonic and sufficiently general [G.S][W].

THEOREM ([C], [W]) : Let (*) be a commutative diagram as above. Suppose that

a) $\tilde{f}$ is analytic

b) for any $x \in S^2$, $d\tilde{f}(T_x) \subset S_{\tilde{f}(x)}$

Then f is harmonic. Conversely suppose that f is harmonic and sufficiently general. Then there exists a unique $\tilde{f}$ having properties a) and b) such that (*) is commutative. This $\tilde{f}$ is the Calabi lifting of f.

In view of the theorem above, the strategy to describe all harmonic maps from S^2 to S^{2n} is the following:

1st step: Describe the space $M(n)$ of all the analytic maps from $S^2 = \mathbb{C}P^1$ to $QG(\mathbb{C}^{2n+1}) = P(S^{2n})$.

2nd step: Describe inside $M(n)$ the subspace $H(n)$ of those maps having property b).

3rd step: Study the map from $H(n)$ to the space Harm(n) of all harmonic maps from S^2 to S^{2n}, given by composition with $\pi : P(S^{2n}) \to S^{2n}$. The theorem above says that this composition map is bijective on some open set.

III. THE VARIETIES $M_d(n)$

On $QG(\mathbb{C}^{2n+1})$ we have:

1) the trivial quadratic vector bundle

$$E = \mathbb{C}^{2n+1} \times QG(\mathbb{C}^{2n+1}) ,$$

2) the canonical subbundle $I \subset E$ maximal isotropic at each point of $QG(\mathbb{C}^{2n+1})$,

3) an exact sequence

$$(*) \qquad 0 \longrightarrow I \longrightarrow E \longrightarrow Q \longrightarrow 0 ,$$

where Q is the quotient bundle of rank $n+1$.

Let $f : S^2 = \mathbb{C}P^1 \longrightarrow QG(\mathbb{C}^{2n+1})$ be an analytic map. Pulling back the exact sequence (*) by f we get

$$0 \longrightarrow I(f) \longrightarrow E \longrightarrow Q(f) \longrightarrow 0$$

where E is again the trivial quadratic vector bundle on $\mathbb{C}P^1$, and $I(f)$ is maximal isotropic at each point. According to a theorem of Birkoff we have

$$Q(f) = \bigoplus_{i=1}^{n} \mathcal{O}_{\mathbb{C}P^1}(\alpha_i)$$

where

$$0 \leq \alpha_1 \leq \cdots \leq \alpha_{n+1}$$

Call $\alpha = (\alpha_1,\ldots,\alpha_{n+1})$ the *type* of f and $d = \sum_1^{n+1} \alpha_i$ the *degree of* f. Since the degree is a topological number, the space $M(n)$ is the disjoint union of the subspaces $M_d(u) = \{f \in M(n) \mid \deg f = d\}$. For a given d we have a partition

$$M_d(n) = \bigcup_{\alpha} M_{d,\alpha}(n)$$

$\deg \alpha = d$, according to the type.

Using general theorems of algebraic geometry we know that the $M_d(n)$ are quasi-projective algebraic varieties and that for each α, $M_{d,\alpha}(n)$ is a locally closed algebraic subvariety of $M_d(n)$. For any $\alpha = (\alpha_1,\ldots,\alpha_n)$ set

$$\delta(\alpha) = \sum_{i,j} (\alpha_i - \alpha_j - 1)^+ \quad \text{where } X^+ = \mathrm{Sup}(X,0),\ X \in \mathbb{R}.$$

<u>THEOREM</u>: 1) $M_d(n)$ *is empty if* d *is odd. If* d *is even,* $M_d(n)$ *is smooth, connected of dimension* $\frac{n(n+1)}{2} + dn$.

2) $M_{d,\alpha}(n)$ *is empty if* $\alpha_{n+1} > \frac{d}{2}$. If $\alpha_{n+1} \leq \frac{d}{2}$ *then* $M_{d,\alpha}(n)$ *is a* (locally closed) *connected submanifold of codimension* $\delta(\alpha)$.

3) *If* $M_{d,\alpha}(n) \cap \overline{M_{d,\alpha}(n)} \neq \emptyset$ *then* $M_{d,\alpha}(n) \subset \overline{M_{d,\alpha}(n)}$. *The* $M_{d,\alpha}(n)$'s *make an analytically locally trivial stratification of* $M_d(n)$.

The difficult points to prove in the theorem are the non emptyness and connectedness in 1) and 2) and the assertion 3). For a generalization see [R].

IV. THE SPACES $H_d(n)$

Denote by $H_{d,\alpha}(n)$ the space $H_d \cap M_{d,\alpha}(n)$. It is a closed algebraic subvariety of $M_{d,\alpha}(n)$.

Set

$$H_d(n) = \bigcup_{\alpha} H_{d,\alpha}(n) \; .$$

The space $H_d(n)$ is a closed algebraic subvariety of $M_d(n)$.

Heuristic calculation supports the following conjecture:

If $d \geq n+1$, *the space* $H_d(n)$ *is of dimension* $n^2 + d$.

The variety $H_d(n)$ is in general extremely complicated. It may have several irreducible components. I do not know examples where it has several connected components. Let us look at the example $n = 2$ i.e. at the harmonic maps from S^2 to S^4. We then have the following facts :

1) For any $d = 2\delta$, $M_d(2)$ consists of the parametrized rational curves in $\mathbb{C}P^3$ of degree δ.
2) The space $H_d(2)$ is made of the parametrized rational curves in $\mathbb{C}P^3$ of degree δ and *self dual*.
3) In equation, $H_d(2)$ is the variety of the coefficients of 4 polynomials P_0, P_1, P_2, P_3 of degree δ, in one variable, such that

   $$W(P_0, P_1) + W(P_2, P_3) = 0$$

   where W is the *Wronskian* $(W(f,g) = f'g-g'f)$.
4) The space $H_2(2)$ corresponds to the isometric equatorial immersions of S^2 in S^4. It is of complex dimension 6, smooth and connected.
5) The space $H_4(2)$ corresponds to the maps

   $$f : S^2 \xrightarrow{\rho} S^2 \xrightarrow{\Psi} S^4$$

   where ρ is a two sheeted analytic covering of degree 2 and Ψ an isometric equatorial immersion. It is smooth connected of dimension 8.
6) The space $H_6(2)$ is connected and has two irreducible components C^1 and C^2 of dimension 10. The space C^1 corresponds to maps

   $$f : S^2 \xrightarrow{\rho} S^2 \xrightarrow{\Psi} S^4$$

where ρ is a three-sheeted covering and Ψ an equatorial isometry. The space $C^2 - C^1 \cap C^2$ is the space of self dual twisted cubics in $\mathbb{C}P^3$. The components C^1 and $C^2 - C^1 \cap C^2$ are smooth.

Finally there are some interesting questions concerning the topology of the harmonic immersions $S^2 \longrightarrow S^4$.

1) They are knotted

2) They have a non trivial normal bundle

3) They have self intersection

The degree of the normal bundle is d-2. The number of self intersection points (cleverly defined) is $1 - \frac{d}{2}$. I do not know how to compute the π_1 of the complementary of the image of S^2.

V. Bibliography

[C] E. Calabi, Quelques applications de l'analyse complexe aux surfaces d'aire minima in Topics in complex manifolds. Presses de l'Université de Montréal 1968.

[D] M. Dubois Violette, Structures Complexes an dessus des variétés. Applications. Séminaire de Mathématique Physique de l'Ecole Normale Supérieure. 1978-1980. To appear.

[G-S] V. Glaser and R. Stora, Regular solutions of the $\mathbb{C}P^n$- models and further generalizations. Preprint.

[R] A. Ramanathan, Deformation of Principal Bundles on Projective line. Inventiones. In print.

[W] J.C. Wood, Some Aspects of Harmonic Maps from a Surface to Complex Projective Space. Proceedings of the Belgian Contact Group on Differential Geometry, Leuven 1981.

COMPLETELY INTEGRABLE HAMILTONIAN SYSTEMS
AND THE SEPARATION OF VARIABLES

P.Winternitz

CRMA, Université de Montréal
Montréal, Québec, Canada H3C 3J7

ABSTRACT

A group theoretical approach to the separation of variables is applied to the Hamilton-Jacobi and Laplace-Beltrami equation in the hermitian hyperbolic space HH(2). Symmetry reduction by maximal abelian subgroups of the isometry group SU(2,1) leads to completely integrable systems defined in a Minkowski space and involving nontrivial interactions.

I. Introduction

The purpose of this contribution is: 1. To summarize some recently obtained results on the separation of variables in the Hamilton-Jacobi and Laplace-Beltrami equations in certain spaces of nonconstant curvature (complex projective spaces CP(n) and hermitian hyperbolic spaces HH(n)). 2. To show how separation of variables for free Hamiltonians in these n dimensional complex spaces generates completely integrable Hamiltonian systems with nontrivial interactions in an n dimensional real space. 3. To point out the role of the subgroups, in particular, the maximal abelian subgroups of the isometry group of the considered space in the separation of variables and generation of completely integrable systems.

The results on separation of variables in CP(n) and HH(n) were obtained in collaboration with C.P.Boyer and E.G.Kalnins[1,2]. The algebraic and group theoretical characterization of the separation of variables goes back quite far[3,4] and has been greatly developed by W.Miller Jr. and collaborators[5,6]. The relationship between subgroups of Lie groups and separation of variables has been emphasized mainly for spaces of constant curvature[7]. Work on classifying the maximal abelian subgroups of all classical Lie groups is progressing[8,9].

In this talk we shall concentrate on the 4-real dimensional space HH(2), for which the isometry group is the noncompact group SU(2,1). The results can be generalized to HH(n). The corresponding compact space CP(n) with isometry group SU(n+1) has been treated in complete generality[1] and the results there are much simpler. Our physical motivation lies in the generation of nontrivial relativistic completely integrable systems on one hand and in the importance of real 4-dimensional (pseudo)-Riemannian spaces like CP(2) and HH(2) in general relativity theory on the other[10,11]. From the mathematical point of view we were interested in seeing how separation of variables works in spaces with nonconstant curvature but with large symmetry groups. The crucial role of nonorthogonal coordinates is of particular interest.

2. Formulation of the Problem

In general we consider a Riemannian or pseudo-Riemannian space V_n with metric

$$ds^2 = g_{ik}(\vec{x})dx^i dx^k \tag{1}$$

and write the Hamilton-Jacobi equation (HJE)

$$H = g^{ik}p_i p_k = g^{ik} \frac{\partial W}{\partial x^i} \frac{\partial W}{\partial x^k} = E \tag{2}$$

and Laplace-Beltrami equation (LBE)

$$\Delta\psi = \frac{1}{\sqrt{g}} \frac{\partial}{\partial x^i} (\sqrt{g}\, g^{ij} \frac{\partial}{\partial x^j} \psi) = E\psi \tag{3}$$

$$g = \det(g_{ij})$$

(summation over repeated indices is understood). For the HJE(2) and the LBE(3) we are interested in additive separation

$$W = \sum_{i=1}^{n} W_i(x^i, \lambda_1, \ldots, \lambda_n), \tag{4}$$

and multiplicative separation

$$\psi = \prod_{i=1}^{n} \psi_i(x^i, \lambda_1, \ldots, \lambda_n), \tag{5}$$

respectively. In both cases the λ_i are separation constants and each of the functions W_i and ψ_i depends on one variable x^i only. A coordinate system which allows separation (5) for the LBE will also yield separation (4) in the HJE; the opposite is only true for certain types of spaces[12] (including CP(n) and HH(n)).

Separation of variables is related to complete integrability via the following classical result[13]: To every separable coordinate system for the HJE in V_n there corresponds a family of n linearly independent second order Killing tensors T_i in involution:

$$T_i = a_i^{jk}(\vec{x})p_j p_k, \qquad [T_i, H] = 0$$

$$[T_i, T_k] = 0, \qquad i,k = 1,\ldots,n \tag{6}$$

$$T_n \equiv H$$

where [,] denotes the Poisson bracket. The separated solution (4) is characterized by the equations

$$H(\vec{x},\vec{p}) = E = \lambda_n, \quad T_k(\vec{x},\vec{p}) = \lambda_k, \quad k = 1,\ldots,n-1 \tag{7}$$

where λ_k are the separation constants.

Thus: separation of variables implies the existence of n (quadratic) integrals of motion in involution and hence complete integrability of the Hamiltonian system.

The converse is in general not true: the existence of n second order Killing tensors in involution implies the separation of variables only if certain further algebraic conditions on the coefficients $a_i^{jk}(\vec{x})$ are satisfied[12].

Our approach consists of the following: given the space V_n and its metric tensor $g_{ik}(\vec{x})$, find all sets of n 2nd order Killing tensors T_i ($T_n = H$) in involution, classify them into orbits under the action of the isometry group G, choose a representative of each orbit, identify those orbits that lead to the separation of variables and find the separable variables in each case. Certain specific types of Killing tensors play particularly important roles. Thus, Killing tensors that are squares of Killing vectors

$$T_i = (L_i)^2, \quad L_i = a_i^k(\vec{x})p_k, \quad [H,L_i] = 0 \tag{8}$$

correspond to ignorable variables (variables that do not figure in the metric tensor, figure additively in the function W and as exponentials in ψ). Killing tensors that are Casimir operators (invariant operators) of a Lie subgroup of G also play a special role and lead to particularly simple types of coordinates.

Let us now realize the above program for the space HH(2) described below.

3. The Space HH(2), Its Isometry Group SU(2,1) and Complete Sets of Killing Tensors

Consider the standard basis $\{e_0, e_1, \ldots, e_n\}$ in $\mathbb{C}^{n+1}$ and introduce the SU(n,1) invariant hermitian form

$$\begin{aligned} F(x,y) &= -\bar{x}_0 y_0 + \sum_{k=1}^{n} \bar{x}_k y_k \\ F(gx,gy) &= F(x,y), \quad g \in U(n,1). \end{aligned} \tag{9}$$

The group U(n,1) acts transitively on the real hypersurface $M \subset \mathbb{C}^{n+1}$ defined by

$$F(y,y) = -1 \tag{10}$$

The space HH(n) is obtained by a projection $\pi: M \to HH(n)$ as the space of U(1) orbits: $y \to e^{i\theta} y$. We thus obtain the diffeomorphism $U(n,1)/U(n)xU(1) \sim HH(n)$.

In addition to homogeneous coordinates $(y_0, y_1, \ldots, y_n) \sim e^{i\theta}(y_0, y_1, \ldots, y_n)$ we introduce affine coordinates on HH(n):

$$z_k = \frac{y_k}{y_0}, \quad k = 1, \ldots, n. \tag{11}$$

The space HH(n) can then be identified with an open unit ball in $\mathbb{C}^n$:

$$z \in \mathbb{C}^n, \quad \sum_{k=1}^{n} \bar{z}_k z_k < 1. \tag{12}$$

The real part of the invariant metric F(y,y) projects onto a metric on

HH(n), namely the noncompact form of the well known Fubini-Study metric[14].

$$ds^2 = -\frac{4}{c}\frac{[1-(\bar{z},z)](d\bar{z},dz) + (\bar{z},dz)(z,d\bar{z})}{[1-(\bar{z},z)]^2} \tag{13}$$

where we have put $(\bar{x},y) = \Sigma_{k=1}^{n} \bar{x}_k y_k$ and $c < 0$ is the constant holomorphic sectional curvature.

From now on we restrict ourselves to the case $n = 2$. The Hamiltonian (2) associated with the Fubini-Study metric (13) is

$$H = 4(1-|z_1|^2-|z_2|^2)[(|z_1|^2-1)p_1\bar{p}_1+(|z_2|^2-1)p_2\bar{p}_2+z_1\bar{z}_2p_1\bar{p}_2+\bar{z}_1z_2\bar{p}_1p_2]. \tag{14}$$

A basis for the Lie algebra of the isometry group SU(2,1) is realized by 8 linearly independent Killing vectors. In terms of the homogeneous coordinates y and the affine coordinates z such that a basis is:

$$\begin{aligned}
X_1 &= -y_2p_{y_1}+y_1p_{y_2}+c.c = -z_2p_{z_1}+z_1p_{z_2} + c.c\\
X_2 &= i(y_2p_{y_1}+y_1p_{y_2})+c.c = -i(z_2p_{z_1}+z_1p_{z_2})+c.c\\
X_3 &= i(y_1p_{y_1} - y_2p_{y_2})+c.c = i(-z_1p_{z_1}+z_2p_{z_2}) + c.c\\
X_4 &= y_1p_{y_0}+y_0p_{y_1}+c.c = (z_1^2 - 1)p_{z_1}+z_1z_2p_{z_2} + c.c\\
X_5 &= i(-y_1p_{y_0}+y_0p_{y_1})+c.c = i[(z_1^2+1)p_{z_1} + z_1z_2p_{z_2}] + c.c\\
X_6 &= y_2p_{y_0}+y_0p_{y_2}+c.c = z_1z_2p_{z_1} + (z_2^2 - 1)p_{z_2} + c.c\\
X_7 &= i(-y_2p_{y_0}+y_0p_{y_2})+c.c = i[z_1z_2p_{z_1}+(z_2^2+1)p_{z_2}] + c.c\\
X_8 &= \frac{i}{\sqrt{3}}(2y_0p_{y_0} - y_1p_{y_1} - y_2p_{y_2})+c.c = i\sqrt{3}(z_1p_{z_1}+z_2p_{z_2}) + c.c
\end{aligned} \tag{15}$$

(c.c denotes complex conjugation). We shall also need an alternative basis: $Y_1 = \frac{3}{2}(X_3+(1/\sqrt{3})X_8)$, $Y_2 = X_1-X_6$, $Y_3 = X_2-X_7$, $Y_4 = -X_5 + \frac{1}{2}(X_3-\sqrt{3}X_8)$, $Y_5 = Y_4$, $Y_6 = \frac{1}{2}(X_3-\sqrt{3}X_8)$, $Y_7 = X_1$, $Y_8 = X_2$. The SU(2,1) quadratic Casimir operator is

$$C_2 = X_1^2 + X_2^2 + X_3^2 - X_4^2 - X_5^2 - X_6^2 - X_7^2 + X_8^2. \qquad (16)$$

Our task now is to represent all triplets of second order Killing tensors $\{T_1, T_2, T_3\}$ (see (6)).

This is greatly simplified by two results valid for certain classes of spaces including CP(2) and HH(2), namely:

1. Every symmetric second order Killing tensor is a bilinear combination of Killing vectors:

$$T_i = \sum_{a,b=1}^{\infty} A^i_{ab} X_a X_b, \qquad A^i_{ab} = A^i_{ba} = \text{const}. \qquad (17)$$

(see Delong[15]).

2. Every separable coordinate system will involve precisely two ignorable variables[1]. Hence two of the Killing tensors, say T_1 and T_2 will be squares of Killing vectors (elements of su(2,1) as in[15]).

$$T_1 = L_1^2 = (\sum_{\alpha=1}^{8} a_\alpha X_\alpha)^2, \qquad T_2 = L_2^2 = (\sum_{\alpha=1}^{8} b_\alpha X_\alpha)^2. \qquad (18)$$

Since L_1 and L_2 commute, this leads us to the problem of finding all maximal abelian subalgebras (MASAs) of su(2,1)[8,9,16]. Four classes of these MASAs exist, namely: the compact Cartan subalgebra $\{X_3, X_8\}$, the noncompact Cartan subalgebra $\{X_3 + (1/\sqrt{3})X_8, X_5\}$, a decomposable non-Cartan subalgebra $\{Y_1, Y_4\}$, and the maximal abelian nilpotent subalgebra[9] $\{Y_3, Y_4\}$. We also need to know all subalgebras of su(2,1) with second order Casimir operators. The relevant algebras and their invariants are:

$$\text{su}(2): \{X_1, X_2, X_3\}, \quad I_0 = X_1^2 + X_2^2 + X_3^2$$

$$\text{su}(1,1): \{X_4, X_5, X_3 - \sqrt{3}X_8\}, \quad I_1 = X_4^2 + X_5^2 - \tfrac{1}{4}(X_3 - \sqrt{3}X_8)^2$$

$$\text{or } \{X_6, X_7, X_3 + \sqrt{3}X_8\}, \quad I_2 = X_6^2 + X_7^2 - \tfrac{1}{4}(X_3 + \sqrt{3}X_8)^2 \qquad (19)$$

$$A_{4,10}: \{Y_1, Y_2, Y_3, Y_4\}, \quad I_3 = 4Y_1Y_4 + 3(Y_2^2 + Y_3^2)$$

($A_{4,10}$ is a four dimensional solvable Lie algebra[17]).

It is now an easy matter to find the most general Killing tensor T_3 in involution with T_1 and T_2 for each MASA of su(2,1).

The result is[2]:

$$
\begin{aligned}
&\text{I.} && T_1 = X_3^2, \quad T_2 = X_8^2, \quad T_3 \equiv Q_I = aI_0+bI_1+cI_2 \\
&\text{II.} && T_1 = \tfrac{9}{4}(X_3+(1/\sqrt{3})X_8)^2, \quad T_2 = X_5^2, \quad T_3 \equiv Q_{II} = aI_1+b(X_1X_6+X_2X_7) \\
&\text{III.} && T_1 = Y_1^2, \quad T_2 = Y_4^2, \quad T_3 \equiv Q_{III} = aI_3+bI_1 \\
&\text{IV.} && T_1 = Y_3^2, \quad T_2 = Y_4^2, \quad T_3 \equiv Q_{IV} = aI_3+b(Y_1Y_3-3Y_2Y_5-6Y_4Y_8)
\end{aligned}
\tag{20}
$$

where a,b,c are real constants. The classification can be further refined by allowing for linear combinations of T_3 with the Casimir operator (16) and by simplifying T_3 in each case with the help of the normalizer of T_1 and T_2 (an external normalizer of $\{T_1,T_2\}$ exists in cases III and IV; Cartan subalgebras are by definition self-normalizing). The final result is: Q_I splits into four cases:

$Q_1 = I_0$, $Q_2 = I_1$,

$Q_3 = I_1+\mu I_2$ with $0 < \mu < 1$,

$Q_4 = I_1+\mu I_2$ with $-1 \leq \mu < 0$.

Q_{II} splits into two cases

$Q_5 = I_1$, $Q_6 = Q_{II}$ with $b=1$, $a \geq 0$;

Q_{III} splits into 4 cases

$Q_7 = I_1$, $Q_8 = I_3$, $Q_9 = I_1+I_3$, $Q_{10} = I_1-I_3$;

Q_{IV} splits into two cases

$Q_{11} = I_3$, $Q_{12} = Y_1Y_3 - 3Y_2Y_5 - 6Y_4Y_8$.

Thus, we have obtained 12 classes of triplets: 6 of them of subgroup type, namely those involving the subgroup Casimir operators Q_1, Q_2, Q_5,Q_7,Q_8 and Q_{11}. We now proceed to establish a one-to-one correspondance between these 12 classes and separable coordinates on HH(2).

4. Separable Coordinates on HH(2) and Integrable Relativistic Systems

To proceed further we use an adaptation of the method of reduction of phase space in classical mechanics by ignorable variables[18], i.e. reduction by an Abelian symmetry group. We proceed in three steps, making use of the homogeneous coordinates y_0, y_1, y_2.

1. Choose a basis L_1, L_2 for the considered MASA of su(2,1) and express L_i in terms of the coordinates y_i and moments p_{y_i} as in (15). Introduce the ignorable variables x, y and ρ by putting

$$L_1 = p_x, \quad L_2 = p_y, \quad X_0 \equiv y_0 p_{y_0} + y_1 p_{y_1} + y_2 p_{y_2} + c.c = p_\rho \tag{21}$$

(X_0 complements su(2,1) to u(2,1)). Solve equations (21) to express y_μ in terms of x, y, ρ and three essential real variables s_0, s_1, s_2 appearing as integration constants and satisfying

$$|y|^2 = |y_0|^2 - |y_1|^2 - |y_2|^2 = s_0^2 - s_1^2 - s_2^2 = 1 \, . \tag{22}$$

2. Express the Hamiltonian H and the remaining Killing tensor $T_3 = Q$ in terms of s_μ, x and y (setting $p_\rho = 0$). We obtain

$$H = -I_{12}^2 + I_{01}^2 + I_{02}^2 + f_1(s_\mu) p_x^2 + f_2(s_\mu) p_y^2 + f_3(s_\mu) p_x p_y \tag{23}$$

$$Q = \sum_{\mu\nu,\mu'\nu'} A_{\mu\nu,\mu'\nu'} I_{\mu\nu} I_{\mu'\nu'} + h_1(s_\mu) p_x^2 + h_2(s_\mu) p_y^2 + h_3(s_\mu) p_x p_y \tag{24}$$

where

$$I_{12} = s_1 p_{s_2} - s_2 p_{s_1} \, , \quad I_{01} = s_0 p_{s_1} + s_1 p_{s_0} \, , \quad I_{02} = s_0 p_{s_2} + s_2 p_{s_0} , \tag{25}$$

the $A_{\mu\nu,\mu'\nu'}$ are constants and f_i, h_i are functions of the essential variables s_μ. These constants and functions have a specific form for each different MASA. The problem of separating variables in the free HH(2) Hamiltonian has thus been reduced to that of separating variables on the O(2,1) hyperboloid (22) for an O(2,1) Hamiltonian (23) with a potential given by the functions $f_i(s_\mu)$ (we put $p_x = c_1$, $p_y = c_2$, c_i=const.). The "kinetic energy" part of H is O(2,1) invariant.

3. Introduce separable coordinates on the O(2,1) hyperboloid(22), compatible with the Killing tensor Q(24) and the potential in(23). We recall that 9 types of separable coordinates exist for free Hamiltonians on O(2,1) hyperboloids and cones[4,19].

Steps 1 and 2 lead to the following results[2]:

I. Compact Cartan subalgebra:

$$y_0 = s_0 e^{i\frac{3\rho-\alpha_1-\alpha_2}{3}}, \quad y_1 = s_1 e^{i\frac{3\rho+2\alpha_1-\alpha_2}{3}},$$

$$y_2 = s_2 e^{i\frac{3\rho-\alpha_1+2\alpha_2}{3}}$$

$$H = -I_{12}^2+I_{01}^2+I_{02}^2+\frac{1}{s_1^2}\,p_{\alpha_1}^2+\frac{1}{s_2^2}\,p_{\alpha_2}^2-\frac{1}{s_0^2}\,(p_{\alpha_1}+p_{\alpha_2})^2$$

$$Q_I = a\left[I_{12}^2+\left(1+\frac{s_2^2}{s_1^2}\right)p_{\alpha_1}^2+\left(1+\frac{s_1^2}{s_2^2}\right)p_{\alpha_2}^2\right]+$$

$$+\, b\left[I_{01}^2+\left(-1+\frac{s_0^2}{s_1^2}\right)p_{\alpha_1}^2+\left(-1+\frac{s_1^2}{s_0^2}\right)(p_{\alpha_1}+p_{\alpha_2})^2\right]+$$

$$+\, c\left[I_{02}^2+\left(-1+\frac{s_0^2}{s_2^2}\right)p_{\alpha_2}^2+\left(-1+\frac{s_2^2}{s_0^2}\right)(p_{\alpha_1}+p_{\alpha_2})^2\right]$$

II. Noncompact Cartan subalgebra:

$$y_0 = e^{i\frac{3\rho-\alpha}{3}}(is_0\,\mathrm{ch}u+s_1\,\mathrm{sh}u), \quad y_1 = e^{i\frac{3\rho-\alpha}{3}}(is_1\,\mathrm{ch}u-s_0\,\mathrm{sh}u),$$

$$y_2 = e^{i\frac{3\rho+2\alpha}{3}}\,is_2$$

$$H = -I_{12}^2+I_{01}^2+I_{02}^2-\frac{s_0^2-s_1^2}{(s_0^2+s_1^2)^2}\,p_u^2+\left[\frac{s_0^2-s_1^2}{(s_0^2+s_1^2)^2}-\frac{1}{s_2^2}\right]p_\alpha^2$$

$$+\,\frac{4s_0s_1}{(s_0^2+s_1^2)^2}\,p_u p_\alpha$$

III. Orthogonally decomposable non Cartan MASA:

$$y_0 = e^{i\frac{3\rho-\alpha}{3}}[s_0+i(s_0-s_1)t], \quad y_1 = e^{i\frac{3\rho-\alpha}{3}}[s_1+i(s_0-s_1)t]$$

$$y_2 = e^{i\frac{3\rho+2\alpha}{3}} s_2$$

$$H = -I_{12}^2 + I_{01}^2 + I_{02}^2 + \frac{1}{s_2^2} p_\alpha^2 + \frac{s_0+s_1}{(s_0-s_1)^3} p_t^2 + \frac{2}{(s_0-s_1)^2} p_\alpha p_t$$

IV. Maximal Abelian nilpotent subalgebra:

$$y_0 = e^{i\rho}[(s_0-s_1)(u - \frac{i}{2}t^2)+s_2 t-is_0]$$

$$y_1 = e^{i\rho}[(s_0-s_1)(u - \frac{i}{2}t^2)+s_2 t-is_1]$$

$$y_2 = e^{i\rho}[-is_2-(s_0-s_1)t]$$

$$H = -I_{12}^2+I_{01}^2+I_{02}^2 + \frac{1}{(s_0-s_1)^2} p_t - \frac{4s_2}{(s_0-s_1)^3} p_u p_t + \frac{3s_2^2-s_0^2-s_1^2}{(s_0-s_1)^4} p_u^2 .$$

The expressions for Q_{II}, Q_{III} and Q_{IV} are given in Ref.2.

Step 3 of the algorithm leads us to 12 separable coordinate systems in HH(2) and hence to 12 completely integrable Hamiltonian systems. In each case the kinetic energy term is O(2,1) invariant and the potentials are such that the corresponding Hamilton-Jacobi equations and Schrödinger equations separate. The formulas are too long to reproduce here[2] so we only give the simplest case as an example. It corresponds to the choice of the compact Cartan subalgebra and to spherical coordinates on the O(2,1) hyperboloid. In this case the affine coordinates in HH(2) are

$$z_1 = \tanh A \cos B e^{i\alpha_1} \qquad z_2 = \tanh A \sin B e^{i\alpha_2}$$

and the Hamiltonian and integral of motion are

$$H = p_A^2 + \frac{1}{\mathrm{sh} A^2} Q_1 - \frac{1}{\mathrm{ch}^2 A} (p_{\alpha_1} + p_{\alpha_2})^2$$

$$Q_1 = p_B^2 + \frac{1}{\cos^2 B} p_{\alpha_1}^2 + \frac{1}{\sin^2 B} p_{\alpha_2}^2 .$$

Putting

$$H = E, \quad Q_1 = c_3, \quad p_{\alpha_1} = c_1, \; p_{\alpha_2} = c_2$$

we have reduced the problem of solving the HJE to quadratures. Putting $H\psi = E\psi$, $Q_1\psi = c_3\psi$, $p_{\alpha_i}\psi = c_i\psi$ (with $p_x = \partial/\partial x$) we have reduced the problem of solving the Schrödinger equation to that of solving ordinary linear differential equations (the solutions are exponentials in α_1 and α_2, hypergeometric functions in A and B).

5. Conclusions

We have shown that the group theoretical approach to separation of variables works well for HH(2) and in particular that the MASAs of the isometry group provide a key to the construction of separable coordinate systems. All separable coordinate systems involve two ignorable variables and all are nonorthogonal. Symmetry reduction by Abelian subgroups reduces the free HH(2) Hamiltonian to a separable O(2,1) Hamiltonian with an interaction term. A possible interpretation of the Laplace-Beltrami operator in this case is that it corresponds to a Klein-Gordon type particle in an external potential field. All of the obtained potentials are "kinematically" induced (like the centrifugal potential $\ell(\ell+1)/r^2$ in the euclidean space Hamiltonian in spherical coordinates).

Work is in progress on the extension of the results to HH(n) for n arbitrary and on the inclusion of nontrivial potentials in HH(n) that allow the separation of variables in at least one of the systems in which the free Hamiltonian separates[20]. The obtained integrable systems are being further investigated in order to find trajectories, wave functions etc. and to inderpret these systems physically, e.g. in terms of the integrable interactions between several particles[21].

REFERENCES

0- Work supported in part by the Natural Sciences and Engineering Research Council of Canada and the "Fonds FCAC pour l'aide et le soutien à la recherche du Gouvernement du Québec".

1- C.P.Boyer, E.G.Kalnins, and P.Winternitz, Preprint CRMA-1064 (1981) Montreal (to be published).

2- C.P.Boyer, E.G.Kalnins, and P.Winternitz, Preprint CRMA-1104 (1982), Montreal, (to be published).

3- P.Winternitz and I.Friš, Yad.Fiz.1, 889 (1965) {Sov.J.Nucl.Phys. 1, 636 (1965)}.

4- P.Winternitz, I.Lukač and Ya.A.Smorodinskii, Yad.Fiz. 7, 192 (1968) {Sov.J.Nucl.Phys. 7, 139 (1968)}.

5- W.Miller,Jr. Symmetry and Separation of Variables, Addison-Wesley, Reading, Mass. 1977.

6- E.G.Kalnins and W.Miller Jr., Research Report 104, Waikato, New Zealand 1982 (contains an extensive list of references).

7- W.Miller Jr., J.Patera, and P.Winternitz, J.Math.Phys. 22, 251 (1981).

8- J.Patera, P.Winternitz, and H.Zassenhaus, Math.Rep.Ac.Sci. (Canada) 2, 231, 237 (1980).

9- J.Patera, P.Winternitz, H.Zassenhaus, Preprint CRMA-1099 (1982) and to be published.

10- G.W.Gibbons and C.N.Pope, Comm.Math.Phys. 61, 239 (1978).

11- C.P.Boyer, Hadronic,J. 4, 2 (1981).

12- E.G.Kalnins and W.Miller Jr., SIAM J. Math.Analysis 11, 1011 (1980).

13- L.P.Eisenhart, Ann.Math. 35, 284 (1934).

14- S.Kobayashi and K.Nomizu, Foundations of Differential Geometry, Vol.2, Interscience, New York, 1969.

15- P.Delong, Ph.D. Thesis, U. of Minnesota, 1982.

16- J.Patera, P.Winternitz and H.Zassenhaus, J.Math.Phys. 15, 1378 (1974).

17- J.Patera, R.T.Sharp, P.Winternitz, and H.Zassenhaus, J.Math.Phys. 17, 986 (1976).

18- R.Abraham and J.E.Marsden, Foundations of Mechanics, Benjamin, Reading, 1978.

19- C.P.Boyer, E.G.Kalnins and W.Miller, Jr., J.Math.Phys. 19, 20 (1978).

20- L.Gagnon and P.Winternitz, to be published.

21- F.Calogero, J.Math.Phys. 12, 2191 (1969).

Conformally Invariant Pure Spinor Models

M.Arık and M.Hortaçsu

Physics Department
Boğaziçi University
Istanbul- Turkey

F.Gürsey proposed a pure spinor model long ago[1] which is described by the Lagrangian

$$\mathcal{L} = \bar{\psi} i \partial\!\!\!/ \psi + g^{4/3} (\bar{\psi}\psi)^{4/3} \tag{1}$$

F.Kortel found classical solutions to this model[2]. These solutions were later interpreted as instantons and merons[3]. These are of the form

$$\psi_{cl} = \frac{a \pm i \gamma x}{(a^2 + x^2)^{3/2}} C \tag{2}$$

$$\psi_{cl} = \frac{1}{(x^2)^{3/4}} \left(1 + \frac{i\gamma x}{(x^2)^{1/2}} \right) C, \text{ C constant spinor} \tag{3}$$

The existence of these solutions may suggest that perhaps this is a completely integrable system, since,for instance, $(\bar{\psi}\psi)^2$ in two dimensions would have similar solutions.

We did not investigate this aspect of the model, but tried to quantize it and show that one gets a renormalizable field theory out of the above Lagrangian[4]. We used the path integral method of quantization. Two auxiliary fields are introduced to the functional integral to write the equivalent Lagrangian as

$$\mathcal{L} = \bar{\psi} i \partial\!\!\!/ \psi + \lambda (g \psi\psi - \phi^3) + g \bar{\psi}\psi\phi \tag{4}$$

One has to modify this Lagrangian and add the Faddeev-Popov term coming from the constraint

$$g \bar{\psi}\psi = \phi^3 \tag{5}$$

This is written by the introduction of two ghost fields c, c* as

$$\mathcal{L} = \bar{\psi} i \partial\!\!\!/ \psi + \lambda (g\bar{\psi}\psi - \phi^3) + g\bar{\psi}\psi\phi + i\, c^* c \phi^2 \tag{6}$$

We redefine our fields

$$\phi \to \Phi = \phi + \lambda \tag{7}$$

$$\lambda \to \lambda \quad , \tag{8}$$

integrate over ψ and ψ and obtain

$$\mathcal{L}_{eff} = -i \ \mathrm{Tr} \ \ln \ (i\not\partial + g\Phi) + i \int d^4x \ \{ \lambda(\lambda-\Phi)^3 + ic^*c(\lambda-\Phi)^2 \} \tag{9}$$

The tadpole conditions,

$$\left.\frac{\partial S_{eff}}{\partial \Phi}\right|_{vac.} = \left.\frac{\partial S_{eff}}{\partial \lambda}\right|_{vac.} = \left.\frac{\partial S_{eff}}{\partial c}\right|_{vac} = 0 \tag{10}$$

are satisfied for

$$< \Phi > = < c > = < \lambda > = 0 \tag{11}$$

This is the only consistent solution we could find which would result in a renormalizable theory. Note also that this solution respects the γ_5 - invariance of the classical Lagrangian, $\psi \to \gamma_5\psi$, $\Phi \to -\Phi, \lambda \to -\lambda$. The inverse propagator for Φ is given by, in the dimensional regularization scheme,

$$\left.\frac{\partial^2 S_{eff}}{\partial \Phi^2}\right|_{vac} = \frac{ig^2 \ \mathrm{Tr}}{(2\pi)^4} \int \frac{d^4p}{\not p(\not p + \not q)} = \frac{g^2}{4\pi^2} (\ln \Lambda) \ q^2 \tag{12}$$

If we choose

$$g^2 = \frac{4\pi^2}{\ln\Lambda} \tag{13}$$

This term is convergent and we obtain the propagator of a mass zero scalar particle.

$$\frac{\partial^2 S_{eff}}{\partial \lambda^2} = 0 \tag{14}$$

$$\left.\frac{\partial^2 S_{eff}}{\partial\lambda\partial\Phi}\right|_{vac.} = 0 \tag{15}$$

$$\left.\frac{\partial^2 S_{eff}}{\partial c \ \partial c^*}\right|_{vac.} = 0 \tag{16}$$

So the ghost and λ- fields do not propagate. We end up in a renormalizable model with Feynman rules

 $= i/\not p$ (17)

 $= i/p^2$ (18)

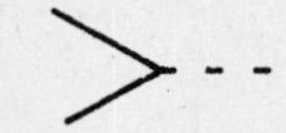 $= g$ (19)

$$g^2 = \frac{4\pi^2}{\ln\Lambda} \qquad (12)$$

g^2 is not a free parameter of the theory but it is fixed to make the theory renormalized.

One can calculate several processes. These calculations also verify that no additional ultraviolet divergences remain in the theory. One may give internal symmetry to the model. Then nothing essentially changes. Then one can calculate, for instance, the $e^+e^- \to$ all process. This work is in progress.

One can mimick this construction for other models. An example is when one takes the Lagrangian[5)]

$$\mathcal{L} = \bar\psi i\not\partial\psi + m\bar\psi\psi + g((\bar\psi\gamma_\mu\psi)(\bar\psi\gamma^\mu\psi))^{2/3} \qquad (20)$$

Although this model does not seem to be locally gauge invariant, the non-gauge invariant part decouples. So we get a model whose composite particle is the gluon . The effective Feynman rules are the same as QCD (with and without the ghost)[6)].

Since the Feynman rules are the same as QCD, the expression for the running coupling constant should also be the same. So we equate

$$\frac{e^2}{4\pi} = \frac{3\pi}{C_f\ln\Lambda^2} = \frac{12\pi}{(11C_f-4C_a)\ln\Lambda^2} \qquad (21)$$

C_f = Casimir operator for the fermion representation
C_a = Casimir operator for the adjoint representation

This equation yields

$$\frac{C_f}{C_a} = \frac{11}{4} \qquad (22)$$

One can find solutions for groups that satisfy this equality. Some simple solutions are given below :

E_6, SO(10) → 11 families in the spinor representation ;

SU(8) → 4 vector + 3 antisymmetric tensor representation

If the latter is broken in the pattern SU(8)→SU(5)xSU(3), here SU(5) is that of Georgi-Glashow, we obtain three conventional families plus their mirrors.

Many other solutions for the models proposed in Ref. 5 and Ref. 6 will be published elsewhere.

References

1.F.Gürsey, Nuovo Cimento, 3, 988 (1956)
2.F.Kortel, Nuovo Cimento, 4, 210 (1956)
3.K.G.Akdeniz, Lett. Nuovo Cimento, 33,40 (1982)
4.K.G.Akdeniz, M.Arık, M.Durgut, M.Hortaçsu, S.Kaptanoğlu,N.K.Pak, Phys.Lett.,116B, 34 (1982)
5.K.G.Akdeniz, M.Arık, M.Durgut, M.Hortaçsu, S.Kaptanoğlu,N.K.Pak, Phys.Lett.,116B, 41 (1982)
6.K.G.Akdeniz, M.Arık, M.Hortaçsu, N.K.Pak, I.C.T.P. Trieste preprint, IC/82/98 (1982)

"EXACT SOLVABILITY IN CHIRAL AND GAUGE THEORIES"

L.Dolan
The Rockefeller University, New York, New York 10021

ABSTRACT

Kac-Moody subalgebras are discussed in the context of the principal chiral models and SU(N) gauge theory. For self-dual gauge fields, a new set of conserved currents is given here as a guide in the search for the $D_\mu F_{\mu\nu} = 0$ hidden invariance.

I- Introduction

An infinite set of conserved commuting charges is often a signal for exact solvability. In the two-dimensional principal chiral models and in various formulations of the SU(N) gauge theory, such an infinite parameter symmetry algebra has been identified[1] as a Kac-Moody subalgebra, modulo local gauge transformations in the gauge theory.

Kac-Moody Lie or associated affine algebras have been of much interest recently in the mathematics literature. First, it was observed that a generating function for a representation could be given by the vertex operator of the dual string, an alternative picture for strong interactions to the non-abelian gauge theory[2]. Second, the orbit method in the theory of representations of Lie groups elucidated the inverse scattering method for integrable systems; and a Kac-Moody algebra was seen to be relevant to the periodic Toda lattice[3]. Third, a general class of explicit solutions for K-dV was constructed with use of the observation that its scattering data was related to a Kac-Moody algebra[4].

Although the chiral models have not been exactly integrated, the S-matrix for the non-linear sigma model is calculable by imposing Kac-Moody charge conservation on the matrix elements[5]. In general, a linear problem is constructed from an infinite parameter invariance. But in this case, the scattering data is constant in time and thus not useful in solving the initial value problem. Nevertheless we see that important non-perturbative information is gained from the new symmetry.

Also, from the mathematician's point of view, i.e. an interest in constructing representations of Kac-Moody algebras, the Luscher-Pohlmeyer non-local charges[6] and their principal chiral generalization can now be identified as a new representation of the subalgebra.

With this background, and the relationship between chiral theories, gauge theories[7] and strings[8], it seems reasonable that non-abelian SU(N) theory should also carry a representation of the infinite dimensional Kac-Moody symmetry. Extra symmetry would be extremely useful in formulating a non-perturbative solution of the string interactions. Its realization deserves considerable attention.

In Section II, the chiral model is reviewed. In Section III, a Kac-Moody-like subalgebra is identified on loop space and in the local four-dimensional self-dual sector. The first Kac-Moody transformation is identified as a particular local gauge transformation: it is trivial to generalize to the full Yang-Mills theory. The associated Noether current can be used to construct a new set of conserved currents on the self-dual set. The currents of the full Yang-Mills theory when found may reduce to these expressions for self-dual fields.

II- Principal Chiral Models

The Kac-Moody algebra associated with a finite parameter semi-simple Lie algebra G is $G \otimes C[t,t^{-1}] + C_c$. The commutation relations are

$$[M_a^{(n)}, M_b^{(m)}] = \varepsilon_{abc} M_c^{(n+m)} + n\delta_{n,-m}\delta_{ab}P$$

$$[P, M_a^{(n)}] = 0 \tag{1}$$

Here $n,m = -\infty, \ldots -1,0,1,\ldots\infty$ ε_{abc} are the structure constants of G. The central extension $(n\delta_{n,-m}\delta_{ab}P)$ vanishes for $n,m = 0,1,\ldots\infty$. What occurs naturally in the chiral and gauge models is half of a Kac-Moody algebra, the subalgebra $G \otimes C[t]$:

$$[M_a^{(n)}, M_b^{(m)}] = \varepsilon_{abc} M_c^{(n+m)} \tag{2}$$

Here $n,m = 0,1,\ldots\infty$.

The two-dimensional Euclidean chiral model is $\mathcal{L}(x)=1/16\mathrm{tr}\partial\mu g\partial\mu g^{-1}$. The field g is the element of some group whose Lie algebra is G. The equations of motion are $\partial\mu(g^{-1}\partial\mu g) = 0$. There exists an infinite set of infinitesimal transformations $\Delta_g^{(n)}g$ which shift $\mathcal{L}(x)$ by a total divergence, leave the equations of motion invariant, and generate Noether charges which close half of a Kac-Moody algebra.

$$\Delta_a^{(n)}\, g = -g\,\Lambda_a^{(n)}$$

$$\Lambda_a^{(n+1)}(x,t) = \int_{-\infty}^{x} dy\,\{\partial_o\Lambda_a^{(n)}(y,t) + [A_o(y,t),\Lambda_a^{(n)}(y,t)]\}$$

$$A_\mu \equiv g^{-1}\partial\mu g$$

$$\Lambda_a^{(o)} = T_a \text{ the generators of G.} \tag{3}$$

$$M_a^{(n)} = -\int d^2x\,\Delta_a^{(n)}\,g(x,t)\,\frac{\delta}{\delta g(x,t)}$$

$$[M_a^{(n)}, M_b^{(m)}] = \varepsilon_{abc}\,M_c^{(n+m)} \tag{4}$$

For the general chiral models there are also "left-multiplications" $\tilde\Delta_a^{(n)} g = \tilde\Lambda_a^{(n)} g$, $\tilde\Lambda_a^{(n+1)} = \int_{-\infty}^{x} dy\,\{\partial_o\tilde\Lambda_a^{(n)} + [\tilde A_o,\tilde\Lambda_a^{(n)}]\}$,

$\tilde A_\mu = g\partial\mu g^{-1}$, $\tilde\Lambda_a^{(0)} = \tilde T_a$. For the non-linear sigma model, these give rise to the same constants of motion as $\Delta_a^{(n)}g$. For $\tilde T_a = T_a$, they also form half of a Kac-Moody algebra. For $\tilde T_a = g(-\infty,t)\,T_a g^{-1}(-\infty,t)$, they can be combined with $\Delta_a^{(n)}g$ to extend the algebra to $G \otimes C[t,t^{-1}] \oplus G$ on shell for these constant time non-local transformations[9].

III- SU(N) Gauge Theory

The loop space Yang Mills equations are $\frac{\delta}{\delta\xi_\mu}(\psi^{-1}\frac{\delta\psi}{\delta\xi_\mu}) = 0$. where $\psi = P\,e^{\oint A.d\xi}$. For a set of infinitesimal symmetry transformations $\psi \to \psi + \Delta^{(n)}\psi$, $\Delta^{(n)}\psi = -\psi S^{(n)}$, then $\frac{\delta}{\delta\xi_\mu}(\frac{\delta}{\delta\xi_\mu}S^{(n)} + [\psi^{-1}\frac{\delta\psi}{\delta\xi_\mu}, S^{(n)}]) = 0$. In two dimensions, the $S^{(n)}$ can be defined iteratively $\frac{\delta}{\delta\xi_\mu}S^{(n)} + [\psi^{-1}\frac{\delta\psi}{\delta\xi_\mu}, S^{(n)}] = \varepsilon_{\mu\nu}\frac{\delta}{\delta\xi_\nu}S^{(n+1)}$ with $S^{(o)} = T$. In three dimensions, Polyakov has proposed

$$\frac{\delta}{\delta\xi_\mu} S^{(n)} + [\psi^{-1} \frac{\delta\psi}{\delta\xi_\mu}, S^{(n)}] = \varepsilon_{\mu\nu\lambda} \dot{\xi}_\nu \frac{\delta}{\delta\xi_\lambda} S^{(n+1)}$$ where S is $S([\xi]_1)$.

In four dimensions, on the self-dual sector,

$$S^{(n)} = - \int_0^1 ds\, \psi^{-1}_{x:\xi(s)} \Delta^{(n)} A_\mu(\xi(s))\, \dot{\xi}_\mu(s) \psi_{\xi(s):x}$$ where

$\Delta^{(n)} A_\mu$ is given by the following.

In four Euclidean dimensions, the self-dual equations are

$F_{\mu\nu} = 1/2\, \varepsilon_{\mu\nu\alpha\beta}\, F_{\alpha\beta} \equiv \tilde{F}_{\mu\nu}$. Here

$F_{\mu\nu} \equiv \partial_\mu A_\nu - \partial_\nu A_\mu + [A_\mu, A_\nu]$, $A_\mu \equiv A^a_\mu (x) T^a$

T^a are the antihermitain generators of SU(N) with structure constants $[T^a, T^b] = \varepsilon_{abc} T^c$. Real solutions of SU(N) correspond to real $A^a_\mu(x)$. The symbol $\varepsilon_{\mu\nu\alpha\beta}$ is totally antisymmetric, $\varepsilon_{1234} = 1$. With the change of variables $\sqrt{2}\, y = x_4 - ix_3$, $\sqrt{2}\, \bar{y} = x_4 + ix_3$, $\sqrt{2}\, z = x_2 - ix_1$ $\sqrt{2}\, \bar{z} = x_2 + ix_1$, the self-dual equations become

$$F_{y\bar{y}} + F_{\bar{z}z} = 0 , \quad F_{y\bar{z}} = 0 , \quad F_{\bar{y}z} = 0. \tag{5}$$

The infinitesimal transformations are

$$\begin{aligned} \Delta^{(n)} A_y &= D_y\, \Omega^{(n)} & \Delta^{(n)} A_{\bar{z}} &= D_{\bar{z}}\, \Omega^{(n)} \\ \Delta^{(n)} A_{\bar{y}} &= -D_{\bar{y}}\, \Omega^{(n)} & \Delta^{(n)} A_z &= -D_z\, \Omega^{(n)} . \end{aligned} \tag{6}$$

Here

$$\Omega^{(n)} = -1/2\, (\bar{D}^{-1} \Lambda^{(n)}\, \bar{D} + D^{-1}\, \Lambda^{(n)\dagger}\, D)$$

$$D(\bar{z},y,z,\bar{y}) = Pe^{\int_{-\infty}^{\bar{z},y,z,\bar{y}} \{dy' A_y + d\bar{z}' A_{\bar{z}}\}}$$

path has z, $\bar{y}$ fixed

$$\bar{D}(\bar{z},y,z,\bar{y}) = Pe^{\int_{-\infty}^{\bar{z},y,z\bar{y}} \{d\bar{y}' A_{\bar{y}} + dz' A_z\}} \tag{7}$$

path has $\bar{z}$, y fixed

$$\partial_z \Lambda^{(n+1)} = \mathcal{D}_y \Lambda^{(n)} \qquad \partial_{\bar z} \Lambda^{(n+1)\dagger} = \tilde{\mathcal{D}}_{\bar y} \Lambda^{(n)\dagger}$$
$$\partial_{\bar y} \Lambda^{(n+1)} = -\mathcal{D}_{\bar z} \Lambda^{(n)} \qquad \partial_y \Lambda^{(n+1)\dagger} = -\tilde{\mathcal{D}}_y \Lambda^{(n)\dagger} \tag{8}$$

,

$$\mathcal{D}_y \equiv \partial_y + [\,P^{-1}\partial_y P,$$
$$\mathcal{D}_{\bar y} \equiv \partial_{\bar y} + [\,P\partial_{\bar y} P^{-1}, \tag{9}$$

$P \equiv D\bar{D}^{-1}$, $\Lambda^{(o)} = T^e \rho^e = T$, ρ^e are constant infinitesimal parameters.

These transformations leave $A^a_\mu(x)$ real and if $A_\mu(x)$ is a solution to $F_{\mu\nu} = \tilde{F}_{\mu\nu}$, so is $A_\mu + \Delta^{(n)} A_\mu$. They close half of a Kac-Moody algebra up to local gauge transformations:

$$[Q_a^{(n)} , Q_b^{(m)}] = \varepsilon_{abc}\, Q_c^{(n+m)} + \int d^4x\, D_\mu [\Omega_a^{(n)}, \Omega_b^{(m)}] \frac{\delta}{\delta A_\mu} \tag{10}$$

Here $Q_a^{(n)} = - \int d^4x\, \Delta_a^{(n)} A_\mu(x) \frac{\delta}{\delta A_\mu(x)}$. The larger set of $\Delta^{(n)} A_\mu$ plus gauge transformations, $\delta_{\rho,\lambda} A_\mu = \rho_b^n\, \Delta_b^{(n)} A_\mu - D_\mu \lambda$, closes to generate finite transformations which form a group.

The extension of $\Delta^{(n)} A_\mu$ off the self-dual sector is difficult because invariance and locality is tied to Eq.(5). (From Eq.(7), D and $\bar{D}$ are path dependent for $F_{\bar z y}$, $F_{\bar y z} \neq 0$ respectively). We note however that the first Kac-Moody transformation $\Delta^{(o)} A_\mu$ can be written as $\Delta^{(o)} A_\mu = D_\mu\, \bar{\Omega}^{(o)}$ where $\bar{\Omega}^{(o)} = -1/2\ (\bar{D}^{-1} T \bar{D} + D^{-1} T\, D)$, i.e. it is a particular local gauge transformation, an invariance already present in the full Yang Mills theory. This is similar to the chiral theory where $M_a^{(o)}$ are the isospin generators. Since a gauge transformation leaves the Yang Mills Lagrangian density invariant, a conserved Noether current can be constructed. For $\bar{\Omega}^{(o)}$ it is $J_\mu = -2\mathrm{tr}\, F_{\mu\nu} D_\nu \bar{\Omega}^{(o)}$ corresponding to the conserved matrix currents j_μ where $J_\mu = -1/2 \mathrm{tr}\ j_\mu T$:

$$j_{\bar z} = \Box (P^{-1} \partial_{\bar z} P) \qquad j_z = \Box (P \partial_z P^{-1})$$
$$j_y = \Box (P^{-1} \partial_y P) \qquad j_{\bar y} = \Box (P \partial_{\bar y} P^{-1}). \tag{11}$$

From the integrability condition of Eq.(8) and the Brezin et al, method[10], a set of conserved matrix currents can be identified:

$J_{\bar{z}}^{(n)} = \mathcal{D}_{\bar{z}}\Lambda^{(n)}$, $J_z^{(n)} = \tilde{\mathcal{D}}_z\Lambda^{(n)\dagger}$, $J_{\bar{y}} = \tilde{\mathcal{D}}_{\bar{y}}\Lambda^{(n)\dagger}$, $J_y^{(n)} = \mathcal{D}_y\Lambda^{(n)}$.

These are not derived as Noether currents and Eq.(11) does not appear naturally in this set. What is more natural is $j_\mu^{(n)} = \Box J_\mu^{(n)}$. These may serve as a guide to generalize Eq.(6) off the self-dual sector, since for n = 0 or Eq.(11) the full Yang Mills analog exists. For example, $\Delta^{(1)}A_\mu$ shifts $\mathcal{L}(x) = -1/2 \text{ tr } F_{\mu\nu}F_{\mu\nu}$ by $\Delta^{(1)}\mathcal{L} = \partial_\mu\Lambda_\mu$, where $\Lambda_\mu = -\text{tr } \partial_\mu([V_{\bar{z}},V_y] + [V_z,V_{\bar{y}}])T$ and $V_{\bar{z}} = P^{-1}\partial_{\bar{z}}P$, $V_y = P^{-1}\partial_y P$, $V_z = P\partial_z P^{-1}$, $V_{\bar{y}} = P\partial_{\bar{y}}P^{-1}$. So a conserved current is $\Box\bar{j}_\mu$ where

$\bar{j}_\mu = \partial_\mu\chi + 1/2\ [V_\mu,\chi], \mu = y,\bar{z}$ and $\bar{j}_{\bar{y}} = -\bar{j}_y^\dagger$, $\bar{j}_{\bar{z}} = -\bar{j}_z^\dagger$;

$\partial_{\bar{y}}\chi = -V_{\bar{z}}$, $\partial_z\chi = V_y$; and the associated charge for $\Box\bar{j}_\mu$ equals that of $j_\mu^{(1)}$ up to a commutator with T and functions of $Q^{(o)}$.

REFERENCES:

1- L.Dolan Phys. Rev. Lett. 47, 1371 (1981) and Phys. Lett. B113, 387 (1982).

2- I.B. Frenkel and V.G.Kac, Inv. Math. 62, 23 (1980); J.Lepowsky and R.L.Wilson, Commun. Math. Phys. 62, 43 (1978).

3- B.Kostant, Inv. Math. 48, 101 (1978); M.Adler and P.V. Moerbeke, Adv. in Math. 38, 267 (1980).

4- E.Date et. al., RIMS preprint 362 (1981) (Kyoto).

5- M.Luscher, Nucl. Phys. B135, 1 (1978).

6- M.Luscher, K.Pohlmeyer, Nucl. Phys. B137, 46 (1978).

7- A.M.Polyakov, Phys. Lett. 82B, 247 (1979); Y.Brihaye, D.B.Fairlie, J.Nuyts, R.G. Yates, J.Math. Phys. 19, 528 (1978).

8- See J.Schwartz, Phys. Rep. 8C, 269 (1973) for a review.

9- L.Dolan, Proceedings of the AMS Summer Seminar at the University of Chicago July 6-16, 1982. These lectures present in more detail some of the material given in this talk.

10- E.Brezin et al., Phys. Lett. B82, 442 (1979).

AN INVERSE SCATTERING TRANSFORM TECHNIQUE

FOR STATIONARY AXI-SYMMETRIC EINSTEIN-MAXWELL FIELDS

Ahmet Eriş and Metin Gürses*

Physics Department, METU, Ankara, Turkey

Recently Belinsky and Zakharov[1] (BZ) integrated the vacuum Einstein equations for space-times with two commuting Killing vectors and constructed exact N-soliton solutions. Their technique was later extended by Aleksejev[2] to construct the soliton configurations of Einstein-Maxwell fields.

In this note we show that, provided one performs the appropriate symmetric space reductions, the (BZ) technique can also be applied to the SU(2)/U(1) formulation of Einstein vacuum[3], and to the SU(3)/SU(2)xU(1) formulation[4] of Einstein-Maxwell field equations. The two systems can be treated in a unified way by introducing a Hermitian matrix P satisfying

$$d[\rho({}^{*}dP)P^{-1}] = 0 \quad , \tag{1}$$

and is given in terms of the appropriate Ernst potentials $\epsilon_o(\rho,z)$ and $\epsilon(\rho,z)$, $\Phi(\rho,z)$ by:

$$P = P_V = \frac{1}{\epsilon_o+\bar{\epsilon}_o}\begin{bmatrix} 2 & i(\bar{\epsilon}_o-\epsilon_o) \\ i(\bar{\epsilon}_o-\epsilon_o) & 2\epsilon_o\bar{\epsilon}_o \end{bmatrix}, \quad P=P_{ev}=F^{-1}\begin{bmatrix} 1 & \sqrt{2}\Phi & i(\bar{\epsilon}-\epsilon+2\Phi\bar{\Phi})/2 \\ \sqrt{2}\bar{\Phi} & -(\epsilon+\bar{\epsilon}-2\Phi\bar{\Phi})/2 & -i\epsilon\bar{\Phi}\sqrt{2} \\ i(\bar{\epsilon}-\epsilon-2\Phi\bar{\Phi})/2 & i\sqrt{2}\Phi\bar{\epsilon} & \epsilon\bar{\epsilon} \end{bmatrix} \tag{2}$$

with $\epsilon+\bar{\epsilon} = -2(F+\Phi\bar{\Phi})$. Here and in the sequel d stands for the exterior derivative, (*) for the "Hodge" dual operation in E^2 with cartesian coordinates (ρ,z), a bar over a quantity for complex conjugation and (†) for Hermitian conjugation. For both cases the matrix P satisfies $(\gamma P)^2 = I$, displaying the symmetric property of the respective configuration spaces for the fields[5], where γ is an involution ($\gamma^2 = I$) taken as

$$\gamma_V = \begin{bmatrix} 0 & i \\ -i & 0 \end{bmatrix} \quad , \quad \gamma_{ev} = \begin{bmatrix} 0 & 0 & i \\ 0 & -1 & 0 \\ -i & 0 & 0 \end{bmatrix} . \tag{3}$$

Defining a generalized exterior derivative operator $D \equiv d-(\partial f/\partial\lambda)^{-1}(df)\partial/\partial\lambda$ where λ is the complex spectral parameter independent of (ρ,z) and $f(\lambda,\rho,z)$ is any scalar function with the property $\lim_{\lambda\to 0} D = d$, the linear eigenvalue problem associated with the field equation (1) is written as:

$$D\Psi = -\Omega\Psi \quad , \qquad D\Omega + \Omega\wedge\Omega = 0 \quad , \tag{4}$$

where $\Psi(\lambda,\rho,z)$ is a complex matrix (2x2 for the vacuum and 3x3 for the electrovacuum case) and $-\Omega(\lambda,\rho,z) = a(dP)P^{-1} + \rho b^*(dP)P^{-1}$ with the complex scalar functions $a(\lambda,\rho,z)$ and $b(\lambda,\rho,z)$ satisfying $\lim_{\lambda\to 0} a = 1$, $\lim_{\lambda\to 0} b = 0$. Equations in (4) also imply $a^2 + \rho^2 b^2 - a = 0$ and $Da = \rho^* Db$ so that the curvature of the connection 1-form Ω vanishes when restricted to the solution submanifold. The limiting values of a and b imply $\lim_{\lambda\to 0} \Omega = -(dP)P^{-1}$ which enables us to identify $P = \Psi(o,\rho,z)$. The linear eigenvalue problem given above is relevant for principal chiral fields. For cases under consideration one must further perform a reduction to ensure the $(\gamma P)^2 = I$ property. This can be done by following Zakharov-Mikhailov[6] and Mikhailov-Yarimchuk[7], by restricting the connection to satisfy

$$\gamma\Omega(\lambda) + \Omega^\dagger(\bar{\lambda})\gamma = \Omega(\lambda)P - P\Omega^\dagger(\bar{\lambda}) = 0 \quad . \tag{5}$$

Equations (4) and (5) together constitute Einstein field equations.

We are now in a position to apply the (BZ) technique to the system given above. For this purpose we need the knowledge of a particular solution P_o in terms of which we construct the corresponding Ω_o and Ψ_o. Any transformation of the form $\Psi = \chi\Psi_o$ defines a new matrix Ψ leading to a new P provided that the matrix $\chi(\lambda,\rho,z)$ satisfies

$$D\chi = \chi\Omega_o - \Omega\chi \quad , \tag{6}$$

while the reduction to a symmetric space requires that

$$\Psi^{-1}(\lambda) = \gamma\Psi^\dagger(\bar{\lambda})\gamma \quad , \qquad P\gamma\Psi(\tau) = \Psi(\lambda)J \quad , \tag{7}$$

and

$$\chi^{-1}(\lambda) = \gamma\chi^\dagger(\bar{\lambda})\gamma \quad , \qquad P\gamma\chi(\tau) = \chi(\lambda)P_o\gamma \; , \tag{8}$$

where J is a matrix with $DJ = 0$ and $\tau:\lambda \to \tau(\lambda,\rho,z)$ is a fractional linear transformation on the complex λ-plane with $\tau^2 = 1$. Under this mapping a and b transform as $a(\tau,\rho,z) = 1-a(\lambda,\rho,z)$, $b(\tau,\rho,z) = -b(\lambda,\rho,z)$. For the N-soliton configuration it is assumed that the matrix χ has the form

$$\chi = I + \sum_{i=1}^{N} \frac{R_i}{\lambda-\mu_i} \tag{9}$$

where μ_i are the roots of the equation $f(\mu_i,\rho,z) = -\omega_i$ (constants) and R_i's are independent of λ. The remaining task is to determine the matrices R_i by solving eqn.(6) at the poles $\lambda = \mu_i$ such that they are compatible with conditions given by eqs.(7) and (8). This procedure can be summarized as follows:

1. The first eqn. in (8) requires that χ^{-1} should have poles at $\lambda = \bar{\mu}_i$ and

$$R_k + \sum_{i=1}^{N} \frac{R_k \gamma R_i^{\dagger} \gamma}{\mu_k - \bar{\mu}_i} = 0 \quad . \tag{10}$$

2. R_k's are degenerate matrices and must be of the form $R_k = n_k m_k^{\dagger}$ (for the vacuum case), $R_k = n_k m_k^{\dagger} + s_k t_k^{\dagger}$ (for the electrovacuum case) where n_k, m_k, s_k and t_k are column vectors.

3. Eqn.(6) evaluated at the poles $\lambda = \mu_k$ reduces to $(DR_k - R_k\Omega_o)\chi^{-1} = 0$ which implies $m_k = \gamma\Psi_o(\bar{\mu}_k)\gamma m_{ok}$, $t_k = \gamma\Psi_o(\bar{\mu}_k)\gamma t_{ok}$ where m_{ok} and t_{ok} are constant vectors.

4. Eqn.(10) reduces to a system of linear algebraic equations for the vectors n_k and s_k determining them completely in terms of m and t.

5. Eqn.(8) evaluated at the poles $\lambda = \mu_k$ reads.

$$R_k P_o[I + \sum_{i=1}^{N} \frac{R_i^{\dagger}}{\tau(\mu_k) - \bar{\mu}_i}] = 0 \quad . \tag{11}$$

Compatibility of the above solution for R_i with this equation requires that: a) There exists no odd numbered soliton solutions. b) For the solutions the poles are related pairwise as $(\mu_1,\tau(\mu_1))$, $(\mu_2,\tau(\mu_2),..$ $..,(\mu_i,\tau(\mu_i))...$ c) The matrix J appearing in eqn.(7) relates the corresponding constant vectors of the pairs as $J\gamma m_{10} = \gamma m_{20}$, $J\gamma m_{30} = \gamma m_{40}$ .... $J\gamma m_{io} = \gamma m_{i+1,0}$ ...., which implies $\gamma P_o m_1 = m_2$, $\gamma P_o m_3 = m_4$,...., $\gamma P_o m_i = m_{i+1}$.... . d) The form of the matrix χ given in eqn. (9) does not quarantee that its determinant is unity, hence the final step in the determination of the new solution P is to scale $\chi(o,\rho,z)$ by $[\mathrm{Det}\chi(o)]^{-1/n}$, n = 2,3 for vacuum and electrovacuum cases respectively.

Once a two soliton solution is obtained explicitly the 2N soliton solution can also be constructed using successive applications of the 2-soliton "operator". Since the reduction process decreases the number of arbitrary constants appearing in the solution to half, it seems reasonable to assume that our 2N soliton solution corresponds to Aleksejev's N soliton solution.

It is a great pleasure to thank the Centre for Theoretical Physics at Trieste for kind hospitality extended to us where part of this work is prepared.

* Present address:
Applied Mathematics Department, TBTAK. Marmara Scientific and Industrial Research Institute, GEBZE/KOCAELİ, TURKEY.

REFERENCES AND FOOTNOTES

[1] V.A.Belinsky and V.E.Zakharov, Sov.Phys. JETP, 48, 935(1978); 50, 1(1979).

[2] G.A.Aleksejev, Abstracts GR9, Jena, 1, 1(1980).

[3] L.Witten, Phys.Rev. D19, 718(1979).

[4] M.Gürses and B.C.Xanthopoulos, "Axially Symmetric, Static Self-Dual SU(3) Gauge Fields and Stationary Einstein-Maxwell Metrics" to appear in Phys.Rev.D.

[5] A.Eriş, M.Gürses and A.Karasu, "Poster Presented at the XI. International Colloquium on Group Theoretical Methods in Physics", Istanbul, Turkey (1982).

[6] V.E.Zakharov and A.V.Mikhailov, Sov.Phys. JETP, 47, 1017(1978).

[7] A.V.Mikhailov and A.I.Yarimchuk, "Cylindrically Symmetric Solutions of the Non-Linear Chiral Field Model (σ Model)" CERN Preprint, TH-3150 CERN (1981).

[8] For a comparison of Aleksejev's extension of the (BZ) technique with other existing solution generation techniques we refer the reader to: C.M. Cosgrove, California Ins. of Tech.Preprint No.OAP-619, (1981); D.Kramer, J.Phys. A15, 220(1982).

THE SYMMETRIC SPACE PROPERTY AND THE EMBEDDING PROBLEM FOR STATIONARY AXI-SYMMETRIC EINSTEIN-MAXWELL FIELDS

A.Eriş, M.Gürses and A.Karasu
Physics Department, M.E.T.U., Ankara, Turkey

Stationary axially symmetric (SAS) Einstein-Maxwell field equations in terms of the complex Ernst[1] potentials $\varepsilon(\rho,z)$, $\Phi(\rho,z)$ are

$$(\varepsilon+\bar{\varepsilon}+2\Phi\bar{\Phi})\,\nabla^2\varepsilon = 2(\nabla\varepsilon)\cdot(\nabla\varepsilon+2\bar{\Phi}\nabla\Phi) \quad (1)$$
$$(\varepsilon+\bar{\varepsilon}+2\Phi\bar{\Phi})\,\nabla^2\Phi = 2(\nabla\Phi)\cdot(\nabla\varepsilon+2\bar{\Phi}\nabla\Phi)\,.$$

Here ∇ is the flat space grad operator in cylindrical coordinates (ρ,z,ϕ) and $\nabla^2=\nabla\cdot\nabla$. Eqns. (1) can also be regarded as equations determining harmonic mappings[2] $f:M\rightarrow M'$ where M and M' are two Riemannian monifolds with metrics

$$ds^2 = d\rho^2+dz^2+\rho^2 d\phi^2\,,\quad ds'^2 = F^{-2}|d\varepsilon+2\bar{\Phi}d\Phi|^2 - 4F^{-1}\,d\Phi d\bar{\Phi}, \quad (2)$$

respectively[3]. M' is then called the configuration space for this class of Einstein-Maxwell fields. The configuration space with the metric given in equation (2) has some nice properties and hence consequences which may be summarized as follows:

1. M' is an Einstein space, i.e. $R'_{\mu\nu} = (R'/4)g'_{\mu\nu}$, $R' = -6$.
2. M' is a Riemannian symmetric space, i.e. $R'_{\mu\nu\alpha\beta;\lambda} = 0$.
3. Properties 1 and 2 mean that M' is a harmonic space and hence its metric can be written as $ds'^2 = \eta_{ij}\tau^i \otimes \tau^j$ ($i,j,\ldots = 1,2,\ldots,8$; are group indices), where τ^i are Killing 1-forms satisfying the Maurer-Cartan equations $d\tau^i + (1/2)C^i{}_{jk}\tau^j\wedge\tau^k = 0$ with $C^i{}_{jk}$ being the structure constants of the group SU(2,1) and $\eta_{ij} = -C^n{}_{ik}C^k{}_{jn}$ being the group metric[4]. Invariance of eqns. (1) under SU(2,1) means that one can also write the field equations as $d*(f^*o\tau^i) = 0$ where $f^*o\tau^i$ are the pull-backs onto M of the Killing 1-forms by the map f and (*) denotes "Hodge" dual operation defined with respect to the Riemannian structure on M.
4. Using the 3x3 matrix representation of the generators X_i of SU(2,1) we can define a Lie algebra valued connection 1-form $W = X_i\tau^i$ which, because of the Maurer-Cartan equations, satisfies $dW + W\wedge W = 0$.
5. Vanishing of the curvature of W implies that we can write $W = -dPP^{-1}$ and Determine P up to a constant gauge transformation, in

$$P = F^{-1}\begin{bmatrix} 1 & \sqrt{2}\Phi & i(\bar{\varepsilon}-\varepsilon+2\Phi\bar{\Phi})/2 \\ \sqrt{2}\bar{\Phi} & -(\varepsilon+\bar{\varepsilon}-2\Phi\bar{\Phi})/2 & -i\varepsilon\bar{\Phi}\sqrt{2} \\ i(\bar{\varepsilon}-\varepsilon-2\Phi\bar{\Phi})/2 & i\sqrt{2}\Phi\bar{\varepsilon} & \varepsilon\bar{\varepsilon} \end{bmatrix}$$

agreement with the result of Gürses and Xanthopoulos[5]. This form for W is particularly useful for the application of some inverse scattering techniques for generation of soliton solutions[6].

6. Using P the metric of M' is written as $-ds'^2 = tr.\,[dPP^{-1} \otimes dPP^{-1}]$ displaying clearly the connection of this problem with static axi-symmetric self-dual SU(3) Yang-Mills fields which is equivalent to a principal SU(3) chiral field theory.
The additional property $(\gamma P)^2 = I$ shows that one can identify the (SAS) Einstein Maxwell theory as a SU(3)/SU(2) xU(1) chiral model.

$$\gamma = \begin{bmatrix} 0 & 0 & i \\ 0 & -1 & 0 \\ -i & 0 & 0 \end{bmatrix}$$

7. Defining $X = \gamma P$ one can write $-ds'^2 = tr(dX)^2$ thus embedding M' into a flat space of dimension eight. The embedding constraints are given by the relation $X^2 = I$. This generalizes the work of Matzner-Misner to Einstein-Maxwell case[7].

We would like to thank M.Arık for drawing our attention to symmetric spaces.

[1] F.J.Ernst, Phys. Rev. 168, 1415 (1968).

[2] J.Eells and J.H.Sampson, Am.J.Math. 86, 109 (1964); Y.Nutku, Ann.Inst. H.Poincaré, A21, 175 (1974)

[3] D.Kramer et al. "Exact Solutions of Einstein Field Equations" Cambridge Monograps on Math. Phys. 6, Ed.by E.Schmutzer(1980).

[4] K.Yano, "Curvature and Betti Numbers" Princeton Univ.Press. Annals of Math. Studies No.32, (1953).

[5] M.Gürses and B.C.Xanthopoulos, "Axially Symmetric, Static Self-Dual SU (3) Gauge Fields and Stationary Einstein-Maxwell Metrics" to Appear in Phys. Rev.D.

[6] A.Eriş and M.Gürses, "An Inverse Scattering Transform Technique for Stationary Axi-Symmetric Einstein-Maxwell Fields" Talk presented at The XL. International Colloquium on Group Theoretical Methods in Physics, Istanbul, Turkey, (1982).

[7] R.Matzner and C.W.Misner, Phys.Rev. 154, 1229 (1967).

SELF-DUAL YANG-MILLS AS A TOTALLY INTEGRABLE SYSTEM

Ling-Lie Chau

Physics Department
Brookhaven National Laboratory
Upton, New York
UNITED STATES

ABSTRACT

The characteristics of a totally integrable system for the self-dual Yang-Mills equations are pointed out: the Parametric Bianchi-Bäcklund transformations, infinite conservation laws, the corresponding linear systems, and the infinite dimension Kac-Moody algebra.

INTRODUCTION

It has become increasingly clear that, besides its mathematical beauty, the Yang-Mills theory[1] may provide the key to our understanding of strong interactions. Despite many interesting theoretical and phenomenological observations such as confinement, asymptotic freedom[2], QCD (quantum chromodynamics) perturbative studies and lattice numerical analysis[3], the non-Abelian gauge theory is far from being solved.

In the past few years, with many colleagues of mine, we have investigated whether the beautiful and powerful techniques developed in solving the many so-called totally integrable systems in two dimensions can be used to solve the Yang-Mills fields in four dimensions. We have found, using the J formulation, that the self-dual Yang-Mills fields strikingly possess many of the characteristics of a totally integrable system[4,5]. The parametric Bianchi-Bäcklund transformations[6], infinite conservation laws[7], and the corresponding linear systems[4,5,8]. Recently we have added a new entry, the infinite-dimensional Lie algebra for the "hidden symmetry" of the self-dual Yang-Mills (SDYM) fields[9,10,11]. **It is the Lie algebra $s\ell(N,C) \otimes C(\lambda,\lambda^{-1})$ for the gauge group SL(N,C), and a symmetric-space over the subalgebra $su(N) \otimes R(\lambda)$ for the real gauge fields of SU(N).**

As for the full Yang-Mills system, it was hoped that similar progress can be made after the beautiful loop-space chiral equation was formulated for the Yang-Mills fields[12]. However there have been many difficulties and not much progress has been made[5].

Because of limitation of space, in the following I shall mainly list the results.

I. THE J FORMULATION OF SDYM FIELD IN COMPLEXIFIED E^4 SPACE

In the complexified E^4 space, $\sqrt{2}\, y = x_1 + ix_2$, $\sqrt{2}\bar{y} = x_1 - ix_2$, $\sqrt{2}z = x_3 - ix_4$, $\sqrt{2}\bar{z} = x_3 + ix_4$, the self-dual Yang-Mills equations[3] $F_{\mu\nu} = 1/2\, \varepsilon_{\mu\nu\rho\sigma} F_{\rho\sigma}$, are

$$F_{yz} = 0 = F_{\bar{y}\bar{z}},\ F_{y\bar{y}} + F_{z\bar{z}} = 0$$

The first two equations imply that the gauge potential A_μ can always be written in the following form

$$A_y = D^{-1}\partial_y D,\ A_z = D^{-1}\partial_z D,\ A_{\bar{y}} = \bar{D}^{-1}\partial_{\bar{y}}\bar{D},\ A_{\bar{z}} = \bar{D}^{-1}\partial_{\bar{z}}\bar{D}.$$

For the gauge group SL(N,C), $\det D = \det\bar{D} = 1$. For real SU(N) potentials A_μ, one can show that D and $\bar{D}$ are related, in real coordinate space, as $D^{\dagger} \doteq \bar{D}^{-1}$. Defining a matrix J by

$$J \equiv D\bar{D}^{-1} \tag{1.1}$$

which can be shown to be gauge invariant, and $\det J = 1$ for the gauge group SL(N,C), and that J can be made Hermitian, $J^{\dagger} \doteq \bar{J}$ in the real coordinate space for real SU(N) gauge fields. Now the SDYM equation can be written as

$$\mathbf{B_y} \equiv \mathbf{J^{-1}\partial_y J},\quad \mathbf{B_z} \equiv \mathbf{J^{-1}\partial_z J},\qquad \text{and} \tag{1.2a}$$

$$\boldsymbol{\partial_{\bar{y}} B_y + \partial_{\bar{z}} B_z = 0}\ , \tag{1.2b}$$

which we call the left SDYM-J equation; or equivalently the SDYM equation can be written as

$$\hat{B}_{\bar{y}} \equiv J\partial_{\bar{y}}J^{-1}, \quad \hat{B}_{\bar{z}} \equiv J\,\partial_{\bar{z}}J^{-1} \quad \text{and} \tag{1.3a}$$

$$\partial_y\hat{B}_{\bar{y}} + \partial_z\hat{B}_{\bar{z}} = 0\ , \tag{1.3b}$$

which we call the right SDYM-J equation.

II. TWO PARAMETER BIANCHI-BÄCKLUND TRANSFORMATION

One can easily show that the following transformation is a Bianchi-Bäcklund transformation

$$J^{-1}\partial_y J - J'^{-1}\partial_y J' = \lambda'\ \partial_{\bar{z}}(J^{-1}J') \tag{2.1a}$$

$$J^{-1}\partial_z J - J'^{-1}\partial_z J' = -\lambda'\ \partial_{\bar{y}}(J^{-1}J'), \tag{2.1b}$$

i.e., if J satisfies Eqs. (1.2, 1.3) so does J'.

For Hermitian J and J', they can be shown to satisfy the algebraic constraint $J'J^{-1} - J'^{-1}J = BI$, and $\lambda' = e^{i\alpha}$, where α, β are real.

III. INFINITE NON-LOCAL CONSERVATION LAWS

Consider B_y and B_z of Eqs. (1.2, 1.3) being the first conserved currents,

$$V_y^{(1)} \equiv B_y = \partial_{\bar{z}}\chi^{(1)},\quad V_z^{(1)} \equiv B_z = -\partial_{\bar{y}}\chi^{(1)} \tag{3.1}$$

$\chi^{(1)}$ exists because Eqs. (1.2, 1.3). From this first current we can generate infinite number of them by the following iterative procedure

$$V_y^{(n+1)} = \mathcal{D}_{y\chi}^{\ (n)} = \partial_{\bar{z}}\ \ \chi^{(n+1)},\quad V_z^{(n+1)} = \mathcal{D}_{z\chi}^{\ (n)} = -\partial_{\bar{y}}\chi^{(n+1)} \tag{3.2}$$

Such generated currents $V_y^{(n+1)}$, $V_z^{(n+1)}$ can be shown to be conserved and satisfy Eq. (1.2b), $\chi^{(n+1)}$ can be regarded as conserved charges.

IV. THE LINEAR SYSTEM FOR THE SDYM FIELDS

From these infinite non-local conservation laws, using a method given in Ref.14, we can obtain the following linear differential equations

$$\partial_{\bar{z}}\chi = \lambda\mathcal{D}_{y}\chi \equiv \lambda(\partial_{y} + J^{-1}\partial_{y}J)\chi; \tag{4.1a}$$

$$-\partial_{\bar{y}}\chi = \lambda\mathcal{D}_{z}\chi \equiv \lambda(\partial_{z} + J^{-1}\partial_{z}J)\chi \tag{4.1b}$$

The integrability of these equations gives the left SDYM equations of motion Eq. (1.2). Similarly, for the right SDYM equations Eq.(1.3) we have

$$\partial_{z}\hat{\chi} = -\frac{1}{\lambda}\hat{\mathcal{D}}_{\bar{y}}\hat{\chi} = -\frac{1}{\lambda}(\partial_{\bar{y}} + J\partial_{\bar{y}}J^{-1})\hat{\chi} \tag{4.2a}$$

$$\partial_{y}\hat{\chi} = \frac{1}{\lambda}\hat{\mathcal{D}}_{\bar{z}}\hat{\chi} - \frac{1}{\lambda}(\partial_{\bar{z}} + J\partial_{\bar{z}}J^{-1})\hat{\chi} \tag{4.2b}$$

V. THE KAC-MOODY ALGEBRA FOR THE SELF-DUAL YANG-MILLS FIELDS

V.1. THE CASE OF SL(N,C) SDYM FIELDS

We introduce the following two infinitesimal parametric transformations for the J-field[13]

$$\delta_{\alpha}(\lambda)J = \alpha_{a}\delta_{a}(\lambda)J = -J_{\chi}(\lambda)\ T_{\alpha}\chi(\lambda)^{-1} = \sum_{m=0}^{\infty}\lambda^{m}\alpha_{a}\delta_{a}^{(m)}J \tag{5.1}$$

$$\delta_{\alpha}(-\frac{1}{\lambda})J = \alpha_{a}\delta(-\frac{1}{\lambda})J = \chi(-\frac{1}{\lambda})T_{\alpha}\chi(-\frac{1}{\lambda})^{-1}J = \sum_{m=0}^{\infty}\lambda^{m}\alpha_{a}\delta_{a}^{(m)}J \tag{5.2}$$

where $T_{\alpha} \equiv \alpha^{a}T_{a}$, α^{a}'s are infinitesimal parameters and T_{a}'s are traceless anti-Hermitian matrices satisfying $[T_{a},T_{b}] = C^{c}_{ab}T_{c}$ with C^{c}_{ab} the structure constants of su(N). For complex α, T_{α} span the Lie algebra $s\ell(N,C)$, and for real α, T_{α} span the Lie algebra su(N). Using Eqs.(4.1, 4.2), it is easy to show that $J + \delta_{\alpha}J$, $J + \hat{\delta}_{\alpha}J$ satisfy the self duality equations, Eqs. (1.2, 1.3) respectively. Moreover, it follows from $T_{r}(T_{a}) = 0$ that $\det(J + \delta_{\alpha}J) = 1 = \det(J + \hat{\delta}_{\alpha}J)$. Therefore, these transformations are infinitesimal Bäcklund transformations. We can actually show that they satisfy the same Bäcklund transformations constructed previously in Ref.6, as given in Section II.

After lengthy calculations, we can derive the following infinite algebraic relations

$$[\Delta_a^{(m)}, \Delta_b^{(n)}]J = - C_{ab}^c \Delta_c^{(m+n)} J, \quad - \infty \leq m, \quad n \geq \infty, \tag{5.3}$$

where

$\Delta_a^{(m)} = \delta_a^{(m)}$, for $m > 0$; $\Delta_a^{(m)} = \delta_a^{(0)} + \hat{\delta}_a^{(0)}$, for $m=0$; and

$\Delta_a^{(m)} = (-)^m \hat{\delta}_a^{(-m)}$, for $m \leq 0$. This is the now well-known Kac-Moody algebra $s\ell(N,C) \otimes C(\lambda,\lambda^{-1})$. The important point to note is that it lacks the center of the algebra, which is of the form $c\delta_{ab}\delta_{k,-\ell}$, where c is a constant.

Since the indices m, n in Eq. (5.3) cover all integers, we can resum it into a single commutator in the complementary variables θ and θ'. Multiplying both sides of Eq. (5.3) by $e^{im\theta}\, e^{in\theta'}$, and summing, with the definition

$Q(\theta) \equiv \sum_{m=-\infty}^{\infty} e^{im\theta} Q_a^{(m)}$, we obtain

$$[Q_a(\theta), Q_b(\theta')] = C_{ab}^c \, Q_c(\theta)\, \delta(\theta - \theta'), \tag{5.4}$$

where the variable θ can be identified as (for unimodular λ) $e^{i\theta} = \lambda$, the CP^3 parameter[5,8].

V.2. THE CASE OF REAL SU(N) SDYM FIELDS

For J Hermitian and α_a real, we see that δJ, $\hat{\delta} J$ give new $J' \equiv J + \delta J$, $\hat{J}' = J + \hat{\delta} J$ respectively with $\det J' = 1 = \det \hat{J}'$; but J', $\hat{J}'$ are not Hermitian. From the condition $\hat{\chi}^{-1}(\bar{\lambda}) = \chi^{\dagger}(-1/\lambda)$ we can easily show $[\delta(-1/\lambda)J]^{\dagger} = \delta(\bar{\lambda})J$, therefore we can form two Hermitian transformations

$$\overset{(+)}{\delta}_a(\lambda)J \equiv \delta_a(\lambda)J + \hat{\delta}_a(-1/\lambda)J = \sum_{k=0}^{\infty} \lambda^k \, \overset{(+)}{\delta}_a{}^{(k)} J \tag{5.5}$$

$$\overset{(-)}{\delta}_a(\lambda)J \equiv i[\delta_a(\lambda)J - \delta_a(-1/\lambda)J] = \sum_{k=0}^{\infty} \lambda^k \overset{(-)}{\delta}_a{}^{(k)} J \tag{5.6}$$

where J is restricted to be Hermitian. After lengthy derivation we find the algebra

$$\left[\overset{(+)}{d}_a(m),\ \overset{(+)}{d}_b(n) \right] = C^c_{ab}\ \overset{(+)}{d}_c(m+n),$$

$$\left[\overset{(+)}{d}_a(m),\ \overset{(-)}{d}_b(n) \right] \equiv C^c_{ab}\ \overset{(-)}{d}_c\ (m+n), \qquad 0 < m,\ n < \infty \tag{5.7}$$

$$\left[\overset{(-)}{d}_a(m),\ \overset{(-)}{d}_b(n) \right] = C^c_{ab} \sum_{\ell=0}^{(m+n)} a_\ell\ \overset{(+)}{d}_c(\ell)$$

where

$$\overset{(+)}{d}_a(0) \equiv \overset{(+)}{\delta}_a(0), \quad \overset{(\pm)}{d}_a(1) \equiv \overset{(\pm)}{\delta}_a(1), \text{ and}$$

$$\left[\overset{(\pm)}{d}_a(m),\ \overset{(\pm)}{d}_b(1) \right] \equiv C^c_{ab}\ \overset{(\pm)}{d}_c(m+1)$$

and the coefficients a_ℓ are completely determined in the calculation.

So we see that $\overset{(+)}{d}_a(m)$, $\overset{(-)}{d}_b(n)$ form a symmetric space-like algebra over the $su(N) \otimes R(\lambda)$.

The existence of such infinite dimensional algebra in the 4-dimensional self-dual Yang-Mills theory certainly is a very nice addition to the list of total-integrability characteristics of the theory. However, the full implication of the algebra is yet to be discovered. Recently the infinite-dimensional algebra of the Toda-Lattice has been used to construct S-matrix for the theory[15]. The hope is that similar development can be made for the 4-dimensional Yang-Mills field.

ACKNOWLEDGMENTS

I would like to thank my colleagues, Ge Mo-Lin, A.Sinha and Wu Yong-Shi for their collaborative efforts in this pursuit.

The submitted manuscript has been authored under Contract No. DE-AC02-76CH00016 with the U.S. Department of Energy.

REFERENCES

1- C.N.Yang and R.L.Mills, Phys. Rev. 96 (1954) 191.

2- G. 't Hooft, Nucl. Phys. B33 (1971) 173, and private communication. H.D.Politzer, Phys. Rev. Lett. 30 (1973); D.J.Gross and F.A.Wilzcek, Phys. Rev. Lett. 30 (1973) 1343; H.D.Politzer, Nucl. Phys. B129 (1977) 301.

3- K.Wilson, Phys. Rev. D10 (1974) 2445; A.M. Polyakov, Phys. Lett. 59B (1975) 82; G. 't Hooft, Phys. Rev. D14 (1976) 3432. For recent work, see M. Creutz, Phys. Rev. Lett. 43 (1979); M. Creutz, L.Jacobs, C.Rebbi, Phys. Rev. D20 (1979) 1915.

4- For reviews on the subject, see Ling-Lie Chau Wang, "Bäcklund Transformations, Conservation Laws and Linearization of Self-Dual Yang-Mills and Chiral Fields", Proc. of the Guanzhou (Canton) Conference on Theoretical Particle Physics, Guanzhou, China, 1980; Proc. Int'l. School of Subnuclear Physics, Erice, 1980; Proc. of Int'l. Workshop on High Energy Physics, Protvino, Serpukov, U.S.S.R., 1980.

5- For recent progress, see Ling-Lie Chau, "Comments on the Linear Systems for the Self-Dual Yang-Mills Fields", "Is Yang-Mills Equation a Totally Integrable System?", and "A Systematic Framework of Generating Multi-monopole Solutions", Lectures at the 18th Winter School of Theoretical Physics, Karpacz, Poland, 1981.

6- M.K.Prasad, A.Sinha and L.-L. Chau Wang, Phys. Rev. Lett. 43 (1979) 750.

7- M.K.Prasad, A.Sinha and L.-L. Chau Wang, Phys. Lett 87B (1979) 237.

8- K.Pohlmeyer, Comm. Math. Phys. 72 (1980) 37; L.-L. Chau, M.K. Prasad and A.Sinha, Phys. Rev. D24 (1981) 1574.

9- L.-L. Chau, M.-L. Ge and Y.-S. Wu, Phys. Rev. D25 (1982) 1086.

10- L.-L. Chau, Y.-S. Wu, "More About Hidden-Symmetry Algebra for the Self-Dual Yang-Mills Systems", BNL-Princeton preprint, 1982.

11- L.-L. Chau, M.-L. Ge, A. Sinha, Y.-S, Wu, "Hidden Symmetry Algebra for the Self-Dual Yang-Mills Equations", BNL-Lanzhou-Ohio-Princeton preprint, 1982.

12- A.M.Polyakov, Phys. Lett. 82B (1979) 249, Nucl. Phys. B164 (1979) 1716; Y.S.Wu, Physica Energiae Fortis et Physica Nuclearis 3 (1979) 382; I. Ya. Aref'eva, Lett. Math. Phys. 3 (1979) 241.

13- C.N. Yang, Phys. Rev. Lett. 38 (1977) 1377; S. Ward, Phys. Lett. 61A (1977) 81; D.B.Fairlie, J. Nuyts and R.G. Yates, JMP 19 (1978) 2528.

14- This transformation was inspired by a transformation introduced by B.Y.Hou in the chiral fields, see B.Y.Hou, Yale preprint YTP80-29, 1980 (unpublished), and B.Y. Hou, M.L. Ge and Y.S. Wu, Phys. Rev. D24 (1981) 2238.

15- B.Kostant, Adv. in Math. 34 (1979) 195, Inventiones Math. 48 (1978) 101; M. Adler, Inventiones Math. 50 (1979) 219; A.G. Reyman and M.A. Semenov-Tian-Shansky, Inventiones Math. 54 (1979) 81.

CONSTRAINED HAMILTONIAN SYSTEMS

Kishore B. Marathe*
Department of Mathematics
Brooklyn College, Brooklyn N. Y. 11210 U.S.A.

1. Introduction

In studying generalized Hamiltonian dynamics, Dirac [1] introduced a bracket operation to replace the classical Poisson bracket when dealing with certain constrained Hamiltonian systems. A geometrical formulation of this Dirac bracket as a generalized Poisson bracket on the constraint manifold was given by the author in [2]. Lichnerowicz [3] has discussed the Dirac bracket in the context of Poisson manifolds (see also Sniatycki [4] for another approach).

The constraints influence the evolution of any dynamical system. The classical example of completely integrable Hamiltonian systems illustrates the situation where the constraints completely determine the dynamics. More generally, if we have a Hamiltonian system on a 2n-dimensional symplectic manifold with k first integrals in involution then we can obtain a reduction of the given system to a (2n-2k) dimensional symplectic manifold. In this paper we start from a given Hamiltonian system and consider the influence of constraits of various types on its evolution. In section 2 we give a brief review of Hamiltonian systems. Constraints are introduced in section 3, where evolution of constrained Hamiltonian systems is discussed. Prospects

* The author wishes to thank the U. di Firenze and C.N.R., ITALY for a visiting professorship during the Summer of 1982 when this work was completed. Thanks are also due to Prof. Modugno for his hospitality.

for quantization of constrained systems are discussed in section 4. Proofs of theorems are omited. They will be given elsewhere along with examples and other details.

2. Hamiltonian Systems

In this section we define Hamiltonian systems and indicate some related concepts that we will use in this paper. For a general discussion of Geometrical Mechanics see, for example, Abraham and Marsden[5], Souriau[6] or Godbillon[7]. Let (M, ω) be a symplectic manifold ($\dim M = 2n$, ω is a closed, nondegenerate 2-form on M). Let F denote the isomorphism of TM (tangent bundle of M) and T*M (cotangent bundle) induced by ω and let $V = F^{-1}$. Let $H \varepsilon \Phi(M)$ (C^{∞} functions on M). Then the dynamical system determined by the vectorfield $X_H = V(dH)$ is called the Hamiltonian system corresponding to the Hamiltonian function H and is denoted by the triple (M, ω, H). Its evolution is given by the integral curves of X_H on M. The Poisson bracket $\{f, g\} = \omega(X_f, X_g)$ makes $\Phi(M)$ into a lie algebra. Let G be a Lie group acting on M; let L denote its Lie algebra and let L* denote the dual of L. For B in L, denote by B^the vectorfield on M induced by B under the action of G. If the action of G on M is symplectic (i.e. G preserves the symplectic structure ω) then we call $J: M \to L^*$ a momentum map provided that $X_{j(B)} = B\hat{}$ for every B in L, where $j(B): M \to R$ is defined by $j(B)(x) = J(x).B$.

3. Constrained Hamiltonian Systems

Let C be a Lie subalgebra of dimension s of the Lie algebra $\Phi(M)$. Define $A = \{f \varepsilon C \mid \{f, g\} = 0 \text{ for all } g \varepsilon C\}$. We call C the algebra of constraints for the Hamiltonian H if $\{H, g\} = 0$ for all g in C. Following Dirac we call a constraint f in A a first class constraint and a constraint in C that is not in A is called a second class constraint. If $\dim A = k$ then C/A can be given a Lie algebra structure induced by the Poisson bracket. If $\bar{f}_1, \ldots, \bar{f}_h$ is a basis of C/A $(s - k = h)$ then $\{\bar{f}_i, \bar{f}_j\}$ is well defined. We now assume the following condition to be satisfied by these brackets.

The matrix formed by $\{\bar{f}_i, \bar{f}_j\}$, $1 \leq i,j \leq h$ is nonsingular. (*)
If we consider the case when only the second class constraints are present, then condition (*) reduces to Dirac's condition and leads to the Dirac bracket as shown in [2)]. Condition (*) thus generalizes Dirac's condition to the case when both types of constraints are present. In what follows we assume that $k > 0$.

Let $f_1, f_2, \ldots, f_h; g_1, g_2, \ldots, g_k$, $h + k = s$ be a basis of C such that $g_1, g_2, \ldots, g_k$ form a basis of A. Let us suppose that the differentials of these s functions are linearly independent at each point of the set of zeroes of C. Then this set becomes a submanifold of M of codimension s. We denote this submanifold by N and call it the constraint submanifold determined by C. The Hamiltonian vectorfields $V(dg_j)$, $0 < j \leq k$ when restricted to N define an involutive distribution on N (i.e. a smooth field of k-planes tangent to N and closed under the Lie bracket of vectorfields). Let $A^\wedge$ denote this involutive distribution. The maximal integral submanifolds of $A^\wedge$ determine a foliation of N. We denote the space $N/A^\wedge$ of the leaves of this foliation of N by $N^\wedge$. Let $p: N \to N^\wedge$ denote the natural projection of N onto the quotient $N/A^\wedge = N^\wedge$. Then we have the following theorem.

Theorem 1: $N^\wedge$ is a symplectic manifold of dimension $2n - s - k$ (this is even in view of the condition (*)) with symplectic structure $\omega^\wedge$ which is related to the symplectic structure ω of M by the relation $\omega|_N = p^* \omega^\wedge$. The Hamiltonian system defined by H on M induces a Hamiltonian system $(N^\wedge, \omega^\wedge, H^\wedge)$ on $N^\wedge$.

We observe that the constraint submanifold N is a fiber bundle over $N^\wedge$ with k-dimensional fibers which are the leaves of the foliation of N by $A^\wedge$. These fibers correspond to what Dirac has called the A-spaces determined by the first class constraints. Every point of N lies in one of these A-spaces, which contain all the motions starting from that point. The essential dynamics is then reduced to considering the evolution of the A-spaces or in our terminology to considering the evolution of the reduced Hamiltonian system on the manifold $N^{\wedge}$. Functions on $N^{\wedge}$ correspond to functions on N which are constant on the fibers (or the A-spaces). The first class constraints may, therefore, be called "guage variables" and the algebra of vector fields $A^{\wedge}$ the "guage algebra".

We now consider an alternative method of reduction where we start by factoring out the first class constraints to reduce the problem to that involving only the second class constraints. This system may then be further reduced by the method considered by the author in [2]. We observe that the Lie algebra of Hamiltonian vectorfields X_f , $f \in A$, corresponding to the first class constraints may be regarded as arising from a symplectic group action of a Lie group G on M. The Lie algebra L of G is isomorphic to the algebra of the Hamiltonian vectorfields X_f , $f \in A$. We can now define a momentum map $J: M \to L^*$ and use it to obtain a reduction of the original Hamiltonian system to the manifold M' of the orbits of G on the set of zeroes of J. M' is a symplectic manifold of dimension 2n - 2k with symplectic structure ω' and the induced Hamiltonian H'. If π denotes the canonical projection of $J^{-1}(0)$ onto M', then we have the following relation between ω and ω'. $\omega \,|_{J^{-1}(0)} = \pi^*\omega'$. The algebra of constraints for H' on (M', ω') is isomorphic to C/A which consists of only the second class constraints. The method of reduction discussed in [2] can now be applied to the system (M', ω', H') to reduce it to a Hamiltonian system (M", ω'', H") with dim M" = 2n - 2k - h. The Poisson bracket with respect to ω'' is the Dirac bracket for the system (M', ω', H') as shown in [2]. The relation between the reduced system of theorem 1 and the reduced system considered above is given by the following theorem.

Theorem 2: The reduced Hamiltonian system ($N^\wedge$, $\omega^\wedge$, $H^\wedge$) of theorem 1 is isomorphic to the reduced Hamiltonian system (M'', ω'',H'') in the sense that there is a diffeomorphism $\psi : N^\wedge \to M''$ such that $\psi^*(\omega'') = \omega^\wedge$ and the integral curves of $H^\wedge$ are mapped onto the integral curves of H''.

In our work we started with the algebras of constraints C and A as given and then proceeded with the reduction of the given Hamiltonian system. On the other hand one may start with a submanifold P of M as a constraint submanifold. Clearly not all submanifolds are admissible as constraint submanifolds so that we need some conditions on P for it to be a constraint submanifold. One may then define the algebras of constraints (at least locally) and study the reduction of the given Hamiltonian system. This approach is used in [3]. Our approach emphasizes the constraint algebras from the begining and this brings out the role that different types of constraints play in the reduction.

4. Prospects for quantization

As Dirac has pointed out in [1] the procedure of canonical quantization cannot be applied to Hamiltonian systems with constraints. In fact, in general, we do not have any mathematically satisfactory procedure for passing from the classical to the quantum theory. It seems that we can apply the Kostant-Souriau quantization (see ,for example, Blattner [8] or Guillemin and Sternberg [9] for a discussion of this and related topics) to the reduced Hamiltonian systems of theorem 2. A comparison of the two resulting quantum systems may through some light on the way that constraints behave in quantization. Another approach is to replace the requirement that Poisson brackets correspond to commutators by the requirement that some other classical bracket correspond to commutators. Introduction of the Dirac bracket was motivated in part by such considerations. We propose to discuss these and other aspects of quantization of constrained Hamiltonian systems in a later paper.

BIBLIOGRAPHY

1. Dirac, P.A.M., Can. J. Math. 2 (1950), p. 129.
2. Marathe, K.B., Group Theoretical Methods in Physics: Proc. 5th Int. Coll., Academic Press, New York (1977), p. 371.
3. Lichnerowicz, A., J. Diff. Geo., 12 (1977), p. 253.
4. Sniatycki, J., Ann. Inst. H. Poincarè Sect. A, 20 (1974), p. 365.
5. Abraham, R. and Marsden, J.E., Foundations of Mechanics, 2nd Ed., Benjamin, New York (1978).
6. Souriau, J.-M., Structure des systèmes dynamique, Dunod, Paris (1970).
7. Godbillon, C., Gèometrie differentielle et mècanique analytique, Hermann, Paris (1969).
8. Blattner, R.J., Proc. Symp. Pure Math.(American Math. Soc.), 26 (1973), p. 147.
9. Guillemin, V. and Sternberg, S., Geometric Asymptotics, Am. Math. Soc. Survey, 14 (1977).

BAČKLUND PROBLEM, DIFFERENTIAL ALGEBRA AND GROUP THEORY

Jean-François POMMARET

Département de Mathématiques, Ecole Nationale des Ponts et Chaussées, 28 rue des Saints-Pères, 75007 Paris, France

ABSTRACT

Our purpose is to prove that certain concepts of mathematical physics involving partial differential equations (PDE) can only be clearly understood in the framework of a new branch of mathematics called "differential algebraic geometry". By this way, we give a new setting for the inverse Bäcklund problem and relate it to the differential Galois theory.

1) BÄCKLUND PROBLEM:

Using the standard notations:

$$p = \frac{\partial z}{\partial x}, \quad q = \frac{\partial z}{\partial y}, \quad r = \frac{\partial^2 z}{\partial x \partial x}, \quad s = \frac{\partial^2 z}{\partial x \partial y}, \quad t = \frac{\partial^2 z}{\partial y \partial y}$$

for a surface $\Sigma = \{(x,y,z) \in \mathbb{R}^3 \mid z = f(x,y)\}$

we may state [1]:

CLASSICAL BÄCKLUND PROBLEM: Find out the systems of PDE defining the two surfaces Σ and Σ' in $\mathbb{R}^3$ with local coordinates (x,y,z) and (x',y',z') such that their respective first order contact elements (x,y,z,p,q) and (x',y',z',p',q') are related by four given relations $\Phi^\tau(x,y,z,p,q;x',y',z',p',q')=0 \quad \tau=1,\ldots,4$.

This way to state the problem is hidding in fact a misunderstanding. Let us indeed parametrize the two surfaces with the same parameters u and v. We get:

$$\begin{cases} \dfrac{\partial z}{\partial u} = p\dfrac{\partial x}{\partial u} + q\dfrac{\partial y}{\partial u} \\ \dfrac{\partial z}{\partial v} = p\dfrac{\partial x}{\partial v} + q\dfrac{\partial y}{\partial v} \end{cases} \Rightarrow \begin{cases} p = \dfrac{\partial(z,y)}{\partial(u,v)} \Big/ \dfrac{\partial(x,y)}{\partial(u,v)} \\ q = \dfrac{\partial(x,z)}{\partial(u,v)} \Big/ \dfrac{\partial(x,y)}{\partial(u,v)} \end{cases}$$

and we are simply led to a problem of differential elimination for x,y,z on one side and x',y',z' on the other side, with the new independent variables u and v. The result will not depend on the parametrization and we must therefore obtain intrinsic conditions for the surfaces (constant curvature in the example of Lie-Bianchi).

REMARK: The common parametrization is already done in the cases where $x-x'=0$ and $y-y'=0$ are among the four given relations.

Introducing two fibered manifolds $\pi : \mathcal{E} \to X$ and $\pi' : \mathcal{E}' \to X$ both with their fibered product over X, we may now state:

GENERALIZED BÄCKLUND PROBLEM: Let a system of PDE be given on $\mathcal{E} \times_X \mathcal{E}'$. Find out the resolvent systems of PDE defining the solutions Σ and Σ' induced by the natural projections of $\mathcal{E} \times_X \mathcal{E}'$ onto $\mathcal{E}$ and $\mathcal{E}'$.

The study of this differential correspondence can be done by means of a finite algorithm which is described in our forthcoming book [3]. We understand that the words "Bäcklund transformation" must be absolutely eliminated and that there is another problem, called inverse Bäcklund problem, which consists in finding a differential correspondence over two given resolvent systems.

2) CLASSICAL GALOIS THEORY:

Let $\mathbb{Q} \subset k \subset k' \subset L$ be fields and $\Gamma = \Gamma(L/k) = \mathrm{aut}(L/k)$ be the group of automorphisms of L fixing k. The three following equivalent definitions of a Galois extension L/k with Galois group Γ can be given when $|L/k| = \dim_k L < \infty$:

°) The field of invariants of Γ in L is just k.

°°) L is obtained from k by adjoining all the roots of some polynomial with coefficients in k.

°°°) $L \otimes_k L \simeq L \oplus \cdots \oplus L$ ($|L/k|$ terms)

Only the last definition is crucial for a generalization [3] but has never been used in a systematic way.

The main result is the celebrated:

FUNDAMENTAL THEOREM OF GALOIS THEORY:

°) L/k Galois $\Rightarrow L/k'$ Galois and there is a bijective dual correspondence between intermediate fields and subgroups of the Galois group given by:

$k' \to \Gamma' = \mathrm{aut}(L/k')$, $\Gamma' \to k'$ = field of invariants of Γ' in L.

°°) k'/k Galois $\Leftrightarrow \Gamma' \triangleleft \Gamma$ (normal behaviour)

Let now $k \subset L$ and M be subfields of a given field N. We may introduce $k' = L \cap M$ and denote by (L,M) the composite field of L and M in N, that is to say the smallest subfield of N containing both L and M. We get:

COROLLARY: L/k Galois $\Rightarrow (L,M)/M$ Galois, $\Gamma((L,M)/M) \simeq \Gamma(L/k')$ and $(L,M) \simeq L \otimes_{k'} M$.

3) DIFFERENTIAL ALGEBRA:

If A is a ring, let $Q(A)=\{a/s \mid a,s\in A, sb=0\ \forall 0\neq b\in A \Rightarrow s=0\}$ be the full ring of quotients of A.

DEFINITION: A differential field k is a field with commutative derivations $\partial_1,\dots,\partial_n$ that is to say $\partial_i(a+b)=\partial_i a+\partial_i b$, $\partial_i(ab)=(\partial_i a)b+a(\partial_i b)$, $\forall a,b\in k$, $\forall i=1,\dots,n$

NOTATIONS: We may define the differential ring of differential polynomials in $y^1,\dots,y^m$ as $k\{y\}=\lim\limits_{q\to\infty} k[y_q]$ with $k[y_q]=k[y^k_\mu \mid k=1,\dots,m\, ;\, \mu=(\mu_1,\dots,\mu_n),\, 0\leq\mu_1+\dots+\mu_n\leq q]$, $y^k_0=y^k$ and $d_i y^k_\mu=y^k_{\mu+1_i}$ where $\mu+1_i=(\mu_1,\dots,\mu_{i-1},\mu_i+1,\mu_{i+1},\dots,\mu_n)$. We shall set $k\langle y\rangle=Q(k\{y\})$.

DEFINITION: $\underline{a}\in k\{y\}$ is a differential ideal if $\underline{a}$ is an ideal and $d_i a\in\underline{a}$, $\forall a\in\underline{a}$, $\forall i=1,\dots,n$.

When $\mathfrak{p}\in k\{y\}$ is a prime differential ideal, that is to say if $\mathfrak{p}$ is a differential ideal and $a,b\in k\{y\}$, $ab\in\mathfrak{p}\Rightarrow a$ or $b\in\mathfrak{p}$ we may define a differential extension L/k with $L=Q(k\{y\}/\mathfrak{p})$. The following example illustrate the subtle problem of knowing whenever a differential ideal is prime or not, the solution of which involves the most sophisticate machinery [3]:

EXAMPLE: n=3, m=1; the ideal generated by $P_1\equiv y_{33}-\frac{1}{2}(y_{22})^2$ and $P_2\equiv y_{23}-y_{22}$ in $\mathbb{Q}\{y\}$ is prime but the differential ideal generated by P_1 and P_2 in $\mathbb{Q}\{y\}$ is not prime because

$$d_2P_1-d_3P_2-d_2P_2\equiv(1-y_{22})\,y_{222}$$

EXAMPLE: Any algebraic evolution equation generates a prime differential ideal (KdV).

4) DIFFERENTIAL GALOIS THEORY:

Using the third definition of a Galois extension, we have shown in [3] that the Galois theory is not at all a theory of field automorphisms but a theory of principal homogenous spaces or PHS for finite algebraic groups (for example permutation groups). The only natural generalization is the theory of PHS for algebraic pseudogroups, that is to say groups of transformations defined by systems of algebraic PDE, according to the ideas of E.Vessiot [4] and J.Drach at the beginning of this century [3].

The main fact is that the fundamental theorem is almost the same. In particular, whenever we study differential fields $K\subset L\subset M$, the search for a differential extension M/L of L/K is nothing else than the concept of conservation law.

Indeed, if $M = Q(k\{y,z\}/\mathfrak{q})$ and $L = Q(k\{y\}/\mathfrak{p})$ we must have $\mathfrak{p} = \mathfrak{q} \cap k\{y\}$ because $L \subset M$.

EXAMPLE: $\Phi^{\tau}(x, y_1) = 0$, $z_i = f_i(x, y_1, z)$

Using crossed derivatives one cannot find new PDE in (x, y_1).

As for the corollary, we are led to a very subtle problem:

PROBLEM: How can one construct a composite differential field (L,M) of two abstract differential extensions L/k and M/k as the concept of intersection has no meaning otherwise?

Using the universal property of tensor products, the answer to this problem is nothing else than to determine a prime differential ideal $\mathfrak{r}$ of the differential ring $L \otimes_k M$ and choose $(L,M) = Q(L \otimes_k M/\mathfrak{r})$. Henceforth, the search for a composite differential field is nothing else than the inverse Bäcklund problem. In particular, if $L = M$ and we deal with an auto-correspondence, then we must introduce the tensor product $L \otimes_k L$ and we just find the link with the differential Galois theory, fulfilling our initial purpose and proving at the same time the existence of an infinite number of differential correspondences as we may choose any intermediate differential field k' between k and L .

CONCLUSION

We hope to have convinced the reader that the philosophy of the Bäcklund problem must be revisited within this new framework and that there is no way to avoid the modern formal theory of PDE [2].

BIBLIOGRAPHY

1) E.GOURSAT: Le problème de Bäcklund, Mémorial des Sciences Mathématiques, n°6, Gauthier-Villars, 1925, 53p.

2) J.F.POMMARET: Systems of partial differential equations and Lie pseudogroups, Gordon and Breach, 1978, 426p.

3) J.F.POMMARET: Differential Galois theory, Gordon and Breach, 1982, 760p.

4) E.VESSIOT: Sur la théorie de Galois et ses diverses généralisations, Ann.Sc.Ec.Norm.Sup., 21, 1904, p.9 .

ARE ATOMIC HARTREE-FOCK EQUATIONS LINEARIZABLE?*

Carl E. Wulfman
Department of Physics
University of The Pacific
Stockton, California 95211

Introduction

Recently Kumei and Bluman developed the first systematic and definitive method for determining whether a given non-linear partial differential equation (or system of PDE's) can be linearized by any 1:1 transformation[1]. Their analysis begins with the observation that solutions of linear equations may be superposed. If ψ and ψ' are two such solutions it follows that $(\exp a\psi'\partial/\partial\psi)\psi = (1+a\psi'\partial/\partial\psi++)\psi$ is also a solution. The linear equation consequently admits a Lie group (with parameter a) of superposition transformations. If the equation is transformed into a non-linear equation by a 1:1 transformation $\psi \leftrightarrow \bar{\psi}$ the Lie generator is transformed from $\psi'\partial/\partial\psi$ into some new form $\pi\ \partial/\partial\bar{\psi}$. Kumei and Bluman show that π has a characteristic and definitive form. Considering the inverse transformation they show that given a non linear equation and the knowledge that it admits a Lie transformation group with generator $\pi\ \partial/\partial\bar{\psi}$ one can from π determine the transformation that will linearize the equation and (without resort to this transformation) determine the resulting linear equation. They also establish that if the equation does not admit a Lie group with a generator of the required form, then it is not linearizable by a 1:1 transformation.

As systematic methods are available for finding the generators of the Lie groups admitted by any differential equation,[2] one is now in a position to definitively investigate the linearizability of any given non linear system of differential equations.

*Research supported by NSF Grant CHE 8014165.

The difficulty imposed on atomic, molecular and solid state quantum mechanics by the non-linearity of the Hartree-Fock equations is so well known that it hardly bears mention. It has, for example, led to the widespread use of pseudopotentials in (linear) Schroedinger equations. However, one naturally supposes that a given pseudopotential is approximately constant for a very restricted submanifold of solutions of the relevant H-F equations, e.g., a submanifold for which there is a fixed core of inner shell electrons.

As will be evident shortly, the advance of Bluman and Kumei makes it possible to determine whether the H-F equations of any system are equivalent via a 1:1 mapping to some system of linear differential equations, Schroedinger-like or otherwise. We here determine whether such linearization is possible for the simplest of Hartree-Fock equations - that for two electrons in the same orbital of a two electron atom. It is just for two electrons in the same orbital that one expects the pseudopotential approximation to be of least value, and hence, another linearization method to be of most value.

Investigation

The relevant H-F equations are:

$$(H_o(\vec{r}_1,\vec{p}_1) + \phi(\vec{r}_1)-\varepsilon)\psi(\vec{r}_1) = 0 \qquad (1a)$$

$$(H_o(\vec{r}_2,\vec{p}_2) + \phi(\vec{r}_2)-\varepsilon)\psi(\vec{r}_2) = 0 \qquad (1b)$$

$$\phi(\vec{r}) = \int d^3r'|\psi(\vec{r}\,')|^2/|\vec{r}-\vec{r}\,'| \qquad (1c)$$

$$H_o(\vec{r},\vec{p}) = \frac{1}{2}p^2 - \zeta/r \qquad (1d)$$

Using Poisson's theorem we replace (1c) by

$$\nabla^2\phi(\vec{r}) = -4\pi|\psi(\vec{r})|^2 . \qquad (1e)$$

In searching for a linearizing transformation we may ignore the boundary conditions on (1e) and so allow some freedom in ϕ.

Because of the equivalence of (1a), (1b) we need only consider (1a), (1c), (1e). However because ψ and ψ^* are functionally independent, we must supplement (1a) by its adjoint. Taking advantage of the spherical symmetry of ψ and making the substitutions

$$X = 2^{\frac{1}{2}}r \quad , \quad U = 2\pi^{\frac{1}{2}}r\psi, \quad W = U^*$$

$$Z = \zeta-\varepsilon X \quad , \quad V = r\phi \qquad (2)$$

we obtain after multiplying by X from the left, the system

$$XU_{xx} - (V + Z)U = 0$$

$$XW_{xx} - (V + Z)W = 0 \qquad (3)$$

$$XV_{xx} + UW = 0$$

We now consider a transformation of variables

$$(X,U,W,V,Z) \leftrightarrow (\bar{X},\bar{U},\bar{W},\bar{V},\bar{Z}) \qquad (4)$$

For this transformation to be a diffeomorphism that linearizes the system (3) it is necessary that (3) be left invariant by a continuous group of transformations with operator $\exp a L^{o}$ that when transformed according to (4) has the effect of superposing solutions of the linear target equations. In our case the Lie generator L^{o} will be allowed to have the general form[1]

$$L^{o} = B\partial/\partial U + C\partial/\partial W + E\partial/\partial V + F\partial/\partial Z \qquad (5)$$

$B = B(X,U,V,W,Z)$, and similarly for C,E,F.

The analysis of Kumei and Bluman shows that for L^{o} to be the generator of superpositions in the target space each of the coefficients B,C,E,F must be the product of a function $\rho^{-1}(X,U,V,W,Z)$ and a function which is a general solution of a set of linear PDE's. The latter are the equations being sought, and ρ determines the transformation (4) that will yield them.

Now the system (3) will be invariant under the continuous group of transformations with generator L^{o} IFF whenever the equations (3) are satisfied the are annihilated by the operator L which is the extension of L^{o} that acts in the space of the variables U,V,W,Z and their first and second derivatives with respect to X [1]. One has

$$L = L^{o} + L^{1} + L^{2} . \qquad (6)$$

Here

$$L^{1} = B'\partial/\partial U_{x} + C'\partial/\partial W_{x} + E'\partial/\partial V_{x} + Z'\partial/\partial Z_{x} , \qquad (7a)$$

with

$$B' = \frac{d}{dx}B(x,U(x),V(x),W(x),Z(x)), \qquad (7b)$$

and similarly for C',E',F'.

Also

$$L^2 = B''\partial/\partial U_{xx} + C''\partial/\partial W_{xx} + E''\partial/\partial V_{xx} + F''\partial/\partial Z_{xx} , \tag{8a}$$

with

$$B'' = \frac{d}{dx}B'(X,U(x),V(x),W(x),Z(x), U_x(x),V_x(x),W_x(x),Z_x(x)), \tag{8b}$$

and similarly for C", E", F".

The form of the generator L^o is determined by equations (7), (8) together with the requirement that L annihilate (3) whenever (3) holds:

$$\begin{aligned} 0 &= L(3a) = XB'' - (V+Z)B - U(E+F) \\ 0 &= L(3b) = XC'' - (V+Z)C - W(E+F) \\ 0 &= L(3c) = XE'' + UC + BW \end{aligned} \tag{9}$$

where (in B", C", E") equations (3) require that

$$\begin{aligned} U_{xx} &= X^{-1}(V+Z)U, \\ W_{xx} &= X^{-}(V+Z)W, \\ V_{xx} &= -X^{-1}UW . \end{aligned} \tag{10}$$

Equations (10) imply that the values of the first derivatives U_x, W_x, V_x, Z_x, are not determined by a knowledge of the values of the variables X,U,W,V,Z. This independence implies that (7)-(9) can only be true if the coefficients of each first derivative, and of each linearly independent polynomial in first derivatives, vanishes ident ally. Using this observation one finds that (7)-(10) may be replaced by a system of 75 coupled linear partial differential equations - the Lie determining equations for the coefficients B,C,E,F of L^o. A few typical members of this overdetermined system are:

$$\begin{aligned} &X(B_{xx} + B_z Z_{xx}) + B_u(V+Z)U + B_w(V+Z)W \\ &+ B_v(UW) - (V+Z)B - (E+F)U = 0 \\ &B_{ux} = 0 \\ &B_{uw} = 0 \end{aligned} \tag{11}$$

Solving the set of equations[2] we obtain as the most general L^o admitted by the system (3), the operator

$$L^o = C_1 L_1^o + C_2 L_2^o + C_3 L_3^o ,$$

where the C_1 are arbitrary constants of integration and

$$L_1^o = U\partial/\partial U - W\partial/\partial W \ ,$$

$$L_2^o = \partial/\partial V - \partial/\partial Z \ ,$$

$$L_3^o = X(\partial/\partial V - \partial/\partial Z) \ .$$

Conclusion

No generator L^o is of the required form. Therefore we have definitively established that the H-F equations (1) are not linearizable by a 1:1 transformation of the space of variables $\vec{r},\psi$, ψ^*,ϕ, $Z = \zeta-\epsilon\cdot 2^{\frac{1}{2}}r$, when $\psi(\vec{r})$ is invariant under rotations.

We wish to thank Sukeyuki Kumei for an early communication, and helpful discussion, of the linearization analysis he and George Bluman discovered.

References

1. S. Kumei G.W. Bluman, Siam J. Appl. Math., in press.
2. For a discussion of these methods, examples of their application and further references, c.f.:
   a) C.E. Wulfman, "Dynamical Groups in Atomic and Molecular Physics", in "Recent Advances in Group Theory and their application to Spectroscopy", J.C. Donini, ed. (Plenum, N.Y., 1979).
   b) G.W. Bluman, J. Cole, "Similarity Methods for Differential Equations" (Springer, N.Y., 1974).
   c) C.E. Wulfman, "Systematic Methods for Determining the Lie Groups Admitted by Differential Equations", in "Symmetries in Science", B. Gruber, R.S. Millmann, eds. (Plenum, N.Y., 1980).

OSCILLATOR-LIKE UNITARY REPRESENTATIONS OF NON-COMPACT GROUPS AND SUPERGROUPS AND EXTENDED SUPERGRAVITY THEORIES

M. GÜNAYDIN

Laboratoire de Physique Théorique de l'Ecole Normale Supérieure

24, rue Lhomond - 75231 PARIS CEDEX 05 - FRANCE

ABSTRACT :

A general theory of a unified construction of the oscillator-like unitary irreducible representations (UIR) of non-compact groups and supergroups is presented. Particle state as well as coherent state bases for these UIRs are given and the case of SU(m,p/n+q) is treated in detail. Applications of this theory to the construction of unitary representations of non-compact groups and supergroups of extended supergravity theories, with particular emphasis on $E_{7(7)}$ and OSp(8/4, ℝ) are also discussed.

1. Introduction

My talk consists of two main parts. In the first part I will give a general theory of the construction of oscillator-like unitary irreducible representations (UIR) of non-compact groups and supergroups. The second part will deal with the application of this theory to the construction of the UIRs of the non-compact groups and supergroups of extended supergravity theories (ESGT).

The general construction of the oscillator-like UIRs of ordinary non-compact Lie groups with a Jordan structure was given in a particle state basis in collaboration with C. Saçlıoğlu(1)(2). The extension of this method to the construction of the oscillator-like UIRs of non-compact supergroups in a particle state basis and the formulation of the oscillator-like UIRs of non-compact groups as well as supergroups in a coherent state basis was done in collaboration with I. Bars(3).

Now the use of boson and fermion annihilation and creation operators to construct representations of ordinary Lie groups has a long history in physics. For example the first unitary representation of the Lorentz group was constructed by E. Majorana using boson operators. Later many physicists developed and applied oscillator methods for the construction of the unitary representations of groups of physical interest, most notably Schwinger, Goshen and Lipkin, Dirac, Gürsey, Gell-Mann, Neeman, Biedenharn, Barut, Nambu, Moshinsky,... The references to these authors and many others can be found in the books and review articles listed in reference (4). For ordinary Lie groups (compact as well as non-compact) these oscillator methods use either the boson or the fermion annihilation and creation operators but never a non-trivial mixing of them. However with the emergence of supersymmetry(5), superalgebras(6) and supergroups in physics the situation has changed dramatically. Under supersymmetry bosons and fermions get transformed into each other. Therefore the application of the oscillator method to the representations of supergroups must necessarily involve a non-trivial mixing of boson and fermion operators. The first formulation of the theory of representations of compact supergroups was given using Kac-Dynkin diagrams(7). Later another approach using Young supertableaux was developed(8). The oscillator methods have also been introduced to study the representations of some compact supergroups of physical interest(9).

Oscillator-like representations of ordinary groups have also been studied by the mathematicians starting with the work of Bargmann(10). For references to the mathematical literature we refer to two more recent works by Howe(11) and Kashiwara and Vergne(12) and to Prof. Jakobsen's talk in these proceedings(13). The general theory of the construction of oscillator-like UIRs of non-compact groups in the Fock space of boson operators given in reference (2) follows simply once the crucial underlying algebraic structure, namely the Jordan structure, is identified. It seems that the notion of dual pairs developed by Howe(11) in connection with the oscillator representation of Lie groups corresponds to what we call the Jordan structure. The power of the algebraic approach that was eloquently advocated by Prof. Zuckerman in his talk (14) is once more demonstrated in the passage from the theory of oscillator-like UIRs of ordinary non-compact Lie groups to those of non-compact supergroups as was done in reference (3) by the simultaneous use of boson and fermion operators.

The plan of my talk will be as follows : I begin by giving the construction of the Lie superalgebras of non-compact supergroups $Osp(2n/2m,\mathbb{R})$, $Osp(2m^*/2n)$ and $SU(m,p/n+q)$ in terms of boson and fermion annihilation and creation operators. This is followed by the section on the general theory of the construction of oscillator-like UIRs of non-compact supergroups with a Jordan structure with respect to their maximal compact subsupergroups in the super-Fock space of boson and fermion operators. Since the theory of the UIRs of ordinary groups is a special case of the one for supergroups, I restrict the general discussion to non-compact supergroups. The general theory is illustrated explicitly by constructing the oscillator-like UIRs of $SU(m,p/n+q)$ $(m \neq n \ , p \neq q)$. Then it is shown how to go from the particle state basis to the coherent state basis of a UIR. The supercoherent state basis exists for all the non-compact supergroups G with a Jordan structure. They are in general labelled by complex rectangular supermatrices Z which take values in a "super-bounded domain" and parametrize the "super-Hermitian symmetric space" G/H where H is the maximal compact subsupergroup of $\mathcal{G}$. Using our formalism one can define super-Hilbert spaces of analytic functions of the supervariable $\mathcal{Z}$ on which to realize the same class of UIRs of the non-compact supergroup.

In the second part of my talk, after reviewing the symmetries of the gauged as well as ungauged ESGTs I discuss the applications of our methods to the non-compact groups and supergroups of these theories. In particular I give the oscillator-like unitary representations of the global invariance group $E_{7(7)}$ of the N=8 ESGT in terms of boson operators transforming like the vector fields in it. Since $E_{7(7)}$ does not have a Jordan structure with respect to its maximal compact subgroup SU(8) these unitary representations are reducible. The corresponding coherent state basis is also given. Remarkably enough these coherent states can be labelled by the scalar fields of the N=8 theory. Thus the same unitary representations can be constructed over the Hilbert spaces of functions of the scalar fields. For the N=4 ESGT this corresponds exactly to the construction of the holomorphic discrete series representations of SU(1,1) by Bargmann[10].

This is followed by the construction of oscillator-like UIRs of the non-compact supergroup $Osp(8/4, \mathbb{R})$ which is the invariance group of the trivial vacuum of the gauged N=8 ESGT. The final section is devoted to some work done in collaboration with Ellis, Gaillard and Zumino on the infinite superalgebra generated by non-compact symmetry and supersymmetry generators in the ungauged N = 4 ESGT and its extension to higher N theories, in particular the case of N = 8 and its unitary realizations.

2. Construction of Lie superalgebras in terms of bosons and fermions operators.

In this section we shall construct the Lie superalgebras of some non-compact supergroups following ref. (3). Consider two sets of boson and fermion annihilation and creation operators $a_i, b_i (a^{i\dagger}, b^{i\dagger})$ and $\alpha_\mu, \beta_\mu (\alpha^{\mu\dagger}, \beta^{\mu\dagger})$ satisfying the canonical relations :

$$
\begin{aligned}
&[a_i, a^{j\dagger}] = \delta_i^j = [b_i, b^{j\dagger}] && i,j = 1,2,\cdots m \\
&\{\alpha_\mu, \alpha^{\nu\dagger}\} = \delta_\mu^\nu = \{\beta_\mu, \beta^{\nu\dagger}\} && \mu,\nu = 1,2,\cdots n \\
&[a_i, \alpha_\mu] = 0 = [b_i, \beta_\mu] \\
&[a_i, b_j] = 0 = \{\alpha_\mu, \beta_\nu\} \qquad \text{etc.}
\end{aligned}
\tag{2.1}
$$

By defining a pair of Bose-Fermi operators $\xi_A (\xi^{A\dagger})$ and $\eta_B (\eta^{B\dagger})$ $A, B = 1,2,\cdots,n+m$ whose first m components are the boson operators $a_i (a^{i\dagger})$ and $b_j (b^{j\dagger})$ and whose last n components are the fermion operators $\alpha_\mu (\alpha^{\mu\dagger})$ and $\beta_\nu (\beta^{\nu\dagger})$ respectively, we can represent the canonical relations (2.1) simply as

$$[\xi_A, \xi^{B\dagger}\} = \delta_A^B = [\eta_A, \eta^{B\dagger}\}$$

$$[\eta_A, \xi_B\} = 0 = [\eta_A, \xi^{B\dagger}\} \tag{2.2}$$

where the product $[\ ,\ \}$ is to be understood as an anti-commutator between any two fermion operators and as a commutator otherwise.

$$\xi_i \equiv a_i \quad , \quad \eta_i \equiv b_i \qquad i = 1,2,\ldots m$$

$$m \neq n$$

$$\xi_{m+\mu} \equiv \alpha_\mu \quad , \quad \eta_{m+\mu} \equiv \beta_\mu \qquad \mu = 1,2,\ldots n$$

The bilinear operators $(a_i a^{j\dagger} + b_i b^{j\dagger})$ and $(\alpha_\mu \alpha^{\nu\dagger} + \beta_\mu \beta^{\nu\dagger})$ generate the Lie algebras of $U(m)$ and $U(n)$ under commutation, respectively. The boson fermion bilinears $(a_i \alpha^{\mu\dagger} + b_i \beta^{\mu\dagger})$ and $(\alpha_\nu a^{j\dagger} + \beta_\nu b^{j\dagger})$ close into the generators of $U(m)\times U(n)$ under anti-commutation :

$$\{a_i \alpha^{\mu\dagger} + b_i \beta^{\mu\dagger}, \alpha_\nu a^{j\dagger} + \beta_\nu b^{j\dagger}\}$$
$$= \delta_i^j (\alpha^{\mu\dagger}\alpha_\nu + \beta^{\mu\dagger}\beta_\nu) + \delta_\nu^\mu (a^{j\dagger} a_i + b^{j\dagger} b_i)$$
$$= \delta_\nu^\mu (a_i a^{j\dagger} + b_i b^{j\dagger}) - \delta_i^j (\alpha_\nu \alpha^{\mu\dagger} + \beta_\nu \beta^{\mu\dagger})$$

Thus the bilinear operator $T_A{}^B \equiv \xi_A \xi^{B\dagger} + \eta_A \eta^{B\dagger}$ generate the Lie superalgebra $U(m/n)$ under the Lie superproduct, which is defined as the anti-commutator among any two odd generators and as the commutator otherwise. Odd generators are those bilinear operators that involve only an odd number of fermion operators.

Now the superalgebra $U(m/n)$ generated by $T_A{}^B$ under Lie superproduct can be enlarged to other simple superalgebras by considering additional bilinear operators involving ξ and η simultaneously.

$U(m/n) \longrightarrow OSp(2n/2m, \mathbb{R})$:

The generators $T_A{}^B$ together with the bilinear operators $S_{AB} = \xi_A \eta_B + \eta_A \xi_B$ and $S^{AB} = \xi^{A\dagger}\eta^{B\dagger} + \eta^{A\dagger}\eta^{B\dagger}$ generate the Lie superalgebra $OSp(2n/2m, \mathbb{R})$ under the Lie superproduct. The even part of $OSp(2n/2m, \mathbb{R})$ is the Lie algebra of $O(2n)\times Sp(2m,\mathbb{R})$ where $O(2n)$ is compact and $Sp(2m, \mathbb{R})$ is non-compact with the maximal compact subgroup $U(m)$. The form of the non-compactness (i.e. the Killing form) is determined in a super-Hermitian basis as explained below.

$U(m/n) \longrightarrow OSp(2m^*/2n)$:

The generators $T_A{}^B$ of $U(m/n)$ together with the super anti-symmetric operators

$$A_{AB} = \xi_A \eta_B - \eta_A \xi_B \qquad ; \quad A^{AB} = \xi^{A\dagger}\eta^{B\dagger} - \eta^{A\dagger}\xi^{B\dagger}$$

generate the Lie superalgebra of the non-compact supergroup $OSp(2m^*/2n)$ whose even subgroup is $O(2m)^* \times USp(2n)$. Here $USp(2n)$ is compact and $O(2m)^*$ is non-compact with a maximal compact subgroup $U(m)$.

If, instead of extending the Lie superalgebra $U(m/n)$ generated by $T_A{}^B$ with di-creation and di-annihilation operators, one considers an extension by bilinear operators of the form $\xi^{\dagger}\eta$ and $\eta^{\dagger}\xi$, then one obtains the Lie superalgebras of compact supergroups in the corresponding super-Hermitian basis. For example, the operators

$$(\xi_A \eta^{B\dagger} + \xi_B \eta^{A\dagger})$$

and

$$(\eta_A \xi^{B\dagger} + \eta_B \xi^{A\dagger})$$

extend the superalgebra $U(m/n)$ generated by $T_A^{\ B}$ to the Lie superalgebra of compact $Osp(2n/2m)$. Similarly, the antisymmetric bilinear operators

$$(\xi_A \eta^{B\dagger} - \xi_B \eta^{A\dagger})$$

and

$$(\eta_A \xi^{B\dagger} - \eta_B \xi^{A\dagger})$$

extend $U(m/n)$ to the Lie superalgebra of compact $Osp(2m/2n)$.

If, instead of two, we have R generations of Bose-Fermi operators $\xi(r)$ and $\eta(r)$ that supercommute with each other

$$[\xi_A(r), \xi^{B\dagger}(s)\} = \delta_A^{\ B} \delta_{rs}$$

$$[\eta_A(r), \xi_B(s)\} = 0 = [\xi_A(r), \eta^{M\dagger}(s)\}$$

$$[\eta_A(r), \eta^{B\dagger}(s)\} = \delta_A^{\ B} \delta_{rs} \tag{2.3}$$

then the construction of the above Lie superalgebras can be achieved in exactly the same way by summing over the generation index, i.e.

$$\xi\eta \longrightarrow \vec{\xi}\cdot\vec{\eta} \equiv \sum_r \xi(r)\eta(r)$$

On the Lie superalgebra level this corresponds to the trivial operation of taking a direct sum of R copies of the same Lie superalgebra. However, as we shall see later, the set of unitary irreducible representations (UIRs) of the non-compact supergroups that can be constructed by our method is determined by the number R of "generations".

Consider now the case when the operators ξ and η transform differently, i.e.

$$[\xi_A(r), \xi^{B\dagger}(s)\} = \delta_A{}^B \delta_{rs}$$

$$[\eta^M(r), \eta^\dagger_N(s)\} = \delta^M{}_N \delta_{rs} \tag{2.4}$$

with $A, B = 1, \ldots, m+n$

$M, N = 1, \ldots, p+q$

$r, s = 1, \ldots, R$

Note that the annihilation operators $\xi_A(r)$ and the $\eta^M(r)$ transform covariantly and contravariantly under $SU(m/n)$ and $SU(p/q)$, respectively. Then the generators of $SU(m/n)$ and $SU(p/q)$ given by the bilinears

$$I_A{}^B = \vec{\xi}_A \cdot \vec{\xi}^{B\dagger} - \frac{1}{m-n} \delta_A{}^B (\vec{\xi}^{C\dagger} \cdot \vec{\xi}_C) \quad , \quad m \neq n$$

$$K_M{}^N = -\vec{\eta}^\dagger_M \cdot \vec{\eta}^N + \frac{1}{p-q} \delta_M{}^N (\vec{\eta}^Q \cdot \vec{\eta}^\dagger_Q) \quad , \quad p \neq q \tag{2.5}$$

together with the $U(1)$ generator

$$N = \frac{1}{m-n} \vec{\xi}^{A\dagger} \cdot \vec{\xi}_A + \frac{1}{p-q} \vec{\eta}^M \cdot \vec{\eta}^\dagger_M$$

and the di-creation and di-annihilation operators

$$L^A{}_M = \vec{\eta}^\dagger_M \cdot \vec{\xi}^{A\dagger}$$

$$L_A{}^M = \vec{\xi}_A \cdot \vec{\eta}^M$$

generate the Lie superalgebra $SU(m,p/n+q)$. The even part of the supergroup $SU(m,p/n+q)$ is $S(U(m,p) \times U(n+q) \times U(1))$ where $SU(m,p)$ is non-compact with a maximal compact subgroup $S(U(m) \times U(p) \times U(1))$.

One can continue and construct the generators of other classes of Lie superalgebras as bilinears of Bose-Fermi operators. The examples given above will be sufficient to illustrate our method of constructing the oscillator-like unitary irreducible representations of non-compact supergroups with a Jordan structure.

3. Super-Hermitian basis of Lie superalgebras and the unitary supergroup action.

In the following sections we shall give the construction of oscillator-like UIRs of non-compact supergroups in the super-Fock space of boson and fermion operators.

Before going into this explicit construction, let us explain how we choose the generators of a unitary supergroup action in the super-Fock space. Denoting the generators of the even subgroup by H_a and the odd generators by A_α we can represent the supergroup action in the Fock space by the operator

$$\hat{U}(g) = e^{i\omega^a H_a + i\Theta^\alpha A_\alpha} \tag{3.1}$$

where ω^a and Θ^α are real "bosonic" and "fermionic" (e.g.,Grassmann) parameters, respectively. Since $\hat{U}$ mixes bosons with fermions, the consistency of the transformations requires that Θ^α be taken as anticommuting with the odd generators A_α (15)

$$\{\Theta^\alpha , A_\beta\} = 0$$

This is equivalent to taking the Θ^α as anticommuting with the fermion operators. On the other hand, for $\hat{U}(g)$ to be a unitary operator in the super-Fock space, we must have

$$\hat{U}^\dagger = \hat{U}^{-1} = e^{-i\omega^a H_a^\dagger - i A_\alpha^\dagger \Theta^\alpha}$$

which implies that we choose the generators such that

$$H_a^\dagger = H_a \quad , \quad A_\alpha^\dagger = - A_\alpha \qquad \text{and} \qquad \omega^{a\dagger} = \omega^a \quad , \Theta^{\alpha\dagger} = \Theta^\alpha \tag{3.2}$$

Thus the even generators H_a are Hermitian operators and the odd generators A_α are anti-Hermitian operators. In this basis all the structure constants of the Lie superalgebra are pure imaginary numbers, except for those relating two odd generators to an even one, which are pure real. We shall refer to this basis as the super-Hermitian basis. It will be implicitly assumed that we are working in such a basis and the form of non-compactness of the supergroup (i.e., the Killing form) is to be determined in such a basis.

As special cases of the construction of Lie superalgebras given in the previous section, one obtains the Lie algebras of ordinary groups. The Lie algebras of non-compact groups arise only in those cases when some of the generators are written in terms of di-boson annihilation and creation operators of the form ab and $a^\dagger b^\dagger$. In all other cases one obtains the Lie algebras of compact groups in an Hermitian basis. The Lie algebras constructed as bilinears of fermion operators alone are always of the compact type in the corresponding Hermitian basis.

4. Oscillator-like UIRs of non-compact supergroups with a Jordan structure.

The Lie superalgebras of non-compact supergroups constructed above all have a Jordan structure with respect to the Lie superalgebra of their maximal compact subsupergroups, i.e. they can be decomposed in the form of a vector space direct sum[16][17]

$$L = L^{-1} \oplus L^{0} \oplus L^{+1} \tag{4.1}$$

where L^{0} is the Lie superalgebra of the maximal compact subsupergroup. It contains the generator Q of an Abelian $U(1)$ factor which gives the grading

$$[Q, L^{\mp 1}] = \mp L^{\mp 1}$$

$$L^{0} = Q \oplus H$$

$$[Q, H] = 0$$

and there is a conjugation $\dagger$ in L such that

$$(L^{+1})^{\dagger} \cong L^{-1} \quad ; \quad (L^{0})^{\dagger} \cong L^{0}$$

Under the Lie superproduct we have

$$[L^{+1}, L^{-1}\} \cong L^{0} \quad ; \quad [L^{+1}, L^{+1}\} = 0$$

$$[L^{0}, L^{+1}\} \cong L^{+1} \quad ; \quad [L^{0}, L^{-1}\} \cong L^{-1}$$

Consider now the super-Fock space $\mathcal{F}$ formed by taking a tensor product of the Fock spaces of all the boson and fermion operators that enter in the construction of the Lie superalgebra L. Choose a set of states $|K^{M\ldots}_{A\ldots}\rangle$ in the super-Fock space of the Bose-Fermi operators that are annihilated by all the operators belonging to the L^{-1} space :

$$L^{-1} \, |K^{M\ldots}_{A\ldots}\rangle = 0$$

and which transform as some representation of the maximal compact subsupergroup K generated by L^{0}. Then the infinite set of states obtained by applying the operators of the L^{+1} space on the states $|K^{M\ldots}_{A\ldots}\rangle$ form the basis of a unitary representation of the non-compact supergroup G generated by L i.e.

$$|K^{M\ldots}_{A\ldots}\rangle \,, \quad L^{+} |K^{M\ldots}_{A\ldots}\rangle \,, \quad L^{+} L^{+} |K^{M\ldots}_{A\ldots}\rangle \,, \quad \ldots$$

If the states $|K^{M\ldots}_{A\ldots}\rangle$ transform like an irreducible representation of the maximal compact subsupergroup K then the resulting unitary representation of the non-compact supergroup G is also irreducible. The proof of this statement is identical for ordinary non-compact groups[2] and for supergroups[3]. It follows simply from the Jordan structure and the irreducibility of the starting representation $|K_{A..}^{M..}\rangle$ under K.

Thus this method of constructing UIRs can be applied to all non-compact supergroups G that have a Jordan structure with respect to their maximal compact subsupergroups K. Then each set of vectors $|K_{A..}^{M..}\rangle$ that transform irreducibly under K and are annihilated by the operators of the L^{-1} space determine uniquely an UIR of G. As we shall see explicitly with the example of $SU(m,p/n+q)$, the number of such states $|K_{A..}^{M..}\rangle$ in our super-Fock space depends on the number R of generations of Bose-Fermi operators that enter in the construction of L. In practically all cases there exists an infinite set of such states. In certain exceptional cases there may not exist any such lowest state $|K_{A..}^{M..}\rangle$ annihilated by the L^{-} space. This happens in those cases where the L^{-} space involves di-creation operators as well as di-annihilation operators. An example of this is the construction of the Lie algebra of $SO(12)^{*}$ in terms of boson operators transforming like a singlet and an anti-symmetric tensor representation 15 of its maximal compact subgroup U(6)[2]. This problem does not arise if one constructs $SO(12)^{*}$ generators as bilinears of boson operators transforming like the fundamental representation of U(6)[2]. In these exceptional cases when there are no states in the super-Fock space $\mathcal{F}$ annihilated by L^{-} space our construction leads to reducible unitary representations.

5. Oscillator-like Unitary Irreducible Representations of the non-compact Supergroups $SU(m,p/n+q)$.

To illustrate our method of constructing UIRs of non-compact supergroups we shall consider the case of $SU(m,p/n+q)$ in detail ($m\neq n, p\neq q$)[3]. The Lie superalgebra of $SU(m,p/n+q)$ was constructed in section 2 in terms of annihilation (creation) operators $\vec{\xi}_A$ ($\vec{\xi}^{A\dagger}$) and $\vec{\eta}^M$ ($\vec{\eta}^{\dagger}_M$) that transform covariantly under $SU(m/n)$ and contravariantly under $SU(p/q)$, respectively. $SU(m,p/n+q)$ has a Jordan structure with respect to its subgroup $S[U(m/n)\times U(p/q)]$

$$L = L^{-1} \oplus L^{0} \oplus L^{+1}$$

where
$$L^{0} \cong I_A{}^B \oplus K_M{}^N \oplus N$$

$$L^{-1} \cong L_A{}^M \quad \text{and} \quad L^{+1} \cong L^A{}_M$$

Consider now the super-Fock space $\mathcal{F}$ formed by taking a tensor product of the Fock spaces of all the Bose-Fermi operators $\vec{\xi}$ and $\vec{\eta}$. The vacuum state $|0,0\rangle$ in $\mathcal{F}$ is then simply the tensor product state of all the individual vacua and is annihilated by all the annihilation operators. Clearly any state of the form

$$\left(\xi^{A\dagger}_{(1)}\right)^{\ell_1}\left(\xi^{B\dagger}_{(2)}\right)^{\ell_2}\cdots\left(\xi^{C\dagger}_{(R)}\right)^{\ell_R}|0,0\rangle \quad ; \quad \ell_1,\dots,\ell_R = 1,2,\dots$$

or of the form

$$\left(\eta^{\dagger}_{M}(1)\right)^{k_1}\left(\eta^{\dagger}_{N}(2)\right)^{k_2}\cdots\left(\eta^{\dagger}_{Q}(R)\right)^{k_R}|0,0\rangle \quad ; \quad k_1,\dots,k_R = 1,2,\dots$$

will be annihilated by all the operators $L_A^{\ M} = \vec{\xi}_A \cdot \vec{\eta}^{\,M}$ belonging to the L^- space. From among these states one can project out states that transform like an irreducible representation of the maximal compact subsupergroup $S(U(m/n)\times U(p/q))$ which can then be used for constructing an UIR of $SU(m,p/n+q)$ by repeated applications of the operators $L^A_{\ M}$ belonging to the L^+ space. For example the state $\xi^{A\dagger}|0,0\rangle$ transforms like the irreducible contravariant fundamental representation of $SU(m/n)$. Using the super-Young tableaux technique of reference (8) one can denote it by a dotted superbox

$$\xi^{A\dagger}|0,0\rangle \cong |\boxed{\dot\diagup},0\rangle \qquad A,B,\dots = 1,\dots,m+n$$

where the box $\boxed{\diagup}$ corresponds to the covariant fundamental representation of $SU(m/n)$ and $\boxed{\dot\diagup}$ denotes the contravariant fundamental representation. Then the state $\xi^{A\dagger}_{(1)}\xi^{B\dagger}_{(2)}|0,0\rangle$ corresponds to the supersymmetrical representation $\boxed{\dot\diagup}\boxed{\dot\diagup}$ of $SU(m/n)$. Similarly by applying the operators $\xi^{A\dagger}_{(1)}$ k times on the vacuum $|0,0\rangle$ we create states that transform like the irreducible representation

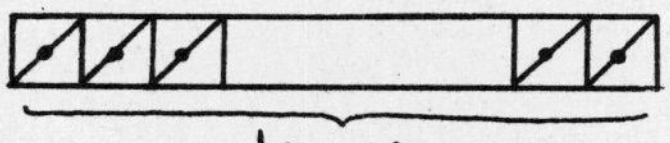

k times $\qquad k = 1,2,\dots$

of $SU(m/n)$. If we have only one set of operators ξ these are the most general irreducible representations of $SU(m/n)$ that can be constructed in our super-Fock space. If we have R sets of operators $\xi(r)$ ($r = 1,\dots R$) then the states

$$\left(\xi^{A\dagger}_{(1)}\right)^{\ell_1}\left(\xi^{B\dagger}_{(2)}\right)^{\ell_2}\cdots\left(\xi^{C\dagger}_{(R)}\right)^{\ell_R}|0,0\rangle$$

transform like the reducible representation

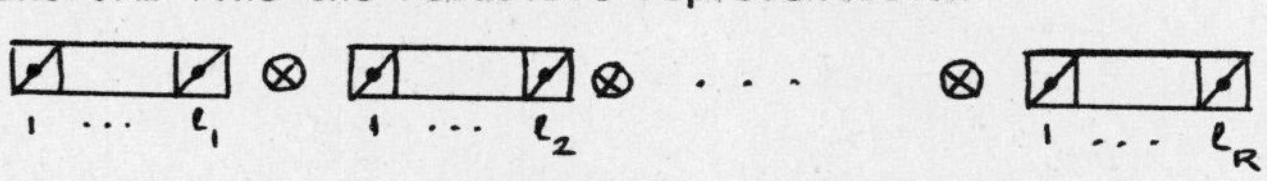

The irreducible representations of $SU(m/n)$ that one can project out of this set can have up to R rows in their supertableaux. Note also that a representation of $SU(m/n)$ ($m \neq n$) corresponding to a supertableau with only dotted (or only undotted) superboxes is irreducible. The same arguments apply to the states created by the multiple action of the operators $\eta^{\dagger}_{M}$ on the vacuum $|0,0\rangle$. In this case one obtains the representations of $SU(p/q)$ that correspond to supertableaux with undotted superboxes only. The irreducible representations obtained by the multiple action of $\xi^{A\dagger}(r)$ (r =1,...R) alone (followed by an appropriate projection operator) transform like the representation

$$[(\dot{r}_1, \dot{r}_2, \dots), 1]$$

of the maximal compact subsupergroup $S(U(m/n)\times U(p/q))$ and those constructed from $\eta^{\dagger}_M(r)$ transform as

$$[1, (r_1, r_2, r_3, \dots)]$$

where (r_1 , r_2,...) or ($\dot{r}_1$, $\dot{r}_2$,...) denote the supertableaux of an irreducible representation which has r_i ($\dot{r}_i$) super(dotted) boxes in its ith row. In general a supertableau can have an arbitrary number of rows[8].

The operators $L^A{}_M = \vec{\xi}^{A\dagger}\cdot\vec{\eta}^{\dagger}_M$ of the L^+ space transform like the representation $[(\dot{1},0,0,\dots),(1,0,0,\dots)]$ of $S(U(m/n)\times U(p/q))$ and

$$(L^A{}_M)^{\ell} \approx [(\dot{\ell},0,0,\dots),(\ell,0,\dots)]$$

Now starting from a "lowest" state $|K^{A\dots}_{M\dots}\rangle$ transforming irreducibly, say like the representation $[(\dot{m}_1,\dot{m}_2,\dots),1]$ or like $[1,(m_1,m_2,\dots)]$ under $S(U(m/n)\times U(p/q))$ and annihilated by the operators $L_A{}^M$ of L^- space we can construct an infinite tower of states by applying powers of the operator $L^A{}_M$ of the L^+ space. They transform under $S(U(m/n)\times U(p/q))$ as

$$(L^A{}_M)^{\ell}|K\rangle = (\vec{\xi}^{A\dagger}\cdot\vec{\eta}^{\dagger}_M)^{\ell}|K\rangle \qquad \ell=1,2,\dots$$

$$\approx [(\dot{\ell},0,\dots),(\ell,0,\dots)]\otimes[(\dot{m}_1,\dot{m}_2,\dots),1]$$

and form the basis of an UIR of the non-compact supergroup $SU(m,p/n+q)$ in the Fock space $\mathcal{F}$. Thus, for each such lowest state $|K\rangle$ we obtain an UIR of $SU(m,p/n+q)$. The remarkable property of these representations is the fact that the "lowest" state $|K\rangle$ uniquely determines the UIR. If we choose the generation number R large enough we can construct any representation of $SU(m/n)$ $(SU(p/q))$ with dotted (undotted) supertableaux only as a lowest state. For the case $n = 0 = q$ they reduce to the oscillator-like UIRs of the non-compact group $SU(m,p)$[2]. For $m = 0 = p$ they give the finite dimensional UIRs of the compact group $SU(n+q)$.

6. Super-coherent state basis of the oscillator-like UIRs of non-compact supergroups.

Above we have given the construction of the oscillator-like UIRs of the non-compact supergroups in a "particle state basis". The states constructed in the super-

Fock space have definite transformation laws under the maximal compact subsupergroup with a well-defined particle number. The proof of the irreducibility of the resulting representation follows from the fact that all the Casimir operators are simultaneously diagonalized and that each irreducible representation of the maximal compact subsupergroup occurs with multiplicity one. To write down the unitary supergroup action in our super-Fock space $\mathcal{F}$ we must exponentiate its generators in the super-Hermitian basis multiplied with their respective parameters. In the case of $SU(m,p/n+q)$ the unitary operator representing the supergroup action in our Fock space can be written as[3]

$$\hat{U}(g) = e^{i\bar{\Psi}\mathcal{M}\psi} \equiv \hat{U}(\mathcal{M})$$

$$\text{where} \quad \psi = \begin{pmatrix} \xi_A \\ \eta^{+}_{M} \end{pmatrix} \quad ; \quad \bar{\Psi} = \psi^{+}\gamma = (\xi^{A+} \quad -\eta^{M}) \tag{6.1}$$

$$\text{with} \quad \gamma = \left(\begin{array}{c|c} 1_{m+n} & 0 \\ \hline 0 & -1_{p+q} \end{array}\right)$$

$\mathcal{M}$ is the supertraceless $(m+n+p+q)\times(m+n+p+q)$ matrix representing the fundamental representation of the Lie superalgebra of $SU(m,p/n+q)$ with the odd and even generators multiplied with their corresponding anti-commuting and commuting "Hermitian" parameters. It can be represented in the form

$$\mathcal{M} = \left(\begin{array}{c|c} \mathcal{H}_{m/n} & -i\nu \\ \hline -i\nu^{+} & \mathcal{H}_{p/q} \end{array}\right) \tag{6.2}$$

where $\mathcal{H}_{m/n}$ and $\mathcal{H}_{p/q}$ are $(m+n)$ and $(p+q)$ dimensional Hermitian matrices representing the superalgebras $U(m/n)$ and $U(p/q)$ multiplied with their respective parameters such that the supertrace of $\mathcal{M}$ vanishes

$$\mathrm{Str}\,\mathcal{M} = \mathrm{Str}\,\mathcal{H}_{m/n} + \mathrm{Str}\,\mathcal{H}_{p/q} = 0$$

ν is a $(m+n)\times(p+q)$ matrix of the form

$$\nu = \begin{array}{c} \\ m \\ n \end{array}\!\!\left(\begin{array}{c|c} \overset{p}{V} & \overset{q}{\lambda} \\ \hline \chi & W \end{array}\right)$$

where V and W are $(m\times p)$ and $(n\times q)$ complex matrices whose entries are all commuting parameters, respectively. λ and χ are $(m\times q)$ and $(n\times q)$ matrices whose entries are anti-commuting complex Grassmann parameters, respectively. The unitarity of $\hat{U}$ follows from $\mathcal{M}^{+} = \gamma\mathcal{M}\gamma$. Thus the operator $\hat{H} \equiv \bar{\Psi}\mathcal{M}\psi$ is Hermitian and $\hat{U}^{+} = \hat{U}^{-1}$.

Under the action of supergroup $SU(m,p/n+q)$ the operators $\xi_A\,(\xi^{A+})$ and $\eta^{+}_{M}\,(-\eta^{M})$ get transformed into each other and together form the covariant (contravariant) fundamental representation of $SU(m,p/n+q)$

$$\hat{U}^{\dagger}(\mathcal{M})\,\psi\,\hat{U}(\mathcal{M}) = e^{i\mathcal{M}}\psi \equiv g\psi$$

$$\hat{U}^{\dagger}(\mathcal{M})\,\bar{\psi}\,\hat{U}(\mathcal{M}) = \bar{\psi}\,e^{-i\mathcal{M}} = \bar{\psi}\,g^{-1} \tag{6.3}$$

$$g \equiv e^{i\mathcal{M}}$$

The supermatrix g can be decomposed as[18]

$$g = t\,h$$

where h is a group element of the maximal compact subsupergroup $S(U(m/n)\times U(p/q))$ and t sits on the coset space $SU(m,p/n+q)/S(U(m/n)\times U(p/q))$. h can be represented as

$$h = e^{i\begin{pmatrix} \mathcal{H}_{m/n} & 0 \\ 0 & \mathcal{H}_{p/q} \end{pmatrix}} \equiv \begin{pmatrix} U_{m/n} & 0 \\ 0 & U_{p/q} \end{pmatrix} \tag{6.4}$$

and t as

$$t = e^{\begin{pmatrix} 0 & \nu \\ \nu^{\dagger} & 0 \end{pmatrix}} = \begin{pmatrix} \dfrac{1}{\sqrt{1-zz^{\dagger}}} & z\dfrac{1}{\sqrt{1-z^{\dagger}z}} \\ \dfrac{1}{\sqrt{1-z^{\dagger}z}}\,z^{\dagger} & \dfrac{1}{\sqrt{1-z^{\dagger}z}} \end{pmatrix} \equiv t(z)$$

where

$$z = \frac{\tanh\sqrt{\nu\nu^{\dagger}}}{\sqrt{\nu\nu^{\dagger}}}\,\nu$$

The corresponding decomposition of the unitary operator $\hat{U}$ is obviously given as

$$\hat{U} = \hat{t}(z)\,\hat{h}$$

$$\hat{t}(z) = e^{i\bar{\psi}\begin{pmatrix} 0 & -i\nu \\ -i\nu^{\dagger} & 0 \end{pmatrix}\psi} = e^{\eta^{T}\nu^{\dagger}\xi - \xi^{\dagger}\nu\eta^{\dagger T}} \tag{6.5}$$

In this case both $\hat{h}$ and $\hat{t}(z)$ are unitary operators. Writing the matrix g representing a general group element of $SU(m,p/n+q)$ in the fundamental representation as

$$g = e^{i\mathcal{M}} = \begin{array}{c} \\ m+n \\ p+q \end{array}\!\!\begin{array}{c} \begin{array}{cc} m+n & p+q \end{array} \\ \left(\begin{array}{c|c} \alpha & \beta \\ \hline \gamma & \delta \end{array}\right) \end{array} \tag{6.6}$$

one finds that the supermatrix Z parametrizing the supercoset space $SU(m,p/n+q)/S(U(m/n)\times U(p/q))$ undergoes a generalized linear fractional transformation under the action of g i.e.

$$g\,t(Z) = t(Z')\;h(Z,g) \tag{6.7}$$

where
$$Z' = (\alpha Z+\beta)(\gamma Z+\delta)^{-1}$$

and
$$h(g,Z) = \left(\begin{array}{c|c} U_{m/n}(g,Z) & 0 \\ \hline 0 & U_{p/q}(g,Z) \end{array}\right)$$

with
$$U_{p/q}(g,Z) = (1-Z'^{\dagger}Z')^{1/2}(\gamma Z+\delta)(1-Z^{\dagger}Z)^{-1/2}$$

and
$$U_{m/n}(g,Z) = (1-Z'Z'^{\dagger})^{1/2}(\alpha+\beta Z^{\dagger})(1-ZZ^{\dagger})^{-1/2}$$

Similarly the unitary operators $\hat{U}(g)$ and $\hat{t}(Z)$ satisfy

$$\hat{U}(g) = e^{i\bar{\Psi}\mathcal{M}\Psi}$$
$$\hat{U}(g)\,\hat{t}(Z) = \hat{t}(Z')\,\hat{h}(Z,g)$$
$$\hat{h}(Z,g) = \exp\left\{\xi^{\dagger}(\ln U_{m/n}(g,Z))\xi - \eta(\ln U_{p/q}(g,Z))\eta^{\dagger}\right\} \tag{6.8}$$

Consider now a set of states $|K^{A\cdots}_{M\cdots}\rangle$ that transform irreducibly under $S(U(m/n)\times U(p/q))$ and is annihilated by the operators L^{M}_{A} of the L^{-} space. We define the supercoherent state $|K^{A\cdots}_{M\cdots}, Z\rangle$ labelled by the rectangular supermatrix Z as

$$e^{\xi^{B\dagger} Z_B{}^N \eta^{\dagger}_N}\,|K^{A\cdots}_{M\cdots}\rangle \equiv |K^{A\cdots}_{M\cdots}, Z\rangle$$

Under the unitary action of the supergroup $SU(m,p/n+q)$ these supercoherent states transform among themselves and form an overcomplete basis of the UIR uniquely determined by the "lowest state" $|K_{M\ldots}^{A\ldots}\rangle$

$$\hat{U}(g)|K_{M\ldots}^{A\ldots}, Z\rangle = R_{M\ldots}^{M'\ldots} |K_{M'\ldots}^{A'\ldots}; (\alpha Z+\beta)(\gamma Z+\delta)^{-1}\rangle R_{A'\ldots}^{A\ldots}$$

where the supermatrices $R_{M\ldots}^{M'\ldots}$ and $R_{A'\ldots}^{A\ldots}$ are related to $S(U(m/n)\times U(p/q))$ rotation of the initial state $|K_{M\ldots}^{A\ldots}\rangle$. For example if the initial state transforms like the covariant fundamental representation of $SU(p/q)$ then

$$\hat{U}(g)|K_M, Z\rangle = [(\gamma Z+\delta)^{-1}]_M^{\ M'} |K_{M'}; (\alpha Z+\beta)(\gamma Z+\delta)^{-1}\rangle$$

The superunitary rotation matrices are uniquely determined by the transformation properties of the Bose-Fermi operators $\vec{\xi}$ and $\vec{\eta}$.

Thus for each "lowest" state $|K\rangle$ transforming irreducibly under the maximal compact subsupergroup $S(U(m/n)\times U(p/q))$ and annihilated by $L_A^{\ M}$ we obtain an UIR in the overcomplete coherent state basis by the above construction. If we expand the coherent states in a particle state basis then the expansion coefficients $\langle K|(L^+)^{\ell}|K,Z\rangle$ are all polynomial functions of Z . Therefore these supercoherent states have the same analyticity properties as the ordinary coherent states[(19)(20)]. This shows in particular that the oscillator-like UIRs we have constructed belong to the so-called "holomorphic discrete series". For ordinary Lie groups in addition to particle and coherent state formulations of the UIRs of the holomorphic series representations there is another formulation over the Hilbert spaces of analytic functions of "complex variables" taking values in a bounded homogeneous domain[(10)(21)(22)]. These complex variables can be taken to be the variables labelling the coherent states. Thus using the coherent state formulation given above one can find the corresponding definition of a "super-Hilbert space" of analytic functions of a "supervariable Z " taking values in a "bounded superdomain " over which to construct the holomorphic discrete series representations.

7. Unitary realizations of the non-compact symmetry groups and supergroups of extended supergravity theories.

In this section we shall discuss the applications of the general theory of UIRs developed in the previous sections to the non-compact groups and supergroups of ESGTs. The ESGTs for N=4-8 have an on-shell global non-compact invariance group G under which the vector field strengths get transformed into their duals and together form a linear representation of G , whereas the scalar fields of the theory transform non-linearly as the coset space G/H where H is the maximal compact subgroup of G[(23,24,25)]. The largest invariance group of these theories on-shell has the form $G_{global} \times H_{local}$ where the local invariance group H_{local} is isomorphic to (but not identical with) the maximal compact subgroup of G_{global}. The fermionic fields ($s = \frac{1}{2}$ or

$s = \frac{3}{2}$) are all singlets under G_{global} and transform as some non-trivial linear representation of H_{local}. In these theories the bound states of a given helicity are expected to fall into unitary representations of the respective non-compact global invariance group G. The theoretical arguments and phenomenological motivations for this expectation can be found in the references (26,27,28,29,30). The oscillator-like unitary representations of these non-compact groups were constructed in references (1,2) using boson operators transforming like the vector fields in the corresponding ESGTs[31,32]. Below we list the representation content of these boson operators and the non-compact groups G for the N=4-8 ESGTs :

	Boson Operators	Representation under H	G
N=4	$\vec{a}_{ij}\ (\vec{b}_{ij})$	$\underline{6}$ of U(4)	SU(4)xSU(1,1)
N=5	$\vec{a}_{ij}\ (\vec{b}_{ij})$	$\underline{10}$ of U(5)	SU(5,1)
N=6	$\vec{v}(\vec{w}) \oplus \vec{a}_{ij}(\vec{b}_{ij})$	$\underline{15}+\underline{1}$ of U(6)	SO(12)*
N=7	$\vec{a}_{ij}(\vec{b}_{ij}) \oplus \vec{v}_i(\vec{w}_i)$	$\underline{21}+\underline{7}$ of U(7)	$E_{7(7)}$
N=8	$\vec{a}_{ij}(\vec{b}_{ij})$	$\underline{28}$ of SU(8)	$E_{7(7)}$

In the case of N=8 ESGT the non-compact invariance group $E_{7(7)}$ does not have a Jordan structure with respect to its maximal compact subgroup SU(8) and the application of our method leads to reducible unitary representations[2]. The generators of $E_{7(7)}$ in terms of the anti-symmetric tensor boson operators $\vec{a}_{ij}$ and $\vec{b}_{ij}$ of SU(8) can be decomposed as $L = T^i_j \oplus V_{ijkl}$ where T^i_j are the generators of SU(8)

$$T^i_j = \vec{a}^{im\dagger} \cdot \vec{a}_{jm} + \vec{b}_{jm} \cdot \vec{b}^{im\dagger} - \frac{1}{8}\delta^i_j \left(\vec{a}^{mn\dagger} \cdot \vec{a}_{mn} + \vec{b}_{mn} \cdot \vec{b}^{mn\dagger}\right) \tag{7.1}$$

and V_{ijkl} are the non-compact generators that transform like the totally anti-symmetric tensor of rank four under SU(8)

$$V_{ijkl} = \vec{a}_{[ij} \cdot \vec{b}_{kl]} + \frac{1}{4}\epsilon_{ijklrstu}\, \vec{a}^{rs} \cdot \vec{b}^{tu}$$

where the bracket $[ij\,kl]$ denotes anti-symmetrization of all four indices.

Now starting from a state $|\psi_A\rangle$ in the Fock space of these boson operators which transforms like an irreducible representation of SU(8) we can generate an infinite set of states by multiple applications of the non-compact generators V_{ijkl}

$$|\psi_A\rangle\ ,\ V_{ijkl}|\psi_A\rangle\ ;\ VV|\psi_A\rangle\ ,\ \cdots \tag{7.2}$$

This infinite set of states forms the basis of a unitary representation of $E_{7(7)}$. This unitary representation is infinitely reducible even though the initial set transforms irreducibly under SU(8)[2].

The boson operators $a_{ij}\ (b_{ij})$ and $b^{ij\dagger}\ (a^{ij\dagger})$ transform among themselves under the action of $E_{7(7)}$ and form the 56 dimensional representation of $E_{7(7)\,global}$. Thus their transformation properties are exactly like the vector field strengths and their duals in the N=8 ESGT[23]. In fact the connection is deeper. For example if we go to the so-called "unitary gauge" which is manifestly free of ghosts, the non-compact invariance group of the theory that preserves this gauge is not $E_{7(7)\,\text{global}}$ but rather another $\tilde{E}_{7(7)}$ which corresponds to the simultaneous action of $E_{7(7)\,\text{global}}$ and $SU(8)_{\text{local}}$. In this gauge the vector field strengths and their duals transform with a 56x56 $\tilde{E}_{7(7)}$ matrix ϑ which is simply the 56-bein in the unitary gauge. Now if we identify the parameters of the non-compact generators V_{ijkl} with the unconstrained 70 scalar fields ϕ_{ijkl} of the N=8 theory, then the action of the non-compact generators on the boson operators a_{ij} (b_{ij}) and $b^{ij\dagger}$ $(a^{ij\dagger})$ give :

$$\hat{t}^{\dagger}(\phi)\begin{pmatrix} a_{ij} \\ b^{ij\dagger} \end{pmatrix} t(\phi) = \vartheta(\phi)\begin{pmatrix} a_{ij} \\ b^{ij\dagger} \end{pmatrix}$$

with

$$\hat{t}(\phi) = e^{i\phi_{ijkl}V^{ijkl}}$$

and

$$\bar{\phi}_{ijkl} \equiv \phi^{ijkl} = \frac{1}{4!}\epsilon^{ijklmnpq}\phi_{mnpq} \quad ; \quad V^{ijkl} = V^{\dagger}_{ijkl} \tag{7.3}$$

where $\vartheta(\phi)$ is the 56-bein in the unitary gauge.[23][33] Interpreting ϕ_{ijkl} as a 28x28 matrix we can write $\vartheta(\phi)$ as

$$\vartheta(\phi) = \begin{pmatrix} \frac{1}{\sqrt{1-y\bar{y}}} & y\frac{1}{\sqrt{1-\bar{y}y}} \\ \bar{y}\frac{1}{\sqrt{1-y\bar{y}}} & \frac{1}{\sqrt{1-\bar{y}y}} \end{pmatrix} \equiv t(y) \tag{7.4}$$

where

$$y_{ij,kl} \equiv \left(\phi\,\frac{\tanh\sqrt{\bar{\phi}\phi}}{\sqrt{\bar{\phi}\phi}}\right)_{ijkl}$$

The fields y are constrained such that $(1-\bar{y}y) > 0$. We can now go from the particle state basis of the unitary representation of $E_{7(7)}$ given above to a coherent state basis where the coherent states are labelled by the scalar field $y_{ij,kl}$. Consider the "initial state" $|\psi_A\rangle$ transforming irreducibly under SU(8) , then the coherent states $|\psi_A ; y\rangle$ defined by

$$|\psi_A ; y\rangle \equiv \hat{t}(y)|\psi_A\rangle = e^{i\phi_{ijkl}V^{ijkl}}|\psi_A\rangle \tag{7.5}$$

transform under the action of $E_{7(7)}$ as

$$\hat{U}(g)\,|\psi_A\,;y\rangle = \hat{U}(g)\,\hat{t}(y)\,|\psi_A\rangle$$
$$= \hat{t}(y')\,\hat{h}(y,g)\,|\psi_A\rangle = \hat{t}(y')\,R(h(y,g))_A^{\ B}\,|\psi_B\rangle$$
$$= R(h(y,g))_A^{\ B}\,|\psi_B\,,(Ay+B)(Cy+D)^{-1}\rangle \tag{7.6}$$

where $y' = (Ay+B)(Cy+D)^{-1}$ and $R(h(y,g))_A^{\ B}$ is the induced SU(8) rotation in the representation defined by $|\psi_A\rangle$. The 28x28 matrices A, B, C, D are defined by the 56 dimensional representation matrix g of $E_{7(7)}$: $g = \begin{pmatrix} A & B \\ C & D \end{pmatrix}$. Thus the coherent states $|\psi_A\,,y\rangle$ form an overcomplete basis of the unitary representation given in the particle basis above. The expansion coefficients $\langle\psi_A|V^n|\psi_A,y\rangle$ n=0,1,2,... of the coherent states in a particle state basis are functions of the scalar field y. Therefore one can equivalently formulate the same class of unitary representations on the Hilbert spaces of functions of y with an $E_{7(7)}$ invariant measure defined in terms of y which is uniquely determined by the scalar product in our Fock space.

Now in ESGTs one has the option of introducing additional couplings, while still preserving all the N supersymmetries, to turn the Abelian vector fields into non-abelian gauge fields of the natural SO(N) symmetry[33]. "Gauging" of the SO(N) symmetry however breaks the non-compact global invariance group of the ESGTs for N=4-8. The gauged ESGTs have local SO(N)xSU(N) symmetry at the Lagrangian level. They have a large cosmological constant and the scalar potential that necessarily accompanies the gauging is unbounded from below. However it has been shown that in spite of the unboundedness of the scalar potential they can have vacua in an anti-de Sitter background geometry which are stable against fluctuations that vanish sufficiently rapidly at spatial infinity[34]. The invariance groups of these vacua are the non-compact supergroups $Osp(N/4,\mathbb{R})$ (N=4-8) whose even subgroup is $SO(N)\times Sp(4,\mathbb{R})$ where $Sp(4,\mathbb{R})$ is isomorphic to the anti-de Sitter group SO(3,2). Thus the particle states (elementary as well as bound) constructed over these vacua must fall into unitary representations of the corresponding supergroup $Osp(N/4,\mathbb{R})$ .The oscillator-like UIRs of these non-compact supergroups can be constructed in a straightforward manner by our methods as explained above. For the case of N=8 the maximal compact subsupergroup of $Osp(8/4,\mathbb{R})$ is U(2/4). By the construction of the Lie superalgebras $Osp(2n/2m,\mathbb{R})$ given in section 2 the corresponding Bose-Fermi operators $\vec{\xi}_A$, $\vec{\eta}_B$ transform like the covariant fundamental representation of U(2/4). The even subgroup of the SU(2/4) is SU(2)xSU(4)xU(1) where SU(2) is the rotation subgroup of the anti-de Sitter group and SU(4) is the subgroup of SO(8). Now if we identify the SU(3)xU(1) subgroup of SU(4) with the color SU(3) and U(1) of electromagnetism, then one can embed an infinite family of color triplet and color singlet states of $S = 1/2$ inside a single irreducible representation of $Osp(8/4,\mathbb{R})$. This shows that if the true vacuum of the gauged N=8 theory has $Osp(8/4,\mathbb{R})$ symmetry, then its bound state spectrum can in principle accomodate the present day particle phenomenology[35]. In this kind of a scenario the gauge fields of SU(3) x U(1) are elementary whereas the weak gauge bosons are composite[36]. The chirality of fermion families is natural if the vacuum of the theory is such that the bound states fall into a single UIR or several UIRs which are not pairwise conjugate[35]. This is a special feature of the oscillator-like UIRs of non-compact groups. In the case of real compact groups or supergroups, a UIR decomposes into pairwise conjugate representations with respect to a complex subgroup. However for the UIRs of non-compact versions of these groups or supergroups, such as $E_{7(7)}$ or $Osp(8/4,\mathbb{R})$, this is not true. In fact the oscillator-like UIRs are intrinsically complex as can be seen from the fact that they can equivalently be realized over Hilbert spaces of analytic functions.

8. Infinite superalgebras in extended supergravity theories.

In ungauged ESGTs for N=4-8 the group of transformations on-shell that preserves the "unitary gauge" for the scalar fields is a non-compact group $\tilde{G}$ which is isomorphic to the global on-shell invariance group G_{gl} but not identical with it. This non-compact group corresponds to a simultaneous action of G_{gl} and $H_{loc.}$. The generators of the non-compact group $\tilde{G}$ do not commute with the supersymmetry generators and the larger superalgebra into which they close is infinite dimensional[27-30,37]. In fact, if the bound states of a given helicity fall into unitary representations of the non-compact invariance group, then the compatibility with supersymmetry even if it is badly broken implies the existence of such a larger superalgebra[1,29]. In the "rigid limit" of spatial infinity where all fields, except for scalar fields and the vierbein, vanish asymptotically the resulting infinite superalgebra for the N=4 case has been studied in detail at the "classical" level by Ellis, Gaillard, Günaydin and Zumino[37]. Except for algebraic complications the structure of the corresponding "rigid superalgebra" for higher N is essentially the same as the N = 4 case. The only novelty appears in the case of N = 8 algebra due to the lack of a Jordan structure of $E_{7(7)}$ with respect to its SU(8) subgroup[38]. In this case, the scalar fields $y_{ij,kl}$ that parametrize the 56-bein in the unitary gauge transform under $\tilde{E}_{7(7)}$ as in equation (7.6) : $\tilde{E}_{7(7)}: y \longrightarrow (Ay+B)(Cy+D)^{-1}$ which reads infinitesimally as :

$$L(\tilde{E}_{7(7)})\, y_{ij,kl} = \omega_{ijkl} - y_{ij,mn}\,\bar{\omega}^{mnpq}\, y_{pq,kl} + y_{ij,mn}\,\bar{\Omega}^{mn}{}_{kl} - \Omega_{ij}^{mn}\, y_{mn,kl}$$

where Ω and ω are the parameters of SU(8) and non-compact generators, respectively. Under the action of supersymmetry generators Q^i with Grassmann spinor parameters ϵ_i we have :

$$\bar{\epsilon}^m Q_m y_{ij,kl} = -2\sqrt{2}\left(\sqrt{1-y\bar{y}}\right)_{ij}^{\ pq} \psi_{pqrs} \left(\sqrt{1-\bar{y}y}\right)^{rs}_{\ kl}$$

$$\psi_{pqrs} \equiv \bar{\epsilon}^{(L)}_{[p} \dot{\lambda}^{(R)}_{qrs]} + \frac{\eta}{24} \epsilon_{pqrstuvz} \bar{\epsilon}^{t}_{(R)} \lambda^{uvz}_{(L)} \qquad ; \eta = \pm 1$$

where λ_{ijk} are the s = 1/2 fields transforming like the 56-dimensional representation of SU(8). Using these actions of $\tilde{E}_{7(7)}$ and taking into account the induced SU(8) rotation of the fields λ_{ijk} under $\tilde{E}_{7(7)}$, one obtains the "rigid" superalgebra generated by $\tilde{E}_{7(7)}$ and supersymmetry generators Q_i [38,39]. The multiple commutator of $E_{7(7)}$ generators with Q_i have the form $f_{ij}{}^{k}(y,\bar{y}) Q_k$ and are to be interpreted as "generalized" supersymmetry generators which generate field dependent supersymmetry transformations. If the bound states of a given helicity fall into unitary representations of $\tilde{E}_{7(7)}$, then the full spectrum of bound states (bosonic as well as fermionic) must correspond to a unitary realization of this algebra generated by $\tilde{E}_{7(7)}$ and Q_i . Referring for details to a forthcoming publication[38] I outline the general method by which to find the unitary realizations of this algebra which applies to all ESGTs (N = 4 - 8)[40]. Consider the coherent state basis $|\psi_A, y\rangle$ of a unitary representation of $\tilde{E}_{7(7)}$ given in the previous section. The helicity λ of the state $|\psi_A, y\rangle$ is the same as the helicity of the initial state $|\psi_A\rangle$. By acting on the states $|\psi_A, y\rangle$ with the supersymmetry generator Q_i one generates states $Q_i|\psi_A, y\rangle$ with helicity $\lambda \pm 1/2$ which form the basis of a unitary representation of $\tilde{E}_{7(7)}$ since the supersymmetry generators undergo an induced SU(8) rotation under the action of $\tilde{E}_{7(7)}$. By multiple action of the Q_i one can generate coherent states of helicities ranging between $\lambda-2$ and $\lambda+2$. For N-extended supergravity the helicity ranges from $\lambda - \frac{N}{2}$ to $\lambda + \frac{N}{2}$. The resulting unitary representations of different helicities are in general reducible even when the initial representation $|\psi_A, y\rangle$ is irreducible which can be the case for N = 4 and 5 as explained in the previous section.

ACKNOWLEDGEMENTS

I would like to thank I. Bars, E. Cremmer, B. de Wit, J. Ellis, M. Gell-Mann, B. Julia, H. Nicolai, E. Onofri, C. Saçlıoğlu, J. Schwarz, R. Stora and G. Zuckerman for many helpful discussions. I would also like to express my gratitude to the CERN Theory Division, where most of the work reported here was done, for its kind hospitality, and J. Prentki for his encouragement.

References

(1) M.Günaydin and C.Saçlioḡlu, Phys.Lett. 108B (1982) 169.
(2) M.Günaydin and C.Saçlioglu, "Oscillator-like UIRs of non-compact Groups with a Jordan Structure and the non-compact Groups of Supergravity", Cern Preprint TH-3209 (Dec 1981), to appear in Comm.Math.Phys.
(3) I.Bars and M.Günaydin, "Unitary Representations of non-compact Supergroups", Cern Preprint TH-3350 (June 1982), to be published.
(4) "Group Theoretical Concepts and Methods in Elementary Particle Physics", ed. F.Gürsey (Gordon and Breach, N.Y., 1964) ; F.J.Dyson, "Symmetry Groups in Nuclear and Particle Physics", (Benjamin Inc., 1966, N.Y.) ; "Group Theory and its Applications", Vols. I-III, ed. E.M.Loebl (Academic Press, N.Y., 1968) ; B.G.Wybourne, "Classical Groups for Physicists", J.Wiley and Sons (N.Y., 1974).
(5) J.Wess and B.Zumino, Nucl.Phys. B70 (1974) 39 ; D.V.Volkov and V.P.Akulov, Phys. Lett. 46B (1973) 109.
(6) For a complete classification of Lie superalgebras see V.G.Kac, Adv.Math. 26 (1977) 8. See also P.G.O.Freund and I.Kaplansky, J.Math.Phys. 17 (1976) 228.
(7) V.G.Kac, "Representations of Classical Lie Superalgebras", in "Differential Geometrical Methods in Mathematical Physics", eds. K.Bleuler, H.R.Petry and A. Reetz (Springer-Verlog, Berlin, 1978).
(8) A.B.Balantekin and I.Bars, J.Math.Phys. 22 (1981) 1149 and 1810 and Yale Preprint YTP-81-24 ; I.Bars, Yale Preprint YTP-81-25 to be published in the proceedings of the "School on Supersymmetry in Physics", Mexico (December 1981).
(9) A.B.Balantekin, I.Bars and F.Iachello, Nucl.Phys. A370 (1981) 284 ; A.B.Balantekin Ph.D. Thesis, Yale Univ. (1982), unpublished.
(10) V.Bargmann, Ann.Math. 48 (1947) 568 and Comm.Pure and App.Math. 14 (1961) 187.
(11) R.Howe, "Classical Invariant Theory", Yale Univ. preprint, unpublished ; R.Howe, "Transcending Classical Invariant Theory", Yale Univ. preprint, unpublished.
(12) M.Kashiwara and M.Vergne, Inventiones Math. 44 (1978) 1.
(13) H.P.Jakobsen, These proceedings.
(14) G.Zuckerman, These proceedings.
(15) We should note that when the generators are represented by finite dimensional ordinary matrices rather than oscillators, $\tilde{\Theta}$ will, of course, commute with those matrices.
(16) I.Bars and M.Günaydin, Journ.Math.Physics 20 (1979) 1977.
(17) M.Günaydin, "Proceedings of the 8th Int.Colloq. on Group Theoretical Methods", Annals of Israeli Physical Society, Vol.3 (1980).
(18) F.Gürsey and L.Marchildon, Phys.Rev. D17 (1978) 2038 and J.Math.Phys. 19 (1978) 942.
(19) For a review of coherent states of Lie groups and their applications, see A.M. Perelomov, Sov.Phys.Usp. 20 (1977) 703.
(20) For the study of the analyticity properties of coherent state representations of Lie groups and further references on the subject, see E.Onofri, Journ.Math.Phys. 16 (1974) 1087.
(21) Harish-Chandra, Amer.J.Math. 77 (1955) 743-77 ; 78 (1956), 1-41, 564-628.
(22) A.W.Knapp, "Bounded Symmetric Domains and holomorphic discrete Series", in "Symmetric Spaces", Marcel Dekker Inc. (1972).
(23) E.Cremmer and B.Julia, Phys.Lett. 80B (1978) 48 ; Nucl.Phys. B159 (1979) 141.
(24) The first non-compact invariance group of this kind was found in the N=4 theory by E.Cremmer, S.Ferrara and J.Scherk, Phys.Lett. 74B (1978) 61.
(25) For a review of the symmetries of ESGTs see E.Cremmer, Proceedings of the Spring School on Supergravity, Trieste (1981) ; B.Julia, Talk at the Johns Hopkins Workshop on Particle Theory (1981), ENS Preprint 81/14, and these proceedings.
(26) B.Zumino, in "Superspace and Supergravity", ed. by S.W.Hawking and M.Roček, Cambridge Univ.Press, Cambridge (1981) p.423.
(27) J.Ellis, M.K.Gaillard and B.Zumino, LAPP preprint TH44/Cern preprint TH-3152 (1981)
(28) J.Ellis, Cern preprint TH-3206 (1981), to appear in the proceedings of the "2nd Europhysics Study Conference on Unification of the fundamental Interactions", Erice, October 1981.
(29) M.Günaydin, Cern preprint TH-3222 (1981) to appear in the proceedings of the "2nd Europhysics Study Conference on Unification of the fundamental Interactions",

Erice, October, 1981.
(30) M.Günaydin, ENS preprint LPTENS 82/27, to appear in the proceedings of the "XXI International Conference on High Energy Physics", Paris, July 26-31 (1982).
(31) The work of references (1) and (2) was originally motivated by the arguments indicating that the bound states of ESGTs (N=4-8) may come in unitary representations of the respective non-compact symmetry groups. For a detailed discussion of the relevance of these unitary representations to ESGTs, see reference (29).
(32) For a review of this construction see C. Saçlioğlu, these proceedings.
(33) The gauged version of the N=8 ESGT was written down by B. de Wit and H.Nicolai, Physics Lett. 108B (1981) 285 ; Nucl.Phys. B208 (1982) 323.
(34) P.Breitenlohner and D.Freedman, M.I.T. Preprints (1982).
(35) M.Günaydin, in preparation.
(36) M.Gell-Mann, Talk presented at the Aspen Workshop on Octonionic Quantum Mechanics (1978), unpublished.
(37) J.Ellis, M.K.Gaillard, M.Günaydin and B.Zumino, in preparation.
(38) M.Günaydin, in preparation.
(39) Aspects of the rigid superalgebra generated by $E_{7(7)}$ and supersymmetry have been independently worked out by M.K. Gaillard, private communication and Berkeley preprint UCB-PTH 82/18 (september 1982)
(40) For a study of the unitary realizations of the N = 4 algebra without using the oscillator methods see reference (37)

GRAVITY, SUPERGRAVITIES AND INTEGRABLE SYSTEMS+

Bernard JULIA
Laboratoire de Physique Théorique de l'Ecole Normale Supérieure[x]
24 rue Lhomond, 75231 Paris cedex 05
FRANCE

ABSTRACT

Around 1968 three wonderful concepts emerged in different places and in seemingly unrelated domains of mathematical physics. They are the Kac-Moody algebras (among them the "affine" Kac-Moody algebras are related to current algebras and to gauge groups over one-dimensional "space-times"), the method of inverse scattering (for nonlinear partial differential equations in two-dimensional space-times), and finally the dual string model which is a two-dimensional field theory describing extended particles moving in a space-time of dimension 26 (10 or 2 if one dresses the string with internal degrees of freedom). In the last two years it was realized that gravity and supergravities provide a three-legged bridge between them and this revived hopes (at least with the author) of breaking the 2-dimensionality constraint for the integrability of interesting nonlinear problems. We shall not here discuss the Yang-Mills self-duality equations for lack of space ; they effectively are reduced to two-dimensions by considering the anti-self-dual null 2-planes. After reviewing the known connections between the 3 concepts listed above, we shall present the table of internal Lie symmetries of the Poincaré (super)-gravities in various numbers of dimensions. Finally, we shall see that a Kac-Moody group (affine type I) plays important roles as a) transformation group of solutions, b) parameter space where fields take their values, c) phase-space.

+Based on an invited talk given at the Istanbul Conf. on Group Theoretical Methods in Physics, Aug. 1982.

[x]Laboratoire Propre du CNRS, associé à l'Ecole Normale Supérieure et à l'Université de Paris-Sud.

1. The 1968 revolutions

General relativists developed many techniques to find solutions to Einstein's vacuum equations (and to the coupled Einstein-Maxwell system) among others. One of them is to assume the existence of (commuting) Killing vectors, consequently one exhibits Lie groups (finite dimensional) of invariances of the partial differential equations. Sometimes invariance is a manifest consequence of the coordinate reparametrization invariance one started with, for example the Matzner-Misner $SL(2,\mathbb{R})$ invariance of the set of solutions that depend only on 2 of the 4 coordinates. More mysteriously one has to use duality transformations to exhibit the famous Ehlers $SL(2,\mathbb{R})$ for stationary solutions of Einstein's equations ($SU(2,1)$ for electrovacs). In 1968 B. Harrison[1)] formulated the problem of combining several Ehlers groups corresponding to several Killing vectors and of studying the resulting group. This program was really launched three years later by R. Geroch[2,3)] who found the conditions for the compatibility of these groups for two commuting Killing vectors and discovered that they generated an infinite dimensional algebra : the Geroch algebra.

a) In the years 1967-68, Mathematics and Physics witnessed three revolutions. V. Kac and R. Moody[4)] independently studied special classes of infinite dimensional Lie algebras, these algebras are now named after them and can be defined by a finite number of generators and relations, they are Lie algebras. We shall focus here on one of their subclasses, the "affine" Kac-Moody algebras ; these are singled out among simple $\mathbb{Z}$ graded Lie algebras generated by their subspaces of grade 0, ± 1 and such that the levels (± 1) are contragredient irreducible representations of the subalgebra of grade 0 by the property of "finite growth": the dimension of the ith level grows like $|i|^{d-1}$, d is the Gel'fand-Kirillov dimension (more precisely one has excluded the algebras associated with the pseudo-groups of Cartan). The affine algebras have a Gel'fand-Kirillov dimension 1 (for a group of diffeomorphisms the Gel'fand-Kirillov dimension is the dimension of the manifold on which the group acts). Among the affine algebras, one distinguishes three types among which only the first one will concern us. It corresponds to the (universal) central extensions of loop algebras by a one-dimensional center c sitting at level 0 (plus sometimes another generator d that "measures the grade" of the others : $[d, x^i] = i\, x^i$ for x^i at level i). The loop algebras are defined in turn as tensor products $\mathcal{G} \otimes \mathbb{R}[t, t^{-1}]$ (polynomials in one variable and its inverse), they are approximately the Lie algebras of gauge groups on the circle $t = e^{i\theta}$ and $\mathcal{G}$ is the Lie algebra of a finite dimensional Lie group G. Let us give a concrete example $\mathcal{A}_1^{(1)}$ is the affine algebra associated with $\mathcal{G} = \mathcal{A}_1 \equiv sl(2,\mathbb{R})$, it is defined by the 17 relations (i, j = 1,2), $A_{ij} = \begin{pmatrix} 2 & -2 \\ -2 & 2 \end{pmatrix}$:

$$[h_i, h_j] = 0,\ [h_i, e_j^{\pm}] = \pm A_{ij} e_j^{\pm},\ [e_i^+, e_j^-] = \delta_{ij} h_i,\ (ad\, e_i^{\pm})^{-A_{ij}+1} e_j^{\pm} = 0 \quad i \neq j \qquad (1)$$

Or equivalently, it is described by its generators τ_A^i (i integer >, =, or < 0 ; A = 1,2,3) and c :

$$\begin{cases} [\tau_A^i, \tau_B^j] = 2\,\varepsilon_{AB}{}^C\, \tau_C^{i+j} + (i\, K_{AB}\, \delta_{i,-j})\, c \\ c : \text{central} \quad , (c \equiv h_1 + h_2) \end{cases} \qquad (2)$$

$K_{AB} = (+2, -2, +2)$ is the Killing form and ε_{ABC} the antisymmetric tensor (indices are lowered with $K_{AB}/2$).

The matrix A_{ij} is called the Cartan matrix and is strictly positive for finite simple Lie algebras, positive semi-definite (one null eigenvalue) for affine algebras, and indefinite for "hyperbolic" algebras (like E_{10}).

If one thinks of n as the number of the Fourier component of a map from the circle to $sl(2\,\mathbb{R})$, eq. (2) is a current algebra where the second term on the right-hand side is the Schwinger term.

Another convergence with Physics arose from ref. [5], where a particularly simple linear representation of $\mathcal{A}_1^{(1)}$ was constructed, it is called the basic representation and some of its matrix generators reminded H. Garland of the dual string model. A very precise connection was established in ref. [6]. But in 1968 the dual models are just being discovered and the string picture will appear only two years later.

b) The dual models are at first S-matrix theories for an infinite number of narrow resonances lying on straight Regge trajectories and satisfying the condition of physical duality[7]. The latter condition is a natural consequence of the string interpretation of the Veneziano model, so we shall quickly review some puzzling features of this model. The free string model is obtained by quantizing the Nambu-Goto action :

$$L = -\frac{1}{2\pi\alpha'}\int_0^\pi d\sigma \int d\tau \sqrt{(\dot{x}\cdot x')^2 - \dot{x}^2 x'^2} \qquad \dot{x} = \frac{dx}{d\tau}\ ,\ x' = \frac{dx}{d\sigma} \qquad (3)$$

where $x^M(\sigma, \tau) = q_0^M + 2\alpha' p^M \tau + i\sqrt{2\alpha'} \sum_{n=1}^{\infty} \frac{(a_n^M \cos n\sigma\, e^{-in\tau} + h.c.)}{\sqrt{n}}$

$M = 1, 2, \ldots, D \qquad \eta^{MN} = (-1, +1, \ldots, +1)$

The state space is $\mathcal{L}^2(\mathbb{R}^D) \otimes \mathcal{F}$, $\mathcal{F}$ Fock space of the oscillators a_n^M but the metric is not positive definite because of the anomalous commutation relation $[a_n^0, a_m^{0+}]_- = -\delta_{n,m}$. It is necessary to use the gauge freedom (reparametrization invariance of the string) to eliminate the ghosts, for example one can ignore some longitudinal modes and use the light cone gauge : $x^+ = \tau$, σ its conjugate harmonic coordinate; but then one can satisfy the Lorentz naive commutation relations only for D-2=24. This number 24 appears in many places in Mathematics : in the theory of modular functions the η function of Dedekind is given by $\eta(q) = q^{1/24} \prod_{n=1}^{\infty} (1 - q^n)$, $q = e^{2i\pi z}$ and η^{24} is a weakly modular form of weight 12, in particular $\eta^{24}\left(\frac{a\ z+b}{c\ z+d}\right) = (c\ z+d)^{12}\ \eta^{24}(z)$; the Leech lattice is a close -packed lattice in 24 dimensions closely related to many sporadic finite groups... In fact $\eta^{-24} = (q - 24q^2 + 252q^3 - 1472q^4 + \ldots)^{-1}$ $= \frac{1}{q} + 24 + 324q + \cdots$ is the mass partition function of the open string (see ref. 8) for example), it contains 1 tachyon, the 24 degrees of freedom of the massless vector A_M etc. Let us now consider the spinning open strings of Ramond, Neveu and Schwarz moving in 10 dimensions. The partition function of the Fermi sector (of G parity +1) is given by: $8\prod_1^{\infty}(1+q^{2m})^8(1-q^{2m})^{-8}$ and for the Bose sector (G=+1) by $\frac{1}{2q}\prod_1^{\infty}(1-q^{2n})^{-8}\left[\prod_1^{\infty}(1+q^{2m-1})^8 - \prod_1^{\infty}(1-q^{2m-1})^8\right]$ their equality was the first evidence for the D=10 supersymmetry of this model[9)]. It was actually discovered by Jacobi, it is the famous algebraic relation between theta functions $\Theta_3^4 = \Theta_2^4 + \Theta_4^4$, which was recently applied to finite group theory in ref. 10). Let us note that the root lattice of E_8 appears there too. The group E_8 was already known to be related to the massless closed string sector of the spinning string model[11)]. We shall explain this below.

The interactions of strings are obtained by splitting and joining them with a coupling constant g, but the S matrix for the sector with mass zero can also be obtained from a local field theory in the limit of small slope $\alpha' \to 0$ (at least in the classical approximation). Open strings dressed with some internal quantum numbers give Yang-Mills theory (supersymmetric Yang-Mills theory, respectively). Some closed strings interact like gravitons and lead to the construction of N=8 supergravity in four dimensions.

c) But let us return to 1967-68 and recall the 3rd revolution.

Stimulated by numerical results, a group of Applied-Mathematicians defined the concept of soliton and invented a method of resolution of non-linear partial differential equations : the method of Inverse Scattering[12]. It is possible to solve special differential systems by a sequence of linear operations (algebraic operations in the n soliton sector). The main tool is the Lax Pair, one must exhibit two differential operators L and P depending on some potential function $u(x,t)$ (the method applies fully in 1+1 dimensions) such that $\left[\frac{\partial}{\partial t} - P, L\right]_- = 0$ is the nonlinear system to be solved. The first model to be studied was the Korteweg de Vries equation for which L is nothing but the Schrödinger operator $-\frac{\partial^2}{\partial x^2} + u$
The idea for localized potentials on the real line is to consider the scattering problem $L\psi = \lambda\psi$ and to change variables from $u(x,t_0)$ to an equivalent set of scattering data for which the time evolution is simple and solvable in terms of elementary functions. For a later time t_1, $u(x,t_1)$ is reconstructed from its scattering data hence the Cauchy problem is solved. The case of almost periodic potentials requires a more sophisticated mathematical apparatus but again the problem can be completely solved (in terms of Abelian functions) for the so-called finite gap potentials, in particular stationary periodic solutions are in this class[13]. Besides the n-solitons and the finite gap potentials a third class of solutions can be explicitly computed : the similarity solutions and their generalizations the "multiphase" solutions of some non linear (2 dimensional) equations[14]. The isospectral problem is replaced by an isomonodromy deformation. It can be a Fuchsian system of "regular singular" ordinary differential equations in one auxiliary parameter which becomes the scattering parameter in the "Garnier" limit and one recovers the previous results. It can be also an irregular singular system (with multiple poles) ; they appear naturally in the study of self-similar solutions.

The main advantage of the last approach is the appearance of an auxiliary space, namely the space of the scattering parameter. As a consequence its generalization leads to solutions of non linear partial differential equations in higher dimensional space-times as well. Typically one has to solve a matrix Riemann-Hilbert discontinuity problem, and this can be done using linear (singular in general) integral equations. The first advocates of the matrix Riemann problem for non linear partial differential equations in recent times were Zakharov and Shabat (around 1976)[15]. Their procedure is summarized in ref. 16), it is in fact an infinite dimensional Bäcklund group of transformations. One starts with one solution of the non-linear problem (the trivial one if it exists) and one generates the others by simply changing the discontinuity matrix (the Riemann data) ; these data are subject to constraints corresponding to the particular class of solutions one is studying.

In particular one is not restricted to fast decreasing potentials and we shall see that σ -models or gravitational plane waves lead to non trivial asymptotics. We should mention for completeness 3 connections between integrable systems and the other two revolutions. In ref. 17) A. M. Polyakov proposed a modified quantization procedure for the string model and found a Liouville theory on the two-dimensional string ; the Liouville scalar field equation must be integrated by Inverse Scattering Procedure[18)]. On the other hand the Kyoto group[19)] used the vertex operator of ref. 5) (similar but not identical to the vertex operator of the string model) to study the invariance groups of solutions of "exactly" integrable systems in 2+1 and 1+1 dimensions and they discovered Affine Kac-Moody groups in the second case. A more abstract analysis of integrable systems using the orbits of Affine Kac-Moody groups had been developed previously in ref. 20) following M. Adler and others. Ref. 21) was in fact very stimulating for ref. 22).

2. The dimensional reductions

One failure of the conventional dual model was its dimensionality : the states of the strings move in 1+9 dimensions. A few people however were not discouraged and tried to make contact with our 1+3 dimensional experience. Two roads were explored : the small slope limit ($\alpha' \to 0$) which leads to effective Yang-Mills theory and Einstein gravity (plus some matter fields)[9)], and the dimensional reduction which was revived by J. Scherk and J.H. Schwarz in order to explain the disappearance of 6 dimensions without spoiling the nice duality properties. In fact the 10 dimensional model admits one (spinorial) local supersymmetry on the string (in two dimensions) but the set of states chosen in ref. 9) and corresponding to the partition functions given above admits a global supersymmetry in 10 dimensions (the Weyl-Majorana spinor has 16 components) ; the zero slope limit of this model is the supersymmetric Yang-Mills theory in 10 dimensions and after reduction to four dimensions it is the N=4 version of the same theory. It was a natural conjecture that the closed string sector which contains gravitation in the bosonic sub-space contains supergravity as well. The problem was to construct the theory in 10 dimensions and to take its small slope limit after which only a finite number of massless fields survive. We refer the reader to ref. 23) for a more detailed discussion. The construction was done actually by assuming local supersymmetry not only in 10 dimensions (with 2x16 generators hence the name N=2) but E. Cremmer, J. Scherk and the author assumed the existence of a supergravity theory in 11 dimensions. It reduces to the other one when one coordinate becomes ignorable. W. Nahm had shown that a global supersymmetry multiplet with the right spins existed in 11 dimensions. The eleven dimensional supergravity does exist and its bosonic part reads:

$$S_{11}^{Bose} = -\int_{\mathcal{M}_{11}} \frac{e}{4} R + \frac{e}{48} F_{MNPQ} F^{MNPQ} - \frac{2e}{(12)^4} \varepsilon^{PQ\cdots Z} A_{PQR} F_{STUV} F_{WXYZ} \quad (4)$$

where $e = \det (e_M{}^A)$

$F_{MNPQ} = 4\, \partial_{[M} A_{NPQ]}$ and $\varepsilon^{PQ\cdots}$ is the totally antisymmetric tensor. The last term has been written so as to suggest a 12 dimensional structure with presumably 2 time directions, which deserves further study. It took a few months of hard work to reduce (4) to four dimensions. More precisely Cremmer and the author rewrote the equations of motion of S_{11} for solutions that depend only on 1+3 coordinates as the Euler-Lagrange equations of another action S'_4 that is manifestly symmetric under the group SO(8). This invariance was expected for N=8 supergravity in 4 dimensions. Naïve suppression of 7 coordinates in S_{11} leads to a dissymmetric Lagrangian L_4 with only $GL(7)$ symmetry. This difficulty is best illustrated in the example of gravitation theory reduced from 4 to 3 dimensions (see ref. 22) and references therein).

Let us consider Einstein's action :

$$\tilde{S}_4 = -\int_{\mathcal{M}_4} \frac{e}{4} R \quad (5)$$

If x^3 is cyclic we can show that the equations of motion can be obtained from the naïve reduction of $\tilde{S}_4$: $(\Delta = g_{33}\ ,\ A_\mu = e_\mu{}^3 / e_3{}^3)$

$$\tilde{S}_3 = \int_{\mathcal{M}_3} -\frac{e}{4} R - \frac{e}{16} F_{\mu\nu} F^{\mu\nu} \Delta^2 - \frac{e}{8} \frac{\partial_\mu \Delta\, \partial^\mu \Delta}{\Delta^2} \quad (6)$$

Ehlers however discovered a hidden SO(2) invariance of the set of three dimensional solutions (strictly speaking in this generality this result is due to Papapetrou). We shall exhibit it by defining a dual potential by the Bäcklund transformation :

$$\partial_\mu B = \varepsilon_\mu{}^{\nu\rho}\, \partial_\nu A_\rho \tag{7}$$

This is called a duality transformation, let us note that it is a special instance of Bäcklund transformation, namely eliminating A_ρ leads to the equation of motion for B and conversely. The equations for B and the other fields are Euler-Lagrange equations for

$$\widetilde{S}'_3 = -\int_{m_3}\left(\frac{\varepsilon}{4}R + \frac{\varepsilon}{8}\frac{(\partial\Delta)^2+(\partial B)^2}{\Delta^2}\right) \tag{8}$$

and $\widetilde{S}'_3$ can be seen to be invariant under $z \to \frac{az+b}{cz+d}$, $z = B+i\Delta$ (a,b,c,d real). Thus $SL(2,\mathbb{R}) \supset SO(2)$ is an invariance of the set of classical solutions of $\widetilde{S}_3$ or $\widetilde{S}'_3$ (they have the same solutions locally). The extension of the SO(7) subgroup of GL(7) to SO(8) involves analogous dualities, S'_4 is even invariant under SL(8).

We mentioned the Matzner-Misner group $SL(2,\mathbb{R})$ in the introduction, its origin is better understood. Quite generally if one starts with a theory that is coordinate reparametrization invariant in D dimensions and one assumes (D-d) commuting Killing vectors, the dimensionally reduced action S_d is invariant under SL(D-d) and sometimes under GL(D-d) for example GL(7) (resp. GL(2)) if (D,d) = (11,4) (resp. (4,2)). SL(D-d) is the group of (internal) volume preserving linear changes of coordinates, it is an internal symmetry after "trivial" dimensional reduction.

In fact the group of bosonic internal invariances of the equations of motion of supergravities is always larger than the expected SL(D-d). For example N = 8 supergravity solutions in four space-time dimensions are exchanged by a non linear action of the non compact group $E_7(+7)$ with maximal compact subgroup SU(8). This theory can be reduced to 3 dimensions, where it is associated to an E_8-invariant action. It is a fascinating question to find the connection with the E_8 mentioned in part 1 (see 6) and the contribution of I. Frenkel to the same volume as 23)). If one reduces it further to 2 dimensions one is then studying (super-) gravitational plane waves or stationary axially symmetric solutions. It is a considerable surprise to discover the appearance of infinite dimensional Lie groups of symmetry which exchange classical solutions among themselves.

We began this lecture by recalling the history of the Geroch algebra. This algebra has been studied extensively since but was not recognized for what it is namely $sl(2\,\mathbb{R})^{(1)}$ until January 1981, see 24). This recognition was suggested by the rules of group disintegration[22)23)] and in particular the remarkable appearance of E_{11-d} invariance groups after reduction of S_{11} to d dimensions. It is not absurd to set d = 2 because E_9 is defined by its Dynkin diagram as $E_8^{(1)}$. This E_9 invariance has not yet been checked explicitly but its N = 0 analogue $sl(2\,\mathbb{R})^{(1)}$ (N is the number of supersymmetries) was studied extensively in the last two years and we shall now describe these works. So we have just seen one of the bridges , dimensional reduction, it connects dual models and supergravities. Further dimensional reduction connects supergravities and integrable systems as well as Kac-Moody algebras.

3. The scattering parameter.

When we first mentioned the Geroch algebra we were in fact describing the finite "presentation" (1), this infinite algebra is generated by the two non-commuting finite Lie algebras of Ehlers and Matzner-Misner. From a computational point of view it might be advantageous to use the description (2) and the loop subalgebra. Actually nobody seems to consider the full $A_1^{(1)}$, because the central charge is a Weyl transformation[24)]. The metric is assumed to be block diagonal :

$$g = \begin{pmatrix} g_{\mu\nu} & 0 \\ 0 & g_{ij} \end{pmatrix} \qquad \begin{matrix} \mu,\nu = 0,1 \\ i,j = 2,3; \end{matrix}$$

The Weyl scaling leaves g_{ij} inert (g_{ij} contains the degrees of freedom of the graviton) and maps $g_{\mu\nu}(x^\rho) \to w\, g_{\mu\nu}(x^\rho)$. The advantages of the loop algebra are first of all its compactness : generating functions are easier to handle than individual operators; secondly the fact that it exponentiates nicely to a loop group, and thirdly a relation $\lambda = f(t)$ between the scattering parameter λ and the loop parameter t .

To develop this idea we must recall some earlier works. In ref. 25) a Lax pair was discovered for gravitation reduced from 4 to 2 dimensions. D. Maison started from Geroch's paper[3)] and used the analogy with the σ-model and the Sine-Gordon equation to find a Lax pair. The scattering parameter appeared however in a complicated fashion and the asymptotics of the scattering problem was non standard. Belinski and Zakharov found another Lax pair, both pairs are related (Maison unpublished). The latter authors avoided the problem of asymptotic behaviour at infinity by using the Riemann problem and computed 1 and 2 soliton solutions ;

in the absence of "radiation" these solutions can be constructed algebraically once one knows the solution of the Lax problem for the particular (trivial for example) background metric solution one starts with. For example one can obtain the Kerr solution as a stationary 2 soliton solution from ordinary flat space time. Two peculiarities of the Belinski-Zakharov pair of operators are that they possess moving poles $\lambda_i(x)$ and that they involve the differential operator with respect to the scattering parameter : $\partial/\partial\lambda$. It turned out that this solution generating procedure leads frequently to singular solutions but the similarity with the Geroch algebra manipulations was revealed by the powerful computations of 26) and 27). Hauser and Ernst found independently the Riemann-Hilbert method ; they started from the Geroch algebra and exponentiated the infinitesimal transformations by this method. Their work culminated with a proof of a conjecture of Geroch (with a mild technical restriction) namely the transitivity of the Geroch group on the set of stationary axially symmetric solutions of Einstein's equations (see ref. 28) for the asymptotically flat case). So the loop algebra $SL(2,\mathbb{R}) \times \mathbb{R}\left[t,\frac{1}{t}\right]$ acts transitively on classical solutions.

In supergravities however we have seen that Dynkin diagrams grow naturally by dimensional reduction and the presentation (1) is more natural than (2). To be more precise we shall study the symmetries of N = 6 supergravity as an example. In four dimensions the equations of motion (and the set of solutions) are invariant under $SO^*(12)$, it has been partially checked that their reduction to 3 dimensions is $E_7(-5)$ invariant and it is known that the theory is $SU^*(6)$ invariant in 5 dimensions[29]. Let us draw the Dynkin diagram of $E_7(-5)$:

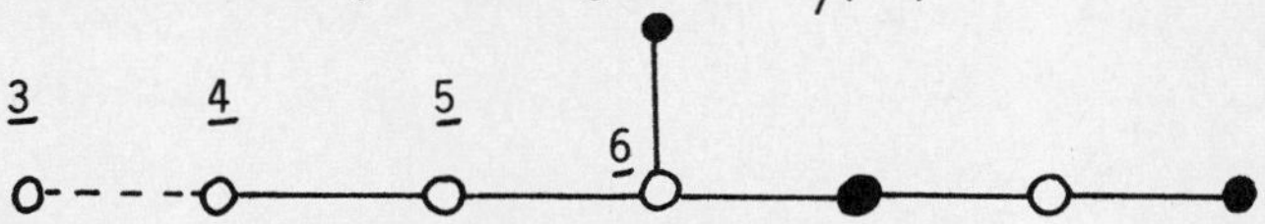

Adding the dot 3 defines the Dynkin diagram of $E_7(-5)^{(1)}$. The black circles correspond to the replacement of h_i, e_i, f_i at those locations by $h'_i = ih_i$, e'_i, f'_i generating su(2) instead of $sl(2,\mathbb{R})$, this notation is due to Tits and Satake. In 3 dimensions all degrees of freedom are spinless bosons (one sets to zero all Fermi fields) ; they form in fact the generalized σ-model for the coset space $E_7(-5)/SO(12) \times SO(3)$ i.e. maps from space time to this coset space. The action is invariant under left action of a global (= space time independent) $E_7(-5)$ element. Somehow the graviton must have been converted into some of these fields. One rule has been checked case by case namely the disintegration of $E_7(-5)$ into the internal invariance group in d dimensions times SL(d-2) ; it has been argued that SL(d-2)/SO(d-2) is precisely the graviton field : a traceless symmetric transverse matrix in the linearized approximation. More precisely the splittings always occur in the "regular" form (preserving Cartan subalgebras) and in d dimensions the Dynkin diagram of the subgroup is obtained by adding the extra root of $E_7^{(1)}$ and taking away the root labelled <u>d</u> . Let us note that if N = 6 supergravity is the dimensional reduction of a 6 dimensional theory, $E_7(-5)$ must contain SL(4) and it does (the diagram of SL(4) is o——o——o). This 6-dimensional theory awaits construction.

The appearance of $G^{(1)}$ in this discussion with G the internal symmetry group in 3 dimensions seems to be valid for N = 0,1,...,8; and the Matzner-Misner SL(2,$\mathbb{R}$) will always appear upon reduction to 2 dimensions. I believe that it will be easier to really understand the appearance of the affine algebras from dimensional reduction than from the loop algebra approach ; it is well known that Lax pairs are always guessed and that the introduction of a spectral parameter is not straightforward. The study of supergravities suggests many conjectures about this : for example it has been shown that there is a Lax-Pair for Einstein Maxwell plane waves as well, but this is the bosonic sector of N = 2 supergravity. N extended supergravities $3 \leqslant N \leqslant 8$ should admit analogous pairs and associated loop algebras $G \times \mathbb{R}\left[t , \frac{1}{t}\right]$. In 24) another conjecture was formulated ; it is based on the observation of a σ -model structure SL(2 $\mathbb{R}$)/SO(2) inside the gravitation problem in 2 dimensions and on the existence of an infinite non-Abelian algebra of non local charges for the SO(3)/SO(2) (or SL(2 R)/SO(2)) σ-model[30]. It is natural to expect, in this S^2 σ-model, a loop algebra (the central charge does not act on g_{ij}) of on-shell symmetries and a relation with the charges of ref. 30).

In fact a stronger result was subsequently proved in ref. 31) and ref. 32) for pure chiral models in the group G, namely the existence of an off-shell symmetry under the loop algebra $G \times \mathbb{R}\left[t , \frac{1}{t}\right]$ such that the associated Noether currents are combinations of the non-local currents generalizing 30). An off-shell group is important for the quantum theory, it was already partially conjectured in 33). In four dimensions such a current had also been studied in 34) for the usual Maxwell duality between electricity and magnetism. It is an important problem to extend these formulas to E_7 and N = 8 supergravity in order to have a symmetry of the full set of field configurations. The connections between the Riemann-Hilbert approach and the transformation group pictures were studied most carefully in ref. 35).

We would like to distinguish three possible roles for groups of "symmetries". We have discussed at length symmetries of the equations of motion and symmetries of the action. A group can also be the space in which fields take their values : in pure σ models, or in generalized σ models G/H, when one describes the coset space as a fiber bundle over itself without choosing a section, one then has global G invariance and gauge (local) H invariance. For dimensions higher than 3 the group G plays both roles and one can write manifestly covariant equations of motion. In dimension 2 however the first attempt at finding manifestly symmetric equations under $G^{(1)}$ was ref. 24) ; some progress has been made in particular in identifying the finite co-dimension subalgebras of $G^{(1)}$. Clearly one must enlarge $G^{(1)}$: for example $A_1^{(1)}$ does not contain the full Matzner-Misner $GL(2,\mathbb{R})$ but only its $SL(2,\mathbb{R})$ subgroup (work in progress). Finally the full phase space can be a group orbit under the co-adjoint action [20),21)]; the group acts by canonical transformations. This last point of view has attracted some attention very recently and led to some surprises[36)]. It is the most powerful approach and it deserves more study.

In conclusion we recall the importance of seeing the scattering parameter as a third coordinate, for example in the three wave problem (ref. 37)), and our conjecture that $G^{(1)}$ (or even the over-extended group $G^{\hat{}\hat{}}$ obtained by adding one more root to the extended Dynkin diagram of G) could be an invariance group in any dimension, $G^{(1)}$ being an internal symmetry in dimension 2 and $G^{\hat{}\hat{}}$ in dimension 1 (see refs. 22)23)).

Acknowledgements

It is a pleasure to thank C. Cosgrove for useful discussions and I. Frenkel, V. Kac, W. Nahm and J. Tits for sharing their insights of affine Lie algebras.

References

1) B.K. Harrison, J. Math. Phys. 9 (1968) 1744.
2) R. Geroch, J. Math. Phys. 12 (1971) 918.
3) R. Geroch, J. Math. Phys. 13 (1972) 394.
4) V. Kac, Funct. An. and Appl. 1 (1967) 82 , and Math. USSR Izvestija 32 (1968) 1271 ; R. Moody, Bull. Am. Math. Soc. 73 (1967) 217, and J. of Algebra 10 (1968) 211.
5) J. Lepowski and R. Wilson, Comm. Math. Phys. 62 (1978) 43.
6) G. Segal, Comm. Math. Phys. 80 (1981) 301 ; I. Frenkel and V. Kac, Inventiones 62 (1980) 23.
7) G. Veneziano, Nuovo Cim. 57A (1968) 190.
8) W. Nahm, Nucl. Phys. B114 (1976) 174 ; J. Scherk, Rev. Mod. Phys. 47 (1975) 123.
9) F. Gliozzi, J. Scherk and D. Olive, Nucl. Phys. B122 (1977) 253.
10) I. Frenkel, J. Lepowski and A. Meurman, talk at the Chicago SIAM Workshop, July 1982.
11) E. Cremmer and B. Julia, Nucl. Phys. B159 (1979) 141.
12) C.S. Gardner, J.M. Greene, M.D. Kruskal, R. M. Miura, Phys. Rev. Lett. 19 (1967) 1095.
13) B.A. Dubrovin, V.B. Matveev and S.P. Novikov, Russian Math. Surveys 31 (1976) 59.
14) For a review see H. Flaschka and A.C. Newell, Comm. Math. Phys. 76 (1980) 65.
15) V.E. Zakharov and A.B. Shabat, Funct. Anal. and Appl. 13 (1979) 13.
16) For a review of 2 dimensional problems see A.V. Mikhailov, CERN preprint TH.3194,(1981).
17) A.M. Polyakov, Phys. Lett. 103B (1981) 207.
18) J.L. Gervais and A. Neveu, Nucl. Phys. (to appear).
19) E. Date, M. Jimbo, M. Kashiwara and T. Miwa, RIMS 362 (July 1981) ; see also G. Segal and G. Wilson, Oxford preprint (in preparation).
20) A.G. Reyman and M.A. Semenov-Tian-Shansky, Inventiones Mat. 63 (1981) 423 ; and V.G. Drinfeld and V.V. Sokolov, Doklady Acad. Nauk. USSR 258 (1981) 457.
21) M. Adler, Inventiones Mat. 50 (1979) 219.
22) B. Julia, in Superspace and Supergravity, ed. S. Hawking and M. Rocek, Cambridge 1981, p. 331 (C.U.P.).
23) B. Julia, "Kac-Moody Symmetry of Gravitation and Supergravity Theories" (to be published by A.M.S. in Proc. Chicago Meeting, July 1982).
24) B. Julia, "Infinite Lie Algebras in Physics", Proc. 5th Johns Hopkins Workshop on Particle Theory, Baltimore, May 1981, p. 23.
25) D. Maison, J. Math. Phys. 20 (1978) 871 ; V.A. Belinsky and V.E. Zakharov, Sov. Phys. JETP 48 (1978) 985 and 50 (1979) 1.
26) W. Kinnersley and D.M. Chitre, J. Math. Phys. 18 (1977) 1538.
27) I. Hauser and F.J. Ernst. See for example a review by the first author in the Proc. Coyococ 1980 Conference of this series, Lecture Notes in Physics 135, Springer.
28) B.C. Xanthopoulos, J. Math. Phys. 22 (1981) 1254.
29) E. Cremmer, see ref. 22).
30) M. Lüscher and K. Pohlmeyer, Nucl. Phys. B137 (1978) 46.
31) L. Dolan, Phys. Rev. Lett. 47 (1981) 1371.
32) Wu Yong-Shi, Nucl. Phys. B211 (1983) 160.
33) R.P. Zaikov, Dubna preprints E2-80-118, 197 and with B.L. Markowsky E2-80-654 ; L. Dolan and A. Roos, Phys. Rev. D22 (1980) 2018.
34) S. Deser and C. Teitelboim, Phys. Rev. D13 (1976) 1592.

35) K. Ueno and Y. Nakamura, Phys. Lett. 117B (1982) 208 ; C. Cosgrove, J. Math. Phys. 23 (1982) 615.
36) A.C. Davies, P.J. Houston, J.M. Leinaas and A.J. Macfarlane, CERN preprint TH 3372.
37) V.E. Zakharov, in Lecture Notes in Physics 153 (Springer) p. 190.

Ref.TH.3499-CERN

FREE GRADED DIFFERENTIAL SUPERALGEBRAS *)

P. van Nieuwenhuizen +)

CERN -- Geneva

ABSTRACT

Two theorems of D. Sullivan on the structure of differential algebras are extended to the algebras mentioned in the title and a few applications of non-trivial cohomology classes to the gauging of extended groups are given. The applications are due to R. D'Auria, L. Castellani, P. Fré, F. Giani, K. Pilch and the author, and are discussed in more detail in the author's talk at the 1982 Chicago Meeting of the American Mathematical Society on group-theoretical methods in physics.

*) Invited talk at the August 1982 Istanbul meeting on group-theoretical methods in physics.

+) On leave from the Institute for Theoretical Physics, State University of New York at Stony Brook, N.Y.

Ref.TH.3499-CERN
20 December 1982

1. INTRODUCTION

Differential algebras have become an important tool in the construction of local Lagrangian field theories. They define the group or supergroup or even generalized supergroup ("groups" with arbitrary antisymmetric tensor gauge fields, see below), from which one hopes to obtain a local Lagrangian field theory. The task of obtaining a local Lagrangian field theory from a given (extended super) group is usually called "the gauging of groups". The first results were obtained by MacDowell and Mansouri [1] and by Chamseddine and West [2] for simple supergravity, by Townsend and the author for N = 2 extended supergravity [3] and by Kaku, Townsend and the author for simple conformal supergravity [4]. In all these cases the base manifold was ordinary Minkovski spacetime and one had to impose certain constraints on the curvatures by hand (just like in superspace supergravity), something which really should come out of the method by itself.

The meaning of these constraints became clear [4,5] and in particular in extended conformal supergravities new constraints were deduced in a systematic fashion [6].

There exists another method, the so-called "group manifold approach", according to which one takes an extended supergroup and following general rules, one turns the crank, without imposing constraints. Then either the program stops, or out comes *the action* of a local Lagrangian field theory. This program was originally proposed by Ne'eman and Regge [7], while it has been developed in particular by D'Auria and Fré [8]. At this point the program is completely geometrical as far as the construction of the action is concerned, but the construction of the transformation rules under which the action is invariant is incomplete in the following sense. When no auxiliary fields are known, the transformation rules follow directly only when all fields are on-shell (= satisfy their equations of motion) but to find their off-shell form one must add to them arbitrary terms proportional to equations of motion and fix those by requiring that the action be invariant. (This is not exactly the same as one does in the so-called component approach because here the action is already known.) When one knows the auxiliary fields, on the other hand, the derivation of the transformation rules is again completely geometrical (it is then equivalent to the independence of the action from the particular hypersurface M chosen in the group manifold on which the action is defined [9]).

Below we will discuss the general form of the differential algebras used; in particular, we shall extend two theorems due to D. Sullivan [10] on ordinary differential algebras to differential superalgebras. Then we shall show examples of differential algebras which are used in applications. This discussion is based on work done with D'Auria, Fré, Castellani, Giani and Pilch [11]. For a review of how to obtain actions from a differential algebra we refer to two sets of lectures [9], while the issue of the transformation laws in the group manifold approach will be discussed in a forthcoming article by D'Auria, Fré, Townsend and the author [12]. What follows is self-contained, but it is only an element in the larger program of the group manifold approach.

2. DEFINITION OF FREE GRADED DIFFERENTIAL SUPERALGEBRAS

The differential algebras we consider are graded superalgebras. This means that they contain forms which have a grade k ($1 \leq k < \infty$) and are bosonic or fermionic (sometimes one also uses the words even and odd). Thus we have really a $Z_\infty \otimes Z_2$ grading. The algebras are generated by a finite number of generators. In an algebra one can add and multiply. We will need to add bosonic p-forms only to bosonic p-forms (idem for fermionic p-forms), but we will multiply any form with any other form. Multiplication is denoted, as usual, by the wedge symbol, and the product of a p-form with a q-form is a (p + q) form, which is bosonic when the p-form and q-form are both bosonic or fermionic, and which is fermionic if the p-form is bosonic (fermionic) while the q-form is fermionic (bosonic). The field over which the algebra is defined is the real number system. Obviously, multiplication of a given p-form by a real number does not change its grade nor its bose or fermi property.

The (anti)commutation relations of these forms are as for ordinary forms, except that one gets an extra minus sign when two fermionic forms are interchanged. Denoting a bosonic p-form by b_p and a fermionic p-form by f_p, we thus have

$$b_p \wedge b_q = (-)^{pq} b_q \wedge b_p \,, \quad b_p \wedge f_q = (-)^{pq} f_q \wedge b_p \,, \quad f_p \wedge f_q = (-)^{pq+1} f_q \wedge f_p \tag{1}$$

Except for these (anti)commutation properties, there are no other relations between the forms, so that we are dealing with free differential algebras. (In general relativity one uses these days also non-free differential algebras, see for example Harrison's talk at the Marcel Grossman meeting in Shanghai, 1982.)

The differential operator d satisfies Leibniz' rule dd = 0, and maps a bosonic p-form into a bosonic p + 1 form, and a fermionic p-form into a fermionic p + 1 form. When in Leibniz' rule d passes a form it acts as if it were a bosonic 1-form; for example

$$d\left(x_p \wedge x_q\right) = (dx_p) \wedge x_q + (-)^p x_p \wedge dx_q \tag{2}$$

independently of whether x_p is bosonic or fermionic.

Let us stress that we do not consider the fermionic forms as forms with a negative grade: all forms have positive grade p with $p \geq 1$. This will be useful when we prove certain theorems by induction. For algebras with 0-forms see [15].

AN EXAMPLE OF A DIFFERENTIAL ALGEBRA

Consider the following differential algebra

$$\begin{aligned}
d\omega^m{}_n &= -\omega^m{}_k \wedge \omega^k{}_n && (m,k=0,3) \\
dV^m &= -\omega^m{}_n \wedge V^n + \bar{\psi}\gamma^m \wedge \psi && (\omega^{mn} = -\omega^{nm}) \\
d\psi^a &= -\tfrac{1}{4}\,\omega^{mn}(\gamma_{mn})^a{}_b \wedge \psi^b && (a=1,4) \\
dA &= \bar{\psi}\gamma_{mn} \wedge \psi \wedge V^m \wedge V^n
\end{aligned} \tag{3}$$

The $\omega^m{}_n$ and V^m are bosonic 1-forms, the ψ^a are fermionic 1-forms and A is a bosonic 3-form. The γ^m are Dirac matrices satisfying

$$\{\gamma^m, \gamma^n\} = 2\eta^{mn} \quad , \quad \eta^{mn} = (-,+,+,+) \tag{4}$$

and $\bar{\psi} = \psi^T C$ where $C\gamma^m C^{-1} = -(\gamma^m)^T$ and $C^T = -C$. (For a detailed discussion of the charge conjugation matrix C in arbitrary dimensions, Majorana spinors, etc, see Ref. [13]).

These forms can be interpreted as

$$\omega^{mn} = dx^\Lambda \omega_\Lambda{}^{mn} \, , V^m = dx^\Lambda V_\Lambda{}^m \, , \psi^a = dx^\Lambda \psi_\Lambda{}^a \tag{5}$$

where x^Λ are either Minkovski coordinates or group-manifold coordinates or superspace coordinates. The x^Λ and $\omega_\Lambda{}^{mn}$ are bosonic or fermionic such that ω^{mn} is always bosonic. We shall never need to use x^Λ and dx^Λ but only work with forms; thus our results apply equally well to the group manifold, to superspace or to Minkovski space.

The differential algebra in (3) should be *consistent*. By this we mean that since dd = 0, (if $d = dx^\Lambda \partial_\Lambda$, dd = 0 follows) also d acting on the right-hand side in (3) must vanish. For ordinary Lie algebras in terms of forms

$$db_1^a = f^a{}_{bc} \, b_1^b \wedge b_1^c \tag{6}$$

this is equivalent to the Jacobi identities for $f^a{}_{bc}$, but for extended Lie algebras (with p-forms where $p \geq 2$) this is a generalization of the Jacobi identities to the case of "extended groups".

The consistency of (3) can be shown as follows. For $d\omega^m{}_n$ one finds consistency at once

$$\begin{aligned} &-d\omega^m{}_k \wedge \omega^k{}_n + \omega^m{}_k \wedge d\omega^k{}_n = \\ &\omega^m{}_\ell \wedge \omega^\ell{}_k \wedge \omega^k{}_n - \omega^m{}_k \wedge \omega^k{}_\ell \wedge \omega^\ell{}_n = 0 \end{aligned} \tag{7}$$

Also for $d\psi^a$ there is no complication. In matrix notation

$$-\frac{1}{4} d\omega^{mn}\gamma_{mn}\wedge\psi + \frac{1}{4}\omega^{mn}\gamma_{mn}\, d\psi = \frac{1}{4}\omega^{m}{}_{k}\wedge\omega^{kn}\gamma_{mn}\wedge\psi - \frac{1}{16}\omega^{mn}\wedge\omega^{k\ell}\gamma_{mn}\gamma_{k\ell}\wedge\psi \tag{8}$$

Now $\omega^{mn}\wedge\omega^{k\ell} = -\omega^{k\ell}\wedge\omega^{mn}$ and

$$\frac{1}{2}[\gamma_{mn},\gamma_{k\ell}] = \eta_{nk}\gamma_{m\ell} + 3 \text{ terms due to } m\leftrightarrow n,\ k\leftrightarrow \ell \tag{9}$$

and again consistency follows. For dV^m we get

$$\begin{aligned}
&-d\omega^{m}{}_{n}\wedge V^n + \omega^{m}{}_{n}\wedge dV^n - 2\bar\psi\gamma^m\wedge d\psi = \\
&= \omega^{m}{}_{k}\wedge\omega^{k}{}_{n}\wedge V^n + \omega^{m}{}_{n}\wedge\left(-\omega^{n}{}_{\ell}\wedge V^\ell + \bar\psi\gamma^n\wedge\psi\right) + \frac{1}{2}\bar\psi\wedge\gamma^m\gamma_{k\ell}\,\omega^{k\ell}\wedge\psi \\
&= \omega^{k\ell}\wedge\left(\delta^m_k\,\bar\psi\gamma_\ell\wedge\psi - \frac{1}{2}\bar\psi\gamma^m\gamma_{k\ell}\wedge\psi\right)
\end{aligned} \tag{10}$$

Now $\bar\psi\gamma^m\gamma_{k\ell}\wedge\psi = \delta^m_k\bar\psi\gamma_\ell\wedge\psi - \delta^m_\ell\bar\psi\gamma_k\wedge\psi$ [13], so that also the consistency of $d\psi$ is shown.

The most interesting case is the consistency of dA. In

$$-2\bar\psi\gamma_{mn}\wedge d\psi\wedge V^m\wedge V^n + 2\bar\psi\gamma_{mn}\wedge\psi\wedge dV^m\wedge V^n \tag{11}$$

we can replace d by $\mathcal{D}$ where $\mathcal{D}$ is the SO(3,1) Lorentz covariant derivative because the extra terms cancel. From (3)

$$\mathcal{D}\psi = 0 \quad , \quad \mathcal{D}V^m = \bar\psi\gamma^m\wedge\psi \tag{12}$$

Hence consistency requires that

$$\bar\psi\gamma_{mn}\wedge\psi\wedge\bar\psi\gamma^m\wedge\psi \tag{13}$$

This identity indeed holds; it is equivalent to the identity for gravitino fields (as opposed to forms) which supergravity practitioners know very well [13]

$$\bar\psi_\mu\gamma_{mn}\psi_\nu\bar\psi_\sigma\gamma^m\psi_\rho\,\epsilon^{\mu\nu\rho\sigma} = 0 \tag{14}$$

It is usually proved by laborious Fierz rearrangements, but a more group-theoretical technique exists which is simpler in the important applications of 10 or 11 dimensional models [14].

It is clear from this example why differential algebras form a starting point for the gauging of "extended groups". In d = 11 (d = dimension) supergravity a 3-index antisymmetric tensor appears, in addition to an elevenbein e_μ^m and a gravitino field ψ_μ^a (a = 1, 32, and m,μ = 0, 10 in d = 11). Ordinary Lie algebras could not accommodate $A_{\mu\nu\rho}$; one would begin with, say $A_\mu{}^{\nu\rho}$ (or the 1-form $A^{\nu\rho}$), and a posteriori $A_\mu{}^{\nu\rho}$ should become totally antisymmetric by some mechanism. Forms allow one to start straight away with a totally antisymmetric $A_{\mu\nu\rho}$ (or the 3-form A).

4. BIANCHI IDENTITIES AND COVARIANT DERIVATIVES

We can define curvatures by bringing the right-hand sides of (3) to the left. For example

$$R(\omega)^{mn} \equiv d\omega^{mn} + \omega^m{}_k \wedge \omega^k{}_n \, , \; R(A) \equiv dA - \bar{\psi}\gamma_{mn} \wedge \psi \wedge V^m \wedge V^n \tag{15}$$

The generalized left-invariant forms are those forms for which all curvatures vanish (the classical vacuum). Physical fields are those forms for which the curvatures do not vanish.

Consider a differential algebra with a finite number of generators, and let the subset of p-form generators be labelled by an index a. Thus the generators consist of $(x_1^a, x_2^a, \ldots x_N^a)$. Let the curvatures be given by

$$R(x_k^a) \equiv dx_k^a + \sum \frac{1}{p} C\left(\begin{matrix} a \\ k \end{matrix}\middle| \begin{matrix} \ell_1 \ldots \ell_p \\ a_1 \ldots a_p \end{matrix}\right) x_{\ell_1}^{a_1} \wedge \ldots \wedge x_{\ell_p}^{a_p} \tag{16}$$

where the sum runs over all possible terms and where we assume that the generalized structure constants C have the same symmetry as induced by permuting the various x's in the wedge product. Of course $\ell_1 + \ldots \ell_p = k + 1$ but in principle all possible partitions can occur. This system is assumed to be consistent: if all $R(x_k^a) = 0$ then also $dR(x_k^a) = 0$.

If the $R(x^k_a)$ are non-vanishing, they still satisfy Bianchi identities. Consistency leads to the simple result

$$dR(x^a_k) = \sum C\left(\begin{smallmatrix}a\\k\end{smallmatrix}\middle|\begin{smallmatrix}\ell_1 \ldots \ell_p\\ a_1 \ldots a_p\end{smallmatrix}\right) R(x^{a_1}_{\ell_1}) \wedge x^{a_2}_{\ell_2} \wedge \ldots x^{a_p}_{\ell_p} \tag{17}$$

This defines the covariant derivative ∇ in the adjoint representation of the extended group, and in particular

$$\nabla R(x^a_k) \equiv 0 \quad \text{(Bianchi identity)} \tag{18}$$

Notice that in $\nabla R(x^a_k)$ one finds in principle all $R(x^b_\ell)$ with $\ell < k$. To define the covariant derivative in the coadjoint representation we consider a set of d-k forms V^{d-k}_a (one V^{d-k}_a per generator x^a_k) and define for arbitrary d

$$d\left(T^a_k \wedge V^{d-k}_a\right) = (\nabla T^a_k) \wedge V^{d-k}_a + (-)^k T^a_k \wedge \nabla V^{d-k}_a \tag{19}$$

where ∇T^a_k is the covariant derivative in the adjoint representation. Later we shall show that ∇ contains a piece D which is the covariant derivative w.r.t. the algebra generated by the 1-form generators alone.

5. DIFFERENTIAL ALGEBRAS SPLIT

Any free graded differential superalgebra A can be decomposed into a contractible algebra C and a minimal algebra M according to D. Sullivan. The contractible algebra consists of *pairs* of k and k + 1 forms (both bosonic or both fermionic) satisfying

$$dx^a_k = x^a_{k+1} \quad , \quad dx^a_{k+1} = 0 \tag{20}$$

In the minimal algebra dx^a_k is equal to a sum of *products* of forms (or equal to zero) but never equal to a single k + 1 form generator. Denoting the algebras *generated* by all p-forms with $p \leq k$ by C^k and M^k we thus have

$$dC^k \subset C^{k+1} \quad , \quad dM^k \subset M^k \wedge M^k \tag{21}$$

Before proving this theorem, let us give an example. Let

$$db_1^1 = 0 , df_1^1 = f_2 + b_1^1 \wedge f_1^2 , df_1^2 = f_2 + b_1^1 \wedge f_1^1 , df_2 = b_1^1 \wedge f_2 \tag{22}$$

Clearly this algebra is consistent.

Redefining the generators

$$\hat{f}_2 = f_2 + b_1^1 \wedge f_1^2 , \hat{f}_1 = f_1^1 , \tilde{f}_1 = f_1^2 - f_1^1 , \tilde{b}_1 = b_1^1 \tag{23}$$

we find that the algebra splits

$$\begin{aligned} d\hat{f}_1 &= \hat{f}_2 , \quad d\hat{f}_2 = 0 \\ d\tilde{f}_1 &= \tilde{b}_1 \wedge (f_1^1 - f_1^2) = -\tilde{b}_1 \wedge \tilde{f}_1 , \quad d\tilde{b}_1 = 0 \end{aligned} \tag{24}$$

We will now prove the theorem straightforwardly at the level of 1, 2 and 3-forms. By then the general inductive proof will become clear, but we shall not present the latter here. Seeing what goes on at the 1, 2 and 3-form level will convince the reader that the theorem holds and is much more understandable.

<u>Proof for 1-forms</u>

The most general expression for dx_1^a where a runs over all independent generators of grade 1 (bosonic or fermionic) contains terms $M^a{}_b x_2^b$. Redefining a maximal set of independent $M^a{}_b x_2^b$ as x_2^a, and taking linear combinations of the x_1^a

$$dx_1^a = x_2^a + f^a{}_{bc} x_1^b \wedge x_1^c \tag{25}$$

By redefining $\hat{x}_2^a = x_2^a + f^a{}_{bc} x_1^b x_1^c$ we find

$$dx_1^a = \hat{x}_2^a \quad \text{or} \quad dx_1^a = f^a{}_{bc} x_1^b \wedge x_1^c \tag{26}$$

By taking linear combinations of the x_1^a we can achieve that the correspondence $dx_1^a = \hat{x}_2^a$ is 1-1. Let us denote the x_1^a, which satisfy $dx_1^a = \hat{x}_2^a$ by $\hat{x}_1^a$ and the rest by $\tilde{x}_1^a$. Thus

$$x_1^a = (\hat{x}_1^a , \tilde{x}_1^a) , d\hat{x}_1^a = \hat{x}_2^a , d\tilde{x}_1^a = f^a{}_{bc} x_1^b \wedge x_1^c \tag{27}$$

To show that at the level of 1-forms the algebra splits into a contractible part ($d\hat{x}_1^a = \hat{x}_2^a$) and a minimal part, we must show that the x_1^b and x_1^c in the expressions for $d\tilde{x}_1^a$ are of the kind $\tilde{x}_1$ and not $\hat{x}_1$. The proof is easy. Since $dd\tilde{x}_1^a$ should yield zero, due to dd = 0, we have

$$dd\tilde{x}_1^a = 0 = f^a{}_{bc}\left(dx_1^b \wedge x_1^c - x_1^b \wedge dx_1^c\right) \tag{28}$$

Noting that

$$d(\hat{x}_1^b \wedge \hat{x}_1^c) = \hat{x}_2^b \wedge \hat{x}_1^c - \hat{x}_1^b \wedge \hat{x}_2^c \tag{29}$$

$$d(\hat{x}_1^b \wedge \tilde{x}_1^c) = \hat{x}_2^b \wedge \tilde{x}_1^c - \hat{x}_1^b \wedge (f^c{}_{pq}\, x_1^p \wedge x_1^q) \tag{30}$$

$$d(\tilde{x}_1^b \wedge \tilde{x}_1^c) = (d\tilde{x}_1^b) \wedge \tilde{x}_1^c - \tilde{x}_1^b \wedge d\tilde{x}_1^c = \text{terms with three } x_1\text{'s} \tag{31}$$

it is clear that the three kinds of terms cannot help each other in cancelling. In fact,

$$f^a{}_{bc}\left(\hat{x}_2^b \hat{x}_1^c - \hat{x}_1^b \wedge \hat{x}_2^c\right) \tag{32}$$

never vanishes, because if $\hat{x}_2$ and $\hat{x}_1$ are both fermionic then $f^a{}_{bc}$ is symmetric in (b,c), while if one or both of $\hat{x}_2$ and $\hat{x}_1$ are bosonic then $\hat{x}_2$ and $\hat{x}_1$ commute and in that case $f^a{}_{bc}$ is antisymmetric in (bc). Hence, there cannot be terms with $\hat{x}_1 \wedge \hat{x}_1$ in $d\tilde{x}_1^a$.

The terms with $\hat{x}_2\tilde{x}_1$ in $d\tilde{x}_1^a$ are non-vanishing and must cancel by themselves. Since they only come from $d(\hat{x}_1\tilde{x}_1)$, also the $\hat{x}_1\tilde{x}_1$ terms in $d\tilde{x}_1$ must be absent. Hence, at the level of 1-forms, the algebra indeed splits: $\vec{x}_1 = (\vec{\tilde{x}},\vec{\hat{x}})$, $d\tilde{x}_1^a = f^a{}_{bc}\tilde{x}_1^b \wedge \tilde{x}_1^c$, $d\hat{x}_1^a = \hat{x}_2^a$.

<u>Proof for 2-forms</u>

We first look for 2-form generators (even or odd) for which dx_2^a contains a 3-form generator x_3 on the right-hand side. Thus $dx_2^a = x_3^a$ + more. By redefining x_3^a into $\hat{x}_3^a$, we obtain that $dx_2^a = \hat{x}_3^a$ and thus $d\hat{x}_3^a = 0$; moreover, we can make the connection $(x_2^a, \hat{x}_3^a)$ again 1-1 by taking linear combinations of the x_2^a. Let us denote these pairs by $(\hat{\hat{x}}_2, \hat{x}_3)$ since the notation $\hat{x}_2$ was already reserved for $d\hat{x}_1 = \hat{x}_2$. The rest of the x_2 we denote by $\tilde{x}_2$. Thus

$$x_2^a = \left(\hat{x}_2^a, \hat{\hat{x}}_2^a, \tilde{x}_2^a\right), \ d\hat{x}_1^a = \hat{x}_2^a, \ d\hat{\hat{x}}_2^a = \hat{x}_3^a \tag{33}$$

We must again show that on the right-hand side of $d\tilde{x}_2^a$ one only finds $\tilde{x}_1 \wedge \tilde{x}_1 \wedge \tilde{x}_1$ or $\tilde{x}_1 \wedge \tilde{x}_2$ but never terms with $\hat{x}_1$ or with $\hat{x}_2$ or $\hat{\hat{x}}_2$. If we can show this then we have accomplished the decomposition into a contractible and minimal part also at the level of 2-forms.

We have in general

$$d\tilde{x}_2^a = f^a_{(2)bc}\, x_1^b \wedge x_2^c + x_1 \wedge x_1 \wedge x_1 \text{ terms} \tag{34}$$

Clearly x_2^c cannot be $\hat{\hat{x}}_2^c$ since otherwise the $\hat{x}_3^a$ terms in $dd\tilde{x}_2^a$ would not cancel.

Suppose the x_2^c were $\hat{x}_2^c$, and x_1^b were $\hat{x}_1^b$. In that case consistency would require that $f^a_{(2)bc}\, \hat{x}_2^b \hat{x}_2^c = 0$, hence $f^a_{(2)bc}$ is super-antisymmetric (antisymmetric except when both x_2^b and x_2^c are fermionic in which case $f^a_{(2)bc}$ is symmetric in bc). The super-antisymmetric part of $f^a_{(2)bc}\, \hat{x}_1^b \hat{x}_2^c$ can be written as a total derivative

$$f^a_{(2)bc}\, \hat{x}_1^b \wedge \hat{x}_2^c = -\frac{1}{2}\, d\left(f^a_{(2)bc}\, \hat{x}_1^b \wedge \hat{x}_1^c\right) \tag{35}$$

and by redefining $\tilde{x}_2^{a\prime} = \tilde{x}_2^a + \frac{1}{2} f^a_{(2)bc}\, \hat{x}_1^b \hat{x}_1^c$ we can eliminate all $\hat{x}_1 \hat{x}_2$ terms in $d\tilde{x}_2$. Thus, although the bosonic and fermionic forms behave exactly opposite ($x_1^a x_1^a$ is non-zero when x_1^a is fermionic but vanishes when x_1^a is bosonic), the notion of super-(anti)symmetry covers both cases.

Suppose next that there is a term $f^a_{(2)bc}\, \tilde{x}_1^b \hat{x}_2^c$ in $d\tilde{x}_2^a$. It can be written as

$$f^a_{(2)bc}\, \tilde{x}_1^b \hat{x}_2^c = -d\left(f^a_{(2)bc}\, \tilde{x}_1^b \wedge \hat{x}_1^c\right) + \tilde{x}_1 \tilde{x}_1 \tilde{x}_1 \text{ terms} \tag{36}$$

and by redefining $\tilde{x}_2^a$ once more, also such terms could be eliminated.

Hence at this point,

$$d\tilde{x}_2^a = f^a_{(2)bc}\, x_1^b \wedge \tilde{x}_2^c + g^a_{\ bcd}\, x_1^b \wedge x_1^c \wedge x_1^d \tag{37}$$

and we must show that all x_1 are $\tilde{x}_1$. This is easy; if there were one or more $\hat{x}_1$, the $\hat{x}_2 x_1 x_1$ or $\hat{x}_2 \tilde{x}_2$ terms in $dd\tilde{x}_2^a$ would not cancel. Hence, also at the level of 2 forms the algebra splits into a contractible and minimal part.

$$\begin{gathered} x_2^a = \left(\hat{x}_2^a, \hat{\hat{x}}_2^a, \tilde{x}_2^a\right), \quad d\hat{x}_1^a = \hat{x}_2^a, \quad d\hat{\hat{x}}_2^a = \hat{x}_3^a \\ d\tilde{x}_2^a = f^a_{(2)bc}\, \tilde{x}_1^b \wedge \tilde{x}_2^c + g^a_{\ bcd}\, \tilde{x}_1^b \wedge \tilde{x}_1^c \wedge \tilde{x}_1^d \end{gathered} \tag{38}$$

<u>Proof for 3-forms</u>

As for x_2, we decompose the 3-forms x_3^a into three classes

$$\begin{gathered} x_3^a = \left(\hat{x}_3^a, \hat{\hat{x}}_3^a, \tilde{x}_3^a\right), \quad d\hat{\hat{x}}_2^a = \hat{x}_3^a, \quad d\hat{\hat{x}}_3^a = \hat{x}_4^a \\ d\tilde{x}_3^a = f^a_{(3)bc}\, x_1^b \wedge x_3^c + g^a_{\ bc}\, x_2^b \wedge x_2^c + h^a_{\ bcd}\, x_2^b \wedge x_1^c \wedge x_1^d + x_1 x_1 x_1 x_1 \text{ terms} \end{gathered} \tag{39}$$

Consistency ($dd\tilde{x}_3^a = 0$) forbids $x_3 = \hat{\hat{x}}_3$. If $x_3 = \hat{x}_3$, then redefinition of $\tilde{x}_3^a$ in $d\tilde{x}_3^a$ can eliminate $\hat{x}_3 x_1$ terms. The x_1 in the $\tilde{x}_3 x_1$ term cannot be an $\hat{x}_1$, since $\tilde{x}_3 \wedge \hat{x}_2$ would not cancel ($d\hat{\hat{x}}_2 \wedge \hat{x}_2$ could only yield $\hat{x}_3 \wedge \hat{x}_2$, not $\tilde{x}_3 \wedge \hat{x}_2$). Hence the first term in $d\tilde{x}_3^a$ is a product of generators of the minimal part of the algebra.

Now the $x_2 \wedge x_2$ terms. None of these x_2 can be a $\hat{\hat{x}}_2$ since in $dd\tilde{x}_3^a = 0$ the $\hat{x}_3 \hat{x}_2$ terms would not cancel. Again the notion of supersymmetry of $g^a_{\ bc}$ is helpful in treating all cases. Suppose both of these x_2 were an $\hat{x}_2$, then $\hat{x}_2 \wedge \hat{x}_2 = d\hat{x}_1 \wedge \hat{x}_2 =$ $= d(\hat{x}_1 \wedge \hat{x}_2)$ could be eliminated by redefining $\tilde{x}_3^a$. Similarly one could eliminate $\hat{x}_2 \wedge \tilde{x}_2$ terms because although we get now an extra term, namely $-\hat{x}_1 \wedge d\tilde{x}_2$, this extra term lies in the h-sector or in the 4 x_1 sector. Going on to the $x_2 \wedge x_1 \wedge x_1$ terms, $\hat{\hat{x}}_2$ is ruled out by consistency, as are the combinations $\hat{x}_2 \hat{x}_1 \hat{x}_1$ and $\hat{x}_2 \hat{x}_1 \tilde{x}_1$. The $\hat{x}_2 \tilde{x}_1 \tilde{x}_1$ can be removed by redefinition. Again the x_1's in the term with h cannot be $\hat{x}_1$ since $\tilde{x}_2 \hat{x}_2 x_1$ would not cancel. As to the 4 x_1 terms, also these must 4 $\tilde{x}_1$ terms.

This concludes the first three steps in the iterative proof that the algebra decomposes into a contractible and minimal part. We covered both cases of bosonic and fermionic forms. There are, of course, important differences (for example, $x_1^a \wedge x_1^a$ is non-vanishing for odd forms) but to cover both, the notion of super-(anti)symmetry turned out to be useful.

6. COHOMOLOGY CLASSES IN MINIMAL ALGEBRAS

Let A now be a minimal algebra and denote all generators of grade p by x_p^a. The x_p^a can be bosonic or fermionic. Dropping tildas from now on, we have

$$dx_p^a = (M_p)^a{}_b \wedge x_p^b + a_p \tag{40}$$

where the matrices M_p are 1-forms and a_p is a p + 1 form which is generated by the generators of grade p - 1 and less

$$(M_p)^a{}_b = f^a_{(p)\,cb}\, x_1^c \tag{41}$$

Consistency implies that (in matrix notation)

$$(dM_p)\wedge x_p - M_p \wedge dx_p + da_p = (dM_p - M_p \wedge M_p)\wedge x_p + (d - M_p)\, a_p = 0 \tag{42}$$

The terms with generators of grade p must vanish separately (we have a free algebra) and hence

$$dM_p - M_p \wedge M_p = 0, \quad (d - M_p)\, a_p = 0 \tag{43}$$

It follows that $D(M_p) \equiv d - M_p$ is nilpotent, just like d

$$D(M_p)\, D(M_p) = 0, \quad D(M_p)\, x_p = a_p \tag{44}$$

This implies two things. The matrices $f_{(p)c}$ for given p form a representation of the Lie superalgebra defined by the 1-forms x_1^c

$$[f_{(p)a}, f_{(p)b}] = f_{(p)c}\, f^c{}_{ab} \;;\; dx_1^c = f^c{}_{ab}\, x_1^a \wedge x_1^b \tag{45}$$

Furthermore, a_p is closed under $D(M_p)$ but not necessarily exact in A^{p-1} [it is, of course, exact in A^p by definition: $a_p = D(M_p)x_p$]. Thus the a_p are elements of a cohomology class

$$a_p \in H^{p+1}(A^{p-1}, M_p) \tag{46}$$

In words: the a_p are p + 1 forms (bosonic or fermionic) constructed from the generators with grade p - 1, which are closed with relation to the derivative $d - M_p$ (where the M_p form a representation of A^1).

The physical relevance of non-trivial cohomology classes (forms which are closed but not exact) is that one can introduce new higher-order forms into the differential algebra and still maintain consistency. Indeed, let

$$a_p^a = \sum C\left(\begin{matrix} a \\ p \end{matrix} \middle| \begin{matrix} p_1 \cdots p_\ell \\ a_1 \cdots a_\ell \end{matrix}\right) x_{p_1}^{a_1} \wedge \ldots \wedge x_{p_\ell}^{a_\ell} \tag{47}$$

where $p_1 + \ldots + p_\ell = p + 1$ and all p_i have $p_i \leq p - 1$. If $D(M_p)a_p = 0$ we can add to the algebra a new p form y_p^a as follows

$$D(M_p)\, y_p^a = a_p^a \tag{48}$$

and consistency would hold due to $D(M_p)a_p = 0$.

In practice one begins with an ordinary Lie superalgebra (usually the super-Poincaré or super de Sitter algebra), finds non-trivial cohomology classes, and then one adds new higher-grade forms to the system. One example was given in (3): the 4-form [11]

$$a_3 = \bar{\psi}\gamma_{mn} \wedge \psi \wedge V^m \wedge V^n \tag{49}$$

is an element of $H^4(A^1, I)$. Numerous other examples can be found in Refs. [11,14]; a discussion is given by the author in the proceedings of the Chicago conference 1982.

7. TRIVIALIZING COHOMOLOGY CLASSES

An interesting development in field theory is the following possibility [14]: given a p-form generator y_p with p > 1 in a differential algebra, can one *add* new k-forms with k < p to the algebra such that if y_p is represented by a product of forms with grade < p, then dy_p is consistent identically. Consider again (3), and write [11]:

$$A = \alpha B^{mn} \wedge V_m \wedge V_n + \beta \bar{\psi} \gamma^m \wedge \eta \wedge V_m + \gamma \bar{\psi} \gamma_{mn} \wedge \psi \wedge B^{mn} + \delta B^{mk} \wedge B_{kl} \wedge B^l{}_m \tag{50}$$

The new forms are here a bosonic 1-form $B^{mn} = -B^{nm}$ and a fermionic 1-form η. Their differential relations (= Cartan-Maurer equations, since they are 1-forms) we take as

$$\mathcal{D} B^{mn} = \bar{\psi} \gamma^{mn} \wedge \psi \tag{51}$$

$$\mathcal{D} \eta = \gamma^m \psi \wedge V_m + \epsilon \gamma^{mn} \psi \wedge B_{mn} \tag{52}$$

where $\mathcal{D}$ is the Lorentz covariant derivative. The consistency of (51, 52) follows from $\mathcal{D}\psi = 0$ and the identities in d = 4

$$\bar{\psi} \gamma^m \wedge \psi \wedge \gamma_m \psi = 0 \ , \quad \bar{\psi} \gamma^{mn} \psi \wedge \gamma_{mn} \psi = 0 \tag{53}$$

Let us now turn to A in (50). We can compute dA in two ways: from (3) or by explicitly working out d on the right-hand side of (50), using (3), (51), and (52). We can arrange both results to be identical by fixing α, β, γ, δ, and ε appropriately. (There is actually a 2-parameter class of solutions.)

In d = 11 supergravity similar things happen. There one must introduce in addition to η and B^{ab} a five-index bosonic 1-form $B^{a_1 \ldots a_5}$, and one finds two discrete solutions. The 1-forms (the original ones plus the new ones) define a new superalgebra, which has now two "supersymmetry charges" Q^a and $\hat{Q}^a$, corresponding to ψ and η. The corresponding superalgebra has the following structure [14]

$$\{Q, Q\} \sim B^{\cdot\cdot} + B^{\cdot\cdot\cdot\cdot\cdot} + P \ ; \ \{Q, \hat{Q}\} = \{\hat{Q}, \hat{Q}\} = 0 \tag{54}$$

$$[Q, P \text{ or } B^{\cdot\cdot} \text{ or } B^{\cdot\cdot\cdot\cdot\cdot}] \sim \hat{Q} \ ; \ [\hat{Q}, P \text{ or } B^{\cdot\cdot} \text{ or } B^{\cdot\cdot\cdot\cdot\cdot}] = 0 \tag{55}$$

Thus $\hat{Q}^a$ is a kind of fermionic central charge: it commutes with as many generators as possible (being a spinor it does not commute with the Lorentz generators).

The idea to reformulate the theory with antisymmetric tensors into a theory with only ordinary gauge fields has not yet been worked out. In particular, whether both theories are equivalent is not known, but it would be interesting in either case.

It is not always possible to add new 1-forms such that a p-form can be represented by a product of 1-forms. We quote here a counter example given to us by D. Sullivan. The proof (if correct) is undoubtedly equal to his, but we were only able to construct the proof in the form below.

<u>Theorem</u>: in a simple ordinary Lie algebra there are non-trivial cohomology classes which remain non-trivial, no matter how many new 1-forms one adds to the differential algebra.

<u>Proof</u>: let the simple Lie algebra S have generators $x_1^1, \ldots x_1^n$ satisfying

$$dx_1^a = f^a{}_{bc}\, x_1^b \wedge x_1^c \tag{56}$$

Consistency implies that $f^a{}_{b[c} f^b{}_{k\ell]} = 0$, in other words, the Jacobi identities. The structure constants $f^a{}_{bc}$ are taken to be totally antisymmetric (S is simple).

The following form is closed

$$a_n = x_1^1 \wedge \ldots\ldots \wedge x_1^n \tag{57}$$

because a totally antisymmetric tensor in n dimensions with n + 1 indices vanishes. This form is not exact: if $a_n = db_{n-1}$ then

$$b_{n-1} = \sum \alpha_k\, x_1^1 \wedge \ldots \wedge x_1^{k-1} \wedge x_1^{k+1} \wedge \ldots \wedge x_1^n \tag{58}$$

However, db_{n-1} vanishes always because in dx_1^ℓ one never finds x_1^ℓ on the right-hand side (due to the antisymmetry of f_{abc}).

Let us now add an arbitrary number of new 1-forms λ^a, satisfying consistent Cartan-Maurer equations

$$d\lambda^a = g^a{}_{bc}\lambda^b\wedge\lambda^c + h^a{}_{bc}\lambda^b\wedge x^c + k^a{}_{bc}x^b\wedge x^c \tag{59}$$

We shall prove that one can never find a b_{n-1} constructed from the x^a and λ^a such that $db_{n-1} = a_n$.

We begin by noting that *any* ordinary Lie algebra (semisimple or not) can always be decomposed into a semidirect sum of a solvable part P and a semisimple part; the latter is, of course, the direct sum of simple parts

$$A = P \ni S_1 \oplus \ldots\ldots \oplus S_n \tag{60}$$

Let the generators of P be denoted by P_A and those of S_i by $S_{i,A}$. Then

$$[P, P] \sim P \;;\; [P, S_i] \sim P \;;\; [S_i, S_j] \sim \delta_{ij} S_j \tag{61}$$

The generators of the Lie algebra corresponding to (56) are linear combinations of the P_A's and $S_{i,A}$'s, and from (56) and (61) it follows that

$$\begin{aligned} &X_a = \hat{P}_a + \hat{S}_a \,, \quad [X_a, X_b] = X_c f^c{}_{ab} \\ &[\hat{S}_a, \hat{S}_b] = \hat{S}_c f^c{}_{ab} \\ &[\hat{S}_a, \hat{P}_b] + [\hat{P}_a, \hat{S}_b] + [\hat{P}_a, \hat{P}_b] = \hat{P}_c f^c{}_{ab} \end{aligned} \tag{62}$$

Thus the $\hat{S}_a$ in X_a form a simple Lie algebra. Since a solvable Lie algebra has no simple subalgebra, the $\hat{S}_a$ can be identified with one of the factors S_i in (60).

Let us now go over to the dual Lie algebra (the algebra in terms of 1-forms). We have

$$x_1^a = \hat{p}_1^a + \hat{s}_1^a \,, \quad d\hat{s}^a_{1,i} \sim \hat{s}_{1,i}\wedge\hat{s}_{1,i} \,, \quad d\hat{p}_1 \sim \hat{p}_1\wedge\hat{p}_1 + \hat{p}_1\wedge\hat{s}_1 \tag{63}$$

Suppose we could find a b_{p-1} as a sum of products of p_1^a and $s^a_{1,i}$ 1-forms, such that $db_{n-1} = a_n$. In a_n one has a term $\hat{s}_1\wedge\ldots\hat{s}_n$. In b_{n-1} one would have terms with and without p_1^a 1-forms, but dp_1^a produces at least one p_1^a, so the terms with p_1^a cannot produce the term $\hat{s}_1\wedge\ldots\,\hat{s}_n$. However, nor can the terms without p_1^a

forms, because of the same argument as given below (58). Hence, even in the larger algebra generated by x_1^a and λ_1^a, the closed form a_n remains non-exact.

Acknowledgements

The author is grateful to D. Sullivan for showing and explaining him the two theorems in the text, and to R. Stora and B. Julia for discussions.

REFERENCES

[1] S. MacDowell and F. Mansouri, Phys. Rev. Lett. 38, 739 (1977).

[2] A.H. Chamseddine and P.C. West, Nucl. Phys. B129, 39 (1977).

[3] P.K. Townsend and P. van Nieuwenhuizen, Phys. Lett. 67B, 439 (1977).

[4] M. Kaku, P.K. Townsend and P. van Nieuwenhuizen, Phys. Lett. 69B, 304 (1977) and Phys. Rev. Lett. 39, 1109 (1977) and especially Phys. Rev. D 17, 3179 (1978).

[5] P. van Nieuwenhuizen in the Proceedings of the 1980 Cambridge Nuffield Workshop.

[6] B. de Wit, in "Supergravity 82", Cambridge Univ. Press, S. Ferrara and J.G. Taylor editors and references therein.

[7] Y. Ne'eman and T. Regge, Phys. Lett. 74B, 54 (1978) and Riv. del Nuovo Cimento 1, 1 (1978).

[8] R. D'Auria and P. Fré, Nucl. Phys. B173, 456 (1980);
P. Fré, Nucl. Phys. B186, 44 (1981);
R. D'Auria, P. Fré and A.J. Da Silva, Nucl. Phys. B196, 205 (1982).

[9] P. van Nieuwenhuizen in Proceedings 1982 Erice School on (Super)gravity in higher dimensions and exact solutions, World Publishing Cy, V. de Sabata editor.
R. D'Auria, P. Fré and T. Regge in "Supergravity 81", Cambridge Univ. Press S. Ferrara and J.G. Taylor editors.

[10] D. Sullivan, "Infinitesimal Computations in Topology", reprints to be obtained from IHES, Bures sur Yvette, France.

[11] L. Castellani, P. Fré, F. Giani, K. Pilch and P. van Nieuwenhuizen, Phys. Rev. D26, 1481 (1982) and Ann. of Phys. (to be published).

[12] R. D'Auria, P. Fré, P.K. Townsend and P. van Nieuwenhuizen, CERN preprint TH 3495.

[13] See, for example, P. van Nieuwenhuizen, Physics Report 68, 189 (1981), pages 361-365.

[14] R. D'Auria and P. Fré, Nucl. Phys. B201, 101 (1982).

[15] P.K. Townsend and P. van Nieuwenhuizen, Ecole Normale Supérieure, preprint LPTENS 82/15.

RELATIVISTIC WAVE EQUATIONS FROM SUPERGROUP QUANTIZATION

V.Aldaya and J.A.de Azcárraga

Departamento de Fisica Teórica, Facultad de Ciencias Fisicas
Universidad de Valencia, Burjasot (Valencia), Spain

ABSTRACT

A formalism of geometric quantization recently introduced which is based on the consideration of Lie groups which are central extensions by U(1) is applied to the relativistic case by using the N=2 super Poincaré group with a central charge.

1- Introduction

In a previous paper[1] we have developed a group theoretical method of geometric quantization based on a group, the "quantum group". The procedure is based on the analysis of the symmetry of several quantum non relativistic (galilean) systems. The analysis reveals a common underlying structure -a central extension $\tilde{G}$ of a Lie group G by U(1)- which allows us to base the quantization procedure of the classical system (with symmetry group $G=\tilde{G}/U(1)$) on the quantum group $\tilde{G}$(*). Broadly speaking, the U(1) extension is responsible of the modification of the commutation relations $[K^i, P_j] = 0$ (which become $[q^i, p_j] = i\,\delta^i_j$ in $\tilde{G}$), of the continuity equation as a conservation law associated with the U(1) symmetry[2] and of the Schrödinger equation. This last is given through a vector field which is part of the characteristic module C_θ associated with the vertical component θ of the canonical 1-form defined on $\tilde{G}$ (**). The simplest case, that of the free quantum galilean particle, corresponds to G = Galilei group; then $\tilde{G}$, which is parametrized by the mass[3], is $\tilde{G}_{(m)}$, $G = \tilde{G}_{(m)}/U(1)$. In the case of a particle with interaction, G is a group which reduces to the Galilei group when the contraction (interaction) parameter is set to zero and which preserves the symplectic cohomology.

(*) The procedure is not, however, limited to quantum systems with classical analogs.

(**) The verticality is defined by the principal bundle $(\tilde{G} \xrightarrow{\pi} G, U(1))$.

It is clear that the above procedure cannot be directly applied to the relativistic case, since the Poincaré group $\tilde{\mathcal{P}}$ allows only for a direct product extension by U(1). At the same time, the commutator $[K^i, P_j]$ is already non-zero in $\tilde{\mathcal{P}}$, so that it is not possible to think of $\vec{K}$ as being associated which the position operator (in galilean mechanics, $\vec{K}/m = \vec{x}$). This difficulty to carry on the above quantization program is bypassed if one moves to consider supersymmetry groups and, more precisely, the N=2 superPoincaré group which can be extended by a central charge[4]. This implies, we note in passing, that fermions are <u>necessary</u> to perform the quantization in the above sense (*). Indeed, a quantization based on the superGalilei group can be also carried out, but it is <u>not</u> necessary in the sense that the U(1)-extended Galilei group $\tilde{G}_{(m)}$ already allows for quantization in our scheme.

2- Derivation of the Klein-Gordon and Dirac equations from the U(1) extended N=2 superPoincaré.

The generalization of the group manifold approach to quantization to the case of super Lie groups requires the appropriate extension of the Cartan calculus to the case of odd variables[5]; this does not introduce any essential difficulty. We shall now apply the procedure to the U(1)-extended N=2 superPoincaré, in which we shall not consider the Lorentz part since it is not relevant for the U(1) central extension; we shall accordingly treat the boosts as external parameters. The starting group law will be the following

$$\tilde{g}' * \tilde{g} = \{x'^{\mu} + x^{\mu} - 1/2\,[\theta'^{\alpha}_{i}(\sigma^{\mu})_{\alpha\dot{\beta}}\,\bar{\theta}^{\dot{\beta}}_{i} - \theta^{\alpha}_{i}(\sigma^{\mu})_{\alpha\dot{\beta}}\,\bar{\theta}'^{\dot{\beta}}_{i}]\,;$$

$$\theta'^{\alpha}_{i} + \theta^{\alpha}_{i}\,;\ \bar{\theta}'^{\dot{\alpha}}_{i} + \bar{\theta}^{\dot{\alpha}}_{i}\,;\ \zeta'\zeta \exp i\zeta_{(m)}(g',g)\} \qquad (1)$$

$$\xi_{(m)}(g',g) \equiv -\,m/2\,[\theta'_{\alpha i}\varepsilon_{ij}\theta^{\alpha}_{j} + \bar{\theta}'_{\dot{\alpha} i}\varepsilon_{ij}\bar{\theta}^{\dot{\alpha}}_{j}] - 1/2\,[\theta^{\alpha}_{i}(\sigma p)_{\alpha\dot{\beta}}\,\bar{\theta}'^{\dot{\beta}}_{i} - \theta'^{\alpha}_{i}(\sigma p)_{\alpha\dot{\beta}}\,\bar{\theta}^{\dot{\beta}}_{i}]$$

where $g = (x^{\mu}, \theta^{\alpha}_{i}, \bar{\theta}^{\dot{\alpha}}_{i})$, $\tilde{g} = (g,\zeta)$, $\zeta \in U(1)$ and all repeated (latin, i,j=1,2 and spinorial $\alpha,\beta = 1,2$) indices are summed.

The left invariant vector fields, which include the "covariant derivatives" are obtained from (2.1) with the result

---

(*) Quantization is here used in the sense of first quantization (derivation of the wave equations).

$$X^L_{(\mu)} = \frac{\partial}{\partial x^\mu} \; ; \; \Xi = i\zeta \frac{\partial}{\partial \zeta} \quad \text{(U(1) central vector field)}$$

$$D^L_{(\alpha i)} = \frac{\partial}{\partial \theta^\alpha_i} + \frac{1}{2} (\sigma^\mu)_{\alpha\dot\beta} \bar\theta^{\dot\beta}_i \frac{\partial}{\partial x^\mu} - \frac{1}{2} [m\varepsilon_{ij}\theta_{\alpha j} + (\sigma p)_{\alpha\dot\beta} \bar\theta^{\dot\beta}_i] \Xi \tag{2}$$

$$D^L_{(\dot\alpha i)} = \frac{\partial}{\partial \bar\theta^\alpha_i} + \frac{1}{2} \theta^\beta_i (\sigma^\mu)_{\beta\dot\alpha} \frac{\partial}{\partial x^\mu} - \frac{1}{2} [m\varepsilon_{ij}\bar\theta_{\dot\alpha j} + \theta^\beta_i (\sigma p)_{\beta\dot\alpha}] \Xi$$

and generate the graded Lie algebra of the N=2 superPoincaré with a central charge parametrized (as in the Galilei case) by the mass. Following the general formalism[1] the (super) wavefunction is defined on the group manifold and satisfies the U(1)-equivariance condition $\Xi \,.\, \Phi = i\Phi$. The (super) wave function must satisfy the condition $X \,.\, \Phi = 0$ for all vector fields of the characteristic module C_θ of the vertical canonical 1-form θ ($\theta(\Xi) = 1$, θ (any other) = 0) which is given by

$$-md\theta^\alpha_i \varepsilon_{ij}\theta_{\alpha j} - m\bar\theta_{\dot\beta i}\varepsilon_{ij} d\bar\theta^{\dot\beta}_j + [\theta^\beta_i (\sigma p)_{\beta\dot\alpha} d\bar\theta^{\dot\alpha}_i - d\theta^\alpha_i (\sigma p)_{\alpha\dot\beta} \bar\theta^{\dot\beta}_i] + \frac{d\zeta}{i\zeta} \tag{3}$$

C_θ is generated by the vector fields X which satisfy $i_X \theta = 0$, $i_X d\theta = 0$; for instance, by

$$\{ X^L_{(\mu)} , C_\alpha \equiv \frac{\partial}{\partial\theta^\alpha_1} - \frac{1}{m} (\sigma p)_{\alpha\dot\beta} \frac{\partial}{\partial\bar\theta_{\dot\beta 2}} , \bar C_{\dot\alpha} \equiv \frac{\partial}{\partial\bar\theta^{\dot\alpha}_1} - \frac{1}{m} (\sigma p)_{\gamma\dot\alpha} \frac{\partial}{\partial\theta_{\gamma 2}} \}; \tag{4}$$

in deriving the expressions for C_α, $\bar C_{\dot\alpha}$ the condition $p^2 = m^2$ is obtained. The reduction of the (super) wavefunction is obtained by adding the conditions $X \,.\, \Phi = 0$ where X now corresponds to the "polarizations" in the geometric quantization language. The fact that all the restrictions have to be consistent implies that polarizations and characteristic vector fields have to generate a horizontal subalgebra; thus, a maximal set of constraints on Φ is obtained by imposing $X \,.\, \Phi = 0$, $X \in \{ X^L_{(\mu)} , C_\alpha, \bar C_{\dot\alpha} , D^L_{(\alpha 2)}, \bar D^L_{(\dot\alpha 1)}\}$. $\Xi . \Phi = i\Phi$, $X^L_{(\mu)} . \Phi = 0$, $D^L_{(\alpha 2)} \,.\, \Phi = 0$ and $\bar D^L_{(\dot\alpha 1)} . \Phi = 0$ give

$$\Phi(\theta_i^{\alpha}, \bar{\theta}_i^{\dot{\alpha}}, p_\mu, x^\mu, \zeta) =$$

$$\exp \frac{i}{2} (m\bar{\theta}_i^{\dot{\alpha}} \bar{\theta}_{\dot{\alpha}2} - \theta_1^{\beta}(\sigma p)_{\beta\dot{\alpha}} \bar{\theta}_1^{\dot{\alpha}} - m\, \theta_2^{\alpha}\theta_{\alpha 1} + \theta_2^{\alpha}(\sigma p)_{\alpha\dot{\beta}} \bar{\theta}_2^{\dot{\beta}}).$$

$$\cdot \mathfrak{F} (\theta_1^{\alpha}, \bar{\theta}_2^{\dot{\alpha}}, p_\mu)\, \zeta. \tag{5}$$

Expanding the superfield[6] $\mathfrak{F}$ in terms of fields depending on p,

$$\mathfrak{F} = \phi(p) + \theta_{\alpha 1}\, \psi_1^{\alpha}(p) + \bar{\theta}_2^{\dot{\alpha}}\, \psi_{\dot{\alpha}2}(p) +$$

$$+ m\, \theta_1^{\alpha}(\sigma_\mu)_{\alpha\dot{\beta}}\bar{\theta}_2^{\dot{\beta}}\, V^{\mu} + \frac{m}{2}\, \phi'(p)\theta_{\alpha 1}\theta^{\alpha 1} + \frac{m}{2}\phi''(p)\, \bar{\theta}_2^{\dot{\alpha}}\, \bar{\theta}_{\dot{\alpha}2}, \tag{6}$$

the condition $C_\alpha \cdot \Phi = 0$ shows that (6) includes a Dirac field $\psi = (\psi_{\alpha 1}, \bar{\psi}_2^{\dot{\beta}})$ satisfiying the Dirac equation (note that in the above formulae $p^2 = m^2$) and two Klein-Gordon fields ϕ and ϕ' (it turns out that $\phi''(p) = -\phi'(p)$ and $V^{\mu}(p) = \frac{p_\mu}{m}\phi'(p)$). This is the basic N=2 matter hypermultiplet.

To conclude, we remark that it is interesting that in order to reconcile relativistic invariance with quantization in the above sense one is forced to include (Z_2)-graded Lie groups and accordingly fermions into the scheme, in agreement with the well known result that the spin-statistics theorem is naturally incorporated in the framework of relativistic field theory.

REFERENCES

1- V.Aldaya and J.A.de Azcarraga, J.Math.Phys. 23 1297 (1982). See also V.Aldaya and J.A.de Azcarraga, Lett.NuovoCim, 33, 341 (1982); V.Aldaya, J.A.de Azcarraga and K.B.Wolf, Physica 114A, 468 (1982)

2- V.Aldaya and J.A.de Azcarraga, J.Math.Phys. 22 1425 (1981)

3- V.Bargmann, Ann.Math. 59, 1 (1954)

4- See R.Haag,J.Lopuszanski and M.Sohnius, Nucl.Phys. B88, 257 (1975)

5- For the Cartan calculus with anticommuting variables see. e.g. R.Hermann, Quantum and Fermion Differential Geometry, Interdisciplinary Mathematics vol.XVI, Math.Sci.Press, Ma.(1977).

6- For general reviews on supersymmetry and on Salam-Strathdee superspace see, e.g., P.Fayet anf S.Ferrara, Phys.Rep. 32C, 249 (1977) and A.Salam and J.Strathdee, Forts. der Phys. 26, 57 (1978).

GENERALIZED KERR-SCHILD TRANSFORMATION

A.H.Bilge and M.Gürses[†]
Physics Department, M.E.T.U., Ankara,Turkey

If (M,g) is a solution of the Einstein field equations with a physical source then the generalized Kerr-Schild transformation is defined as $\tilde{g}_{\mu\nu} = g_{\mu\nu} + 2V\ell_\mu\ell_\nu$, where ℓ_μ is a null vector and V is a scalar function. The resulting geometry $(M,\tilde{g})$ is called the generalized Kerr-Schild space-time. If $(M,\tilde{g})$ satisfies the Einstein field equations then the resulting differential equations are of the form of backlund transformations which are mostly linear in function V. Generalized Kerr-Schild space-times have been partly studied by Xanthopoulos[1] for vacuum and by Taub[2] for pure radiation cases. In this work we use the generalized Kerr-Schild transformation for two purposes; to investigate how the geometrical properties of (M,g) transform, and to find new solutions of the Einstein field equations for several sources.

Let $\zeta^a = (\ell,n,m,\bar{m})$ and $\tilde{\zeta}^a = (\tilde{\ell},\tilde{n},\tilde{m},\tilde{\bar{m}})$ be the null tetrad bases one-forms corresponding to the space-times (M,g) and $(M,\tilde{g})$ respectively, then the GKS transformation is simply given as:

$$\tilde{\ell} = \ell \qquad \tilde{n} = n + V\ell , \quad \tilde{m} = m , \quad \tilde{\bar{m}} = \bar{m} , \tag{1}$$

Under this transformation Newman-Penrose (NP) spin coefficients, trace-free Ricci, Ricci scalar and Weyl spinors transform respectively as follows:

$$\tilde{\kappa} = \kappa , \quad \tilde{\sigma} = \sigma , \quad \tilde{\rho} = \rho , \quad \tilde{\epsilon} = \epsilon , \quad \tilde{\tau} = \tau , \quad \tilde{\pi} = \pi$$

$$\tilde{\lambda} = \lambda + \bar{\sigma}V , \quad \tilde{\mu} = \mu + \rho V , \quad \tilde{\beta} = \beta + 1/2\kappa V , \quad \tilde{\alpha} = \alpha + 1/2\bar{\kappa}V ,$$

$$\tilde{\gamma} = \gamma + 1/2(D+2\bar{\epsilon}+\rho-\bar{\rho})V , \quad \tilde{\nu} = \nu + (\bar{\delta}+2\alpha+2\bar{\beta}-\pi-\bar{\tau})V+\bar{\kappa}V^2 , \tag{2}$$

$$\tilde{\Phi}_{00} = \Phi_{00} + 2\kappa\bar{\kappa}V \equiv \Phi_{00} + S_{00}$$

$$\tilde{\Phi}_{10} = \Phi_{10} + 1/2D(\bar{\kappa}V)+1/2\kappa\bar{\sigma}\, V+1/2\bar{\kappa}(D+3\epsilon+\bar{\epsilon}-\bar{\rho})V \equiv \Phi_{10} + S_{10}$$

$$\tilde{\Phi}_{20} = \Phi_{20}+(D+3\epsilon-\bar{\epsilon}-2\rho)(\bar{\sigma}V)+\bar{\kappa}(\bar{\delta}+2\alpha+2\bar{\beta}-\pi-\bar{\tau})V \equiv \Phi_{20} + S_{20}$$

$$\begin{aligned}\tilde{\Phi}_{11} = \Phi_{11}&+1/4D(D+2\bar{\epsilon}+\rho-\bar{\rho})V+1/2V\, D\epsilon+1/4\delta(\bar{\kappa}V)-1/4\bar{\delta}(\kappa V)\\ &+\kappa(2\alpha+2\bar{\beta}-\pi-\bar{\tau}+V\bar{\kappa}+\bar{\delta})V+1/2(\epsilon+\bar{\epsilon}-\rho+\bar{\rho})(2\bar{\epsilon}+\rho-\bar{\rho}+D)V\\ &+\epsilon(\epsilon+\bar{\epsilon}+D)V-[\rho^2+\epsilon(\rho-\bar{\rho})-\sigma\bar{\sigma}+1/2\bar{\kappa}(\tau+\bar{\pi}+\bar{\alpha}-\beta)\\ &+1/2\kappa(\bar{\tau}+\pi+\bar{\beta}-\alpha)]V \equiv \Phi_{11} + S_{11}\end{aligned}$$

$$\begin{aligned}\tilde{\Phi}_{12} = \Phi_{12}&\; 1/2\delta(2\bar{\epsilon}+\rho-\bar{\rho}+D)V-1/2\Delta(\kappa V)+1/2(-\tau+\bar{\alpha}+\beta+V\kappa)\; x\\ &(D+\rho-\bar{\rho}+2\bar{\epsilon})V+(\gamma\kappa-\rho\tau)V+\sigma(2\alpha+2\bar{\beta}-\bar{\tau}-\pi+\bar{\kappa}V+\bar{\delta})V\\ &+\epsilon(2\bar{\alpha}+2\beta-\tau-\bar{\pi}+\kappa V+\delta)V+\beta(\bar{\epsilon}-\epsilon-\bar{\rho})V+1/2\kappa(\gamma-\bar{\gamma}-\mu)V\\ &+1/2\kappa(\bar{\epsilon}-\epsilon-\bar{\rho})V^2-\alpha\sigma V-1/2\bar{\kappa}(\bar{\lambda}+\sigma V)V \equiv \Phi_{12} + S_{12}\end{aligned}$$

$$\tilde{\Phi}_{22} = \Phi_{22}+\delta(2\alpha+2\bar{\beta}-\bar{\tau}-\pi+\bar{\kappa}V+\bar{\delta})V-\Delta(\rho V)+VD(\mu+\rho V)$$
$$-2\mu\rho V-\rho^2V^2-\lambda\sigma V-\bar{\sigma}(\bar{\lambda}+\sigma V)V-(\gamma+\bar{\sigma})\rho V$$
$$-(\mu+\rho V)(D+\epsilon+\bar{\epsilon})V+\pi(2\bar{\alpha}+2\beta-\tau-\bar{\pi}+\kappa V+\delta)V$$
$$+2\nu\kappa V-(\tau-3\beta-\bar{\alpha}-2V\kappa)(2\alpha+2\bar{\beta}-\bar{\tau}-\pi+\bar{\kappa}V+\bar{\delta})V$$
$$\equiv \Phi_{22} + S_{22} \qquad (3)$$
$$6\tilde{\Lambda} = 6\Lambda-1/2D(2\bar{\epsilon}-3\rho-\bar{\rho}+D)V-VD\epsilon+1/2\delta(\bar{\kappa}V)-1/2\bar{\delta}(\kappa V)$$
$$+\kappa(2\alpha+2\bar{\beta}-\bar{\tau}-\pi+\bar{\kappa}V+\bar{\delta})V-1/2(\rho-\bar{\rho}+\epsilon+\bar{\epsilon})(2\bar{\epsilon}+\rho-\bar{\rho}+D)V$$
$$-\epsilon(\epsilon+\bar{\epsilon}+D)V-\rho(\rho+2\bar{\rho}-2\epsilon-2\bar{\epsilon})V-\epsilon(\rho-\bar{\rho})V-\sigma\bar{\sigma}V$$
$$+1/2\bar{\kappa}(\tau+\bar{\pi}-\bar{\alpha}+\beta)V+1/2\kappa(\bar{\tau}+\pi+\alpha-\bar{\beta})V \equiv 6\Lambda+6S_{\Lambda}, \qquad (4)$$
$$\tilde{\Psi}_0 = \Psi_0 + 2\kappa^2V \quad , \quad \tilde{\Psi}_1 = \Psi_1 + 1/2(D+\epsilon+3\bar{\epsilon}+3\rho-2\bar{\rho})(\kappa V)-1/2\bar{\kappa}\sigma V$$
$$3\tilde{\Psi}_2 = 3\Psi_2+1/2(D+\epsilon+\bar{\epsilon}+\rho-\bar{\rho})(D+2\epsilon+2\bar{\epsilon}+3\rho-\bar{\rho})V-1/2\delta(\bar{\kappa}V)+1/2\bar{\delta}(\kappa V)$$
$$-1/2[4\sigma\bar{\sigma}+\bar{\kappa}(\tau+\bar{\pi}+\beta-\bar{\alpha})+\kappa(\bar{\tau}+\pi+\alpha-\bar{\beta})]V$$
$$+2\kappa(\bar{\delta}+2\alpha+2\bar{\beta}-\bar{\tau}+\pi+\bar{\kappa}V)V$$
$$\tilde{\Psi}_3 = \Psi_3-1/2\Delta(\bar{\kappa}V)+VD(\alpha+1/2\bar{\kappa}V)+1/2(\bar{\delta}+\bar{\beta}-\bar{\tau}+\alpha+\bar{\kappa}V) \times$$
$$(D+2\bar{\epsilon}+\rho-\bar{\rho})V+(\rho+\epsilon)(\bar{\delta}+2\alpha+2\bar{\beta}-\pi-\bar{\tau}+\bar{\kappa}V)V-(\tau+\beta)\bar{\sigma}V$$
$$-1/2\kappa V(\lambda+\bar{\sigma}V)+1/2\bar{\kappa}(\gamma+\bar{\gamma}-\bar{\mu})V$$
$$\tilde{\Psi}_4 = \Psi_4+(\bar{\delta}+3\alpha+\bar{\beta}+\pi-\bar{\tau}+2\bar{\kappa}V)(\bar{\delta}+2\alpha+2\bar{\beta}-\pi-\bar{\tau}+\bar{\kappa}V)V-\Delta(\bar{\sigma}V)$$
$$-[\bar{\sigma}(\mu+\bar{\mu}+3\gamma-\bar{\gamma})+(\lambda+\bar{\sigma}V)(3\bar{\epsilon}-\epsilon+3\rho-\bar{\rho})-2\nu\bar{\kappa}]V \quad , \qquad (5)$$

The above transformations (2-5) imply the following (transformation of the properties of the geometry (M,g)):

a) If $\kappa=0$, i.e., ℓ_μ is a geodesic null vector, and if (M,g) is algebraically special, then $(M,\tilde{g})$ is also algebraically special. If $\kappa\neq 0$ $(M,\tilde{g})$ may not be algebraically special. If (M,g) is algebraically special but ℓ_μ is not a proncipal null direction of the Weyl tensor of (M,g) then $(M,\tilde{g})$ is not necessarily algebraically special.
b) If the null vector ℓ_μ is a principal null direction of the both Weyl tensors of (M,g) and $(M,\tilde{g})$ then ℓ_μ must be a geodesic null vector, or if ℓ_μ is a repeated principal null direction of the Weyl tensor of (M,g) then ℓ_μ must also be a repeated principal null direction of the Weyl tensor of $(M,\tilde{g})$.
c) If the tetrad frame $\zeta^a=(\ell,n,m,\bar{m})$ propagates parallely along ℓ in (M,g) then the tetrad frame $\tilde{\zeta}^a=(\ell,\tilde{n},m,m)$ propagates parallely along ℓ in $(M,\tilde{g})$.
d) If ℓ_μ is a geodesic null vector then the mixed component of the Einstein tensor in $(M,\tilde{g})$ is linear in V.
e) The local Lorentz transformations which changes the tetrad bases ζ^a leave the metric g invariant. The four parameter subgroup of

these transformations (Preserving ℓ-direction) leaves the metric $\tilde{g}$ invariant as well. The remaining two parameter transformation, $n'=n$, $\ell'=\ell+sm+\bar{s}\bar{m}+s\bar{s}n$, $m=m+\bar{s}n$, where s is a complex function, is very useful while searching for new solutions of Einstein's equations.

f) If $\xi = A\ell+Bn+C\bar{m}+\bar{C}m$ is a Killing vector both in (M,g) and $(M,\tilde{g})$ then

$$\Delta(BV)-VD(A+BV)+(\gamma+\bar{\gamma})BV+(A+BV)(D+\epsilon+\bar{\epsilon})V$$

$$+\bar{C}(\delta+2\bar{\alpha}+2\beta-\bar{\pi}-\tau)V+C(\bar{\delta}+2\alpha+2\bar{\beta}-\pi-\bar{\tau})V = 0 \quad ,$$

$$[A(D+\epsilon+\bar{\epsilon})+B(\gamma+\bar{\gamma})]V = 0 \quad .$$

g) Let ξ be a strongly asymptotically constant vector field both on (M,g) and $(M,\tilde{g})$, and S be a unit two sphere where intersects the future null infinity $\mathscr{I}^+$. Then the components of the total Bondi energy-momenta P^B and $\tilde{P}^B$ of the isolated systems described by (M,g) and $(M,\tilde{g})$ respectively, in the direction of ξ on $\mathscr{I}^+$ are related by

$$\eta(\tilde{P}^B,\xi) = \eta(P^B,\xi) + 1/4\pi \int_S B_o V_o \, dS \quad ,$$

where $V_o = V|_{\mathscr{I}^+}$, $B_o=\ell_\mu\xi^\mu|_{\mathscr{I}^+} \geq 0$, and η denotes the flat Minkowski inner product[3]. If $P^B = 0$ then $V_o \geq 0$.

h) If both (M,g) and $(M,\tilde{g})$ are asymptotically flat vacuum space-times, then their Weyl tensors must be algebraically special. GKS transformation does not preserve algebraic generality and asymptotic flatness simultanously for vacuum space-times.

The transformation of the Einstein tensor is found as:

$$\tilde{G}_{\mu\nu} = G_{\mu\nu}+[2S_{22}+8VS_{11}+2V^2\Phi_{00}+2V^2S_{00}+4V(\Phi_{11}+3\Lambda)]\ell_\mu\ell_\nu$$

$$+2(S_{11}-3S_\Lambda)\tilde{g}_{\mu\nu}+2S_{00}\, n_\mu n_\nu-2(2S_{11}+V\Phi_{00}+VS_{00})(\ell_\mu n_\nu+\ell_\nu n_\mu)$$

$$-2(S_{21}+V\Phi_{10}+VS_{10})(\ell_\mu m_\nu+\ell_\nu m_\mu)-2(S_{12}+V\Phi_{01}+VS_{01})(\ell_\mu\bar{m}_\nu+\ell_\nu\bar{m}_\mu)$$

$$-2S_{01}(n_\mu\bar{m}_\nu+n_\nu\bar{m}_\mu)-2S_{10}(n_\mu m_\nu+n_\nu m_\mu)+2S_{20}m_\mu m_\nu+2S_{02}\,\bar{m}_\mu\bar{m}_\nu \,, \qquad (6)$$

where $S_{ij} \equiv \tilde{\Phi}_{ij}-\Phi_{ij}$ and $S_\Lambda = \tilde{\Lambda} - \Lambda$ are given by Eqs.(3) and (4). Talbot[4] has studied the case when both (M,g) and $(M,\tilde{g})$ are vacuum and algebraically special space-times. He considered that ℓ_μ is a repeated principal null direction of the Weyl tensor of (M,g). The case when ℓ_μ is not a principle null direction remains to be studied. Robinson-Robinson[5], Hughston[6], and Stephani[7] have tried to use GKS transform to obtain a new solution for Einstein's field equations with a pure radiation. By use of Eq.(6) we show that one can utilize GKS transform to find a solution of Einstein's field equations with a null fluid. The procedure is a follows;,Equations to be satisfied are:

$$\kappa = 0,\ 2S_{11}+V\Phi_{00} = 0 \quad , \quad S_{21}+V\Phi_{10} = 0$$

$$(D-2\bar{\rho}+3\bar{\epsilon}-\epsilon)(\sigma V) = 0 \quad , \qquad (7)$$

then

$$\tilde{G}_{\mu\nu} = G_{\mu\nu} + (p+e)\ell_\mu\ell_\nu - p\tilde{g}_{\mu\nu}$$

where

$$p+e = 4VS_{11} + 2S_{22} + 4V(\Phi_{11} + 3\Lambda)$$

$$p = 2(S_{11} - 3S_{\Lambda})$$

of course, the positivity of both p and e should be required. If the background geometry is an Einstein-space then $\tilde{g}_{\mu\nu}$ solves the following field equations:

$$\tilde{G}_{\mu\nu} = (p+e-2V\lambda)\ell_\mu \ell_\nu - (p-2V\lambda)\tilde{g}_{\mu\nu}$$

$$= (\tilde{p}+\tilde{e})\ell_\mu \ell_\nu - \tilde{p}\tilde{g}_{\mu\nu}$$

where $\tilde{p} = p-2V\lambda$, $\tilde{e}=e$ and since $\Phi_{ij} = 0$ which implies $S_{11} = S_{12} = 0$ then $\tilde{p} = (6S_\Lambda + 2V\lambda)$, $\tilde{e} = 2S_{22} + 6S - 2V\lambda$. The function V satisfies the following second order linear differential equations ($\kappa = 0$):

$$S_{11} = S_{21} = 0 \quad , \quad (D-2\bar{\rho}+3\bar{\varepsilon}-\epsilon)\sigma V = 0 \quad .$$

An interesting application of the GKS transformation and some new exact solutions of Einstein field equations with a null fluid distribution are given as a poster in this meeting.

We would like to thank Tekin Dereli, Rahmi Güven and Martin Walker for discussions.

† Present Address: Applied Mathematics Department, TÜBİTAK, Marmara Scientific and Industrial Research Institute, P.K.21 GEBZE/KOCAELİ, TURKEY.

REFERENCES

1 Basilis C.Xanthopoulos, J.Math.Phys. 19, 1607(1978).

2 A.H.Taub, Annals of Physics, 134, 326(1981).

3 Martin Walker, "On the positivity of total gravitational energy at retarded times "Lectures at the 1982 Les Houches Summer School on gravitational radiation. Max-Planck Preprints, MPA 19, Garching bei München, May 1982.

4 C.S.Talbot, Commun.Math.Phys. 1345(1969).

5 I.Robinson and J.R.Robinson, Int.J.Theor.Phys. 2, 231(1969).

6 L.P.Hugston, Int.J.Theor.Phys. 4, 267(1971).

7 H.Stephani, J.Phys. A12, 1045(1979).

8 A.H.Bilge and M.Gürses, "Adding a Λ-term to pp-wave solutions of Einstein field equations", Poster presented at the XI.International Colloquium on Group Theoretical Methods in Physics, Istanbul, Turkey (1982).

ADDING A Λ-TERM TO PP-WAVE SOLUTIONS OF THE EINSTEIN FIELD EQUATIONS

A.H.Bilge and M.Gürses
Department of Physics, M.E.T.U.,Ankara,Turkey

As a simple application of the generalized Kerr-Schild (GKS) Transformation[1] we start with a conformally flat Einstein space

$$\ell = \phi^{-1}du \quad , \quad n = \phi^{-1}dv \quad , \quad m = \phi^{-1}dz \tag{1}$$

where $\phi = \phi_o + az + \bar{a}\bar{z} + bu + cv$, the cosmological constant $\Lambda = -a\bar{a} + bc = -\lambda/6$ $(G_{\mu\nu} = \lambda\ g_{\mu\nu})$, and a is an arbitrary complex, b and c are arbitrary real constants. For this background geometry $\Phi_{ij} = 0$, $(i,j = 0,1,2)$, $\Psi_A = 0$, $(A = 0\text{-}4)$ and $D = \phi\partial_v$, $\Delta = \phi\partial_u$ and $\delta = -\bar{\phi}\partial_z$. The Newman-Penrose spin coefficients are simply given by:

$$\kappa = \lambda = \nu = \sigma = 0 \ , \qquad \rho = -2\epsilon = c \ , \ -\mu = 2\gamma = b \tag{2}$$

$$-2\alpha = 2\bar{\beta} = \pi = -\bar{\tau} = a$$

By use of the GKS transformation $\tilde{g}_{\mu\nu} = g_{\mu\nu} + 2V\ell_\mu\ell_\nu$ we want to construct some new exact solutions of the Einstein field equations with a null-fluid distribution. For this purpose we take $\tilde{\Phi}_{00} = \tilde{\Phi}_{01} = \tilde{\Phi}_{11} = 0$ which imply

$$\tilde{G}_{\mu\nu} = 2\tilde{\Phi}_{22}\ell_\mu\ell_\nu + 6\tilde{\Lambda}\tilde{g}_{\mu\nu}$$

where

$$\tilde{\Phi}_{22} = (\delta + 2\bar{a})\bar{\delta}V + a\delta V - c\Delta V - 2bcV \tag{3}$$

$$3\tilde{\Lambda} = cDV - 3c^2V + 3\Lambda$$

When $c \neq 0$ the function V must satisfy the following differential equation:

$$\delta DV - c\delta V + \bar{a}DV = 0 \tag{4}$$

and $D\tilde{\Lambda} = 0$. The Weyl spinors are given as:

$$\tilde{\Psi}_0 = \tilde{\Psi}_1 = 0 \ , \quad \tilde{\Psi}_2 = cDV \quad , \quad \tilde{\Psi}_3 = c\delta V \ , \ \tilde{\Psi}_4 = (\bar{\delta} + a)\bar{\delta}V \ . \tag{5}$$

If both DV and δV are different than zero then the space-time is of Petrov-Pirani type II. If $\delta V = 0$ but $DV \neq 0$ then the type is D. If $DV = 0$ which implies $\delta V = 0$ $(c \neq 0)$ then the space-time becomes conformally-flat. If the constant $c = 0$ then V satisfies two differential equations:

$$(\delta + \bar{a})DV = 0 \quad , \quad D^2V = 0 \tag{6}$$

and $\tilde{\Lambda} = \Lambda$. $\tilde{\Phi}_{22}$ and $\tilde{\Psi}_A$ $(A = 0\text{-}4)$ reduce to

$$\tilde{\Phi}_{22} = (\delta + \bar{a})\bar{\delta}V + (\bar{a}\bar{\delta} + a\delta)\ V$$

$$\tilde{\Psi}_0 = \tilde{\Psi}_1 = \tilde{\Psi}_2 = \tilde{\Psi}_3 = 0 \quad , \quad \tilde{\Psi}_4 = (\bar{\delta} + a)\bar{\delta}V \tag{7}$$

Hence the space-time for this case is of type N. When $DV = b = 0$ the only non-zero spin coefficients are $-2\alpha=2\bar{\beta}=\pi=-\bar{\tau}=a$. Our aim is now to interpret the source of $(M,\tilde{g})$ as:

a. Empty space-time with a cosmological constant $(\tilde{\Phi}_{22}=0)$:Defining d and ^{*1} as the exterior differential operator and the volume four-form we have $d^*dV=0$ and $\tilde{\Psi}_4=\phi^3\Psi_4{}^0(u)$, $\tilde{G}_{\mu\nu} = \lambda\tilde{g}_{\mu\nu}$.

b. Null Yang-Mills field with a cosmological constant: $\tilde{\Phi}_{22} = K\Phi_2{}^a\bar{\Phi}_2{}^a$ where K is the gravitational coupling constant and Yang-Mills field two form $F^a = -\bar{\Phi}_2{}^a \ell\wedge\bar{m} - \Phi_2{}^a \ell\wedge m$ and a is the internal group index. Yang-Mills field equations imply $\Phi_2{}^a = \phi^2\Phi_2{}^{0a}(u)$ and gravitational field equations give

$$d^*dV = -K\,\rho_{Y-M}\,{}^*1 \quad , \quad \rho_{Y-M} = 2\phi^2\Phi_2{}^{0a}(u)\bar{\Phi}_2{}^{0a}(u) \tag{8}$$

and $\psi_4 = \phi^3[K\,a\Phi_2{}^a\bar{\Phi}_2{}^a + \Psi_4{}^0(u,z)]$. The potential one-form $A = \mathcal{A}^a\ell$ in which $\Phi_2{}^a = -(\bar{\delta}\mathcal{A}^a + a\mathcal{A}^a)$ satisfies

$$d^*d\mathcal{A}^a + 2\lambda/3\,\mathcal{A}^{a*}1 = 0 \tag{9}$$

which is formally a massive Klein-Gordon equation in (M,g).

c. One component neutrino field with a cosmological constant. Let $\psi^A = \psi o^A$, the component of ψ^A along ι^A is taken to be zero[2], then $T^{neut.}_{\mu\nu} = \rho_N \ell_\mu \ell_\nu$ where $\rho_N = 2i(\psi\Delta\bar{\psi}-\bar{\psi}\Delta\psi)$. The neutrino field equation gives $\psi = \phi^{3/2}\psi_0(u)$ and the gravitational field equations reduce to

$$d^*dV = K\rho_N{}^*1$$

It is interesting that the Weyl spinor field Ψ_4 in part (a), Yang-Mills field $\Phi_2{}^a$ in part (b) and the neutrino field ψ in part (c) have a common property that they behave like, field = $\phi^{s+1}f_s(u)$ where s is the spin of the fields and $f_s = (\psi_4{}^0(u)$ for $s = 2$, $\Phi_2{}^{0a}(u)$ for $s = 1$, and $\psi^0(u)$ for $s = 1/2)$. Such a behaviour is the consequence of the conformal invariance of the massless spin-field equations. In all cases the energy density is of the form $2\tilde{\Phi}_{22} = \phi^4\varepsilon(u)$, where ε is a function of u. When the cosmological constant vanishes which implies $a = 0$, then the space-times (M,g) and $(M,\tilde{g})$ becomes the flat Minkowski and pp wave geometries respectively[3,4].

REFERENCES

1 A.H.Bilge and M.Gürses, "Generalized Kerr-Schild Transformation Talk presented at the XI. International Colloquium on Group Theoretical Methods is Physics, Istanbul, Turkey (1982).

2 J.Wainwraight, J.Math.Phys. 12, 828(1970).

3 J.Ehlers and W.Kundt, "Gravitation, an introduction to Current Research", Wiley, New York, London.Ed. L.Witten (1962).

4 M.Gürses, J.Phys. A14, 1957(1981).

ASPECTS OF A SPIN(1,4) GAUGE THEORY WITH KALUZA-KLEIN SYMMETRY*

T.Dereli

Physics Dept., Middle East Technical Univ.

Ankara, Turkey

R.W.Tucker

Dept. of Physics, Univ. of Lancaster

Lancaster, U.K.

We describe the interaction of Dirac spinors with gravitation and electromagnetism in terms of a 5-dimensional non-Riemannian geometry with a one-parameter group of isometry. This is a generalization of one of the oldest attempts of unification by Kaluza and Klein[1]. In fact the so-called dimensional reduction technique has become very popular recently in the various attempts made to find a viable extended supergravity[2]. We analyse our model in terms of co-ordinate independent concepts and extensive use is made of the calculus of differential forms.

* Research supported by the NATO Scientific Collaboration Grant No.RG.160.81

The dynamics of the theory is determined by a variational principle from an action 5-form which is invariant under SO(1,4) induced gauge transformations. Next we specialise to spaces that admit a Killing vector. (Kaluza-Klein symmetry) The integral curves of this vector field are chosen to be closed and the resulting stratification of the 5-dimensional manifold by these curves is used to identify a projection onto a 4-dimensional submanifold which is identified as space-time. The unique solution for the 5-torsion and SO(1,4) valued connection is used to construct the reduced action 4-form and some of the space-time interactions are displayed. We included a Jordan-Thiry scalar in the 5-frame ansatz in order the field equations obtained by varying the reduced action 4-form be consistent with the 5-field equations. However, we explicitly display the reduced spinor field equation which contains an additional coupling to the Jordan-Thiry scalar field.

Details will be published.[3]

(1) A.Lichnerowicz, Théories relativistes de la gravitation et de l'électromagnétisme, (Masson et Cie, 1955, Paris)

(2) J.Scherk, J.H.Schwarz, Nucl.Phys. B153 (1979)61

E.Cremmer, B.Julia, Nucl.Phys. B159 (1979)141

(3) T.Dereli, R.W.Tucker, Nucl.Phys.B (1982)

COMPOSITE GRAVITY AND COMPOSITE SUPERGRAVITY

Jerzy Lukierski
International Centre for Theoretical Physics, Trieste, Italy,
and
Institute for Theoretical Physics, University of Wroclaw,
ul. Cybulskiego 36, Wroclaw, Poland.

ABSTRACT

It is known that the composite YM H-gauge theory can be constructed from σ-fields taking values in a symmetric Riemannian space $\frac{G}{H}$. We extend such a framework to graded σ-fields taking values in supercosets. We show that from supercoset σ-fields one can construct composite gravity, and from supercoset σ-superfields the composite supergravity models.

1. The composite gauge potentials describing the internal symmetry group H were constructed from the σ-fields which take values in larger internal symmetry group $G \supset H$ [1),2)]. Usually one assumes that $(H, K \equiv \frac{G}{H})$ is the Riemannian symmetric pair, i.e. the algebra $g = h \oplus k$ is Z_2-graded. The composite H-connections A^k are defined by the following Cartan one-form:

$$\mathcal{A} = K_2^{-1}\, dK = A + E = A^k h^k + e^r K^r \quad . \tag{1}$$

Using Cartan-Maurer equation for G one gets the formula for composite H-curvature

$$F = dA - \frac{1}{2} A \wedge A = \frac{1}{2} E \wedge E \quad . \tag{2}$$

If we observe that $A = A_\mu\, dx^\mu$ and $F = F_{\mu\nu}\, dx^\mu \wedge dx^\nu$ one can conclude that

$\frac{G}{H}$ σ-fields	⟶	composite YM H-gauge theory	(A)

The known examples of such a construction are the composite U(n) potentials obtained from $\frac{U(n+m)}{U(n)\times U(m)}$ σ-fields [3)-5)], $SP(1) \simeq SU(2)$ composite gauge fields in HP(n) σ-models [6),7)], or SU(8) gauge fields constructed from $\frac{E_7}{SU(8)}$ σ-fields describing scalar sector of N = 8 supergravity [8)].

2. Such a scheme can be supersymmetrized in three ways [9)]. The first already known way is obtained by <u>supersymmetric extension of the space-time co-ordinates</u>, i.e. by the replacement of σ-fields by σ-superfields.

In such a formulation the composite gauge superfields are defined again by the formula (1), but the differentials dK and one-forms A are expanded in even and odd differentials. For D = 4 "flat" Salam-Strathdee superfields one can write

$$dK = D_\alpha K \; d\theta^\alpha + \partial_\mu K \; e^\mu \quad , \tag{3}$$

$$A = A_\alpha \; d\theta^\alpha + A_\mu \; e^\mu \quad , \tag{4}$$

where $A_\alpha = G^{-1} D_\alpha G$, $A_\mu = G^{-1}\partial_\mu G$, $e_\mu = dx_\mu - \frac{1}{4}\bar{\theta}\gamma_\mu d\theta$ and $D_\alpha = \frac{\partial}{\partial\theta_\alpha} - \frac{1}{4}\bar{\theta}\gamma_\mu \frac{\partial}{\partial x_\mu}$. If we observe that $de^\mu = -\frac{1}{4} d\bar{\theta}\,\gamma^\mu\, d\theta$ the formulae for field strength components (3) look as follows:

$$F_{\alpha\beta} = D_\alpha A_\beta - \{A_\alpha, A_\beta\} - \frac{1}{4}(\gamma^\mu)_{\alpha\beta} A_\mu \quad , \tag{5a}$$

$$F_{\alpha\mu} = D_\alpha A_\mu - \partial_\mu A_\alpha - [A_\alpha, A_\mu] \quad , \tag{5b}$$

$$F_{\mu\nu} = \partial_\mu A_\nu - \partial_\nu A_\mu - [A_\mu, A_\nu] \quad , \tag{5c}$$

where

$$F = F_{\alpha\beta}\, d\theta^\alpha \wedge d\theta^\beta + F_{\alpha\mu}\, d\theta^\alpha \wedge e^\mu + F_{\mu\nu}\, e^\mu \wedge e^\nu \quad . \tag{6}$$

Because the coset structure implies F = 0 , using (2) and (5a), one gets

$$A_\mu = -(\gamma_\mu)^{\alpha\beta}\left(D_\alpha A_\beta - \{A_\alpha, A_\beta\} - \frac{1}{2} E_\alpha \wedge E_\beta\right) \tag{7}$$

Substituting (7) in (5b) one obtains the composite field strength superfield $F_{\alpha\mu}(x,\theta)$ which provides the composite SSYM Yang-Mills action (see,e.g. Ref.10). The formulation of D = 4 composite SSYM theory has been proposed independently in Refs.11 and 12. The relation (A) is generalized as follows:

$$\boxed{\frac{G}{H}\ \sigma\text{-superfields} \longrightarrow \text{composite SSYM H-gauge theory}} \qquad \text{(B)}$$

3. In order to describe the composite D = 4 gravity we should construct the composite vierbein e_μ^a which transforms in tangent space as a four-vector under local SL(2,C) spin group and is a world vector under general co-ordinate transformations. Such a composite vierbein we obtain from the Cartan one-form on the supercoset

$$K = \frac{Osp(1;4)}{SL(2,C)} = \frac{Sp(4)}{SL(2,C)} \times \frac{OSp(1;4)}{Sp(4)}$$

which we parametrize as follows (see also Refs.13-16):

$$\tilde{K} = \exp V^a(x)\,P_a \quad \exp \psi^\alpha(x)\,Q_\alpha \quad , \qquad (8)$$

where

- V^a describes local space-time co-ordinates,
- ψ^α describes the spinor Goldstones.

The composite SL(2,C) connections ω^{ab} and composite vierbein e_μ^a are defined in a very analogous way to (1)

$$\mathcal{A} = \tilde{K}^{-1}\,d\tilde{K} = \Omega + E + \Phi = \omega^{ab}M_{ab} + e^aP_a + \Phi^iK_i \quad , \qquad (9)$$

where M_{ab} are Lorentz generators. The Cartan structure equations

$$R = d\Omega - \Omega\wedge\Omega \qquad T = dE - \Omega\wedge E \qquad (10)$$

provide after inserting Ω and E from (9) the formulae for composite SL(2,C) curvature and torsion. The gravity described by any action which is the function of vierbeins, SL(2,C) curvature and torsion is made composite by inserting respective composite formulae for e_μ^a , R and T (see Ref.17).

In general the supergroup OSp(1;4) can be replaced by a larger one (e.g. $\tilde{G}$ = OSp(4;N)). We have therefore the relation

$\frac{\tilde{G}}{SL(2,C)}$ σ-fields	$\longrightarrow$	composite gravity with fermionic fundamental fields	(C)

4. The composite N = 1 D = 4 supergravity is obtained by inserting the composite achtbein superfields $E_A^{\ B}$ into the well-known supergravity action formula

$$S = \int d^4x\, d^4\theta \quad \mathrm{sdet}(E_B^{\ A}) \quad . \tag{11}$$

The composite supervierbeins are obtained as functions of Goldstone superfields by considering the Cartan one-form on the supercoset

$$\tilde{\tilde{K}} = \frac{\tilde{G}}{SL(2,C)} = \frac{OSp(1;4)}{SL(2,C)} \times \frac{\tilde{G}}{Osp(1;4)}$$

parametrized as follows:

$$\tilde{\tilde{K}} = \exp(V^a(x,\theta)\, P_a + \eta^\alpha(x,\theta)\, Q_\alpha)\, \exp \xi^A(x,\theta)\, L_A \quad , \tag{12}$$

where (V^a, η^α) parametrize the local superspace co-ordinates, ξ^A describe Goldstone superfields living in the supercoset $S = \frac{\tilde{G}}{OSp(1;4)}$. The Cartan one-form

$$\tilde{\tilde{K}}^{-1}\, d\tilde{\tilde{K}} = \omega^{ab} M_{ab} + E^a P_a + E^\alpha Q_\alpha + \Phi^A L_A \tag{13}$$

defines the composite achtbein superfields $E_B^{\ A}$ (A = (a,α), B = (μ,β)) through the expansion into the covariant differentials $d\theta^\alpha$, e^μ

$$\begin{aligned} E^a &= E^a_\beta\; d\theta^\beta + E^a_\mu\; e^\mu \\ E^\alpha &= E^\alpha_\beta\; d\theta^\beta + E^\alpha_\mu\; e^\mu \; . \end{aligned} \tag{14}$$

For the choice $\tilde{G}$ = OSp(2;4) the composite achtbeins and composite supergravity have been introduced in Ref.17.

In order to obtain composite supergravity it is necessary to consider supercoset σ-superfields as fundamental ones, i.e.

$$\frac{\tilde{G}}{SL(2,C)} \ \sigma\text{-superfields} \longrightarrow \text{composite supergravity} \qquad (D)$$

The composite gravity and composite supergravity satisfy the torsion constraints modified in comparison with "elementary" gravity and supergravity models. We expect that torsion constraints describe only the effective theory valid at sufficiently large distances. For the support of this conjecture we mention that in "non-geometric" composite gravity of Amati and Veneziano [18] the non-vanishing torsion can be neglected in the sufficiently low energy region.

REFERENCES

1) M.A. Semenov-Tjan-Saenski and L.D. Faddeev, Westnik Leningr. Univ. 13, 81 (1977) (in Russian).
2) A.P. Balachandran, A. Stern and G. Trahern, Phys. Rev. D19, 2416 (1978).
3) J. Fröhlich, Lecture at Bielefeld Symposium, December 1978, IHES preprint (1979).
4) M. Dubois-Violette and Y. Georgelin, Phys.Lett. 82B, 251 (1979).
5) I. Bars in Proc. of 8th Int. Colloquium on Group Theor. Methods in Elem. Particle Phys., Kiriet Amavin, Israel, March 1979, published in Ann. of Isr. Phys. Soc. 3, 58 (1980).
6) J. Lukierski in Proc. of Summer Inst. of Field-Theor. Methods in Elem. Particle Phys., August 1979, Kaserslautern, published by Springer, p.309 (1980).
7) F. Gürsey and H.C. Tze, Ann. Phys. 128, 29 (1979).
8) E. Cremmer and B. Julia, Nucl.Phys. B159, 141 (1979).
9) J. Lukierski in Proc. of Symposium on Differential-Geometric Methods, Aix-en-Provence, September 1979, published by Springer-Verlag; Lecture notes in Math. Vol. 835, 225 (1980).
10) M. Roček in Superspace and Supergravity, Eds. S.W. Hawking and M. Roček (Cambridge Univ. Press, 1981), p.71.
11) I. Bars and M. Günaydin, Phys.Rev. D22, 1403 (1980).
12) J. Lukierski and B. Milewski, Phys. Lett. 93B, 91 (1980).
13) D.P. Akulov and D.W. Volkov, Teor. Math. Fiz. 18, 39 (1974).
14) D.W. Volkov and W.A. Soroka, Teor. Mat. Fiz. 20, 291 (1974).
15) F. Gürsey and L. Marchildon, Phys. Rev. D17, 2038 (1978).
16) N.S. Baaklini, J. Phys. G 4, 1 (1978).
17) J. Lukierski, CERN preprint TH-3386, August 1982.
18) D.Amati and G. Veneziano, Phys. Lett. 105B, 358 (1981); CERN preprint TH.3197 (to appear in Nucl. Phys. B).

LABELLING OF IRREDUCIBLE REPRESENTATIONS OF SUPER LIE ALGEBRAS

H. Ruegg *
Département de Physique Théorique
Université de Genève
1211 Genève 4, Switzerland

ABSTRACT

We give the relations between highest weights and tensors for irreducible representations.

1. INTRODUCTION

This report is based on work done with I. Bars and B. Morel(1).

The irreducible representations (IR) of the Lie algebra of SU(M) can be labelled in different ways. One is to give the highest weight whose components are specified on a Dynkin diagram. They correspond to eigenvalues of the operators of the Cartan subalgebra. The remaining weights are obtained by applying lowering operators. Another one is to introduce covariant tensors whose symmetry properties are given by a Young tableau. One can also introduce contravariant and mixed (traceless) tensors. In this way, one gets easily all the basis states of an IR.

For superalgebras, or graded Lie algebras, the product is defined either by a commutator, or an anticommutator. The labelling of IR is analogous to ordinary Lie algebras. Kac(2) has generalized the Dynkin approach. Bars and Balantekin(3) have introduced super Young tableaux. All types of tensors are necessary.

We give the dictionary between the two labellings. While Kac-Dynkin diagrams characterize uniquely all IR, it is found that pure covariant and pure contravariant tensors correspond to IR, while for mixed tensors this is only the case provided M, N are sufficiently large compared to the number of boxes. Otherwise, the representation is reducible, but indecomposable. Supertableaux are also useful to compute the decomposition of direct product representations.

2. THE SUPERALGEBRA SU(M/N) AND KAC-DYNKIN DIAGRAMS

The super Lie algebra SU(M/N) is simple for $M \neq N$. It consists of an even ("Bosonic") part, the subalgebra SU(M) x SU(N) x U(1), and an odd ("Fermionic") part, which transforms as the representation $(M,N^*) + (M^*,N)$ of the even part. The Cartan subalgebra consists of the $M + N - 1$ mutually commuting generators H_i, the $M - 1$ first ones belonging to SU(M), the $N - 1$ last ones to SU(N), H_M playing a special role. The generator Q of U(1) is a linear combination of H_i (see below). To each H_i corresponds a simple root α_i, a "raising" operator E_i^+ and a "lowering" operator E_i^-. We shall need the commutation relations : $[H_i, E_j^\pm] = \pm a_{ij} E_j^\pm$, $i,j = 1 \ldots M + N - 1$, where a_{ij} are the elements of the Cartan matrix of SU(M/N) given by Kac[2]. They coincide with that of SU(M+N) except for $a_{MM} = o$; $a_{MM+1} = +1$. The even generators obey the commutation relations $[E_j^+, E_i^-] = \delta_{ij} H_j$, $i,j \neq M$ and the two odd generators the anticommutation relation $[E_M^+, E_M^-]_+ = H_M$. The remaining odd generators are commutators of $E_M^\pm$ with $E_i^\pm$ $(i \neq M)$. The generator Q of U(1) is given by

$$Q = \sum_{k=1}^{M-1} \frac{kH_k}{M} + H_M - \sum_{\ell=1}^{N-1} \frac{(N-\ell)}{N} H_{M+\ell}, \quad \text{from this one gets}$$

$$[Q, E_M^\pm] = \left(\frac{1}{M} - \frac{1}{N}\right) E_M^\pm$$

The state $|\Lambda\rangle$ in the representation space of an IR with highest weight Λ is defined[2] by $E_i^+ |\Lambda\rangle = o$, $i = 1 \,..\, M + N - 1$, $H_i |\Lambda\rangle = a_i |\Lambda\rangle$ The numbers a_i are non negative integers for $i \neq M$. a_M may be any real number. An IR of SU(M/N) is thus defined by the values a_i of the highest weight, which are the labels of the Kac-Dynkin diagram :

a_1 a_2 a_M a_{M+N-1}

o-o- - - - o-⊗-o- - - - - - - o

The part without ⊗ decomposes into ordinary Dynkin diagrams for SU(M) and SU(N). ⊗ corresponds to the odd root α_M, or to the special generator H_M. If we define $Q|\Lambda\rangle = q_\Lambda|\Lambda\rangle$, we find

$$q_\Lambda = \text{maximum for } M < N, \text{ minimum for } M > N \qquad (1)$$

3. SUPERYOUNG TABLEAUX AND RELATIONS TO KAC-DYNKIN DIAGRAMS

BB[3] assign to the covariant tensors $t_{AB}\dots$ resp. mixed tensors $t_{AB\dots}^{A'B'\dots}$ the Young supertableaux :

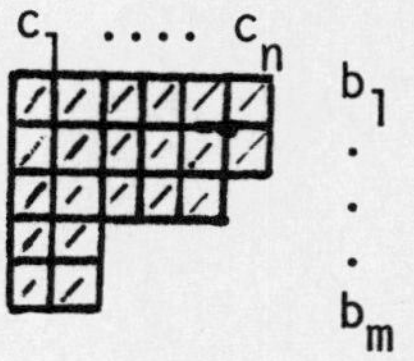

resp.

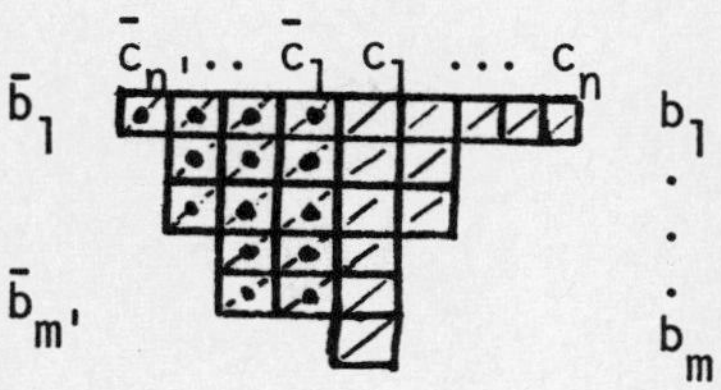

$b_i (i = 1, \dots, m)$ counts the boxes in the row i, $c_j (j = 1, \dots, n)$ counts the boxes in the column j, with the conditions :
$b_1 \geq b_2 \geq \dots \geq b_m > 0$, $c_1 \geq c_2 \geq \dots \geq c_n > 0$. Similarly for $\bar{b}_i$ and $\bar{c}_j$.
The tensor indices take the values $A, B, A', B', \dots = 1, 2, \dots, M + N$.
For the first M values, the symmetry properties given by the supertableau are those of a usual tableau. For the last N values, symmetrization and antisymmetrization must be interchanged. This is called supersymmetrization[3]. The decomposition of the fundamental IR (of dimension $M + N$) of SU(M/N) into IR of SU(M) x SU(N) x U(1) is ⧄ = (□,1)$_{q = \frac{1}{M}}$ + (1,□)$_{q = \frac{1}{N}}$, the more general rule is given in ref. (3). There is no limitation on the number of rows. One must introduce contravariant tensors $t^{A'B'\dots}$ inequivalent to covariant tensors. They correspond to the conjugate representations. The fundamental conjugate IR is denoted by ⊡ = (⊡,1)$_{q = -\frac{1}{M}}$ + (1,⊡)$_{q = -\frac{1}{N}}$ where ⊡ denotes a conjugate IR of SU(M)[4]. A general mixed tensor $t_{AB\dots}^{A'B'\dots}$ has zero supertrace[3].

The relations between Kac-Dynkin diagrams and supertableaux is obtained if one finds the highest weight determined by a given tensor. For this, it is enough to find the extremum value of q(see eq. (1)) counting boxes and using the q-values of the two fundamental IR. Taking into account supersymmetrization, one finds the dictionary[1] between Kac-Dynkin labels a_i and Young labels b_j, c_k : $a_i = b_i - b_{i+1}$,
$i = 1 \dots M - 1$, $a_M = b_M + c_1'$, $a_{M+j} = c_j' - c_{j+1}'$, $j = 1 \dots N - 1$,
$c_j' = (c_j - M)\,\Theta(c_j - M)$, $\Theta(x) = 1, 0$ if $x > 0$, $x < 0$, $b_{M+1} \leq N$,

$$a_{M-i} = \bar{b}'_i - \bar{b}'_{i+1},\ i = 1 \,..\, M-1,\ \bar{b}'_i = (\bar{b}_i - N)\,\Theta(\bar{b}_i - N),$$

$$a_M = -\bar{c}_N - \bar{b}'_1,\ a_{M+N-j} = \bar{c}_j - \bar{c}_{j+1},\ j = 1 \,\ldots\, N-1,\ \bar{c}_{N+1} \leq M.$$

4. DISCUSSION

One sees that the dictionary is not one to one. One can convince oneself that pure covariant or pure contravariant tensors belong to IR, but give not all IR. Hence one needs mixed tensors. Typical and atypical IR have been discussed in ref. 1. The adjoint IR is given by the mixed supertableau  . There are mixed tableaux corresponding to reducible but indecomposable representations. For example, for SU(1|2), this is the case for  . The reason is that the corresponding tensor t^{AB}_{C} cannot be made supertraceless.

Supertableaux are useful for computing products of IR. The usual rules can be applied. Sometimes, indecomposable representations appear, and a more detailed discussion is required.

5. REFERENCES

(1) I. Bars, B. Morel and H. Ruegg, CERN Preprint, Ref. TH 3333 (1982).

(2) V.G. Kac, Lecture Notes in Math. 676, 597-626 (Springer Verlag (1978).

(3) A.B. Balantekin and I. Bars, J. Math. Phys. 22, 1149 (1981) and 22 1980 (1982).

(4) R.C. King, J. Math. Phys. 11, 280-293 (1970).

* Partially supported by the Swiss National Science Foundation.

TENSORIAL PROPERTIES OF INCOMMENSURATE CRYSTALS

A. Janner and B.W. van Beest
INSTITUTE FOR THEORETICAL PHYSICS
CATHOLIC UNIVERSITY OF NIJMEGEN
Toernooiveld 6525 ED Nijmegen

ABSTRACT

The problem of characterizing the symmetry of tensorfields describing physical properties of incommensurate crystals is discussed in relation with the covariance group of the physical phenomena involved, and the superspace group symmetry of the crystals considered.

The invariance with respect to superspace group of tensorfields defined in the superpace (which is (3+d)-dimensional euclidean) leads to necessary and sufficient conditions for the physical tensorfields to reflect crystal symmetry in the incommensurate case also.

1. The Problem

The number of independent components of a tensor(field) describing a physical property of a crystal may be greatly reduced by the symmetry of the crystal considered: some components may even be zero because of that symmetry. Indeed according to the Curie principle the "effects" have at least the symmetry of the "causes". In a more specific and precise form this means that the tensorfield has to be invariant with respect to the symmetry group of the crystal, i.e. its space group. In the case of macroscopic crystal properties, this tensorfield reduces to a constant tensor and the symmetry group to a crystallographic point group.

These standard views are explicitely based on the crystal lattice periodicity and implicitely on the covariance group of the physical phenomena considered. The latter being non-relativistic and involving electromagnetic properties, this group is the intersection of the Galilei and of the Poincaré group, which is the so-called Newton Group, i.e. the direct product of the Euclidean group in three dimensions E(3) and the time reversal group Θ. Time reversal operations lead to "magnetic" symmetry groups and are not relevant for the aspects we want discuss and shall, therefore, not be considered here.

The problem we are concerned with is connected with the existence of incommensurate crystal phases missing 3-dimensional lattice periodicity: their euclidean symmetry is therefore not a space group and in general very low. Despite that, the ideal crystal structure is highly ordered as revealed by the diffraction pattern, and one is forced to conclude that, this euclidean symmetry is not the relevant one. (See refs 1 to 3; in ref. 4 one can find a more extended list of references.)

The ordering which makes that such a condensed phase can still be considered a crystal is reflected in the Fourier spectrum of the crystal density function. The necessary and sufficient property is the Fourier wave vectors span a 3-dimensional Z-module M^* of rank (3+d). The normal (commensurate) crystal appears to be a special case only, for which the "internal dimension" d is zero. In that case, M^* is a 3-dim. lattice, the well-known reciprocal lattice.

In the superspace approach Euclidean crystallography is recovered by considering M^* as the orthogonal projection of a lattice Σ^* in a (3+d)-dim. euclidean space, the <u>superspace</u>. The Fourier expansion of the density is extended accordingly, and one constructs a scalar density function in the superspace (describing a <u>supercrystal</u>) from which by intersection with the (real) space one gets the crystal density back.

The euclidean symmetry of the supercrystal is then a (3+d)-dim. space group (the <u>superspace group</u>) and such a group is considered to be the relevant symmetry group of the incommensurate crystal. Amazing enough such extended groups seem to be the appropriate ones even in the case of some commensurate crystals.

Coming back to crystal tensorial properties, the discussion of these would be rather elementary if one would have a natural extension to superspace of all physical phenomena considered, and, accordingly, an extension of their covariance group. Although this approach seems to have promising features, it is not a straightforward one.

Furthermore, one can feel reluctant to engage himself in such a direction because even if a crystal is incommensurate, no new physical laws are involved, and the system still obeys to non-relativistic mechanics and to electromagnetism.

It is possible not to take into account incommensurability either using a more or less well defined average symmetry, or by treating commensurate and incommensurate crystal phases in a fundamentally different way: the former by considering their space group symmetry, the latter by describing deviation from that by means of space group representations.

The present paper follows the superspace group approach, despite the lack of an explicitely worked out covariance group. It has, therefore, a preliminary character only which admits alternative possibilities, but shows how useful this approach is for dealing with tensorial properties of incommensurate crystals in a way that reduces to the standard formulation for normal crystals.

2. Recalling Superspace Symmetry

A crystal can be described by a scalar density function $\rho(\vec{r})$ defined on the 3-dim. euclidean space V which has a Fourier series expansion with wave vectors $\vec{k}$ generating a Z-module M* of rank (3+d):

$$\rho(\vec{r}) = \sum_{\vec{k} \in M^*} \hat{\rho}(\vec{k})\, e^{i \vec{k} \vec{r}} \tag{1}$$

The vectors $\vec{k}$ can thus be written as:

$$\vec{k} = \sum_{\nu=1}^{3+d} Z_\nu \vec{a}^{\nu}_{*} \tag{2}$$

with $\{\vec{a}^*\}$ a (free) basis of M* and Z_ν integral coefficients. One now considers M* as the orthogonal projection of a lattice Σ^* in a (3+d)-dim. euclidean space V_S having V as fixed subspace:

$$V_S = V \oplus V_I \tag{3}$$

The subspace V_I is called "internal" whereas $V = V_E$ is also denoted as "external". Accordingly the elements r_s of V_S can be decomposed in external and in internal components:

$$r_s = (\vec{r},\vec{r}_I) \quad \text{with} \quad \vec{r} \in V \quad \text{and} \quad \vec{r}_I \in V_I \tag{4}$$

The embedding of M* in V_S is defined by a 1-to-1 relation between a basis of M* and one of Σ*:

$$\vec{a}_*^{\nu} \leftrightarrow a_*^{\nu} \qquad \nu = 1,2,\ldots 3+d \tag{5}$$

so that

$$k_s = (\vec{k},\vec{k}_I) = \sum_{\nu=1}^{3+d} Z_\nu\, a_*^{\nu} \tag{6}$$

has an external component:

$$\vec{k} = \sum_{\nu=1}^{3+d} Z_\nu\, \vec{a}_*^{\nu} \qquad \text{(same coefficients)} \tag{7}$$

The supercrystal density $\rho_s(r_s)$ is obtained from the crystal density $\rho(\vec{r})$ by the identification of the corresponding Fourier components:

$$\hat{\rho}_s(k_s = (\vec{k},\vec{k}_I)) \overset{\text{def}}{=} \hat{\rho}(\vec{k}) \tag{8}$$

Accordingly one has:

$$\rho_s(r_s) = \sum_{k_s \in \Sigma^*} \hat{\rho}_s(k_s)\, e^{i\, k_s\, r_s} \tag{9}$$

with $k_s r_s = \vec{k}\,\vec{r} + \vec{k}_I\,\vec{r}_I$.
Note that the crystal is then the (real) space intersection of the supercrystal:

$$\rho_s(r_s = (\vec{r},0)) = \rho(\vec{r}) \tag{10}$$

The euclidean symmetry group G_S of ρ_s is a (3+d)-dim. space group called superspace group (see ref. 3 for more details):

$$\rho_s(g_s^{-1}\, r_s) = \rho_s(r_s) \quad \text{for any } g_s \in G_S. \tag{11}$$

The group G_S is reducible, i.e.

$$G_S \subset E(3) \otimes E(d). \tag{11}$$

Writing the elements of G_S as $g_s = \{R_S|t_s\}$ one has $R_S = (R,R_I)$ and $t_s = (\vec{t},\vec{t}_I)$ for the rotational and the translational part, respectively. The elements $R \in O(3)$ and $R_I \in O(d)$ form the crystallographic point groups K_E and K_I, respectively. One can show that the point group K_S of G_S is a subdirect product of K_E and K_I:

$$K_S = (K_E,K_I) \subset K_E \otimes K_I \tag{12}$$

Furthermore, and because of the incommensurability K_S and K_E are isomorphic

$$K_S \simeq K_E \tag{13}$$

and the usual 3-dimensional classification in 32 crystal classes can be retained. Because of the reducibility the elements g_s belonging to G_S can also be written as:

$$g_s = (g_E,g_I) \in (G_E,G_I) = G_S \tag{14}$$

with $g_E \in E(3)$ and $g_I \in E(d)$. The group G_E is a space group, but this is not the case for G_I. Very often G_E is the symmetry group of the (high temperature) commensurate phase which undergoes the incommensurate phase transition: and in addition G_E describes the symmetry of an averaged structure. The symmetry (11) expressed in terms of the Fourier coefficients $\hat{\rho}(\vec{k})$ takes the simple form:

$$\hat{\rho}(\vec{k}) = \hat{\rho}(R\vec{k})\, e^{i(R_S k_s)t_s} \tag{15}$$

for any $g_s \in G_S$ noted as above.

3. Crystal Tensorial Properties

Consider a crystal tensorial property described by a tensorfield

$T(\vec{r})$ on V. This tensorfield has to share the symmetry of the crystal. In the case that this symmetry is a 3-dim. space group G this invariance condition can be expressed by:

$$(R \,.\, T)(g\vec{r}) = T(\vec{r}) \tag{16}$$

for any $g = \{R|\vec{t}\} \in G$. In particular for a rank zero tensor one recovers the same symmetry condition as for $\rho(\vec{r})$.

In the case of an incommensurate crystal one first embeds the crystal as supercrystal in V_S determining the superspace group G_S. One then considers tensorfield $T_S(r_s)$ on V_S invariant with respect to G_S:

$$(R_S \,.\, T_S)(g_s r_s) = T_S(r_s) \tag{17}$$

for $g_s = \{R_S|t_s\} \in G_S$.
Expressing now T_S in terms of tensor product of vector fields:

$$v_s(r_s) = (\vec{v}(r_s), \vec{v}_I(r_s)) \tag{18}$$

one can decompose T_S into an external component T_E, a mixed one T_M and an internal one T_I:

$$T_S = T_E + T_M + T_I \tag{19}$$

where T_E, T_M, T_I are the contributions obtained from tensor products of the type

$\vec{u} \otimes \ldots \otimes \vec{v}$, $\vec{u} \otimes \ldots \otimes \vec{v}_I$ or $\vec{u}_I \otimes \ldots \otimes \vec{v}$, and $\vec{u}_I \otimes \ldots \otimes \vec{v}_I$, respectively.

In particular the restriction of T_E to the (real) subspace V defines a tensorfield $T(\vec{r})$.

$$T_E(r_s = (\vec{r},0)) = T(\vec{r}). \tag{20}$$

We claim that tensorfields $T(\vec{r})$ obtained in this way have the good crystal symmetry and are the admitted ones for describing the tensorial properties of incommensurate crystals. The symmetry condition (17) for the external component simplifies to:

$$(R \, . \, T_E)(g_s r_s) = T_E(r_s) \tag{21}$$

This implies that the Fourier series expansion involves the same Z-module M* as the crystal density and that the Fourier coefficients for T_E and for T are correspondingly the same:

$$\hat{T}_E(k_s = (\vec{k},\vec{k}_I)) = \hat{T}(\vec{k}) \tag{22}$$

$$\text{for } T(\vec{r}) = \sum_{\vec{k} \in M^*} \hat{T}(\vec{k}) \, e^{i \vec{k} \vec{r}} \tag{23}$$

$$\text{and } T_E(r_s) = \sum_{k_s \in \Sigma^*} \hat{T}_E(k_s) \, e^{i k_s r_s} \tag{24}$$

Therefore one can express the superspace symmetry condition directly in terms of the physical tensors $\hat{T}(\vec{k})$ by:

$$\hat{T}(\vec{k}) = (R \, . \, \hat{T})(R\vec{k}) \, e^{i(R_S k_s) t_s} \tag{25}$$

for any $\vec{k} \in M^*$ and $g_s = \{R_S | t_s\} \in G_S$, where $R_S = (R, R_I)$.

In the case of a macroscopic tensorial property these tensors reduce to constant ones and the symmetry condition simply becomes:

$$R \, . \, T = T \qquad \text{any } R \in K_E \tag{26}$$

One sees that the external point group takes over the role of the point group of the normal case. As in the non-symmorphic space group case where the point group is not the point symmetry of the (microscopic) crystal structure, but only of a kind of average now also the point group K_E is a symmetry only in an averaged sense, and only after having taken in an account internal transformations also. Indeed one can easely give examples of different symmetry conditions for tensors of incommensurate crystals having a same averaged space group, but a different superspace group symmetry.

4. Final Remarks

One can have the impression that the full superspace embedding of the tensor as presented here is superfluous, as the "physical" one only involves T_E and not T_I or T_M. The point is that despite the fact that if the interpretation of T_E is a fairly straightforwards one, whereas this is not the case for T_I and T_M, these latters also carry structural and physical information. The situation is quite analogous to that one finds by extending real functions describing physical properties to functions defined in the complex plane. Even "non-physical" regions of the complex plane carry useful physical information.

5. References

[1] P.M. de Wolff, Acta Cryst A30 (1974), 777

[2] A. Janner and T. Janssen, Phys. Rev. B15 (1977), 643

[3] A. Janner and T. Janssen, Physica 99 (1979), 47

[4] A. Janner, Proc. of the Tenth Int. Conf. on Group Theoretical Methods in Physics, Canterbury, 1981. To appear in Physica 114A.

BIFURCATIONS AND SYMMETRY CHANGES IN CRYSTALS

T. Janssen
Institute for Theoretical Physics, University of Nijmegen, Holland

J.A. Tjon
Institute for Theoretical Physics, University of Utrecht, Holland

Abstract

Bifurcation theory and the theory of discrete area-preserving mappings are both related to models for structural phase transitions in crystals. Here it is shown how to apply these theories to a lattice problem. On the other hand the latter throws a new light on the former theories. Special attention is paid to the role of symmetry of the system.

1. Introduction

The study of structural phase transitions in crystals, where there is a transition from a phase with a certain space group symmetry to one with another space group, has a long history. In more recent years especially transitions towards an incommensurate phase have aroused much interest. An incommensurate phase is a crystal phase where next to the three periodicities of a basic lattice there are additional and incommensurate periodicities present, for example in the form of displacement or occupation probability wavess. The reason for the occurrence of an incommensurate phase is, in principle at least, understood for certain classes of systems, e.g. conductors with charge density waves and magnetic systems. The origin of incommensurability in insulators has recently been studied on simple one-dimensional models in which competitive interactions and anharmonic forces are important. The equations which determine the ground state

Presented at the XIth International Colloquium on Group-Theoretical Methods in Physics, Istanbul, August 23-28, 1982

in these models can be seen as nonlinear functional equations. Like in similar partial differential equations solutions may give rise to bifurcations if the parameters, which are assumed to be temperature dependent, are changed. These bifurcations may sometimes be identified with phase transitions.

This point of view is not new. Several years ago this was realized already in the study of phase transitions in the context of Landau theory. In the model studies we are concerned with here, there is a direct relation with bifurcations of discrete symplectic mappings which have received a lot of interest in the study of Hamiltonian systems, turbulence and the onset of stochasticity.

Usually the symmetry group of a system changes at a phase transition. In the Landau theory of phase transitions this fact plays an important role. This aspect has already been studied by many group theorists. Also in the theory of bifurcations of solutions of differential equations the symmetry of the problem has been used to simplify the equations. We shall use these results here to study the symmetry changes at the phase transitions in models and in particular the symmetry of orbits of discrete mappings.

First we shall discuss the relation between models for structural phase transitions and general bifurcation theory. Then we shall speak about discrete symplectic mappings, the symmetry of orbits and the very interesting universal behaviour of bifurcation series. Finally we shall use a combination of these approaches for the study of a specific model in two dimensions.

2. Models for structural phase transitions.

Recently a number of one-dimensional models has been studied to investigate phase tranitions in crystals, in particular towards incommensurate phases [1),2)]. We generalize this approach to the case of a crystal in d dimensions. For simplicity we assume that there is only one kind of particles, at or near the points of a lattice, such that the particles may be labeled by $n \in Z^d$. The crystal is described by a function u which can represent displacements from the points of Z^d, in which case u is a function from Z^d to R^d, or an internal variable. Then u is a real function. Again for simplicity we consider scalar functions u only here. As an example u_n may represent the torsion angle of a molecule at n with an internal degree of freedom .

The potential energy of the crystal is assumed to be a sum of pair potentials:

$$V(u) = \sum_{ns}\sum \Phi_s(u_n - u_{n-s}) + \Phi_0(u_n), \tag{1}$$

where $n,s \varepsilon Z^d$ .The summation over s gives the interactions with various neighbours. A term $\Phi_0(u_n)$ may represent a local energy, e.g. an interaction with an external potential or the internal energy of the particle at n.

The equilibrium values for u_n satisfy

$$\partial V/\partial u_n = \Phi_0'(u_n) + \sum_s (\Phi_s'(u_n - u_{n-s}) - \Phi_s'(u_{n+s} - u_n)) = 0. \tag{2}$$

These equations can be viewed as a (nonlinear) mapping from the space of real functions on Z^d into itself:

$$F(\alpha,u) = 0, \tag{3}$$

where α denotes a set of parameters appearing in the potentials Φ_s. We choose our variables in such a way that $u_n=0$ is a solution with $V=0$.

The Fréchet derivative of F is a linear operator $F_u(\alpha,u)$ in the space of functions on Z^d. It is defined by

$$F_u(\alpha,u)f = \lim_{t\to 0} \frac{F(\alpha,u+tf) - F(\alpha,u)}{t}. \tag{4}$$

From general bifurcation theory it is known that if u^0 is a solution of eq.(3) for which $F_u(\alpha_c,u^0)$ has an eigenvalue zero a new solution may bifurcate for α past α_c.

For the lattice model the operator F is given by (2). Then for a function $f:Z^d \to R$ one has

$$[F_u(\alpha,u)f]_n = \sum_{n'} \frac{\partial^2 V}{\partial u_n \partial u_{n'}} f_{n'}. \tag{5}$$

If u^0 is a solution of (3) the equations of motion for small displacements $f_n e^{-i\omega t}$ are given by

$$m\omega^2 f_n = \sum_{n'} \frac{\partial^2 V}{\partial u_n \partial u_{n'}} f_{n'}. \tag{6}$$

Hence the kernel M of $F_u(\alpha_c,u^0)$ is nontrivial if there is an eigenmode frequency equal to zero. In solid state theory such a mode for which the frequency goes to zero as a function of the parameters is known as a soft mode. So the soft mode theory of phase transitions corresponds to the bifurcation theory.

Suppose $F_u(\alpha,0)$ has an eigenvalue zero for $\alpha=\alpha_c$. Then the kernel M is nontrivial. One can then reduce the nonlinear equations in an infinite dimensional space to a finite number of algebraic equations by the Ljapunov-Schmidt procedure. The construction is as follows [3]. Consider the projection P of the function space on M. One may write for an arbitrary function $u=u_1+u_2$ with $u_1=Pu$ and $u_2=(1-P)u$. Then eq.(3) may be written as

$$(1-P)F(\alpha,u_1+u_2)=0 \tag{7}$$
$$PF(\alpha,u_1+u_2)=0 \tag{8}$$

From the implicit function theorem it follows that for given u_1 eq.(7) has a unique solution in a neighbourhood of α_c and u^0. So one may write $u_2=u_2(u_1)$. Substitution in (8) gives an equation for u_1 in M which is supposed to be finite-dimensional. The equation

$$PF(\alpha,u_1+u_2(u_1))=0 \tag{9}$$

is called the bifurcation equation.

A very simple example is the following. The equations for an extremum of the function

$$V=y^2/2+(\alpha+y)x^2/2+x^4/4 \tag{10}$$

are $(\alpha+y)x+x^3=y+x^2/2=0$. Hence

$$F(x,y)=((\alpha+y)x+x^3,y+x^2/2) \tag{11}$$

$$F_u=\begin{pmatrix} \alpha+y+3x^2 & x \\ x & 1 \end{pmatrix}$$

For the solution x=y=0 the kernel M is nontrivial for $\alpha_c=0$. It coincides with the x-axis. Eq.(7) can be solved uniquely from $(1-P)F(x,y)=y+x^2/2=0$. The bifurcation equation (9) is 1-dimensional:

$$(\alpha-x^2/2)x+x^3=0$$

Hence for $\alpha>0$ there is only the trivial solution. For $\alpha<0$ the extrema are x=y=0 and $x^2=-2\alpha, y=\alpha$. This case is so simple because eq.(7) can be solved easily. In general the derivation of the bifurcation equations is much harder. Nevertheless, knowing that there are bifurcation equations in a space with the dimension of M can serve as a basis for further investigation, as we shall see below.

3. One-dimensional models and symplectic mappings.

For one-dimensional mappings with short range interactions the equations (3) can, alternatively, be viewed as discrete volume preserving mappings. Since these have recently been shown to have interesting properties and because, in principle, most of the physics of incommensurate phase transitions is already present in these models, we shall discuss this case in some detail.

We specialize eq.(1) for d=1 as follows:

$$V=\sum_n(\alpha(u_n-u_{n-1})^2/2-(u_n-u_{n-2})^2/2 \tag{12}$$

$$+\delta(u_n-u_{n-3})^2+(u_n-u_{n-1})^4/4).$$

involving harmonic interactions with 1st, 2nd and 3rd neighbours and an anharmonic term. The eq.(3) can be written more concisely by introducing a new function $x:Z\to R$ as $x_n=u_n-u_{n-1}$. The equations for x are

$$(\alpha-2+3\delta)x_n+x_n^3-(1-2\delta)(x_{n+1}+x_{n-1})+\delta(x_{n+2}+x_{n-2})=f \tag{13}$$

for some arbitrary constant f, independent of n. For convenience we take f=0.

If one introduces a vector v_n in R^4 with components $(x_{n+1},x_n,x_{n-1},x_{n-2})$, (13) can for $\delta\neq0$ be written as

$$v_n\to v_{n+1}=Sv_n \tag{14}$$

because with (13) x_{n+2} may be expressed in the components of v_n. Actually (14) is related to (3), but where the latter is an equation in an infinite-dimensional space, the former acts in R^4. The derivative of S is

$$DS=\begin{pmatrix} A & B & A & -1 \\ 1 & 0 & 0 & 0 \\ 0 & 1 & 0 & 0 \\ 0 & 0 & 1 & 0 \end{pmatrix} \tag{15}$$

where $A=(1-2\delta)/\delta$, $B=(2-\alpha-3\delta-3x_n^2)/\delta$. The matrix DS leaves invariant a symplectic form and has determinant +1. Therefore, the mapping S is volume-preserving and symplectic.

If $\delta=0$ one can express x_{n+1} in x_n and x_{n-1} via eq.(13). If one introduces a vector v_n with components (x_n,x_{n-1}), (13) is again a transformation of the form (14), this time with

$$DS = \begin{pmatrix} \alpha-2+3x_n^2 & -1 \\ 1 & 0 \end{pmatrix} \qquad (16)$$

S is an area-preserving symplectic mapping.

The volume-preserving mappings are of interest for Hamiltonian mechanics, since the Poincaré map of a system with N degrees of freedom is such a discrete symplectic transformation in R^{2N-2}. These mappings appear in the study of stellar dynamics, but also in such seemingly unrelated subjects as the turbulence in fluid mechanics, the motion of charged particles in storage rings and the chaotic behaviour of systems in statistical mechanics.

An orbit under S may be periodic, i.e. for some N and v one has $v=S^N v$. The N points of such an orbit form an N-cycle. Each of its points is a fixed point of S^N. For the crystal it corresponds to a periodic function u. Hence the crystal still has translation symmetry and forms a superstructure. For a nonperiodic orbit the translation symmetry of the crystal is lost.

The fixed points of S^N (N=1,2..) are characterized by the linearized mapping DS^N. Because S is a real symplectic tranformation it follows that, if λ is an eigenvalue of DS, also λ^* and λ^{-1} are eigenvalues. For the 2-dimensional mapping a fixed point is either elliptic (both eigenvalues on the unit circle), hyperbolic (both eigenvalues real$\neq\pm1$) or parabolic ($\lambda=\pm1$). The character is already completely determined by $T=\mathrm{Tr}DS^N$:it is elliptic if $|T|<2$, hyperbolic if $|T|>2$ and parabolic if $T=\pm2$. Near an elliptic point other points move around it under S^N with mean angular velocity ϕ, where $2\cos\phi=T$. In special cases ,when N=2,3 or 4, the points do not remain necessarily near the fixed point, but for longer periods an elliptic fixed point is always stable. For the 4-dimensional mapping one can make a similar classification, but we shall not discuss that here in detail.

If the parameters (α and δ) are varied the trace T will change. In particular the type of a fixed point may change if T goes through ±2. A stable point may become unstable.

This type of stability, however, is not the one we are interested in primarily. For the minimum of V we have to require that the crystal configuration corresponding to a cycle is stable under small displacements:if u_n^0 corresponds to a cycle, the equations of motion for small displacements $f_n \exp(i\omega t)$ are given by (6). One has the following

lemma[4]: a cycle is elliptic and DS^N has an eigenvalue e^{ik} if and only if there is an eigenmode with wave vector k and frequency zero for the corresponding crystal configuration.

Hence if for varying parameters a hyperbolic cycle corresponding to a stable crystal configuration changes into an elliptic cycle, the corresponding crystal becomes unstable. Of course this can be described again as a soft mode: when the eigenvalues approach the unit circle, one of the eigenmode frequencies goes to zero. If this happens for the ground state, a new ground state grows out of the original one. Because in R^2 the two eigenvalues are either both on the real axis or both on the unit circle, the transition has to take place at $\lambda=\pm 1$. Hence in this case the soft mode wave vector is always 0 or π. In R^4 this is different, because the 4 eigenvalues may collide at an arbitrary point of the unit circle. Hence in that case there is a soft mode at an arbitrary wave vector (Fig.1).

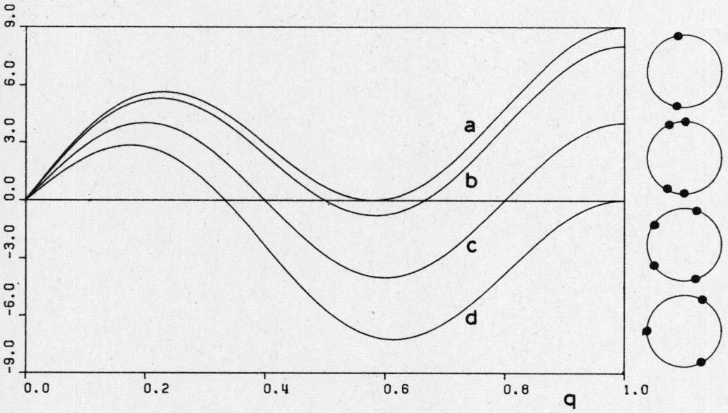

Fig.1
Dispersion curves $\omega(q)^2$ and eigenvalues of DS on the unit circle for the trivial solution (q in units π). α=1.25(a),1.0(b),0(c) and -1.0(d),resp.;δ=1.

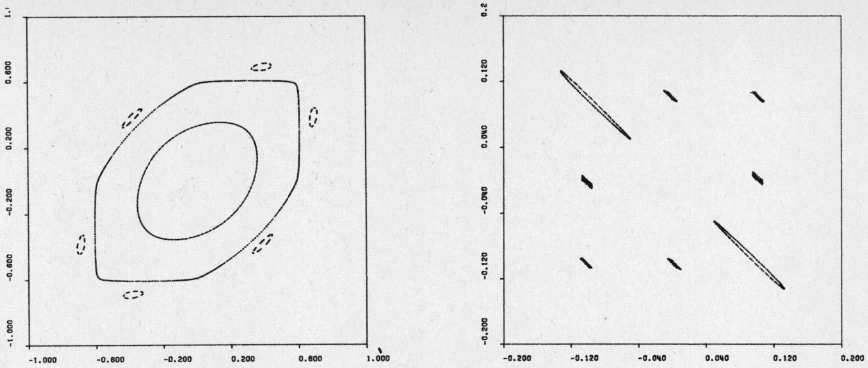

Fig.2 Left:orbit around 6-cycle bifurcated at $\alpha=3,\delta=0$. Right:orbits around a 2- and a 6-cycle born at $\alpha=-1,\delta=1$.Projection on x_n, x_{n-1} plane

This bifurcation is, however, not the only one. If the parameters vary further and the eigenvalues move over the unit circle, each time there is an eigenvalue $\lambda=\exp(2\pi is/N)$(s,N coprime integers)one or more cycles with an N times as large period may split off from the original one. There are several types of bifurcations which we shall not discuss in detail. Two examples are given in Fig.2. One can interpret this result in terms of dispersion curves: as soon as for a rational wave vector there is an eigenfrequency zero, there appears a new solution to (13). However, although for the mapping space there appear pairs of stable and unstable cycles, the new crystal configurations are generally all unstable (except for example if the ground state becomes unstable). For further variation of the parameters some of the new configurations may become stable. Actually, for decreasing value of α the number of dynamically stable equilibrium configurations becomes infinite.

4. Symmetry of the bifurcating solutions.

As explained in Section 2 the bifurcation equations can be formulated in a finite-dimensional space. In the present case this dimension is the number of eigenvalues on the unit circle. Suppose that the original equation (3) is covariant under a group G. Then Sattinger[3)] has proved the following

<u>Lemma</u>: if (3) is covariant under a group G, then the kernel M carries a representation of G.

Suppose that u^0 is a solution of eqs.(7,8). From Section 2 we know that one may write $u^0=u_1^0+u_2^0$. Then the function $W(u_1)\equiv V(u_1,u_2(u_1))$ has an extremum for $u_1=u_1^0$. Hence eq.(8) amounts to the extremalisation of the function W on M. We now assume that the representation is irreducible. The covariance group G is generated by the translation $t:n\to n+1$, the mirror $m:n\to -n$ and (accidentally for the function V which contains only even powers) the reflection $p:u_n\to -u_n$: it is the direct product of a one-dimensional space group pT with the group C_2 generated by p. The irreducible representations of pT are characterized by a wave vector k (and if k=0 or π by an irreducible representation of the group generated by m). If $k\neq 0$ or π the representation is 2-dimensional: an arbitrary function in M is $z.\cos(kn+\phi)$. The function W depends on z and ϕ and is invariant under G. The value of z does not change under this action. Under G ϕ transforms as follows

$$t:\phi\to\phi+k; \quad m:\phi\to -\phi; \quad p:\phi\to\phi+\pi \tag{17}$$

Consider a rational value of $k=2\pi s/N$. Then the solution corresponds to an N-fold superstructure. For N even this implies that ϕ is left invariant by t^N and by $pt^{N/2}$ and $\phi=l.k/2$ (integer l) also by $t^n m$. For N odd it is invariant under t^N and $\phi=l.k/2$ also by $t^n m$ and $\phi=\pm\pi/2+l.k/2$ by $t^n mp$. Since for these special values of ϕ the function W has reflection symmetry there are extrema at those points. If W has no other extrema the minimum will be among these. Therefore, one can expect that the orbits are of one of the following types:

N twice even: ..b,a,b..c,0,-c..-b,-a,-b.. or ..a,a..b,-b..-a,-a..,

N twice odd : ..b,a,b..c,-c..-b,-a,-b.. or ..a,a..,b,0,-b..-a,-a..,

N odd : ..b,a,b...-c,-c... or ..-a,a..b,0,-b..-a,a...

Numerical calculations in this model have confirmed that this is the case in most situations. The only (trivial) exception found is the case where W does not depend on ϕ. The crystal corresponding to such an orbit with symmetry has itself also a higher symmetry than required for the bifurcation. We shall come back to this point in the last section.

5. Series of bifurcations.

A solution that originates from another one may itself give rise to again new solutions if the parameters are varied. In this way one can build whole family trees of bifurcations.

This same phenomenon occurs in non-length-preserving mappings from an interval on itself. There Feigenbaum[5] has observed a very interesting behaviour of these bifurcations, which are in this case period-doubling. If one denotes the value of the parameter (α in this case) for which a bifurcation takes place from an $N=2^p$ orbit to an $N'=2^{p+1}$ orbit by α_p ,he found that these values form a geometric series, i.e.

$$\lim_{p\to\infty} \frac{\alpha_{p+1}-\alpha_p}{\alpha_{p+2}-\alpha_{p+1}}=\eta \tag{18}$$

exists. Moreover, the most important discovery was that this ratio η does not depend on the specific map, but is a universal constant valid for a large class of mappings. This behaviour has been found also in higher-dimensional mappings. For dissipative systems in which the space contracts to a line and the mapping becomes essentially one-dimensional, one finds for period-doubling series the same exponent $\eta=4.66..$ as in the one-dimensional case. Area-preserving mappings have been studied in two dimensions[6),4)]. In particular, series of period-doubling bifurcations have been studied and again one finds geometric behaviour (Feigenbaum sequences) for the bifurcation values of the parameters and again the ratio is universal. For area-preserving mappings, however, this value is different from the Feigenbaum value:here $\eta=8.721....$

Also for other series of bifurcations with another multiplicity one finds geometric series, but the values of the exponents are different[4]. This same behaviour is found in 4 dimensions:there the exponents are the same as for the area-preserving mappings in R^2.

The motion of points in the neighbourhood of a period 2N fixed point is a scaled down version of the motion around the period N point and the scaling factor has also a limit of universal value. For the crystal configuration this means that in the neighbourhood of an equilibrium configuration may appear new equilibria and if the bifurcation series continues there is an infinite number of such configurations in a finite neighbourhood. Because all these configurations are extrema of the potential this may give rise to a chaotic behaviour. This is only one way to chaos. Another one occurs if for changing parameters the number of hyperbolic fixed points increases:in the neighbourhood of such a point an orbit is chaotic and so is the corresponding crystal configuration. The question is, however, whether such a configuration is dynamically stable.

6. Two-dimensional models.

A second specialization of the general model of Section 2 is one for a 2-dimensional crystal. Such a model may be of interest, because there are several incommensurate crystal phases with more than one modulation wave vector. Moreover, in the study of structural phase transitions this is the lowest dimensionality in which a nontrivial phase transition may occur. Finally, having applied what is known for discrete symplectic mappings in a finite-dimensional space to the lattice problem, we can here investigate what one can learn from the latter for an infinite-dimensional mapping.

We consider a rectangular lattice with particles at the positions (n,m) of a lattice Z^2 and with potential energy

$$V = \sum_{nm} (\alpha u_{nm}^2/2+u_{nm}^4/4+\beta(u_{nm}-u_{n-1\,m})^2/2 \qquad (19)$$

$$+\gamma(u_{nm}-u_{n-2m})^2/2+\rho(u_{nm}-u_{nm-1})^2/2+\tau(u_{nm}-u_{nm-2})^2/2.$$

The nonlinear equation (3) becomes

$$\alpha u_{nm}+u_{nm}^3+\beta(2u_{nm}-u_{n-1m}-u_{n+1m})+\gamma(2u_{nm}-u_{n-2m}-u_{n+2m}) \tag{20}$$

$$+\rho(2u_{nm}-u_{nm-1}-u_{nm+1})+\tau(2u_{nm}-u_{nm-2}-u_{nm+2})=0$$

Instabilities and bifurcations occur similarly to the situation in the linear chain. One can interpret (20) again as a symplectic discrete mapping. For example ,for the simplest case $\gamma=\tau=0$:if one considers an infinite-dimensional vector x_m with components $u_{n,m}$ and a second vector p_m with components $u_{n,m}-u_{n,m-1}$ the mapping $(x_m,p_m)\to(x_{m+1},p_{m+1})$ is a symplectic mapping in an infinite-dimensional space.

The nonlinear equations (20) have always the trivial solution $u_{n,m}=0$. Small oscillations around this solution have a frequency $\omega(q)$ with

$$\omega^2(q)=\alpha+2\beta(1-\cos q_1)+4\gamma(1-\cos^2 q_1)+2\rho(1-\cos q_2)+4\tau(1-\cos^2 q_2) \tag{21}$$

The extrema of the frequency function are situated at $q=(0,0),(\pi,0),(0,\pi),(\pi,\pi)$ or in a point inside the Brillouin zone:

$$\cos q_1=-\frac{\beta}{4\gamma}, \quad \cos q_2=-\frac{\rho}{4\tau} \tag{22}$$

If by variation of the parameters the minimum of the frequency surface reaches zero, the trivial solution becomes dynamically unstable and a new ground state splits off for which the whole ω^2 surface is non-negative. As seen from (22) this happens in the center or in a corner of the Brillouin Zone if $\gamma=\tau=0$. Otherwise also incommensurate values are possible.

If for certain q_0 one has $\omega^2=0$, there is a bifurcation from the trivial solution. The derivative F_u has a kernel carrying a representation of the symmetry group G of the crystal which is the direct product of the 2-dimensional space group pmm and a group of order 2 generated by the reflection p. The generators of G are $(1,a):u_{n,m}\to u_{n+1,m}$; $(1,b):u_{n,m}\to u_{n,m+1}$; $(m_x,0):u_{n,m}\to u_{-n,m}$; $(m_y,0):u_{n,m}\to u_{n,-m}$; $p:u_{n,m}\to -u_{n,m}$. The representation is characterized by the star of q_0.

As discussed in Section 3 the bifurcation equations are equivalent with the minimalisation of a function W on M. If q_0 is not in a special position in the Brillouin Zone, the representation is 4-dimensional and the coordinates are z_1, z_2, ϕ_1, ϕ_2. The star of q_0 consists of 4 vectors $k_1 = q_0 = 2\pi(s_1/N_1, s_2/N_2)$, $k_2 = 2\pi(-s_1/N_1, s_2/N_2)$, $k_3 = -k_1$, $k_4 = -k_2$. Under the generators of the symmetry group the components $(z_1, z_2, \phi_1, \phi_2)$ transform as follows. (t is either a or b).

$$(1,t):(z_1,z_2,\phi_1,\phi_2)\rightarrow(z_1,z_2,\phi_1+k_1t,\phi_2+k_2t), \tag{23}$$

$$(m_x,0):(z_1,z_2,\phi_1,\phi_2)\rightarrow(z_2,z_1,\phi_2,\phi_1),$$

$$(m_y,0):(z_1,z_2,\phi_1,\phi_2)\rightarrow(z_2,z_1,-\phi_2,-\phi_1),$$

$$p:(z_1,z_2,\phi_1,\phi_2)\rightarrow(z_1,z_2,\phi_1+\pi,\phi_2+\pi).$$

The points which are invariant under 2 perpendicular mirrors form again, as in the one-dimensional case, a discrete lattice. These points correspond to extrema of the function W and hence to solutions of eq.(20). The translation symmetry of the solutions is determined by the elements $(1, n_1a+n_2b)$ for which the point (ϕ_1, ϕ_2) is invariant in (23). The total symmetry group is in this case the maximal one compatible with the bifurcation. It should be noticed, however, that the absolute minimum does not necessarily correspond to the highest symmetry. As an example consider Fig.3, where two solutions are given which originate from the trivial solution for $q_0 = 2\pi(1/3, 1/6)$. The first one is a "double-q" solution with $z_1 = z_2$ and (ϕ_1, ϕ_2) invariant under m_x and m_y. The second one is a "single-q" solution with $z_2 = 0$ and ϕ_1 determined by the requirement that the solution has maximal point group symmetry (p2 in this case: ϕ_1 is left invariant by $m_x m_y$). The latter solution with lower point symmetry has the lower energy. Notice, however, that also in this case the phase is fixed by symmetry.

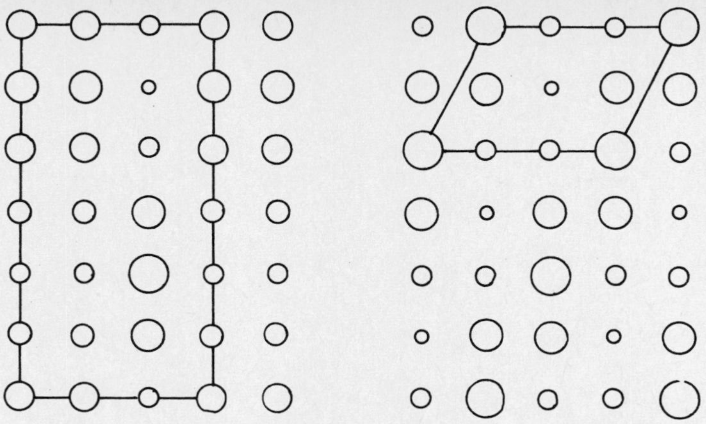

Fig.3 Left:18-fold superstructure with pmm symmetry. Right:6-fold one with p2 symmetry. Both are born from the trivial solution for $q=2\pi(1/3,1/6)$.Diameter of the circles is measure for u.

References

1) T.Janssen and J.A.Tjon,Phys.Rev.B25,3767(1982),B24,2245(1981)

2) A.Bussmann-Holder,H.Buettner and H.Bilz, Ferroelectrics, 36,273(1981),
F.Axel and S.Aubry,J.Phys.C14,5433(1981)

3) D.H.Sattinger,Group Theoretic Methods in Bifurcation Theory, Lecture Notes in Mathematics 762,Springer,Berlin (1979)

4) T.Janssen and J.A.Tjon,Phys.Lett.87A,139(1982),preprint submitted to J.Phys.A (1982)

5) M.J.Feigenbaum,J.Stat.Phys.19,25(1978)

6) J.M.Greene,R.S.McKay,F.Vivaldi,M.J.Feigenbaum,Physica,3D,468(1981)

THE STRUCTURE OF SPACE - GROUPS' UNITARY REPRESENTATIONS

Louis MICHEL
I.H.E.S.
91440 Bures-sur-Yvette
FRANCE

Jan MOZRYMAS
Institute of Theoretical Physics
University of Wrocław
ul.Cybulskiego 36
50-206 Wrocław, POLAND

ABSTRACT

For systems with a symmetry group G, the description of physical phenomena corresponding to a representation of G, depends only on the image of this representation. The classification of the images of the unirreps (unitary irreducible representations) of the little space groups G_k is remarkably simple. The nearly four thousands inequivalent unirreps corresponding to high symmetry wave vectors k have only 37 inequivalent images.

Unitary representations of space groups are a very useful tool for the study of many problems of crystal physics such as: the structure (and labelling) of electronic levels and energy bands, the vibration spectrum and phonon dispersion relations, the selection rules in transitions between quantum states, the symmetry change in a second order phase transition, etc. Under the influence of Wigner, a systematic study of these representations started nearly fifty years ago[1,2,3,4]. Finally, for the last twenty years, more and more complete (and expensive) tables of space groups unirreps (unitary irdeducible representations) appeared[5,6,7,8,9]. (Each of the 230 spacegroups has an infinity of unirreps, so most of them are labelled by continuous parameters which appear explicitly in the tabulated matrix elements). Thanks to these and associate tables (e.g. Clebsch Gordan coefficients, physicists can compute for each case what they need, as efficiently as they wish. There is a drawback in this situation; it can be best understood by an historical comparison: trigonometric function tables were built to satisfy as efficiently the needs of astronomers, land surveyors, physicists... However, for the progress of science, the knowledge of the trigonometric function values has not been sufficient, another type of knowledge was required: their geometrical meaning, their analytic properties, etc. This lecture is devoted to the "other type of knowledge" physicists need concerning space groups unirreps.

Given a representation Γ of a group G, one must first look for its kernel, Ker Γ, which is an invariant subgroup of G and its image, Im Γ, which is the quotient group Im Γ = G/Ker Γ. For physical phenomena concerning the state-vectors of the carrier space of the representation Γ of the symmetry group G, physics feels only Im Γ and forgets Ker Γ. For example, most macroscopic properties of crystals do not depend on the translation T of the space group G but only on the pointgroup P = G/T. How does the physics formalism implement the underline sentence? All physical properties of a system with symmetry group G must be described in term of invariants and covariants of G; the invariants and covariants built on a representation Γ of G depend only on Im Γ.

A classification of group representation images yields therefore a classification of corresponding physical phenomena. To perform this classification we introduce a new equivalence, much weaker than the usual one, between group representations[11]. Given two linear group representations

$$G \xrightarrow{\Gamma} \mathrm{Im}\,\Gamma \rightarrow 1 \ , \quad G' \xrightarrow{\Gamma'} \mathrm{Im}\,\Gamma' \rightarrow 1$$

on the carrier spaces E and E', they are weakly equivalent if there exist an inversible linear map $E \xrightarrow{\gamma} E'$ which transform the set of operators (or matrices) of Im Γ into that of Im Γ'

$$\{\mathrm{Im}\,\Gamma'\} = \gamma\{\mathrm{Im}\,\Gamma\}\gamma^{-1} \tag{1}$$

We can also say that $\gamma_o\Gamma$ is a representation of G whose image Im $\gamma \circ \Gamma$ = Im Γ'. Remark that γ defines an isomorphism $1 \rightarrow \mathrm{Im}\,\Gamma \xrightarrow{\tilde{\gamma}} \mathrm{Im}\,\Gamma' \rightarrow 1$ between the two images. We emphasized that the nature of the physical phenomena "depends only on the image, but the isomorphism $\tilde{\gamma}$ will help us to establish the dictionary for translating the phenomena concerning the state vector of E into those of E'. This two vector spaces need not be distinct: a simple example occurs when $\alpha \in$ Aut G (α is an automorphism of G) and $\Gamma' = \Gamma \circ \alpha$. The two representations Γ' and Γ of the symmetry group G may not be equivalent when Γ is not an inner automorphism, but they are weakly equivalent since their images coincide, (γ = 1). There is a complete translation (i.e. an isomorphism) between the two sets of physical phenomena described by the carrier space of the two representations Γ' and Γ. This translation is the object of the contribution by R.Dirl[12] in these proceedings.

Although our program of classifying the weak equivalence classes of space group unirreps is not completed, we hope that the preliminary results we give here will show its interest. Before explaining our results, we wish to present a relevant remark on the nature of the image of the space group unirrep Γ of G on the vector space E which is involved in the Landau theory[13,14] for the second order phase transition with a spontaneous symmetry breaking from G to its subgroup H. This group H is an isotropy group of the unirrep Γ of G, that of a minimum of a G invariant potential on E. It is easy to show that the Kernel of the representation Γ is the intersection of all G subgroups in $[H]$, the class of G-subgroups conjugated to H . If the transition is to an ordinary crystal state (and not to an incommensurate crystal) H is a space group. This implies that Ker Γ contains a lattice translation group ($\sim Z^3$) so Im Γ is finite. On the contrary, in the case of incommensurate transition, Im Γ is infinite (i.e. enumerable).

We do not need to recall here the notions of group orbits, of cohomology, of free modules since all these notions were used by previous lecturers and no questions were asked concerning their meaning, but it might be appropriate to explain a few basic concepts concerning crystallographic space groups. The translation group T of a space group G is an invariant subgroup isomorphic to Z^3 (and closed in R^3, the translation group of the Euclidean group). Its automorphism Aut T $\sim$ GL(3,Z). A space group G is a discrete closed subgroup of E(3), the 3-dimensional Euclidean group (which is the semi-direct product E(3) = $R^3 \,\square\, O(3)$) so the point group P = G/T has to be a finite group. All possible actions of P on T are given by the distinct injections

$$1 \to P \overset{\Delta}{\to} GL\,(3,Z) \sim \text{Aut } T \qquad (2)$$

i.e. by the conjugation classes of finite subgroups of GL(3,Z). There are 73 such classes and they are called <u>arithmetic classes</u>. For each of the 73 pairs P, Δ one can determine all possible groups extensions G solutions of

$$
\begin{array}{ccccccccc}
 & & & & & & 1 & & \\
 & & & & & & \downarrow & & \\
1 & \longrightarrow & T & \longrightarrow & G & \longrightarrow & P & \longrightarrow & 1 \\
 & & & & & & \downarrow \Delta & & \\
 & & & & & & \mathrm{Aut}\ T & &
\end{array} \qquad (3)
$$

These solutions form an Abelian group $H^2_\Delta(P,T)$, the second cohomolgy group of P in T with action Δ. To the zero element correspond the semi-direct product $T \overset{\Delta}{\square} P$. (The 73 such semi-direct products are called "symmorphic" in crystallography). Let N(P) be the normalizer of P in Aut T $\sim$ GL(3,Z); since this group acts on P and on T, it acts on $H^2_\Delta(P,T)$. Inequivalent but isomorphic extensions form an orbit of N(P). As a result there are 219 isomorphic classes of 3-dimensional space groups. However the traditional classification of space groups established at the end of last century in crystallography corresponds to the orbits of $N_o(P) = N(P) \cap SL(3,Z)$. Since 11 orbits of N(P) split into pairs of orbits of $N_o(P)$ (the so called enantiomorphic pairs) there are therefore 230 crystallographic classes of space groups.

Remark also that the set of translation lattices T form an orbit [GL(3,R): GL(3,Z)]. The action of the subgroup O(n) of GL(3,R) on this orbit yields seven strata (a stratum is the union of all orbits of the same type i.e. with same conjugation class of isotropy groups) corresponding to the seven crystallographic systems. The corresponding seven isotropy groups C_i, C_{2h}, D_{2h}, D_{4h}, D_{3d}, D_{6h}, O_h, are called holohedries and we denote them P_H. They are 14 conjugation classes (respectively 1, 2, 4, 2, 1, 1, 3) of the seven P_H in GL(3,Z). These fourteen actions of P_H of T define the 14 Bravais classes of lattice.

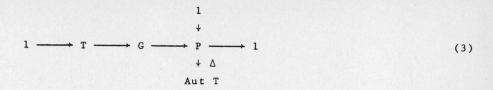

Since T is isomorphic to Z^3, its dual T^*, (i.e. its group of character) is isomorphic to $U(1)^3$ and has the topology of a 3-dimensional torus. T* is the Brillouin zone and its elements are the wave-vectors usually denoted k. (Beware that the use in physics of the wave-vectors may not be always equivalent to their use here as characters of T). The coordinate of k in $U(1)^3 = T^*$ are traditionally given by three real numbers modulo one and the group law of T* is noted additively. Wigner taught us the Frobenius method for determining the unirrep of a group with one Abelian invariant subgroup. In the case of G of (3), consider an orbit G.k of G on T*. Let G_k be an isotropy group of this orbit and Γ^α one of its unirrep. By induction to G one obtains an unirrep $\Gamma^{\alpha}_{G_k}\uparrow G$ of G and this method yields the whole set of inequivalent G unirreps. Theseare therefore labelled by the orbits (physicists say the stars) of G on T^* and the inequivalent irreps of the corresponding G_k. All quoted tables leave the induction $\Gamma^{\alpha}_{G_k}\uparrow G$ to be performed by the users. Here also we discuss only the unirreps Γ^{α} (Remark that each G_k is a space group and any space group can be a G_k, e.g. $G = G_o$ for $k = 0$; then the corresponding G unirrep has T in its kernel and is therefore a unirrep of P).

We denote by Ker k, Im k, the kernel and the image of the unirrep k of T. Given the unirrep Γ^{α} it is natural to apply the Noether isomorphism theorems to G_k and its two invariant subgroups Ker Γ^{α} and Ker k. All results are summarized (and visualized) in the commutative Diagram 1 of exact sequences.

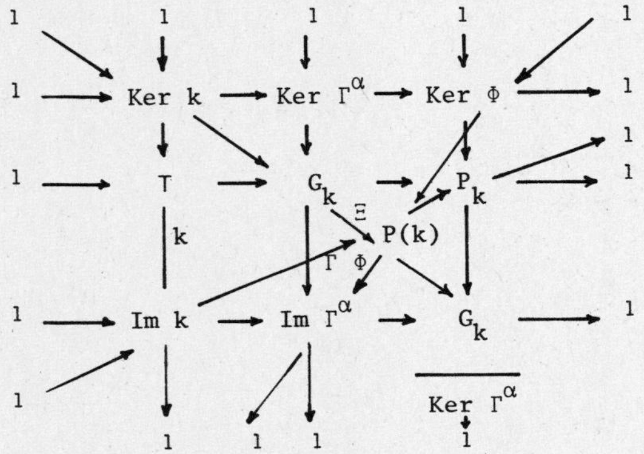

Diagram 1

On this diagram, aligned arrows define an exact sequence of group homomorphisms.

We have used the notation $P_k = G_k/T$ for the "little point group" (i.e. the isotropy group of k in the action of P on T*). Very naturally we are led to consider the quotient group

$$P(k) = G_k/\text{Ker } k = \text{Im } \Xi \tag{4}$$

which seems to have been first introduced by Herring[4] and is sometimes called the "extended little point group".

We note that Im Γ^{α}, the image of the representation Γ^{α} of G_k is also the image of the irrep Φ of P(k); indeed $\Gamma = \Phi \circ \Xi$. It is easy to check that P(k) is a central extension of P_k by Im k

$$P_k = P(k)/\text{Im } k, \quad \text{Im } k < \text{Center of } P(k) \tag{5}$$

For a given arithmetic class, the set of P(k) form the cohomology group $H_o^2(P_k, \text{Im } k)$. To the map $T \xrightarrow{k} \text{Im } k$ corresponds the functorial homomorphism

$$H_\Delta^2(P_k, T_m) \xrightarrow{\tilde{k}} H_o^2(P_k, \text{Im } k) \tag{6}$$

This defines a method of computation of the P(k)'s. Before giving the results of these computation, we wish to make several remarks:

i) Only the images Im Φ of "allowed" irreps Φ_a of P(k) yields Im Γ = Im Φ. These allowed irreps are characterized by "Im k is a subgroup of Im Φ = Im Γ^α" or equivalently

$$\text{"}\Phi \text{ is allowed"} \iff \text{Ker } \Phi \cap \text{Im } k = \{1\} \tag{7}$$

In the literature, the construction of G_k irreps is usually based on the study of projective representation of P_k. Indeed to the irreps $\text{Im } k \xrightarrow{r} U(1)$ of the Abelian group corresponds the homomorphism $T \xrightarrow{r o k} U(1)$ and the functorial homomorphism $H_\Delta^2(P_k, T) \xrightarrow{\tilde{r} o \tilde{k}} H_o^2(P_k, U(1))$. In our opinion, this traditional method inspired from the old Clifford paper[15] is awkward; since $\tilde{r} o \tilde{k}$ factorizes we do believe that our method based on $\tilde{k}$ and P(k) is preferable.

ii) It is true that some authors have considered "auxiliary" groups for building the irreps of G_k (e.g.)[9] where 92 auxiliary groups are used). However these auxiliary groups are not systematically obtained and not all of them are P(k)'s.

iii) We see that "Im $\Gamma^\alpha_{G_k} \uparrow G$ is finite" ⟺ "Im Γ^α finite" ⟺ "Im k finite". Moreover, from Artin's theorem dim $\Gamma^\alpha_{G_k}$ = dim Φ divides the order of "P(k)/Center of P(k)" which divides $|P_k|$, the order of $P_k = P(k)/\text{Im } k$. So the dimension of the induced representation divides $|P|$ (which divides 48), a well known result.

iv) If P_k is cyclic (there are ten cyclic groups among the 32 point groups) its central extension (e.g. P(k)) are Abelian. When it is possible, it is interesting to decompose the P(k)'s into a direct product

$$P(k) = A(k) \times S(k) \tag{8}$$

where A(k) is an Abelian group and S(k) does not contain Abelian factors (we will call it the skeleton of P(k)). Indeed the unirreps of the Abelian groups are one-dimensional and easy to determine: their images are cyclic when the Abelian group is finite. As an example of the simplification introduced by (8) note that for trivial extensions of $H_o^2(P_k, \text{Im } k)$ and this includes the symmorphic groups and all G_k in the kernel of $\tilde{k}$, $P(k) = \text{Im } k \times P_k$ and the 32 P_k's themselves lead to only 4 non-isomorphic skeletons (isomorphic to D_3, D_4, T, O). The unirreps of P(k) are those of S(k) multiplied tensorially by the one dimensional unirreps of A(k).

v) Finally, not only Im k and P_k are the same for all G_k of the same arithmetic class, but there is a strong correlation between P_k and Im k. For a given P_k, and independently of its Bravais class, the structure of Im k is fixed up to few alternatives. We did not find these simple geometrical relations in the literature. They were published with B. Stawski as a poster of the Austin conference[16]. It is useful to reproduce them here (Table 1). Note that for the 22 non - polar groups P_k the image Im k is finite, cyclic and its order divides 6. For the ten polar P_k's, Im k is infinite only if it contains a factor Z^δ with $\delta = 1,2,3$; it is the number of (relatively) irrational components of k in $T^* \sim U(1)^3$.

Here are some results we have already obtained. The zero dimensional strata in the action of the seven P_H on the T* of the 14 Bravais lattice contain 80 orbits which contain 128 k's, the "high symmetry wave vectors" of Herring. They yield 1370 P(k)'s and more than 3800 inequivalent unirreps of C_k's. However these P(k)'s fall into 68 isomorphic classes and there are only 26 distinct skeletons S(k) of order.

order	6	8	12	16	24	32	48	96	Total	
nb of S(k)	1	2	2	6	4	5	4	2	26	(9)

Finally these nearly four thousand tabulated G_k unirreps have only 37 <u>inequivalent images</u>.

TABLE 1: Possible Im k's corresponding to a given P_k (independently from its Bravais lattice); d = dimension of the stratum in the action of the holohedry P_H of P_k on the Brillouin zone T*. The 10 cyclic P_k are underlined. m is an arbitrary positive integer. Z_m is the cyclic group of order m, Z the infinite cyclic group.

d	nb of P_k		P_k	Im k
0	22	13	$\underline{C_i}, C_{2h}, C_{4h}, D_{2h}, \underline{S_6}, C_{6h}, D_{3d}, D_{4h}, D_6, D_{6h}, T_h, O, O_h$	$\{0\}, Z_2$
		6	$D_2, D_{2d}, \underline{S_4}, D_4, T, T_d$	$\{0\}\ Z_2, Z_4$
		3	$D_3, \underline{C_{3h}}, D_{3h}$	$\{0\}, Z_2, Z_3, Z_6$
1	8	2	$C_{6v}, \underline{C_6}$	Z_m, Z
		2	$C_{3v}, \underline{C_3}$	$Z_m, Z, Z_3 x Z$
		4	$C_{4v}, \underline{C_4}, C_{2v}, \underline{C_2}$	$Z_m, Z, Z_2 x Z$
2	1		$\underline{C_s}$	$Z_m, Z_m x Z, Z^2, Z_2 x Z^2$
3	1		$\underline{C_1} = \{1\}$	$Z_m, Z_m x Z, Z_m x Z^2, Z^3$

Their dimension is

dimension	1	2	3	4	6	(10)
nb of images	7	20	6	3	1	

It is worthwhile to compute the isotropy groups of these 37 images and give the generators of their free module of invariants. This is nearly completed. Only for the four P_k's (C_{nv}, n = 2,3,4,6) of one dimensional strata will there appear new skeleton of non Abelian P(k)'s. They are infinite in number but fall into few families, all characterized in reference[17].

We will publish elsewhere, with all relevant details these results on the structure of all G_k unirreps when they will be completed. The complete realization of this program will require to perform the induction $\Gamma^{\alpha}_{G_k} \uparrow G$. Not only, as we will show, the present sets of long tables can be replaced by a sequential set of few short tables easy to implement in computers, but we hope it becomes clear that the "other type of knowledge" about space group unirreps can be useful for a deeper understanding and, in the same time, a simplification of the study of the related physics phenomena.

Another comparable example concern the knowledge of the invariants of a given symmetry group G. With computer help physicists are producing more and more tables of G invariants (for point groups and spacegroups). They can only be incomplete (since there is an infinity of polynomial invariants). However we know since Hilbert[18] that the ring of polynomial invariants of a finite group image is finitely generated. This extend to compact images: this is the case of the topological closure of the non-finite image of space groups (we remark that their smooth invariants are those of this closure). Very recently it can be proven that the smooth invariant - or covariant - functions on the m dimensional carrier space of Im Γ form a <u>free finite dimensional module</u> on a smooth function ring generated by m algebraically independent invariant polynomials[19]. The knowledge of such m polynomials and those of the free module basis gives a much more important knowledge for physicists than any table listing invariant polynomials.

ACKNOWLEDGEMENTS

One of us (L.M.) is very grateful to the organizers of this colloquium for its very stimulating atmosphere.

J.Mozrzymas has not been able to read the last version of this manuscript and is not responsible for the errors it could contain.

REFERENCES

1- F.Seitz, Z.Kristallogr. Kristalgeom. 88 (1934) 433, 90 (1935) 289, 91 (1935) 336, 94 (1930) 100.

2- L.P.Bouckaert, R.Schmoluchowski, E.P.Wigner, Phys.Rev. 50 (1936) 58.

3- F.Seitz, Ann.Math. 37 (1936) 17.

4- C.Herring, J.Franklin Institute 233 (1942), 525.

5- D.K.Faddeyev, Tables of the principal unitary representations of the Fedorov groups Acad.Sci. USSR, Translation Pergamon, New York, 1946.

6- O.V.Kovalev, Irreducible representation of space groups, Izd.Akad. Nauk. Ukraine hoj USSR, translation-Gordon Breach, New York, 1965.

7- S.C.Miller, W.F.Love, Irreducible representations of space groups. Pruett, Boulder Colorado 1967.

8- J.Zak, P.Casher, M.Gluck, Y.G.Gur, The irreducible represenations of space groups, Benjamin, New York 1969.

9- C.J.Bradley, A.P.Cracknell, The mathematical theory of symmetry in solids, Clarendon Press, Oxford 1972.

10- A.P.Cracknell, B.L.Davies, S.C.Miller, W.F.Love, General introduction and Tables of irreducible representations of space groups, Plenum Press, London 1979.

11- L.Michel, J.Mozrymas, Match 10 (1981) 223.

12- R.Dirl, These Proceedings.

13- L.D.Landau, Phys.Z.Soviet 11 (1937) 26, 545.

14- L.D.Landau, E.M.Lifschitz, Statistical Physics, translation Pergamon Oxford 1965.

15- A.H.Clifford, Ann.Math. 38 (1937) 533.

16- L.Michel, J.Mozrymas, B.Stawski, Lecture Notes Phys. 94 (1979) 86, Springer 1979.

17- H.S.M.Coxeter, W.O.J.Moser, Generators and Relations for discrete groups, Springer (1957).

18- D.Hilbert, Math.Ann. 36 (1890) 473.

19- M.Jaric, L.Michel, R.T.Sharp, These Proceedings p. and to appear.

BAND STRUCTURE OF ALMOST PERIODIC POTENTIALS

J.B.Sokoloff

Physics Department
Northeastern University
Boston, Massachusetts 02146, U.S.A.

Recently there has been a good deal of interest in the problem of the nature of electronic states in an almost periodic potential (APP). This interest is due in part to the connection with the problem of two dimensional electronic systems in a magnetic field[1] and to that of electrons in a random potential[2]. It is well known that a two dimensional solid in a magnetic field can be mapped to a one dimensional tight-binding model for an electron in a sinusiodal potential with period which can be incommensurate with that of the lattice[3]. Since the wave vector of the sinusoidal potential is proportional to the number of magnetic flux quanta through a unit cell of the 2-D system, it is usually extremely small, for the values of the fields easily obtainable in the laboratory. It is known, however, that for a two dimensional metallic system, a strong magnetic field can aid in the formation of a Wigner lattice or charge density wave and the unit cell of the Wigner lattice can be large enough to contain a sufficiently large magnetic flux to make the effects observable[4]. Furthermore, regarding the relationship to the random potential problem APP are potentials that lack translational invariance; yet, they are not disordered in the usual sense because there can be distances over which the potential almost repeats[2]. In fact, work by several authors, including one of the present authors, has indeed shown that an APP represents a case intermediate between random and periodic[2,3]. It is well known that in a 1-D disordered system almost all states are localized[5]. In a periodic potential there are regions in energy for which the states are completely extended. In an APP, however, there can be both extended and localized states, with the possibility of a metal-insulator transition at a critical value of the potential strength. This would seem to indicate that if there is a lower critical dimensionality in the APP problem it should be smaller than one. Although there can be extended states in an APP[2,3], the spectrum is not the usual band structure that occurs in the periodic case. Rather, the spectrum is singular, Cantor-set-like, with gaps almost everywhere[6]. Intuitively we can understand this from the fact that a gap can occur at one half of any reciprocal lattice vector, and for an APP every possible linear combination of the periods of these potentials with integer coefficients is a reciprocal lattice vector. Since the periods are incommensurate, by taking linear combinations with appropriate integer coefficients, we can construct a reciprocal lattice vector arbitrarily close to any value, and hence a band gap can occur arbitrarily close to any wave vector. In reference 2b it has been argued that while the spectrum can be Cantor-set like in the region of extended states, most of the gaps are negligible making the band structure appear like that of an ordinary periodic system.

Much of the work on almost periodic systems has centered on the following one-dimensional tight binding model proposed by Aubry and André[2a]:

$$t(f_{n+1} + f_{n-1}) + V_o \cos Q\, na\, f_n = E\, f_n, \qquad (1)$$

where the wave function is given in terms of the coefficient f_n as

$$\phi(x) = \Sigma_n f_n \phi(x-na), \qquad (2)$$

where $\phi(x-na)$ is a Wannier or atomic function centered on lattice site na (n is an integer) and t, V_o, Q and E are the hopping matrix element, potential strength, wave vector of the sinusoidal potential and the energy eigen-value, respectively. We are interested in the case where Qa is an irrational multiple of 2π. The study of this model is important for four reasons: first, it is the simplest possible model of an almost periodic system. Second, Bellisard, et.al.[7], have shown that there exists a mapping from a Schroedinger equation with an array of almost periodically modulated δ-functions to this model, and recently Jose and the present author[8] have shown that a Schroedinger equation with an almost periodically modulated array of potential barriers can be mapped into a generalization of the Aubry model in which the parameter t is also almost periodically modulated. The third reason for studying this model is the tight binding approximation for an electron in a periodic potential and a magnetic field reduces to this model and generalizations of it with t almost periodically modulated[3], and the fourth reason for studying this model is that the problem of a three dimensional crystal with a plane wave modulation with wave vector along a primitive reciprocal lattice vector of the lattice (a very common case in charge and spin density wave systems reduces precisely to the Aubry model in the tight binding approximation. This, can easily be seen if we consider the tight binding approximation equation in the presence of such a modulation:

$$t\Sigma_{\vec{a}}\ f(\vec{R} + \vec{a}) + 2V_o \cos \vec{Q}.\vec{R}\ f(\vec{R}) = Ef(\vec{R}) \tag{3}$$

Here, the wave function is given by

$$\Sigma_{\vec{R}}\ f(\vec{R})\ \phi(\vec{r}-\vec{R}), \tag{4}$$

where $\vec{R}$ is a lattice vector and $\vec{a}$ is a near neighbor lattice vector. The Fourier transformed equation is

$$t\ \Sigma_{\vec{a}}\ e^{i(\vec{k}+n\vec{Q}).\vec{a}}\ g_n + V_o(g_{n+1} + g_{n-1}) = Eg_n, \tag{5}$$

where

$$g_n = \frac{1}{N}\ \Sigma_{\vec{R}}\ e^{i(\vec{k}+n\vec{Q}).\vec{R}}\ f(\vec{R}) \tag{6}$$

If we choose $\vec{Q} = \alpha\vec{G}$, where $\vec{G}$ is one of the primitive reciprocal lattice vectors and α is an irrational number, equation (5) reduces to equation (1). For general directions of Q we also get an Aubry-like model, but the model is more complicated, and as yet has not been studied in detail.

In connection with this model, mathematicians believe that the spectral function is absolutely continuous for $V_o < 2t$ in equation (1)[6]. For $V_o > 2t$, Auron and Simon have proven[6] that the spectral function is singular continuous for Q equal to 2π multiplied by a type of irrational number which is extremely well approximated by rationals. Such a spectral function is neither continuous nor is it a pure point (i.e. delta-function) spectral function, as occurs when all states are exponentially localized. The result implies unusual assymptotic behavior of the wave functions in which they alternately die practically to zero and then become large again[6].

Huberman has recently extended work due to Romerio which attemps to prove the existence of a Bloch theorem for the one dimensional almost periodic problem[9]. There appears to be a contradiction between our results and Romerio's proof of a Bloch theorem for a Schoedinger equation with an APP[9]. DeLange and Janssen[10] in fact consider essentially the same transformation of the system to higher dimensions as is used in Romerio's proof (see also ref. 11). They conclude that if there are localized states, the wave function must be non-analytic in this higher dimensional space, because it is in fact a non-analytic function of the phase of the incommensurate periodic potentials relative to each other. In fact, they present numerical calculations of the wave function based on this model, as a function of position and phase for high order commensurate systems. As the system becomes higher order commensurate the localized wave functions become rapidly varying functions of the phase. We have performed similar calculations on the Aubry model which show that, although when the phase is shifted by $2\pi/N$, where N is the order of commensurability, the energy spectrum does not change, each localized state becomes localized around a new lattice site which is generally quite far from the location of the original site around which the state was localized. This implies that when N becomes infinite (the incommensurate limit), the wave function will be a non-analytic function of the phase. The physical reason for expecting such non-analyticity can be understood as follows: Consider a pair of incommensurate sinusoidal potentials chosen so that their minima coincide at one point; call it the origin. An infinitesimal phase shift will make the potentials coincide at another point which is in general far from the original origin. If we shift the origin to this new point we recover the original problem, but clearly a state localized at a point in the crystal which is a given distance away from the old origin will be shifted to a point which is the same distance from the new origin. Such non-analyticity actually signifies a translational symmetry break[10,11]. Since Romerio's proof depends crucially on the asummption that the wave function is analytic in the phase, the above observation may be the source of the discrepancy.

REFERENCES

1- D.J.Thouless, M.Kohmoto, M.P.Nightingale and M.den Nija, Phys. Rev.Lett. 49, (1982); R.E.Prange, private communications.

2a- S.Aubry and C.André Proc. Israel Physical Society, ed. C.G.Kuper (Adam Hilger, Bristol, 1979), vol.3, p.133; J.B.Sokoloff, Solid State Communications 40, 633 (1981).

2b- Phys. Rev. B23, 2034, 6422 (1981).

2c- Phys. Rev. B22, 5823 (1980).

2d- M.Ya.Azbel, Phys. Rev. Lett. 43, 1954 (1979).

2e- C.M.Soukoulis and E.N.Economuu, Phys. Rev. Lett. 48, 1043 (1982).

2f- D.R.Grempel, S.Fishman and R.E.Prange, to be published.

3- A.Rauh, G.H.Wannier and G.Obermair, Physica Status Solidi (b) 63, 215 (1974); P.G.Harper, Proc. Phys. Soc. (London). A68, 874 (1955); Dr. Hofstadter, Phys. Rev. B14, 2239 (1976); G.H.Wannier, Phys. State Solidi (b) 88, 757 (1978) and additional references contained therein.

4- H.Fukuyama, P.M.Platzman, and P.W.Anderson, Phys. Rev. B19, 5211 (1979); H.Fukuyama, to be published; R.E.Prange, Private Communication; D.C.Tsui, H.L.Stormer, and A.C.Gossard, Phys. Rev. Lett. 48, 1559 (1982).

5- N.F.Mott and W.D.Twose, Adv.Phys. 10, 107 (1960), H. Kunz and B. Souillard Comm. Math Phys. 78, 201 (1980).

6- B.Simon, Advances in Appl. Mathematics (in press); and J.Avron and B.Simon, submitted to Bull. Am. Math. Soc.; M.Ya.Azbel, Zh.Eksp. Teor. Fiz. 46, 929 (1963) (Sov. Phys.-JETP 19, 634 (1963)); Dokl. Akad. Nauk. SSSR 159, 703 (1964) (Sov. Math. Dokl. 5, 1549 (1964)).

7- J.Bellisard, A.Formoso, R.Lima and D.Testard, Phys. Rev. B (May 14, 1982).

8- J.B.Sokoloff and J.V.Jose, Phys. Rev. Lett. 49, (1982); also see erratum in Phys. Rev. Lett (to be published).

9- T.Hogg and B.A.Huberman, unpublished; M.V.Romerio, J. Math. Phys. 12, 552 (1971).

10- C. de Lange and T.Janssen, to be published.

11- P.M. de Wolff, Acta Cryst. A30, 777 (1974); A. Janner and Janssen, Phys. Rev. B15, 643 (1977).

SPACE GROUP REPRESENTATIONS FOR CRYSTAL STRUCTURE TYPES

L.L. BOYLE

University Chemical Laboratory, Canterbury, Kent, England.

1. INTRODUCTION

The usual procedure for specifying a crystal structure type is to state which atoms occupy which sets of equivalent sites within the unit cell of the lattice. These sets were calculated by Wyckoff and are most conveniently found in the *International Tables for X-ray Crystallography, Vol. 1.* [1]

To specify an actual crystal structure one needs to specify not only the crystal structure type but also to specify the metric properties of the structure, i.e. the lattice parameters and the parameters necessary for specifying those Wyckoff sets possessing degrees of freedom. To illustrate these differences it is useful to remember that rock salt and diamond both have face-centred cubic structures but belong to different space groups; that diamond and normal spinels belong to the same space group, O_h^7, but not the same crystal structure type. Different normal spinels will (in a certain temperature range) have the same crystal structure type but not the same crystal structure. From the group-theoretical point of view, however, such matric differences (lattice parameters and parameters used to specify the coördinates of atoms belonging to Wyckoff sets with one, two or three degrees of freedom) are irrelevant unless one starts to determine coördination numbers.

Each point within a Wyckoff set has a different site symmetry group. The site symmetry groups are, however, isomorphic, and are conjugated by the inner automorphisms of the space group. Site symmetry groups are finite and are isomorphic to point groups. The specification of the site symmetry groups of a Wyckoff set by that point group which is isomorphic to each individual site symmetry group in the *International Tables* is a cause of misunderstanding, since for calculations outside the centre of the Brillouin zone with space group representations, the translational parts of the site symmetry group operations are indispensable. The specification of site symmetry elements by Seitz symbols is an origin-dependent calculation, but the physical meaning of each symmetry element is origin-independent.

Some sets of Wyckoff sites are equivalent to each other under the outer automorphisms of the space group, or equivalently, are conjugated by the inner automorphisms of the normalizer of the space group in the Euclidean group of three-dimensional space. [2,3]

2. DEFINITIONS AND PROPERTIES

The purpose of this paper is to define a representation which may be used to characterize any given set of Wyckoff sites. By summation with, if desired, assignation of atomic type present, it can be used to characterize a crystal structure type. The required representation is defined by induction from the scalar representation of the site symmetry group of any point within the Wyckoff set to the space group. Since the various site symmetry groups within the set are conjugate (finite) subgroups of the (infinite) space group this representation is independent of the point chosen. However, since the index of the site group within the space group is infinite, the reducible representation so obtained will be infinite in dimension. It is nonetheless manageable because at any given point in the Brillouin zone it will be finite-dimensional and indeed its dimension is given by the index of the point group isomorphic to the site symmetry group in the point group isomorphic to the factor group of the space group with the translational symmetry group (i.e. lattice). In those physical problems to which this theory may be usefully applied, interest is usually restricted to one or two points of the Brillouin zone and hence these reducible representations may usefully be calculated. Although the components at a given point in the Brillouin zone for two different sets of Wyckoff sites may occasionally be identical, this cannot be true for all points due to the translational symmetry. Such representations are therefore fully characteristic of the Wyckoff set.

3. MANIPULATIONS

The basic processes which can be applied to representations are addition, resolution, subduction, induction, multiplication and power symmetrization. By exploring each in turn we shall find various useful properties of these representations.

(i) Addition

Addition of representations may seem trivial, but it is this process which allows us to build a structure from different types of atoms. Different crystal structure types will have different reducible representations. This method of specification of a structure by a reducible representation provides a means of storing information about the structure in a very usable and useful, if unusual, way. It is particularly well adapted for computer retrieval systems.

(ii) Resolution

Resolution (or reduction when one wishes to resolve as far as irreducible) is the inverse of addition. Usually we resolve a representation into irreducible representations but here we wish to resolve into the basic *reducible* representations corresponding to the individual Wyckoff sets. This process is possible, but instead of having a neat orthogonality formula we have sets of Diophantine equations which must be derived for each space group. Inequalities are also often involved. A sufficient number of such equations or inequalities can always be found to solve the reducibility problem. Consequently we do not need to store a crystal structure type representation in too expanded a form. The sum of the representations for the Na and the Cl atoms in the NaCl structure representation can be resolved back into that for Na and Cl even if the Na and Cl labels are lost.

(iii) Subduction

This is the process of descent in symmetry from a supergroup to a subgroup. On subduction a representation corresponding to a set of atoms will always give another such representation. If the set has decomposed into subsets which are no longer related by symmetry elements in the lower symmetry, the subduced representation is resolvable.

This procedure provides a representation-theoretical basis for the site correlation problem.

(iv) Induction

This is the process of ascent in symmetry from a subgroup to a supergroup. Whilst in the subduction process the number of sites remains constant, in induction it increases by a factor equal to the index of the subgroup in the supergroup. The process enables one to determine which sites or sets of sites are related by the new symmetry elements.

(v) Multiplication

This is the most important in terms of useful applications. It provides a new method for obtaining the symmetries of the inelastic phonon modes of a crystal, i.e. its lattice vibrations at any point of the Brillouin zone, not just at zero wave vector.

The procedure is simply to multiply the crystal structure representation by the vector representation (i.e. that spanned by the translation vectors at the Γ-point) and then subtract the acoustic modes (i.e. that same vector representation). This enables one to identify which atoms are involved in a given motion.

(vi) Power symmetrization

This is the process of making the square, cube, or other power of the representation in such a way that the permutation symmetry with respect to the component parts is recognized.

The calculation of the symmetric part of the square of a crystal structure representation is in practice the most useful. It may be used for finding the symmetries of the sets of connections (i.e. bonds, interactions, etc.) which exist between sets of atoms in the lattice. To formulate this we realise that a bond b_{ij}

is a two-particle function which is symmetric to interchange of i and j. When bonds *between* two different sets of atoms $\{a_i\}$, $\{a_j\}$ are concerned, the symmetries of the sets of bonds $\{b_{ij}\}$ are given by the Kronecker product of the representations corresponding to the atomic sets. When, however, the bonds *within* a given set of atoms are considered, the symmetric part of the square of the representation corresponding to the atomic set must be used and further one must subtract the atomic representation from the result to account for the 'bonds' corresponding to the cases when i=j. (Such a set is clearly just another set of 1-particle scalar functions and hence has the symmetry of the atomic sets.)

Resolution of these reducible representations into bond sets requires, in general, a different set of Diophantine equations.

Complete sets of such representations with the appropriate sets of Diophantine equations for atom and bond sets have been described [4] and calculated for *all* point groups and are available in a Canterbury thesis [5]. Calculation of similar information for the space groups would be an enormous task but is in progress for a few selected representative cases of greater interest. This work is greatly facilitated by the recent publication of the complete sets of *Kronecker Product Tables* for the space groups [6,7] together with tables of symmetrized powers [8].

REFERENCES

[1] Norman F.M. Henry and Kathleen Lonsdale (eds.), *International Tables for X-ray Crystallography, Vol. I: Symmetry Tables*. 3rd edn. (Birmingham: The Kynoch Press, 1969).

[2] L.L. Boyle and J.E. Lawrenson, *The origin dependence of the Wyckoff site description of a crystal structure*, Acta Crystallographica (A), 29, 353-357 (1973).

[3] L.L. Boyle and J.E. Lawrenson, *The dependence of the Wyckoff site description of a crystal structure on the labelling of the axes*, Communications to the Royal Society of Edinburgh (Physical Sciences), 1, 169-175 (1978).

[4] L.L. Boyle, *Atom and Bond representations*, Europhysics Conference Abstracts, 1G, 8 (1976).

[5] Kerie F. Green, *Representation Theory of Finite Groups*, Ph.D. Thesis, University of Kent at Canterbury, (1976) pp. 16-64.

[6] B.L. Davies and A.P. Cracknell, *Kronecker Product Tables, Vol. 2: Wave Vector Selection Rules and Reductions of Kronecker Products for Irreducible Representations of Orthorhombic and Cubic Space Groups*. (New York: Plenum Press, 1979). pp. 141-918.

[7] A.P. Cracknell and B.L. Davies, *Kronecker Product Tables, Vol. 3: Wave Vector Selection Rules for Irreducible Representations of Triclinic, Monoclinic, Tetragonal and Hexagonal Space Groups*. (New York: Plenum Press, 1979). pp. 147-872.

[8] B.L. Davies and A.P. Cracknell, *Kronecker Product Tables, Vol. 4: Symmetrized Powers of Irreducible Representations of Space Groups*. (New York: Plenum Press, 1980). pp. 1-35, 159-542.

SELECTION RULES FOR POLYMERS

Milan Damnjanović[+], Ivan Božović[+&] and Nataša Božović[+&]

[+]Faculty of Science , University of Belgrade
11001 Belgrade , P.O. Box 550
YUGOSLAVIA

[&]Department of Physics, University of California at Berkeley
Berkeley CA 94720
U S A

ABSTRACT

The determination of the selection rules for physical processes in polymers (i.e. of the reduction coeficients for the irreducible representations of the line groups) is reported. Relationships with conservation laws of quasi momentum, quasi angular momentum and parities with respect to the mirror planes are discussed.

Recent discoveries of exciting electronic properties in some polymers and quasi one-dimensional solids have attracted much attention. Symmetries of such systems thus became worth studying and so the line group theory has been developed[1)]. Continuing these efforts we report here on the derivation of the selection rules for the line groups, i.e. of the reduction coeficients for the Kronecker products of their irreducible representations (reps).

These coeficients have been determined both directly and recursively (utilizing the subgroup chain structure of the line groups).

As expected the obtained selection rules can be interpreted as conservation laws for certain physical observables (quasi momentum, quasi angular momentum and parities with respect to the vertical and horizontal mirror planes). As for the quasi momentum, in the case of the line groups isogonal to $\underline{C}_n$ or $\underline{C}_{nv}$ one finds the usuall selection rule: $\langle i|V|f\rangle$ is zero unless

$$k_f = k_i + k_v + Q \qquad (1)$$

where Q=0 for the normal processes and $Q=\pm 2\pi/a$ for the Umklapp processes. However, the other line groups contain elements which convert k into -k ; their reps are in general labelled by pairs {k,-k} and (1) is replaced by:

$$k_f = k_i + k_v + Q \quad \text{or} \quad k_f = k_i - k_v + Q \quad . \tag{2}$$

Analogously, for the rotations through $2\pi/n$ arround the chain axis the selection rule reads:

$$m_f \doteq m_i + m_V \quad , \tag{3}$$

where m is the quantum number of the quasi angular momentum (m is an integer from the interval $(-n/2,n/2]$). In the Ln_p line groups the rotations are coupled with the translations and the quasi angular momentum is conserved in normal processes; however in the Umklapp processes the screw axis requires +p or -p to be added to the right hand side of (3).

Similar results are found for the behaviour of the parities with respect to the mirror planes in the Umklapp processes.

Once the reduction coeficients are known one can readily find the selection rules for different processes in polymers (optical absorption[2]), electron and neutron scattering, two-phonon Raman and infra-red processes).

References

1) I.Božović,M.Vujičić and F.Herbut, 1978 J.Phys.A 11 2133
M.Vujičić,I.Božović and F.Herbut,1977 J.Phys.A 10 1271
I.Božović and M.Vujičić, 1981 J.Phys.A 14 777
M.Damnjanović and M.Vujičić 1982 Phys.Rev. B25
2) I.Božović,J.Delhalle and M.Damnjanović 1981 Int.Journ. of Quant.Chem. XX 1143 .

AUTOMORPHISM SYMMETRIES OF SPACE GROUP SELECTION RULES

R. Dirl
Institut für Theoretische Physik, TU Wien
A-1040 Wien, Karlsplatz 13; Austria

A new class of useful symmetry relations between reduced matrix elements of irreducible tensor operators is derived. These relations are generated by certain auter automorphisms.

1. Unitary representations of automorphism groups

Let G be any finite, compact continuous or countable group. For the sake of simplicity we assume that the centre Z(G) of G is trivial and that the index of G with respect to its automorphism group A(G) is finite. Although we shall primarily be interested in space groups the following discussion will be carried out quite generally. Since due to our assumptions G must be a normal subgroup of A(G), each element $a \in A(G)$ defines an equivalence relation $D^{\lambda}(a(g)) \simeq D^{a(\lambda)}(g)$; $g \in G$. Thereby we adopt the notation D^{λ}, $D^{a(\lambda)}$ for *unirreps* of G and λ, $a(\lambda)$ for equivalence classes of G. Accordingly some unitary matrix $Z^{\lambda}(a)$ must exist that satisfies

$$D^{\lambda}(a(g)) = Z^{\lambda}(a)\ D^{a(\lambda)}(g)\ Z^{\lambda}(a)^{\dagger} \qquad g \in G \tag{1}$$

where $a(g) = a\ g\ a^{-1}$. As usual we call $A^{\lambda}(G) = \{a \in A(G) \mid a(\lambda) = \lambda\}$ the corresponding *little group*, where $Z^{\lambda}(a)$; $a \in A^{\lambda}(G)$ in general forms a projective representation.

With regard to physical applications we assume that a *unitary* representation $U(A(G)) = \{U(a) : a \in A(G)\}$ of A(G) is defined on a separable Hilbert space H. In addition we assume that a G-*adapted* orthonormal basis $\{\Phi^{\lambda}_{j}\}$ of H is given, i.e.

$$U(g)\ \Phi^{\lambda}_{j} = \sum_{k} D^{\lambda}_{kj}(g)\ \Phi^{\lambda}_{k} \qquad g \in G \tag{2}$$

where for the sake of simplicity further state labels are omitted.

A simple manipulation yields

$$\Phi_j^{a(\lambda)} = \sum_k Z_{kj}^{\lambda}(a)\ U(a^{-1})\ \Phi_k^{\lambda} \qquad a \in A(G) \tag{3}$$

Eqs.(3) reduce to identities, if $a \in A^{\lambda}(G)$, but correlate in a generic way G-adapted states that belong to inequivalent unirreps of G, if $a \in A(G) \setminus A^{\lambda}(G)$. It is readily verified that the states (3) transform according to the unirrep $D^{a(\lambda)}$ of G.

Moreover we assume that *irreducible tensor operators* $\{T_j^{\xi}\}$ with respect to G are given. By similar arguments we define corresponding irreducible tensor operators by means of

$$T_j^{a(\xi)} = \sum_k Z_{kj}^{\xi}(a)\ U(a^{-1})\ T_k^{\xi}\ U(a^{-1})^{\dagger} \qquad a \in A(G) \tag{4}$$

whose correct transformation properties with respect to G are easily demonstrated. Eqs.(4) become identities, if $a \in A^{\xi}(G)$, but yield inequivalent irreducible tensor operators, if $a \in A(G) \setminus A^{\xi}(G)$. Now it is well known that according to *Wigner Eckart's Theorem*, matrix elements of irreducible tensor operators factorize into CG-*coefficients* and *reduced matrix elements*.

$$(\Phi_k^{\gamma}\ ,\ T_i^{\alpha}\ \Phi_j^{\beta}) = \sum_w (\alpha i, \beta j \,|\, \gamma w k)^* \, (\gamma \| T^{\alpha} \| \beta)_w \tag{5}$$

For *non-simply reducible* groups G the right hand side of (5) consists of $m(\alpha,\beta;\gamma)$ terms ($m(\alpha,\beta;\gamma)$ = *multiplicity*).

2. Automorphism symmetries of reduced matrix elements

As has been discussed extensively[1), CG-*vectors* of G are symmetry adapted vectors that transform according to unirreps of G. Their components are just the CG-coefficients.

$$\{\vec{C}_k^{\alpha\beta;\gamma w}\}_{ij} = (\alpha i, \beta j \,|\, \gamma w k) \tag{6}$$

For given unirreps of G appertaining CG-coefficients are *unique* up to arbitrary $m(\alpha,\beta;\gamma)$-dimensional unitary transformations. We reported in Ref.2 about new symmetries of CG-coefficients of G that are associated with certain auter automorphisms. These relations are given by

$$Z^{\alpha\beta}(a)\ \vec{C}_k^{\alpha\beta;\gamma w} = \sum_v B_{vw}(a) \sum_j Z_{jk}^{\gamma}(a)\ \vec{C}_j^{\alpha\beta;\gamma v} \qquad a \in A^{\alpha\beta\gamma}(G) \tag{7}$$

$$Z^{\alpha\beta}(a)\ \vec{C}_k^{\alpha\beta;\gamma w} = \sum_j Z_{jk}^{\gamma}(a)\ \vec{C}_j^{a(\alpha),a(\beta);a(\gamma)w} \qquad a \in A(G):A^{\alpha\beta\gamma}(G) \tag{8}$$

where the former were called *symmetry* and the latter *generating* relations. In particular $B(a)\ ;\ a \in A^{\alpha\beta\gamma}(G) = A^{\alpha}(G) \cap A^{\beta}(G) \cap A^{\gamma}(G)$ forms a uniquely defined $m(\alpha,\beta;\gamma)$-dimensional projective representation of the factor group $A^{\alpha\beta\gamma}(G)/G$.

Utilizing the invariance of the scalar product on H, some orthogonality relations of CG-coefficients, Eqs.(4) and Eqs.(7,8) respectively, one obtains immediately the following identities.

$$(\gamma\|T^{\alpha}\|\beta)_w = \sum_v B_{vw}(a)\ (\gamma\|T^{\alpha}\|\beta)_v \qquad a \in A^{\alpha\beta\gamma}(G) \tag{9}$$

$$(\gamma\|T^{\alpha}\|\beta)_w = (a(\gamma)\|T^{a(\alpha)}\|a(\beta))_w \qquad a \in A(G):A^{\alpha\beta\gamma}(G) \tag{10}$$

Obviously both sets of equations present useful relations between reduced matrix elements, assuming G is a proper subgroup of $A^{\alpha\beta\gamma}(G)$ and the latter is one of A(G), respectively. Thus without computing explicitely reduced matrix elements of irreducible tensor operators, *selection rules* for them can be predicted by means of (9) and (10). In particular Eqs.(9) entail vanishing reduced matrix elements, if $B(a)\ ;\ a \in A^{\alpha\beta\gamma}(G)$ does not contain the *trivial* unirrep. In comparison therewith, Eqs.(10) establish that reduced matrix elements belonging to inequivalent unirreps D^{λ}, $D^{a(\lambda)}$; $\lambda = \alpha,\beta,\gamma$ of G must coincide, if $A^{\alpha\beta\gamma}(G)$ is a proper subgroup of A(G) and the group element a does not belong to the triple intersection group.

3. Applications to cubic space groups

It is well known[3)] that the automorphism group of a given space group G is isomorphic to the factor group of the *affine normalizer* N(G) with respect to the *affine centralizer* C(G). As already pointed out we assume that Z(G) and C(G) are trivial in order to ensure that G is a normal subgroup of N(G) = A(G). This situation is realized for many space groups as can be seen from Table 3 of Ref.3.

For instance let us consider the symmorphic face centered cubic space group G = Fm3m(a). Its automorphism group A(G) = Pm3m(a/2) is a symmorphic primitive cubic space group, but whose lattice constant is half of that of G. By virtue of $A(G) = G + (E|\vec{b}_o)G$, where $\vec{b}_o$ is a special primitive translation of Pm3m(a/2), it follows that only one non-trivial auter automorphism for Fm3m(a) exists, namely $(E|\vec{b}_o)$. Denoting the elements of the *representation domain* ΔBZ{Pm3m(a/2)} of Pm3m(a/2) by $\vec{q}$ and those of ΔBZ{Fm3m(a)} by $\vec{k}$, one can show by a representation theoretic argument that $A^{\lambda}(G) = A(G)$ must hold, if the appertaining *little cogroups* coincide (i.e. $P(\vec{k}) = P(\vec{q})$ with $\vec{k} = \vec{q}$), whereas $A^{\lambda}(G) = G$ is always realized, if $P(\vec{k})$ contains $P(\vec{q})$ as subgroup of index two. Obviously only these two situations may occur. For example it holds $P(\vec{k}_{\Gamma}) = P(\vec{q}_{\Gamma}) = O_h$ or $P(\vec{k}_{\Lambda}) = P(\vec{q}_{\Lambda}) = C_{3v}$, whilst e.g. $P(\vec{k}_L) = D_{3d}$ and $P(\vec{q}_{\Lambda}) = C_{3v}$ is a subgroup of index two of the former, where $\vec{k}_L = \vec{q}_{\Lambda}$. Accordingly if $A^{\lambda}(G) = A(G)$ is valid, symmetry relations of type (9) that are generated by the group element $(E|\vec{b}_o)$ must hold for reduced matrix elements, whereas for $A^{\lambda}(G) = G$ relations of type (10) must be expected. When for instance $\lambda = (\vec{k}_L;(\mu,\sigma))$ is considered, where (μ,σ) characterizes unirreps of $P(\vec{k}_L) = C_{3v} \times \{E,I\}$, the group element $(E|\vec{b}_o)$ maps (μ,σ) onto $(\mu,\sigma+1)$, i.e. changes the parity of the unirreps of $P(\vec{k}_L)$.

References

1) R. Dirl; J.Math.Phys.20,659(1979)
2) R. Dirl; Physica 114A,50(1982)
3) H. Burzlaff, H. Zimmermann; Z.Krist.153,151(1980)

INVARIANT FORMULATION FOR THE ZEROS OF COVARIANT VECTOR FIELDS

Marko V. Jarič*, L. Michel and R.T. Sharp**

Institut des Hautes Etudes Scientifiques, 91440 Bures-sur-Yvette, France

ABSTRACT

Invariant formulation for the zeros of covariant vector fields is presented. It is shown that they can be determined at each stratum from certain cannonical equations in terms of invariants and the field components relative to a covariant basis.

1. Introduction

Let G be a finite group acting orthogonally on a carrier space $\mathbb{R}^n$ and let $f: \mathbb{R}^n \longrightarrow \mathbb{R}^n$ be a G-covariant vector field,

$$f(gx) = gf(x), \ \forall g \in G, \ \forall x \in \mathbb{R}^n . \tag{1}$$

Such fields and their zeros play an important role in many areas of physics. Determination of the zeros directly in the carrier space is pursued in another paper in the same proceedings[1]. An invariant, orbit space, approach will be sketched here. More detailed study will be presented elsewhere[2].

2. Orbit space approach

It is well known[3] that there is a finite G-covariant polynomial basis $e_s(x)$ such that every G-covariant polynomial field f(x) can be uniquely decomposed

$$f(x) = \sum_s q_s(\theta) e_s(x), \tag{2}$$

where $q_s(\theta)$ are polynomials in denominator invariants $\theta(x)$. ($\theta(x)$ are n algebraically independent G-invariant polynomials).

It is also well known[4] that at a particular stratum $\sum[L]$ (associated with a class $[L]$ of isotropy subgroups of G) a G-covariant vector field is tangential to the stratum. Therefore, we determine at each $\sum[L]$ the minimal set of basic G-covariant fields (say e'_t, $t=1,\ldots,\dim\sum[L]$) which are linearly independent at $\sum[L]$. The zeros of Eq. (2) can then be obtained from the equations of the stratum[5] and from the equations

$$f\cdot e'_t = \sum_s q_s(\theta)(e_s \cdot e'_t) = 0. \tag{3}$$

In all of these equations x-dependence is only implicite through denominator and numerator invariants (integrity basis). Thus, zeros of f(x) may be determined directly in the orbit space.

The basic fields e'_t and the equations (3) have been determined for all strata of all (finite and infinite) two- and three-dimensional point groups[2].

MVJ acknowledges an Alexander von Humboldt research fellowship and partial support from the Deutsche Forschungsgemeinschaft, Sonderforschungsbereich 161.

* Also at Freie Universität Berlin, Institute for Theoretical Physics; current address: Dept. of Physics, MSU, Bozemann, Montana 59717.

** On leave from Physics Department, McGill University, Montréal, Québec, Canada

References

1. M.V. Jarić, the same proceedings, pp.
2. M.V. Jarić, L. Michel and R.T. Sharp, to be published.
3. See the review by R.P. Stanley, Bulletin Am. Math. Soc. 1, 475 (1979).
4. L. Michel, Lecture Notes in Physics 6, 36 (1970).
5. M. Abud and G. Sartori, Phys. Lett. 104B, 147 (1981); M.V. Jarić preprint IHES/P/82/16 (1982).

ASSOCIATED COREPRESENTATIONS AND SYMMETRY OF CLEBSCH-GORDAN COEFFICIENTS

J.N.Kotzev and M.I.Aroyo

Physics Department, Sofia University, Sofia 1126, BULGARIA

1- The transformation properties of wave functions and operators of systems with magnetic symmetry are determined by Wigner corepresentations (coreps)[1]. Recently we have published full tables of Clebsch-Gordan coefficients (CGC) for the coreps of all 90 antiunitary (AU) Shubnikow point groups[2]. In this paper we will discuss in detail an additional type of symmetry of CGC of coreps, which is related to the operation of association[3]. We have shown that the symmetry of CGC under association considerably decreases the volume of the tables and that it is a necessary condition for the transition from CGC to 3DΓγ symbols (analogous to 3jm symbols).

2- The symmetry of CGC for linear representations under association is discussed in Griffith[3]. Let D^{A_i} be one dimensional corep and D^{α_i} be an arbitrary irreducible corep of AU group A. The Kronecker product $D^{\alpha_i} \otimes D^{A_i}$ is equivalent to an irreducible corep $D^{\alpha_i'}$ which is associated to D^{α_i} by D^{A_i}.

The matrices and the basis functions of the associated corep are connected by CGC $U^{\alpha_i A_i}$

$$D^{\alpha_i'}(g) = U^{\alpha_i A_i^{-1}} D^{\alpha_i}(g) \otimes D^{A_i}(g)\, U^{\alpha_i A_i (*)}, \quad g\varepsilon A \tag{1}$$

$$\psi^{\alpha_i'}_{a_i'} = \sum_a \psi^{\alpha_i}_{a_i} \psi^{A_i}_1 U^{\alpha_i A_i}_{a_i 1, \alpha_i' a_i'} \tag{2}$$

The asterisk in parenthesis means complex conjugation only for the antiunitary operators $g\varepsilon A$.

We will discuss the change of the CGC for coreps when the three coreps D^{α_1}, D^{α_2} and $D^{\alpha_3} \varepsilon D^{\alpha_1} \otimes D^{\alpha_2}$ are associated by three different one-dimensional coreps D^{A_1}, D^{A_2} and $D^{A_3} \cong D^{A_1} \otimes D^{A_2}$. In result we will get CGC which reduce the Kronecker product $D^{\alpha_1'} \otimes D^{\alpha_2''}$

$$U^{\alpha_1'\alpha_2''} = (U^{\alpha_1 A_1} \otimes U^{\alpha_2 A_2})^{-1} (U^{\alpha_1\alpha_2} \otimes U^{A_1 A_2}) \left[\bigoplus_{\alpha_3 \varepsilon \alpha_1 \otimes \alpha_2} (e^{\alpha_1\alpha_2}_{\alpha_3} \otimes U^{\alpha_3 A_3}) \right] \cdot$$

$$\cdot \left[\bigoplus_{\alpha_3} (\omega^{\alpha_1\alpha_2}_{\alpha_3} \otimes E^{\alpha_3}) \right] \tag{3}$$

where $e^{\alpha_1\alpha_2}_{\alpha_3}$ and E^{α_3} are identity matrices and $\dim e^{\alpha_1\alpha_2}_{\alpha_3} = \dim \omega^{\alpha_1\alpha_2}_{\alpha_3} =$ $(\alpha_1\alpha_2|\alpha_3)$ = multiplicity of D^{α_3}. The matrix elements of $\omega^{\alpha_1\alpha_2}_{\alpha_3} =$ $\| \omega^{\alpha_1\alpha_2\alpha_3}_{\Gamma_3'\Gamma_3} \|$ are the so-called inner isoscalar factors (IIF) (in analogy to the outer isoscalar factors of Racah lemma[2]). All our CGC are chosen in such a way that the matrices of IIF for the coreps of all 90 AU Shubnikow point groups are diagonal:

$$\omega^{\alpha_1\alpha_2\alpha_3}_{\Gamma_3'\Gamma_3} = \omega^{\alpha_1\alpha_2\alpha_3}_{\Gamma_3} \delta_{\Gamma_3'\Gamma_3} \tag{4}$$

For the case of generalized CGC for coreps we get the following relation:

$$U^{\alpha_1'\alpha_2''\alpha_3''' \dots \alpha_n^{(n)}} = (U^{\alpha_1 A_1} \otimes U^{\alpha_2 A_2} \otimes \dots \otimes U^{\alpha_n A_n})^{-1} (U^{\alpha_1\alpha_2 \dots \alpha_n} \otimes U^{A_1 A_2 \dots A_n})$$

$$\times \left[\bigoplus_{\alpha} (e^{\alpha_1\alpha_2 \dots \alpha_n}_{\alpha} \otimes U^{\alpha A}) \right]^{(*)} \left[\bigoplus_{\alpha} (\omega^{\alpha_1\alpha_2 \dots \alpha_n}_{\alpha} \otimes E^{\alpha}) \right] \tag{5}$$

And the corresponding relation for the $3D\Gamma^{\alpha}_{\gamma}$ symbol is:

$$V^{\alpha_1'\alpha_2''\alpha_3'''} = (U^{\alpha_1 A_1} \otimes U^{\alpha_2 A_2} \otimes U^{\alpha_3 A_3})^{-1} (V^{\alpha_1\alpha_2\alpha_3} \otimes U^{A_1 A_2 A_3}) \cdot$$

$$\cdot \left[\bigoplus_{\alpha_o} (e^{\alpha_1\alpha_2\alpha_3}_{\alpha_o} \otimes U^{\alpha_o A_4}) \right]^{(*)} \times (\bigoplus_{\alpha_o} \omega^{\alpha_1\alpha_2\alpha_3}_{\alpha_o}) \tag{6}$$

where α_o is the identity corep.

3- The $3D\Gamma_{\gamma}$ symbols calculated on the base of CGC will have the required permutation properties if the starting CGC are connected under association with diagonal IIF matrices[4] (i.e. in the case of coreps the symmetry of CGC under association is a necessary condition for the construction of 3DΓγ symbols with proper permutation properties).

The relation (3) helps us to decrease considerably the volume of the calculation and the tables of CGC. For example all odd coreps $D^{\Gamma^-}_{\alpha}$ of the centrosymmetrical AU group $A \otimes G_i$ are associated to the even coreps $D\Gamma^+_{\alpha}$ by $D\Gamma^-_1$. From (3) we get the following relations for the CGC:

$$U^{\alpha_1^- \alpha_2^-} = U^{\alpha_1^+ \alpha_2^+} = U^{\alpha_1^{\pm} \alpha_2^{\mp}}$$

By association with the pseudoscalar corep we have calculated the CGC for the coreps of all noncentrosymmetrical groups for the case of odd (under space inversion) basis functions.

REFERENCES

1- C.J.Bradley, A.P.Cracknell. The Mathematical Theory of Symmetry in Solids. Oxford Univ. Press, Oxford, 1972.

2- J.N.Kotzve and M.I.Aroyo. J.Phys.A: Math and Gen. 13(1980)2275; ibid 14(1981) 1543; ibid 15 (1982) 711, 725.

3- J.S.Griffith. The Irreducible Tensor Method for Molecular Symmetry Groups, Prentice Ha-l, Inc., Englewood Cliffs, N.J., 1962.

GENERALIZED AND SYMMETRIZED CLEBSCH GORDAN COEFFICIENTS FOR ANTIUNITARY GROUPS

J.N.Kotzev and M.I.Aroyo

Physics Department, University of Sofia, Sofia 1126, BULGARIA

1. The method of irreducible tensorial sets, or the algebra of Wigner-Racah is applied successfully in the spectroscopy of crystals[1]. The generalization of the algebra of Wigner-Racah for the case of corepresentations[2] (coreps) is imposed by the fact that the symmetry of magnetic crystals is described by antiunitary (AU) Shubnikov groups. In a series of papers[3] we have published full tables of Clebsch-Gordan Coefficients (CGC) $U^{\alpha_1\alpha_2}$ for coreps of all AU point groups and tables of the corresponding isoscalar factors (IF)[4,5]. In this paper we introduce the concept of generalized CGC (GCGC) for coreps $U^{\alpha_1\alpha_2\ldots\alpha_n}$ which reduce the Kronecker product of n coreps (1). We find the connection between the GCGC and the CGC in the case of coreps and we discuss the special case of GCGC for n=3, i.e. $U^{\alpha_1\alpha_2\alpha_3}$ (3). Useful relations for the symmetrized GCGC $V^{\alpha_1\alpha_2\ldots\alpha_n}$ (analogous to the 3njm symbols) are derived and 3DΓγ symbols for the coreps of all 90 AU point groups. Racah lemma is reformulated[1,3] for the case of GCGC and 3DΓγ symbols. We discuss a number of essential differences of the algebra of Wigner-Racah for the coreps case.

2. The matrix elements $U^{\alpha_1\alpha_2\ldots\alpha_n}_{a_1a_2\ldots a_n,\alpha\Gamma_\alpha a}=[\alpha_1a_1\alpha_2a_2\ldots\alpha_na_n|\alpha\Gamma_\alpha a]$ are the generalized CGC

$$U^{\alpha_1\alpha_2\ldots\alpha_n^{-1}} D^{\alpha_1}(g) \otimes D^{\alpha_2}(g) \ldots \otimes D^{\alpha_n}(g)\, U^{\alpha_1\alpha_2\ldots\alpha_n(*)} = \bigoplus_\alpha (e^{\alpha_1\alpha_2\ldots\alpha_n} \otimes D^\alpha(g)),\ g\in A \qquad (1)$$

where $e_\alpha^{\alpha_1\alpha_2\ldots\alpha_n}$ are identity matrices with dimension equal to the multiplicity of D^α and Γ_α is the multiplicity index; the asterisk in parenthesis means complex conjugation only for the AU operators. The reduction (1) can be carried out step by step using $U^{\alpha_1\alpha_2}$, $U^{\alpha_{12}\alpha_3}$ ... etc. and in this case we find the relation.

$$[\alpha_1 a_1 \alpha_2 a_2 \ldots \alpha_n a_n \,|(\alpha_{12}\Gamma_{12} \ldots)\, \alpha \Gamma_\alpha a] =$$

$$=\Sigma [\alpha_1 a_1 \alpha_2 a_2 | \alpha_{12}\Gamma_{12} a_{12}] [\alpha_{12} a_{12} \alpha_3 a_3 | \alpha_{123}\Gamma_{123} a_{123}] \ldots \tag{2}$$

$$\ldots [\alpha_{1\ldots n-1}\, a_{1\ldots n-1}\, \alpha_n a_n | \alpha \Gamma_\alpha a]\, \mu^{\alpha_1\alpha_2 \ldots \alpha_n, \alpha}_{\Gamma_{12}\Gamma_{123} \ldots \Gamma_\alpha}$$

Where $\mu^{\alpha_1\alpha_2 \ldots \alpha_n,}_{\Gamma_{12}\Gamma_{123}\ldots}$ are the so-called inner IF and the sum is carried over all intermediate coreps. In matrix notation for n=3 we have

$$U^{\alpha_1\alpha_2\alpha_3} = (U^{\alpha_1\alpha_2} \otimes E^{\alpha_3})[\underset{\alpha_{12}}{\oplus}(e^{\alpha_1\alpha_2}_{\alpha_3} \otimes U^{\alpha_{12}\alpha_3})][\underset{\alpha}{\oplus}(\mu^{\alpha_1\alpha_2\alpha_3}_{\alpha} \otimes E^\alpha)] \tag{3}$$

The Racah lemma can be generalized for the GCGC in the following way

$$\underset{\beta_1\beta_2\ldots\beta_n}{\oplus} (e^{\alpha_1}_{\beta_1} \otimes e^{\alpha_2}_{\beta_2} \otimes \ldots \otimes e^{\alpha_n}_{\beta_n} \otimes U^{\beta_1\beta_2\ldots\beta_n}) =$$

$$= (S^{\alpha_1} \otimes S^{\alpha_2} \times \ldots \otimes S^{\alpha_n})^{-1}\, U^{\alpha_1\alpha_2\ldots\alpha_n} [\oplus (e^{\alpha_1\alpha_2\ldots\alpha_n}_{\alpha}\, S^\alpha)] . (*) \\ . X^{\alpha_1\alpha_2\ldots\alpha_n\,-1} \quad B \subset A \tag{4}$$

where $X^{\alpha_1\alpha_2\ldots\alpha_n}$ are outer IF and S^{α_i} are subduction matrices.

3. The symmetrized CGC $V^{\alpha_1\alpha_2\ldots\alpha_n}$ have convenient permutation symmetry. They can be derived directly from GCGC by separating only those columns of $U^{\alpha_1\alpha_2\ldots\alpha_n}$ which correspond to the identity corep D^{α_o}. For n=3 case we get 3DΓγ matrix $V^{\alpha_1\alpha_2\alpha_3}$

$$V^{\alpha_1\alpha_2\alpha_3} = U^{\alpha_1\alpha_2\alpha_3}_{\alpha_o} = (U^{\alpha_1\alpha_2}_{\alpha_3^*} \otimes E^{\alpha_3})(e^{\alpha_1\alpha_2}_{\alpha_3} \otimes U^{\alpha_3^*\alpha_3}_{\alpha_o})\mu^{\alpha_1\alpha_2\alpha_3,\alpha_o} \tag{5}$$

where the inner IF gives the possibility for optimum choice of the phase.

4. We will discuss some essential peculiarities contributed by the coreps in Racah algebra. One of them is connected with the Kronecker multiplicity. For linear representations we have always $(\alpha\alpha^*|\alpha_o)=1$, i.e. there exists only one invariant for every α, while for the coreps

of type a, b or c we have $(\alpha'\alpha^*|\alpha_o) = 1,4$ or 2 respectively, so

$$(\alpha_1\alpha_2\alpha_3|\alpha_o) = (\alpha_1\alpha_2|\alpha_3^*)(\alpha_3^*\alpha_3|\alpha_o) \qquad (6)$$

i.e. $2D\Gamma\gamma$ symbol (or 1jm Wigner symbol) depends on the multiplicity of α_o in $\alpha_3^* x \alpha_3$ because of the second factor.

$$\begin{pmatrix}\alpha^* & \alpha \\ a^* & a\end{pmatrix}^{\Gamma_o} = [\alpha]^{-1/2}\,[\alpha^* a^* \alpha a|\alpha_o\Gamma_o a_o],\ \Gamma_o = 1,\ldots,(\alpha^*\alpha|\alpha_o) \qquad (7)$$

Another point is that using Racah lemma we can not calculate all $3D\Gamma\gamma$ symbols of a subgroup B A from the corresponding $3D\Gamma\gamma$ symbols of A_i alongside with $V^{\alpha_1\alpha_2\alpha_3}_{\alpha_o}$ we need GCGC $U^{\alpha_1\alpha_2\alpha_3}_{\alpha}$, $\alpha \neq \alpha_o$ but $(D^{\alpha}\downarrow B) = = D^{\beta_o}$.

The symmetrization of all CGC for AU group A under association (see our next paper in this Colloquim) is a necessary condition for the calculation of $3D\Gamma\gamma$ symbols(5).

We have calculated and tabulated $3D\Gamma\gamma$ symbols for all AU point groups and $6D\Gamma$ symbols for same groups using the results of this paper and the tables of Kotzev and Aroyo[3].

REFERENCES

1. P.H.Butler. Point Group Symmetry Applications; Plenum Press, 1981.
2. C.J.Bradley and A.P.Cracknell. The Mathematical Theory of Symmetry in Solids, Oxford Univ. Press. Oxford, 1972.
3. J.N.Kotzev and M.I.Aroyo. J. Phys. A: Math and Gen 13 (1980) 2275; ibid 14 (1981) 1543, ibid 15 (1982) 711, 725.
4. J.N.Kotzev, M.I.Aroyo and M.N.Angelova. JINR, E17-81-376, Dubna, 1981.
5. J.N.Kotzev, M.I.Aroyo and M.N.Angelova. Physica 114A (1982) 533.

GENERALIZED COUPLING COEFFICIENTS FOR SPACE GROUPS

Herbert Willi Kunert

Institute of Physics, Technical University
60 - 965 Poznań, Piotrowo 3, POLAND

ABSTRACT

A practical method for calculating Racah coefficients (Rcs) for crystal space groups is presented. The formulae for Rcs are expressed by products of Clebsch-Gordan coefficients (CGcs). In the following paper this is applied to A-15 O_h^3-Pm3n structure.

1- INTRODUCTION

In order to construct the theory calculating Rcs for crystal space groups we are starting from the theory of CGcs for crystal space groups which has been presented in several places[1,2,3]. In Section 2 we present the formulae for Rcs. Basing on the CGcs we have calculated some Rcs listed in Tables I-IV.

2- RACAH COEFFICIENTS

For the direct products

$$D^{k^1 l_1} \otimes D^{k^2 l_2} \sim \sum_{l_o k^o} \oplus \gamma^o D^{k^o l_o}, \quad D^{k^o l_o} \otimes D^{k^3 l_3} \sim \sum_{l' k'} \oplus \gamma' D^{k' l'} \tag{1,2}$$

We have standard relations for the function basis:

$$\Phi^{k^o, l_o}_{\sigma^o a^o} = \sum_{\sigma^1 a^1 \sigma^2 a^2} U^{k^1 l_1 \otimes k^2 l_2}_{\sigma^1 a^1 \sigma^2 a^2 : \sigma^o a^o l_o k^o \gamma^o} \Phi^{k^1, l_1}_{\sigma^1 a^1} \Phi^{k^2_2, l_2}_{\sigma^2 a^2} \tag{3}$$

$$\Phi^{k', l'}_{\sigma' a'} = \sum_{\sigma^o a^o \sigma^3 a^3} U^{k^o l_o \otimes k^3 l_3 \; x^3 k^3}_{\sigma^o a^o \sigma^3 a^3 : \sigma' a' l' k' \gamma'} \Phi^{k^o, l_o}_{\sigma^o a^o} \Phi^{k^3_3, l_3}_{\sigma^3 a^3} \tag{4}$$

Substituting eq.(3) in eq.(4) we have

$$\Phi^{k',l'\gamma'}_{\sigma' a'} = \sum_{\sigma^o a^o \sigma^3 a^3} \sum_{\sigma^1 a^1 \sigma^2 a^2} U^{k^o l_o \otimes k^3 l_3}_{\sigma^o a^o \sigma^3 a^3 : \sigma' a' l' k' \gamma'} \times$$

$$\times U^{k^1 l_1 \otimes k^2 l_2}_{\sigma^1 a^1 \sigma^2 a^2 : \sigma^o a^o l_o k^o \gamma^o} \Phi^{k^1_1, l_1}_{\sigma_1 a^1} \Phi^{k^2_2, l_2}_{\sigma_2 a^2} \Phi^{k^3_3, l_3}_{\sigma_3 a^3} \tag{5}$$

so that, the Rcs are the products of CGcs:

$$U^{k^1 l_1 \otimes k^2 l_2 \otimes k^2 l_3, \gamma'}_{\sigma^1 a^1 \sigma^2 a^2 \sigma^3 a^3 : \sigma' a' l' k'} = U^{k^o l_o \otimes k^3 l_3, \gamma'}_{\sigma^o a^o \sigma^3 a^3 : \sigma' a' l' k'} \; U^{k^1 l_1 \otimes k^2 l_2, \gamma^o}_{\sigma^1 a^1 \sigma^2 a^2 : \sigma^o a^o l_o k^o} \tag{6}$$

All definitions, symbols, notations and the properties of CGcs can be found in papers[1-3].

3- TABLES

In this article we consider an example for O_h^3 space group. We start from CGcs for $M_{1+} \otimes M_{1+} \sim \Gamma_{1+} \oplus \Gamma_{3+} \oplus M_{1+} \oplus M_{2+}$ for O_h^3 presented in paper[4]. Multiplying both sides of this relation by M_{1+} we have $M_{1+} \otimes M_{1+} \otimes M_{1+} \sim \Gamma_{1+} \otimes M_{1+} \oplus \Gamma_{3+} \otimes M_{1+} \oplus M_{1+} \otimes M_{1+} \oplus M_{2+} \otimes M_{1+}$. Now, using the standard method, we calculate the CGcs for $\Gamma_{1+} \otimes M_{1+}$, $\Gamma_3 \otimes M_{1+}$, $M_{1+} \otimes M_{1+}$ and $M_{2+} \otimes M_{1+}$. Now, using the eqs. (3,4,5) we get the Rcs for the irreducible representations contained in $M_{1+} \otimes M_{1+} \otimes M_{1+}$ and presented in Tables I-IV.

4- DISCUSSION

Using eq.(5) we can calculate the Rcs for crystal space groups. First of all we must calculate CGcs in this method and next the Rcs. Now, we are investigating the other method of calculation of Rcs for space groups. We want to express the Rcs by the products of the matrix elements of the small irreducible representations of the little space groups. We have obtained some results by this method but for some cases we cannot get the correct tables of Rcs and for others we got the correct results. The investigations on this are still underway. The Rcs can be used for calculating the matrix elements of the scattering tensors, in the construction of the invariants in the second-order phase transitions and several other problems.

5- ACKNOWLEDGEMENTS

The author wishes to thank Prof.M.Suffczyński for helpful discussions.

This work was supported under project number MR.1.5.6.03.

TABLE 1

Racah Coefficients for $M_{1+} \otimes M_{1+} \otimes M_{1+}$ in O_h^3

$M_{1+} \otimes M_{1+} \otimes M_{1+} \sim \underset{M_{1+}}{\Gamma_1 \otimes M_{1+}} \oplus \underset{M_{1+}}{\Gamma_{3+} \otimes M_{1+}} \oplus \underset{M_{2+}}{\Gamma_{3+} \otimes M_{1+}}$

$k^1_{\sigma 1} + k^2_{\sigma 2} + k^3_{\sigma 3} = k'_{\sigma'}$	a' 1, σ' 1	1, 2	1, 3	1, 1	1, 2	1, 3	1, 1	1, 2	1, 3
$k_M + k_M + k_M = k_M$	b	0	0	2d	0	0	0	0	0
$5k_M + 5k_M + k_M = k_M$	b	0	0	-d	0	0	ia	0	0
$9k_M + 9k_M + k_M = k_M$	b	0	0	-d	0	0	-ia	0	0
$5k_M + 5k_M + 5k_M = 5k_M$	0	b	0	0	2d	0	0	0	0
$k_M + k_M + 5k_M = 5k_M$	0	b	0	0	-d	0	0	-ia	0
$9k_M + 9k_M + 5k_M = 5k_M$	0	b	0	0	-d	0	0	ia	0
$9k_M + 9k_M + 9k_M = 9k_M$	0	0	b	0	0	2d	0	0	0
$k_M + k_M + 9k_M = 9k_M$	0	0	b	0	0	-d	0	0	ia
$5k_M + 5k_M + 9k_M = 9k_M$	0	0	b	0	0	-d	0	0	-ia

TABLE 2

$M_{1+} \otimes M_{1+} \otimes M_{1+} \sim \underset{M_{1+}}{M_{1+} \otimes M_{1+}} \oplus \underset{M_{1+}}{M_{2+} \otimes M_{1+}} \oplus \underset{M_{2+}}{M_{1+} \otimes M_{1+}} \oplus \underset{M_{2+}}{M_{1+} \otimes M_{2+}}$

	a' 1, σ' 1	1, 2	1, 3	1, 1	1, 2	1, 3	1, 1	1, 2	1, 3	1, 1	1, 2	1, 3
$k_M + 5k_M + 5k_M = k_M$	a^2	0	0	a^2	0	0	$-a^2$	0	0	a^2	0	0
$5k_M + k_M + 5k_M = k_M$	a^2	0	0	$-a^2$	0	0	$-a^2$	0	0	$-a^2$	0	0
$k_M + 9k_M + 9k_M = k_M$	a^2	0	0	$-a^2$	0	0	a^2	0	0	a^2	0	0
$9k_M + k_M + 9k_M = k_M$	a^2	0	0	a^2	0	0	a^2	0	0	$-a^2$	0	0
$k_M + 5k_M + k_M = 5k_M$	0	a^2	0	0	$-a^2$	0	0	$-a^2$	0	0	a^2	0
$5k_M + k_M + k_M = 5k_M$	0	a^2	0	0	a^2	0	0	$-a^2$	0	0	$-a^2$	0
$5k_M + 9k_M + 9k_M = 5k_M$	0	a^2	0	0	a^2	0	0	a^2	0	0	a^2	0
$9k_M + 5k_M + 9k_M = 5k_M$	0	a^2	0	0	$-a^2$	0	0	a^2	0	0	$-a^2$	0
$k_M + 9k_M + k_M = 9k_M$	0	0	a^2	0	0	a^2	0	0	$-a^2$	0	0	a^2
$9k_M + k_M + k_M = 9k_M$	0	0	a^2	0	0	$-a^2$	0	0	$-a^2$	0	0	$-a^2$
$5k_M + 9k_M + 5k_M = 9k_M$	0	0	a^2	0	0	$-a^2$	0	0	a^2	0	0	a^2
$9k_M + 5k_M + 5k_M = 9k_M$	0	0	a^2	0	0	a^2	0	0	a^2	0	0	$-a^2$

TABLE 3

$M_{1+} \otimes M_{1+} \otimes M_{1+} \sim M_{1+} \otimes M_{1+} \; M_{2+} \otimes M_{1+} \oplus M_{1+} \otimes M_{1+} \oplus M_{2+} \otimes M_{1+}$

		Γ_{1+}	Γ_{2+}	Γ_{3+}		Γ_{3+}	
	a'	1	1	1	2	1	2
	σ'	1	1	1	2	1	2
$5k_M + 9k_M + k_M = k\Gamma$		d	d	d	d	d	-d
$9k_M + k_M + 5k_M = k\Gamma$		d	-d	dw	dw*	-dw	dw*
$k_M + 5k_M + 9k_M = k\Gamma$		d	d	dw*	dw	dw*	-dw
$5k_M + k_M + 9k_M = k\Gamma$		d	-d	dw*	dw	-dw*	dw
$9k_M + 5k_M + k_M = k\Gamma$		d	-d	d	d	-d	d
$k_M + 9k_M + 5k_M = k\Gamma$		d	d	dw	dw*	dw	-dw*

In the tables 1, 2, 3 we have $a = 1/\sqrt{2}$, $a^2 = 1/2$, $b = 1/\sqrt{3}$, $d = ab = 1/\sqrt{6}$, $w = -1/2 + i\sqrt{3}/2$ and $i = \sqrt{-1}$. The table 4 present the quasidiagonal structure the matrices of Rcs contained in the tables 1, 2 and 3.

TABLE 4

$M_{1+} \otimes M_{1+} \otimes M_{1+} \sim M_{1+} \oplus M_{1+} \oplus M_{2+} \oplus M_{1+} \oplus M_{1+} \oplus M_{2+} \oplus M_{2+} \oplus \Gamma_{1+} \oplus \Gamma_{2+} \oplus \Gamma_{3+} \oplus \Gamma_{3+}$

A	O	O
O	B	O
O	O	C

where A,B and C are the matrices of Rcs contained in the Tables 1, 2 and 3 respectively.

6. REFERENCES

1- R.Berenson and J.L.Birman, J.Math.Phys. 16, 227, 1975. and R.Berenson, R.Itzkan and J.L.Birman, J.Math.Phys. 16, 236, 1975.

2- J.L.Birman, Theory of Crystal Space Groups and Infrared and Raman Lattice Processes of Insulating Crystals, ind Handbuch der Physik. Encyclopedia of Physics. Vol.XXV/2b, Light and Matter Ib, edited by S.Flügge. Springer-Verlag, Berlin-Heidelberg-New York, 1974.

3- Van den Broek, P.M. and J.F.Cornwell, Phys.Status Solidi b 90, 211, 1978.

4- H.Kunert and M.Suffczynski, J.Physique 40, 199, 1979.

ACTIVE REPRESENTATIONS OF SPACE GROUPS BASED ON THE CUBIC LATTICE

Herbert Willi Kunert

Institute of Physics, Technical University
60 - 965 Poznan, Piotrowo
POLAND

ABSTRACT

We have computed the active representations of space groups Pm3m-O_h^1, Pm3n-O_h^3, Pn3m-O_h^4, Fm3m-O_h^5, Fm3c-O_h^6, Fd3m-O_h^7, Fd3c-O_h^8, Im3m-O_h^9 and Ia3d-O_h^{10}. All single valued representations for high-symmetry wave vectors in the Brillouin zones have been investigated.

1- INTRODUCTION

We have used Landau and Lifshitz criteria and Miller and Love[1] notation. In Table I we give representations for which the reality test and criteria of Landau and Lifshitz are not satisfied. In the last column the active representations are listed explicitly. The results for the representations at the zone centre, Γ, apply equally to all cubic space groups in the series O_h^{1-10}. With these active representations it is possible to determine lower-symmetry groups which could arise by second-order phase transitions from the nine space groups. The space group O_h^2 is not represented in nature.

2- COMPARISON AND DISCUSSION

The comparison and discussion for the active representations of space groups $O_h^{1,3,4}$ have been done in paper[2]. The results for some representations of $O_h^{9,10}$ space groups are in agreement with those of CDML[3]. According to CDML the representations N_{3+} of O_h^9 and N_2 of O_h^{10} are not satisfying the Landau criterion, yet we find that they should be satisfying this criterion.

3- ACKNOWLEDGEMENTS

The author wishes to thank Prof.M.Suffczynski for helpful discussions. This work was supported under project number 62-905.

TABLE I

Active Representations of Space Groups O_h^1, O_h^{3-10}

Space Group	Reality Test	Landau Criterion	Lifshitz Criterion	Active Representations
O_h^1		$\Gamma_{1+,3+,5+}$		$\Gamma_{1-,2\pm,3-,4\pm,5-}$
				$R_{1\pm,2\pm,3\pm,4\pm,5\pm}$
		M_{1+}		$M_{1-,2\pm,3\pm,4\pm,5\pm}$
				$X_{1\pm,2\pm,3\pm,4\pm,5\pm}$
O_h^3	$R_{2,3}$		R_4	$R_1, R_2 \oplus R_3$
		M_{5+}		$M_{1\pm,2\pm,3\pm,4\pm,5\pm}$
			$X_{1,2}$	$X_{3,4}$
O_h^4		M_1	$M_{1,2}$	$M_{3,4}$
				$R_{1\pm,2\pm,3\pm,4\pm,5\pm}$
			$X_{1,2}$	$X_{3,4}$
O_h^5		$X_{1+,4+,5+,}$		$X_{1-,2\pm,3\pm,4-,5-}$
				$L_{1,2,3}$
O_h^6	$L_{1,2}$		$L_{1,2,3}$	
		$X_{1+,4+,5+}$		$X_{1-,2\pm,3\pm,4-,5-}$
O_h^7		$X_{1,3}$	$X_{1,2}$	X_4
				$L_{1\pm,2\pm,3\pm}$
O_h^8	$L_{1,2}$		$L_{1,2,3}$	X_4
		$X_{1,3}$	$X_{1,2}$	
O_h^9		N_{1+}		$N_{1-,2\pm,3\pm,4\pm}$
				$H_{1\pm,2\pm,3\pm,4\pm,5\pm}$
O_h^{10}	$H_{2,3}$		H_4	$H_1, H_2 \oplus H_3$
		N_1	$N_{1,2}$	

4- REFERENCES

1- S.C.Miller and W.L.Love, Tables of Irreducible Representations of Space Groups and Co-representations of Magnetic Space Groups. Pruett Press, Boulder, Colorado. 1967.

2- H.W.Kunert and M.Suffczynski, Physica, North-Holland Publishing Co. 1982, to be published.

3- A.P.Cracknell, B.L.Davies, S.C.Miller and W.F.Love, Kronecker Product Tables, vol 1-4, vol 4. IFI. Plenum, New York, Washington, London, 1979.

REDUCED LARGE N MODELS

ITZHAK BARS
DEPARTMENT OF PHYSICS
YALE UNIVERSITY
NEW HAVEN, CT 06511 USA

The recent reduction technique of Eguchi-Kawai and the quenching ideas of Bhanot et al., Parisi, Gross-Kitazawa are used to construct the large-N reduced and quenched models for continuum QCD and for the Chiral U(N)*U(N) model in any number of dimensions. It is emphasized that the momenta of the reduced model should be frozen in a specified way, in order to agree with the underlying perturbative large-N field theory. The unitary matrix variables of reduced QCD are directly related to gauge invariant string variables that appeared in a reformulation of QCD 5 years ago. Through such a formalism several analogies to the chiral model become evident. The recent progress in evaluating the group integrals for the chiral model, done in collaboration with Gunaydin and Yankielowicz, is reviewed. The exact results in 1-dimension and the semi-classical approximations in arbitrary dimensions are presented. The properties of the vacuum state including its exact symmetries are established for all dimensions.

DERIVATION OF INFINITE-COMPONENT WAVE EQUATIONS FROM FIELD THEORY

A.O. Barut
Department of Physics,
The University of Colorado,
Boulder, Colo 80309

ABSTRACT

Infinite-component wave equations describe composite particles relativistically and nonperturbatively. They have been used in the past phenomenologically to describe mass spectra, magnetic moments, form factors, etc. of atoms, nuclei and hadrons. They are now derived from field theory, hence related to the properties of basic fields. Their solutions therefore provide nonperturbative solutions to the underlying field theory.

1. Introduction

One of the most difficult part of particle physics is the relativistic dynamics of bound states. We try to postpone it as far as possible. Group theory can help in the form of relativistic dynamical groups and infinite component wave equations.

The equation for the electron is

$$(\gamma^{\mu} p_{\mu} - m)\Psi = 0 \qquad (1)$$

This means that all the properties of the electron in interaction with the electromagnetic field A_{μ} is completely described by the minimal coupling of (1) via the substitution $p_{\mu} \to p_{\mu} - eA_{\mu}$. In fact, Eq. (1) is the best established wave equation in particle physics.

Equation (1) most likely also describes neutrino, (although the problem of the magnetic moment of the neutrino is not yet completely settled) and the heavy leptons μ and possibly τ. However, it has been known for a long time that Eq. (1) does not apply to proton, the third of the three stable or almost stable particles. This is because of the large magnetic moment of the proton and its dipole form factor. The mass of the proton is still a great mistery, and the calculation of masses of low lying hadrons is still one of the most important unsolved problem of particle physics, even the neutron-proton mass difference.

The properties of the proton can be, and have been, described and correlated by an equation of the form

$$(\Gamma^{\mu}P_{\mu}+K)\Psi=0. \qquad (2)$$

These properties includes the mass spectrum of the excited states of the proton, the prediction of the dipole form factor of the proton (which has been confirmed up to very high energies), the magnetic moments, transitions moments and form factors, polarizabilities, etc.

The mathematical structure of the Eq. (2) is as follows: Γ_{μ} is a vector operator in a representation $\pi(g)$, not necessarily unitary, of the Lie Algebra g of a dynamical group G. G must contain the Lorentz group $SL(2,C)$ as a subgroup and Γ_{μ} is a 4-vector with respect to this subgroup. K is a scalar operator in $\pi(g)$ with respect to $SL(2,C)$ subgroup. One example of this structure is in fact the electron, Eq. (1), where $G=O(4,2)$ or $SU(2,2)$ and the representation π is the 4-dimensional non-unitary representation of G. For the spinless H-atom G is again $O(4,2)$, but π is an infinite-dimensional ladder representation with lowest spin $j_0=0$. In the case of the H-atom with electron spin, $G=O(4,2)\otimes O(4,2)$ and π is a product of the 4-dimensional and an infinite-dimensional representation. Another example is the proton with $G=O(4,2)$ and π the infinite-dimensional fermion representation with $j_o=1/2$. The representation π may be unitary or non-unitary. But Eq. (2) always induces unitary (in general reducible) representation of Poincarè group.

Equation (2) may have an abstract and difficult appearance but has a very clear physical meaning. In Eq. (1), the wave function Ψ has 4-components $\Psi_{\alpha}(x)$. In Eq. (2) we can either think of Γ_{μ} and K as infinite-dimensional matrices, or better, as functions of differential operators in an auxiliary space with coordinates $\vec{r}$, for example,

$$\{\Gamma_{\mu}(\vec{r},\vec{\nabla}_r)P^{\mu}+K(\vec{r},\vec{\nabla}_r)\}\Psi(\vec{r},X_{\mu})=0, \qquad (2')$$

i.e. Γ_{μ} and K act on the coordinates $\vec{r}$, P_{μ} acts on X_{μ}. If we regard X_{μ} as the coordinates of the center of mass and $\vec{r}$ as the relative coordinates (they may be a set of them $\vec{r}_1$, $\vec{r}_2$, ...), then we see that Eq. (2') describes the motion of a composite object treated as a single relativistic object or particle with infinitely many internal degrees of freedom. Thus the salient features of Eq. (2) are: it is covariant, nonperturbative, contains one time and correct number of internal degrees of freedom.[1]

We now ask the question if Eq. (2) can be derived from field theory?

2. Field Theory Derivation of Wave Equation for Composite Objects

We take as our basic theory the most general electromagnetic interactions of a number of fundamental fermions Ψ_j:

$$L=\sum_j \overline{\Psi}_j(\gamma^\mu i\partial_\mu - m_j)\Psi_j - \frac{1}{4}F_{\mu\nu}F^{\mu\nu} + \sum_j e_j\overline{\Psi}_j\gamma^\mu\Psi_j A_\mu + \sum_j a_j\Psi_j\sigma^{\mu\nu}\Psi_j F_{\mu\nu} \tag{3}$$

Here the last term is the anomalous magnetic moment Pauli-coupling of the fermions. Since we are interested in nonperturbative solutions of this field theory there will be no renormalization infinities, all integrals are finite. We omit the writing of field equations here.

If we solve A_μ from field equations in terms of Ψ's and insert back in (3), the interaction action is

$$A_{int} = \int dx \sum_{jk} e_j e_k \int dy \overline{\Psi}_j(x)\overset{(1)}{\gamma^\mu}\Psi_j(x) D(x-y)\overline{\Psi}_k(y)\overset{(2)}{\gamma_\mu}\Psi_k(y) + \text{magnetic term.}$$

We shall illustrate the method for the electric term above and write the result for the magnetic term. Consider for simplicity two district commuting fields Ψ_1 and Ψ_2. Because of the δ-functions in $D(x-y)$, namely $\delta(x^o-y^o\mp|\vec{x}-\vec{y}|)$ the vector $(x-y)$ is lightlike, consequently we may define a two-body (or bilocal) field by

$$\Phi(\vec{x},\vec{y},t)\equiv\Psi_1(\vec{x},t)\Psi_2(\vec{y},t-|\vec{x}-\vec{y}|) \tag{4}$$

In the free parts of the action we multiply, for example, the first term $\int dx\overline{\Psi}_1(\gamma^\mu p_{1\mu}-m_1)\Psi_1$ by the normalization condition $\int d\vec{y}\Psi^+(y)\Psi(y)=1$, and express them also as a function of $\Phi(x,y,t)$. Then if we vary the action with respect to Φ^+ we obtain the equation

$$\{(\overset{(1)}{\gamma^\mu}p^{(1)}_\mu - m)\otimes\overset{(2)}{\gamma_o}+\overset{(1)}{\gamma_o}\otimes(\overset{(2)}{\gamma^\mu}p^{(2)}_\mu - m_2)+V(\vec{x},\vec{y},t)\}\Phi(\vec{x},\vec{y},t)=0 \tag{5}$$

This is a 16x16-spinor wave equations in two variables $\vec{x}$ and $\vec{y}$ with one time. The potential obtained from (3) for both electric and magnetic couplings is given by[2]

$$\begin{aligned}\overset{(1)}{\gamma_o}V\overset{(2)}{\gamma_o}=\sum_{j\neq k}\{ & e_je_k\frac{1-\vec{\alpha}_j\vec{\alpha}_k}{r} - e_ja_k\vec{\alpha}_j\cdot(\beta\vec{\sigma})_k\wedge\frac{\vec{r}}{r^3} - e_ja_k \mathbb{1}_j(\beta\vec{\alpha})_k\frac{\vec{r}}{r^3} \\ & -a_ja_k\left[\frac{3(\beta\vec{\sigma})_j\cdot\vec{r}(\beta\vec{\sigma}_k)\cdot\vec{r}}{r^5} - \frac{(\beta\vec{\sigma})_j\cdot(\beta\vec{\sigma})_k}{r^3} + \frac{8\pi}{3}(\beta\vec{\sigma})_j\cdot(\beta\vec{\sigma})_k\delta(\vec{r})\right] \\ & +a_ja_k\left[\frac{3(\beta\vec{\alpha})_j\cdot\vec{r}(\beta\vec{\alpha})_k\cdot\vec{r}}{r^5} - \frac{(\beta\vec{\alpha})_j\cdot(\beta\vec{\alpha})_k}{r^3} - \frac{4\pi}{3}(\beta\vec{\alpha})_j(\beta\vec{\alpha})_k\delta(\vec{r})\right]\}\end{aligned} \tag{6}$$

In Eq. (5) we now introduce the center of mass and relative variables

$$r_\mu = x_\mu - y_\mu \qquad \text{and} \qquad P^\mu = p_1^\mu + p_2^\mu$$
$$R_\mu = a x_\mu + (1-a) y_\mu \qquad\qquad p^\mu = (1-a) p_1^\mu - a p_2^\mu$$

These transformations give $[R,P]=iI$ and $[r,p]=iI$ for any a. Because of the light-like condition $r_\mu r^\mu = 0$, there is no relative time, hence $P_o\Phi = 0$ so that $\Phi = \Phi(\vec{r}, R_\mu)$, or in momentum space $\Phi(\vec{p}, P_\mu)$ and we obtain finally

$$\{\Gamma^\mu P_\mu + K\}\Phi(\vec{p}, P) = 0 \tag{7}$$

where

$$\Gamma_\mu = (a\overset{(1)}{\gamma_\mu} \otimes \overset{(2)}{\gamma_o} + (1-a)\overset{(1)}{\gamma_o} \otimes \overset{(2)}{\gamma_\mu})\Gamma$$

$$K = -(\overset{(1)}{\vec{\gamma}} \otimes \overset{(2)}{\gamma_o} - \overset{(1)}{\gamma_o} \otimes \overset{(2)}{\vec{\gamma}}) \cdot \Gamma\vec{p} - \Gamma(m_1 I \otimes \overset{(2)}{\gamma_o} + m_2 \overset{(\Gamma)}{\gamma_o} \otimes I) + \lambda \overset{(1)}{\gamma_\mu} \otimes \overset{(2)}{\gamma^\mu}$$

This is the desired infinite-component wave equation for the two-body electric case (e.g. positronium).

The solutions of (7) provide also solutions of the underlying field theory (1). In order to do this we go to the center of mass of the composite system, $\vec{P}=0$, and solve first the equation $(\Gamma^o P_o + K)\overset{(o)}{\Phi}(\vec{p}) I \equiv W_{rel}\overset{(o)}{\Phi} = 0$ for the mass spectrum of the composite particle, $P_o = M$.

Then if we define the operator U by

$$U(\vec{r}, \vec{P} + W_{rel})U^{-1} = W_{rel},$$

the "moving" solution is given by $\Phi(p, P_\mu) = U^{-1}\overset{(o)}{\Phi}(\vec{p})$.

T he wave operator (7) actually sits in the action

$$A = \int d\vec{x}\, d\vec{y}\, \Psi_1^+(x)\Psi_2^+(y)[\Gamma^\mu P_\mu + K]\Psi_1(x)\Psi_2(y), \tag{8}$$

or

$$A = \int dR\, d\vec{r}\, \Phi^+(\vec{r}, R)[\Gamma^\mu P_\mu + K]\Phi(\vec{r}, R)$$

So that it is possible to integrate over the internal variables and obtain an action for the center of mass variables which would then describe the composite object with many mass states.

For three or more particles the procedure is similar. The composite field is now defined by

$$\Phi(\vec{x}_1,\vec{x}_2,\vec{x}_3,t)=\Psi_1(\vec{x}_1,t)\Psi_2(\vec{x}_2,t-|\vec{x}_1-\vec{x}_2|)\Psi_3(\vec{x}_3,t-|\vec{x}_1-\vec{x}_3|) \quad (9)$$

Note that the action gets contributions from those points only for which

$$t-|\vec{x}_1-\vec{x}_3|=t-|\vec{x}_1-\vec{x}_2|-|\vec{x}_2-\vec{x}_3|.$$

The 3-body equation for the electric coupling is

$$\Big\{(\overset{(1)}{\gamma}{}^{\mu}p_{1\mu}-m_1)\otimes\overset{(2)}{\gamma}_o\otimes\overset{(3)}{\gamma}_o+\overset{(1)}{\gamma}_o\otimes(\overset{(2)}{\gamma}{}^{\mu}p_{2\mu}-m_2)\otimes\overset{(3)}{\gamma}_o+\overset{(1)}{\gamma}_o\otimes\overset{(2)}{\gamma}_o\otimes(\overset{(3)}{\gamma}{}^{\mu}p_{3\mu}-m_3) \quad (10)$$
$$+\frac{\overset{(1)}{\gamma}_\mu\otimes\overset{(2)}{\gamma}{}^{\mu}\otimes\overset{(3)}{\gamma}_o}{|\vec{x}_1-\vec{x}_2|}+\text{cyclic}\Big\}\Phi(\vec{x}_1,\vec{x}_2,\vec{x}_3,t)=0$$

and the passage to the center of mass momentum P_μ and relative coordinates can be made similarly as in Section 2, to arrive at the form (7). Now Γ_μ and K are more lengthy operators.

References

1) For a more detailed discussion of the principles of infinite component wave equations see A.O. Barut, "Dynamical Group for the Motion of Relativistic Composite Systems", in Groups, Systems and Many-Body Physics (edit. P. Kramer et al), Vieweg Verlag (1980); Ch. VI, pp 285-317.

2) A.O. Barut and Bo-wei Xu, Physica Scripta 126, 129 (1982).

MAXIMAL SYMMETRIES ON POTENTIALS AND GAUGE INVARIANCE

J. BECKERS and V. HUSSIN

Physique théorique et mathématique
Institut de Physique au Sart Tilman
Université de Liège - Belgique

ABSTRACT

Subsymmetries of relativistic potentials are classified under the kinematical groups of associated constant electromagnetic fields. A maximal character on symmetries of potentials is obtained leading to maximal symmetries of the corresponding wave equations with interaction.

1. Introduction

A free Klein-Gordon (K.G.) equation does admit the Poincaré group as symmetry group. If the scalar particle of mass m and electric charge e interacts with a constant electromagnetic (elm) field $F^{(0)}$, it is then described by a modified K.G. equation dealing with minimal coupling, i.e. by

$$(D^{\mu}D_{\mu} + m^2)\psi(x) = 0 \tag{1}$$

with

$$D_{\mu} = \partial_{\mu} - ie\, A_{\mu} \quad , \tag{2}$$

the so-called covariant derivatives including the 4-potential $A \equiv (A^{\mu})$ leading to the elm field tensor $F^{(0)} \equiv (F^{\mu\nu}) \equiv (\vec{E},\vec{B})$.

What is the symmetry group of Eq.(1) ? Inside the Poincaré group, the answer is not unique due to possible gauge transformations on A and the gauge invariance of F. Outside the Poincaré group, an answer has been given through compensating gauge transformations[1] and the symmetry group of A which appears as the extension of R by the kinematical group[2] G_F of F.

Here we want to come back on symmetries inside the Poincaré group by discussing symmetries on potentials A. We evidently know that, in such a problem with interaction, the physical quantity is the elm field $F^{(0)}$ and that scalar (V) and vector ($\vec{A}$) potentials leading to such a field $F^{(0)}$ fall into equivalence classes through usual gauge transformations. But the symmetry of the wave equation is connected with the one of the potential A and not with the one of the field $F^{(0)}$. With the results on kinematical groups[2] of $F^{(0)}$, we propose to combine invariance conditions[3] on A in order to get precise answers to the above problem.

2. Symmetries of potentials <u>inside</u> Poincaré.

Under the Poincaré group $(P^\mu,\vec{J},\vec{K})$, there are two kinds of invariant <u>constant</u> elm fields[2], the so-called $F_\parallel^{(0)}$ and $F_\perp^{(0)}$ admitting the kinematical groups (according to some choices[2]) :

$$G_{F_\parallel^{(0)}} \equiv \{P^\mu,J^3,K^3\} \quad , \quad G_{F_\perp^{(0)}} \equiv \{P^\mu,J^1+K^2,J^2-K^1\} \tag{3}$$

as Poincaré symmetry subgroups. Owing to the fact that $F^{(0)} = dA$, we have discussed[3][4] <u>invariant</u> conditions under the Poincaré group on physical A summarized by :

$$\vec{\phi}\cdot\vec{A} - \mathcal{D}V = 0 \quad , \quad \vec{\theta}\wedge\vec{A} - \vec{\phi}V + \mathcal{D}\vec{A} = 0 \tag{4}$$

where

$$\mathcal{D} \equiv (t\vec{\phi} + \vec{x}\wedge\vec{\theta})\cdot\frac{\partial}{\partial\vec{x}} + (\vec{x}\cdot\vec{\phi})\partial_t - a\cdot\nabla \tag{5}$$

$\vec{\phi}$, $\vec{\theta}$, a, being the parameters associated with boosts, spatial rotations and space-time translations respectively.

First, we can easily show that $G_A \subset G_{F^{(0)}}$ if G_A is the symmetry group of A (leading to $F^{(0)}$). Then, through the Patera-Winternitz-Zassenhaus method[5] on subgroup classifications, we can determine up to conjugacy the nonequivalent subgroups of the kinematical groups (3). Finally, we can examine among these subgroups those which are of maximal dimension <u>and</u> admit 4-potentials, solutions of Eqs.(4) and (5) <u>and</u> corresponding to $F^{(0)}$.

These steps lead us to nontrivial G_A^{max} of dimension 3 in the parallel case and of dimension 4 in the perpendicular case. For example we get the only four following G_A^{max} in the parallel case :

$$G_{\parallel,\pm} \equiv \{P^0\pm P^3,J^3,K^3\} \quad , \quad G'_{\parallel,\pm} \equiv \{P^0\pm P^3,K^3,P^1\} \ . \tag{6}$$

The 4-potentials associated with $G_{\parallel,\pm}$ are, for example :

$$A_\parallel = \frac{1}{2}\{\mp E(t\pm z),- By,Bx,- E(t\pm z)\} \ . \tag{7}$$

Introduced in Eqs.(1),(2), those values ensure that the interaction between the scalar particle and the elm field $F^{(0)}$ takes place <u>and</u> that the symmetry group of the K.G. equation is a Poincaré subgroup of maximal dimension given by (6).

3. Symmetries <u>inside</u> and <u>outside</u> Poincaré.

As already mentioned, another notion[1] of (gauge independent) symmetry on potentials has been introduced. It is based on the fact that

$$gA(x) = A(x) + \partial\chi_g(x) \ , \quad \forall\, g \in G_F \tag{8}$$

where $\chi_g(x)$ is the compensating gauge. Then the symmetry group of A (leading to F) is the set $\{(\chi_g, g)\}$ such that

$$(\chi_g, g)A(x) = gA(x) - \partial\chi_g(x) = A(x) \tag{9}$$

defining the extension of R by G_F. In this context, our invariance conditions (4) and (5) write :

$$\vec{\phi}\cdot\vec{A} - \mathcal{D}V = -\partial_t\chi_g \ , \quad \vec{\theta} \wedge \vec{A} - \vec{\phi}V + \mathcal{D}\vec{A} = -\vec{\nabla}\chi_g \tag{10}$$

determining χ_g and showing, in particular, that

$$\chi_g = \text{constant} \iff g \in G_A \subset G_F \ . \tag{11}$$

Let us conclude by noticing that this symmetry group of A leaves also the K.G. equation with interaction invariant and is included in the Maxwell group - the largest symmetry group - obtained by Schrader[6]. We finally have the inclusions :

$$\text{Maxwell group} \supset \text{Extension of R by } G_F \supset G_A^{max} \supset G_A \ , \tag{12}$$

G_F being not contained in such a chain.

4. Remarks.

Minimal elm couplings do enter in other wave equations than the K.G. one : so our considerations also apply there when the spin of the particle is not zero (Dirac equation,...). Moreover, all our developments can be done in a nonrelativistic context (Galilean symmetry) and their interests are transferred on the study of Schrödinger equations with interaction. These developments will appear elsewhere[4].

5. References.

1] A. JANNER and T. JANSSEN, Physica 53,1 (1971); N. GIOVANNINI, Physica A87,546 (1977).

2] H. BACRY, Ph. COMBE and J.L. RICHARD, Nuovo Cimento A67,267 (1970).

3] J. BECKERS and V. HUSSIN, Found.Phys. 11,881 (1981).

4] J. BECKERS and V. HUSSIN, to be published in J.Math.Phys.(1983)

5] J. PATERA, P. WINTERNITZ and H. ZASSENHAUS, J.Math.Phys. 16, 1597 (1975); J.Math.Phys. 17,977 (1976).

6] R. SCHRADER, Forts.Phys. 20,701 (1972).

GHOSTS, ANOMALIES AND THE GEOMETRY OF GAUGE FIELDS

Loriano Bonora

INFN and Istituto di Fisica dell'Università di Padova
Via Marzolo 8 35100 PADOVA (ITALY)

Paolo Cotta-Ramusino

Istituto di Fisica dell'Università di Milano and INFN
Via Celoria 16 20133 MILANO (ITALY)

ABSTRACT

We show that the ghosts in gauge theories are differential forms defined on the Group $\mathcal{G}$ of Gauge Transformations. The problem of finding the possible anomalous terms (as the triangle anomaly) is related to the problem of studying the cohomology of the Lie Algebra of $\mathcal{G}$.

1. INTRODUCTION

It is well known that in gauge theories the effective Lagrangian (with gauge breaking and ghost terms included) is invariant under a special class of transformations (BRS transformations) ([1]). BRS transformations have attracted much interest ever since they were discovered by Becchi, Rouet and Stora. There have recently been some attempts to give a geometrical interpretation to them, as well as to the ghost fields ([2],[3]); but these attempts exhibit unsatisfactory aspects. In this paper we change geometrical framework and try to avoid the flaws of the previous interpretations; moreover we construct a clear setting where the problem of the Adler-Bell-Jackiw anomalies can be dealt with. Our main object will be the action of the group of gauge transormation on the space of connections.

Let us now recall briefly what BRS transformations and anomalous terms are. The infinitesimal BRS transformations are usually written as follows:

$$A(x) \longrightarrow A(x) + \lambda\, s(A(x)) =: A(x) + \lambda Dc(x) \qquad (1.1)$$

$$c(x) \longrightarrow c(x) + \lambda\, s(c(x)) =: c(x) - \frac{1}{2}\lambda\, [c(x), c(x)] \qquad (1.2)$$

$$\bar{c}(x) \longrightarrow \bar{c}(x) + \lambda\, s(\bar{c}(x)) =: \bar{c}(x) + \lambda\, B(x) \qquad (1.3)$$

where $A(x)$, $c(x)$, $\bar{c}(x)$ are respectively the vector potential, the ghost and the antighost fields, D is the covariant derivative, B(x) is an auxiliary field and λ is an anticommuting x-independent parameter. We have also:

$$s(Dc(x)) = -\frac{1}{2}(D[c(x),c(x)]) + [Dc(x), c(x)] = 0 \qquad (1.4)$$

that is $s^2 = 0$ (nilpotent character of the BRS operator).

An integrated anomalous term can be defined as a local functional Q of the gauge fields and the ghosts such that sQ = 0 and there exists no functional Q' such that Q = sQ' (see [4]).

In this paper we don't consider matter fields, even though it would be rather straighforward to extend to them most of the considerations we are going to make here.

2. THE GEOMETRY OF GAUGE FIELDS AND THE GHOSTS.

Let P(M,G) be a principal fiber bundle over a compact, connected, orientable, Riemannian manifold M. G is a compact Lie Group with an invariant inner product defined on its Lie Algebra $\mathfrak{g}$. Consider the bundles AdP =: $P\times_G G$ and adP =: $P\times_G \mathfrak{g}$, where the action of G on itself (respectively on $\mathfrak{g}$) is the adjoint one. Let us denote by $\mathcal{G}$ (respectively by E) the space of C^∞-Sections of AdP (respectively adP): $\mathcal{G}$ is a Group under pointwise multiplication (Group of Gauge Transformations) and E is a Lie Algebra (Gauge Lie Algebra) (see [5]).

For each integer p denote by $\Omega^p(M,adP)$ the space of C^∞-Sections of the bundle $\Lambda^p T^* M \otimes adP$ (locally p-forms on M with values in $\mathfrak{g}$). Combining the bracket operation in $\mathfrak{g}$ with the exterior multiplication for ordinary forms on M, we can give the structure of a graded Lie Algebra to $\Omega^*(M,adP) =: \sum_p \Omega^p(M,adP)$.

For each connection A the covariant derivative d_A is a linear operator: $d_A: \Omega^p(M,adP) \rightarrow \Omega^{p+1}(M,adP)$ and satisfies the following equation:

$$d_A[\varphi, \chi] = [d_A\varphi, \chi] + (-1)^{\text{degree of } \varphi}[\varphi, d_A\chi] .$$

The space $\mathcal{Q}$ of all connections for P(M,G) is an affine space modelled on $\Omega^1(M,adP)$ (see [5]); $\mathcal{G}$ acts on $\mathcal{Q}$ (Gauge transformation of a vector potential). $\mathcal{Q}$, $\mathcal{G}$, E, $\Omega^p(M,adP)$ can be considered as Hilbert manifolds by taking the completion of the various spaces of sections with respect to suitable Sobolev norms (see [6]); $\mathcal{G}$ is then a Hilbert Lie Group and it can be shown that E is its Lie Algebra ([6]).

We are now in a position to explain the geometrical meaning of the ghost field and the BRS transformations.

Consider the E-valued left invariant 1-form ω on $\mathcal{G}$ (Maurer Cartan form) defined by : $\omega(\xi) = \xi$, $\forall \xi \in E$ (we identify the left invariant vector fields on $\mathcal{G}$ with the corresponding elements of E).

We assume that ω is the ghost field (for more details see [7]).

$\forall$ connection A and $\forall \xi \in E$, $A + d_A(\omega(\xi))$ is another connection, which is obtained from A, by performing the infinitesimal gauge transformation given by ξ . If ω is the ghost field, the transformation (1.1) becomes:

$$A \longrightarrow A + d_A\omega \qquad (2.1)$$

which in turn can be seen as representing in a symbolic form the effect of all the possible infinitesimal gauge transformations; that is (2.1) stands for:

$$(A,\xi) \rightsquigarrow A + d_A(\omega(\xi)) = A + d_A\xi \qquad \xi \in E,\ A \in \mathcal{Q} \quad (2.2).$$

Notice that $d_A\omega$ can be considered as a $\Omega^1(M,adP)$-valued 1-form on $\mathcal{G}$ such that: $(d_A\omega)(\xi) = d_A(\omega(\xi))$. If we apply twice the transformation (2.2), antisymmetrizing with respect to the vector fields ξ, $\zeta \in E$ involved, we have:

$$(A,\xi,\zeta) \rightsquigarrow A+d_A\omega(\xi) + d_{A+d_A\omega(\xi)}\omega(\zeta) - d_A\omega(\zeta) - d_{A+d_A\omega(\zeta)}\omega(\xi) =$$
$$= A + d_A[\omega(\xi),\omega(\zeta)] =: A + [d_A\omega,\omega](\xi,\zeta) . \qquad (2.3)$$

Notice that $[d_A\omega,\omega]$ is a 2-form on $\mathcal{G}$ with values in $\Omega^1(M,adP)$ and satisfies the equation:

$$[d_A\omega,\omega] - \frac{1}{2}d_A[\omega,\omega] = [d_A\omega,\omega] + d_A(d_{\mathcal{G}}\omega) = 0 \qquad (2.4)$$

where $d_{\mathcal{G}}$ is the exterior derivative for forms defined on $\mathcal{G}$.

From what we have said before and from the comparison of (1.1)-(1.4) with (2.1)-(2.4) , we can conclude ([7]) :

a) we know that the ghost is an anticommuting scalar field: this is explained by the fact that it is a 1-form on $\mathcal{G}$ (anticommutativity) with values in a space of zero forms on M (scalar character);

b) the transformation (1.1) can be identified with the transformation (2.2);

c) in the transformation (1.2) the operator s can be identified with the exterior derivative for forms on $\mathcal{G}$;

d) eq. (2.4) is the same as eq.(1.4) . That is, applying twice the transformation (2.2) as in (2.3) and adding the term $d_A((d_{\mathcal{G}}\omega)(\xi,\zeta))$, we come back to the same connection where we started from (nilpotent character of the BRS).

Studying the geometrical meaning of the BRS transformations has led us to consider that the relevant object is the action of $\mathcal{G}$ on $\mathcal{A}$. Now, whenever we have an action of a Group on a manifold, we have , in a natural way , a cohomology of the Lie algebra of the Group with coefficients in the algebra of differential forms on the manifold. In our case the coboundary operator is exactly the BRS operator, as we shall see in the next section.

3. COHOMOLOGY OF THE GAUGE LIE ALGEBRA.

Let $\Omega^0(\mathcal{A})$ be the space of zero forms (functionals) on $\mathcal{A}$. $\forall \xi \in E$ we can define the fundamental vector field Z_ξ which acts on $\Omega^0(\mathcal{A})$ in the following way: $(Z_\xi f)(A) = \frac{d}{dt} f(A + t d_A \xi)\Big|_{t=0}$ $\quad A \in \mathcal{A}$.
Notice that $Z_\xi(A) = d_A \omega(\xi)$.

Let $\Gamma^p(E,\mathcal{A})$ be the space of the p-linear skew mappings from $E \times \ldots \times E$ (p times) to $\Omega^0(\mathcal{A})$, which satisfy a suitable continuity condition (see [7]) . In $\Gamma^*(E,\mathcal{A}) =: \sum_p \Gamma^p(E,\mathcal{A})$ we can define a coboundary δ by:

$$\delta : \Gamma^p(E,\mathcal{A}) \to \Gamma^{p+1}(E,\mathcal{A})$$

$$(\delta\psi)(\xi_1,\ldots,\xi_{p+1}) =: \sum_{i=1}^{p+1} (-1)^{i+1} Z_{\xi_i}(\psi(\xi_1,\ldots,\hat{\xi}_i,\ldots,\xi_{p+1})) + \sum_{i<j} (-1)^{i+j} \psi([\xi_i,\xi_j],\xi_1,\ldots,\hat{\xi}_i,\ldots,\hat{\xi}_j,\ldots,\xi_{p+1}),$$

where $\wedge$ indicates omission and $\xi_i \in E, \forall i$.

Let us denote by $H^p(E,\mathcal{A})$ the p-th cohomology group for the complex $(\Gamma^*(E,\mathcal{A}), \delta)$. We have (see [7]) :

a) functionals of the vector potential with p-ghosts are elements of $\Gamma^p(E,\mathcal{A})$ (p-cochains) ;

b) the BRS operator can be identified with the coboundary operator δ ;

c) gauge invariant functionals with no ghosts are elements of $H^0(E,\mathcal{A})$;

d) (integrated) anomalous terms with p-ghosts ($p \geq 1$) are elements of $H^p(E,\mathcal{A})$.

The cohomology of the complex $(\Gamma^*(E,\mathcal{A}),\delta)$ is related to the cohomology of the principal bundle of gauge orbits (see [7]) if we consider not the space of all connections , but only the space of the irreducible ones.

4. LOCAL COHOMOLOGY AND ANOMALIES

The spaces $\Gamma^p(E,\mathcal{Q})$ are too 'large' for field theory ; we need only functionals which are integrals over M of polynomials in the fields, in the ghosts and in their derivatives. So we have to find a suitable sub-complex of the complex $(\Gamma^*(E,\mathcal{Q}),\delta)$ and study the cohomology of this sub-complex (what we have called "local" cohomology in [7]). We now summarize very briefly the steps of our construction and some of our results: full details can be found in [7].

a) We fix a connection A_0 and consider $\eta(A) =: A - A_0$ which belongs to $\Omega^1(M,\mathrm{ad}P)$ whenever $A \in \mathcal{Q}$;

b) we define the space $D^k(\mathcal{Q},A_0,M)$ of all k-forms on M, whose components are, in any coordinates system , polynomials in $\eta(A)$ and its derivatives;

c) we define $S^{p,k}$ to be the space of all p-linear skew local (i.e. decreasing supports) mappings from E x ....x E (p times) to $D^k(\mathcal{Q},A_0,M)$;

d) on $S^{**} =: \sum_{p,k} S^{p,k}$ we consider two coboundary operators: d which is the extension of the exterior derivative for forms on M and δ_{loc} (the BRS operator) which is constructed in the same way as the operator δ in section 3. We have $d\delta_{loc} = \delta_{loc} d$;

e) if n = dim M and ϕ^n_p belongs to $S^{p,n}$, $\int_M \phi^n_p \in \Gamma^p(E,\mathcal{Q})$ and $\int_M \delta_{loc}\phi^n_p = \delta\int_M \phi^n_p$;

f) we prove that $\int_M \phi^n_p \in H^p(E,\mathcal{Q}) \Rightarrow \exists\, \phi^0_{p+n} \in S^{p+n,0}$ such that ϕ^0_{p+n} belongs to the (p+n)-th cohomology group computed w.r.t. the coboundary operator δ_{loc} (we prove more generally that the spectral sequence for the double complex S^{**} is degenerated to the E_2 term) ;

g) under the correspondence mentioned above, the Adler-Bell-Jackiw anomaly and the "anomaly" in two dimension (see [8]) correspond respectively to Ψ^0_5 and Ψ^0_3 given by:

$$\Psi^0_5(\xi_1,\xi_2,\xi_3,\xi_4,\xi_5) = f_3(\xi_1,[\xi_2,\xi_3],[\xi_4,\xi_5]) \qquad \xi_i \in E \ (n=4)$$

and

$$\Psi^0_3(\xi_1,\xi_2,\xi_3) = K(\xi_1,[\xi_2,\xi_3]) \qquad \xi_i \in E \ (n=2),$$

where f_3 and K are respectively the trilinear symmetric invariant map for $\mathfrak{g}$ and the Killing form (we consider that every ad-invariant multilinear symmetric map can be extend to a map from $\Omega^{i_1}(M,adP) \times \Omega^{i_2}(M,adP) \times \ldots\ldots$ to $\Omega^{i_1+i_2+\ldots}(M,adP)$).

REFERENCES

[1] C.Becchi, A.Rouet, R.Stora, Ann. Phys. 98 , 287 (1976)

[2] J.Thierry-Mieg, J. Math. Phys. 21 , 2834 (1980)

[3] M.Quiròs et al. J. Math. Phys. 22 ,1767 (1981)

[4] R.Stora in H. Levy, P. Mitter (eds.) : New developmments in Quantum Field Theories and Statistical Mechanics, Plenum Press (1977)

[5] M.Atiyah , R.Bott , On the Yang Mills equations over Riemann surfaces Harvard preprint

[6] P.K.Mitter, C.M. Viallet , Comm. Math. Phys. 79 , 457 (1981)

[7] L.Bonora, P.Cotta-Ramusino , Some remarks on BRS transformations, anomalies and the cohomology of the Lie Algebra of the group of Gauge transformations Harvard preprint, to be published by Comm. Math.Phys.

[8] L.Bonora , P.Cotta-Ramusino , Phys. Lett. B 107, 87 (1981).

COLOUR ALGEBRAS AND GENERALIZED STATISTICS

H. S. Green
University of Adelaide
SOUTH AUSTRALIA 5001

ABSTRACT

A generalization of Lie groups and algebras is formulated, which includes graded, modular and colour groups and algebras. As an example, a generalization of $gg\ell((n))$ is defined, with its associated Lie algebra and vector operators. The application to the dynamics and statistics of the quark model and similar models of composite particles is outlined.

1. Introduction

In recent years, the problems of confinement and quantum statistics associated with the quark model have directed attention to generalizations of Lie groups and algebras in which the elements of the group manifold are also elements of an associative algebra. For any such group, there is a corresponding generalized Lie algebra. The simplest generalization of this type was introduced into mathematics by Hopf[1)], and has found extensive applications to the study of supersymmetry[2)]. More recently, generalizations of Lie algebras and superalgebras called colour algebras have been described by Rittenberg and Wyler[3)] and Scheunert[4)]. A related generalization was introduced by the author[5)] with the specific application to the quark model in mind. In the following we shall describe a further generalization of this type.

2. Generalized Lie Groups and Algebras

We consider a general Lie group L, such that any point ξ of the group manifold is parametrized by m coordinates ξ^r, which are elements of the group algebra A of a finite group G. If $\{g_\rho\}$ is the canonical basis in G, we write

$$\xi^r = \Sigma_\rho^{(r)} \xi_\rho^r g_\rho \qquad (1)$$

where the summation is restricted to one of the conjugate classes $G_{(r)}$ of G and the complex coefficients ξ_ρ^r are chosen so that

$$\xi^r \eta^s = u_{sr} \eta^s \xi^r \qquad (2)$$

where $|u_{rs}| = 1$, and $u_{rs} u_{sr} = 1$, so that $u_{rr} = \pm 1$.

If $e(\xi)$ is the element of L corresponding to the point ξ, and $e(o)$ is the identity, it follows in the usual way that $e(\xi) e(\eta) = e(\omega)$, where

$$\omega^t(\xi,\eta) = \xi^t + \eta^t + c_{rs}^t \xi^r \eta^s + \ldots \qquad (3)$$

correct to terms linear in ξ^r or η^s. Also, if e_r are the generators of L,

$$e_r \xi^r e_s \eta^s - e_s \eta^s e_r \xi^r = e_t (c_{rs}^t \xi^r \eta^s - c_{sr}^t \eta^s \xi^r) .$$

Thus we may introduce a general Lie product, defined by

$$[e_r, e_s] \equiv e_r e_{s,r} - u_{rs} e_s e_{r,s} = C_{rs}^t e_t ,$$

$$C_{rs}^t = c_{rs}^t - u_{rs} c_{sr}^t , \qquad (4)$$

where $e_{s,r}$ is defined by

$$\xi_r e_s = e_{s,r} \xi_r . \qquad (5)$$

It follows directly from (4) that the structure constants C_{rs}^t satisfy

$$C_{rs}^t = - u_{rs} C_{sr}^t ,$$

$$u_{qr} C_{rs}^t C_{pq}^s + u_{rp} C_{ps}^t C_{qr}^s + u_{pq} C_{qs}^t C_{rp}^s = 0 . \qquad (6)$$

It is evident that the subscripts (r) and (s) of u_{rs} can be interpreted as grading vectors. If C_{rs}^t does not vanish, we write $(t) = (r + s)$, and it follows from (6) that the addition so defined is commutative and associative. The null vector (o) is associated with ξ^o, which is just a multiple of the identity g_o of G. If the order of the conjugate class $G_{(r)}$ corresponding to (r) is k, (kr) is equivalent to (o), and we say that (r) is of order k. If (s_1), (s_2), ... form a basis of the grading vector space, and are of order $k_1, k_2, \ldots$, we can write $(r) = (\Sigma_i r_i s_i)$, where r_i is an integer (mod k_i). If m_r is the number of coordinates associated with the same grading vector (r), we denote the grading by $(m) = (m_1/m_2/\ldots)$.

If the generators e_r commute with all coordinates ξ^s, or if each $e_{r,s}$ is proportional to e_r, the general Lie algebra reduces to a color algebra or superalgebra; if $u_{rs} = \pm 1$ for all r and s, the algebra is a modular algebra; if both of these conditions are satisfied, we have a simply graded Lie algebra. The relations between these algebras have been studied recently by Kleeman[6).

Most properties of ordinary Lie algebras have a straightforward generalization to this context. Representations of the generators e_r (not completely reducible in general) are obtained by writing $e_r = C^t_{sr} e^s_t$, where the e^s_t are generators of $gg\ell((m))$, with grading vectors $(t - s)$, and satisfy

$$[e^p_q, e^r_s] = \delta^r_q e^p_s - u_{q-p,s-r} \delta^p_s e^r_q . \tag{7}$$

A contravariant vector a^r satisfies

$$[e^p_q, a^r] = \delta^r_q a^p . \tag{8}$$

There is also a generalization $osp((m))$ of the orthosymplectic group. For the colour algebras, Jarvis and Green[7)] have shown how to determine the Casimir invariants in terms of highest weights, and have obtained shift vectors and characteristic identities by methods analogous to those used earlier by Bracken and Green[8)] for ordinary Lie algebras. These results generalize without substantial modification to the present context.

3. Dynamics and Statistics of Subparticles

There is an interesting application of general Lie groups to composite models of the observable particles, such as the well known quark model of the hadrons, and the rishon or quip model of quarks and leptons. It is usual to base such models on a gauge group such as SU(3) or SU(3) SU(3), but the different species of subparticles are sufficiently described in terms of representations of a finite group G, such as C_3 or $C_3 \times D_3$. The structure of the composite particle is specified by a set of collective space-time coordinates

$$x^\lambda = \Sigma'_r x^\lambda_r \xi^r , \tag{9}$$

where the summation Σ'_r is over the basis of an abelian subalgebra A_x of the group G, including the identity g_o. Then x^λ_o may be interpreted as the coordinates of the centre of mass and the other x^λ_r are complex relative coordinates chosen so that if the eigenvalues of ξ^r in a particular subparticle state are ξ^r_j, the space-time coordinates of the subparticle are $\Sigma'_r x^\lambda{}_r \xi^r{}_j$. The subscript j may specify properties of the particle such as colour, hypercolour and flavour.

A generalized Poincaré group P(G) can be defined, consisting of transformations which leave $ds^2 = dx^\lambda dx_\lambda$ unchanged, where x^λ is the collective coordinate defined in (9). This group obviously has the ordinary Poincaré group as a subgroup, but includes transformations of the basis of the algebra A_x, and is therefore a general Lie group of the type which we have described. Representations of P(G) are carried by a generalized field variable $\phi(x)$, which is not necessarily reducible with respect to the Lorentz group and therefore may have as components both spinors $\psi(x)$ and four-vectors $A_\lambda(x)$ depending on the collective coordinates. If we expand this field variable in terms of a set of suitably normalized functions $V^r(x)$ and $V_r(x)$, corresponding to positive and negative frequencies respectively:

$$\phi(x) = \Sigma_r[a_r V^r(x) + a^r V_r(x)] \quad (10)$$

the coefficients a_r and a^r can be interpreted as annihilation and creation operators for subparticles of various spins, momenta, colours etc. We require only that $u_{rr} = \pm 1$ for integral spin and $u_{rr} = -1$ for half integral spin. Their algebraic properties are sufficiently specified by the requirement that $e^r{}_s = a^r a_s$ should be generators of $g\ell((m))$, as defined by (7), in a Fock-like representation such that $a_r|o\rangle = 0$, where $|o\rangle$ is the vacuum state vector. It can be inferred from this that a^r is a contravariant vector operator satisfying (8), and that a_r is the corresponding covariant operator, so that

$$[a_s, a^r] = \delta^r{}_s \quad (11)$$

where [,] is of course the generalized Lie product defined in (4). Submaximal factors of the characteristic identity satisfied by the matrix $\hat{e}$ with elements $e^r{}_s u_{ss}$ can be used to project from a_r and a^r components corresponding to particles with a particular spin, colour, etc.

A product $a_r a_s \ldots a_v$ of creation operators is associated with a grading vector $(r + s + \ldots + v)$ determined by the subscripts, and of particular importance are those products, called modules, which are associated with the null grading vector (o). Such modules wither commute or anticommute with one another and therefore represent composite particles satisfying Bose or Fermi statistics[5]. Although the subparticles do not satisfy Bose or Fermi statistics, causality will not be violated because they cannot appear in isolation. A gauge theory can be constructed in which the requirement of containment for the subparticle is automatically satisfied. Thus the general Lie groups may be used to formulate a field theory in which many different kinds of particles are represented by a single field variable and which has several attractive features well adapted to the composite models of the observable particles of physics.

4. References

1) H. Hopf, Ann. of Math. 42, 22 (1941).

2) See V. Rittenberg, Lecture Notes in Physics 79 (Springer, Berlin, 1975).

3) V. Rittenberg and D. Wyler, J. Math. Phys. 19, 2193 (1978).

4) M. Scheunert, J. Math. Phys. 20, 712 (1979).

5) H. S. Green, Aust. J. Phys. 28, 115 (1975); 29, 483 (1976).

6) R. Kleeman, Aspects of Modular Quantization, Univ. of Adelaide preprint (1982).

7) P. D. Jarvis and H. S. Green, J. Math. Phys. 20, 2115 (1979); also Univ. of Adelaide preprint (1982).

8) A. J. Bracken and H. S. Green, J. Math. Phys. 12, 2107 (1971).

THE BRANCHING RULE OF WEYL AND THE QUANTUM NUMBER $b(\lambda_1, \lambda_2, \lambda_3)$

G.IOMMI AMUNATEGUI

INSTITUTO DE FISICA, U.C.V.
Casilla 4059, Valparaiso
CHILE

Following the line of thought displayed in the work of Antoine, Speiser and Oakes[1] we have transcribed the branching rule given by H.Weyl[2] for $SU_5 \rightarrow SU_4$ in geometrical language[3]. As is well known some experiments[4,5,6] suggest the existence of a new family of hadrons which in the quark picture would require an additional quark carrying "beauty".

This geometrical transcription allows us to introduce a quantum number $b(\lambda_1, \lambda_2, \lambda_3)$ which can be considered as "beauty" for class 0 representations of SU_5, i.e., if $(\Lambda_1, \Lambda_2, \Lambda_3, \Lambda_4)$ denotes an irreducible representation of SU_5, $C = \Lambda_1 + 2\Lambda_2 + 3\Lambda_3 + 4\Lambda_4 = 0 \pmod 5$ defines class 0 representations. The minimum value of b is given by $b_{min} = -\frac{1}{5}(4\Lambda_1 + 3\Lambda_2 + 2\Lambda_3 + \Lambda_4)$. Moreover, the number b corresponding to the representations which belong to the SU_4 content of $(\Lambda_1, \Lambda_2, \Lambda_3, \Lambda_4)$ may be derived from b_{min}. As an example, the SU_4 content of the representation (1,1,1,1) of SU_5 consists of the lattice points belonging to two cubes in the SU_4 Cartan-Stiefel diagram.

SU_4 decomposition of the representation (1,1,1,1) of SU_5

1st. cube	$b(\lambda_1, \lambda_2, \lambda_3)$	2nd. cube	$b(\lambda_1, \lambda_2, \lambda_3)$
(1,0,1)	0	(0,1,1)	-3
(2,0,1)	1	(1,1,1)	-2
(1,0,2)	-1	(0,1,2)	-4
(2,0,2)	0	(1,1,2)	-3
(1,1,0)	-1	(0,2,0)	-4
(1,1,1)	-2	(1,2,0)	-3
(2,1,0)	0	(0,2,1)	-5
(2,1,1)	-1	(1,2,1)	-4

REFERENCES

1. J.P.Antoine, D.Speiser, R.J.Oakes Phys.Rev. 141 1542 (1966).
2. H.Weyl, "The Theory of Groups and Quantum Mechanics" Dover Pub. Inc. New York(1950)
3. I. Amunategui, to appear, Ann. Soc. Scient. Bruxelles (1982)
4. S.W.Herb et al. Phys. Rev. Lett. 39, 252 (1977).
5. W.R. Innes et al. Phys. Rev. Lett. 39, 1240 (1977).
6. A. Martin, TH. 3162-CERN (1981).

ON A NEW REGULARIZATION METHOD FOR FEYNMAN DIAGRAMS

G.A.Kerimov and Yi.A.Verdiyev
Institute of Physics of the Academy of Sciences
of the AzSSR, Baku 73
U.S.S.R.

ABSTRACT

A new regularization procedure, in which the spins of the particles are used as regulating parameters, has been developed.

1. Introduction

Regularization procedures often introduce a new parameter into the theory, e.g., a mass, Λ or μ , or, perhaps, the variable dimension, n, of space-time. If the bare Lagrangian of a theory possesses a certain invariance or covariance, we hope that the regularization procedure will not break the invariance or covariance. For instance, we always deal with Lorentz-covariant theories and would hardly think of using a non-Lorentz-covariant regularization.

In papers[1,2] we have presented a new regularization procedure, in which the spins of the particles are used as regulating parameters. This regularization procedure may be applied unambiguously to any arbitrary-spin theories.

2. The Continuous Spin Method

Following Weinberg[3], we suppose that the interaction among three particles of spin j_1, j_2 and j_3 is given by

$$H(x)= g \sum_{\sigma_1 \sigma_2 \sigma_3} \begin{pmatrix} j_1 & j_2 & j_3 \\ \sigma_1 & \sigma_2 & \sigma_3 \end{pmatrix} \varphi_{\sigma_1}^{(1)}(x)\, \varphi_{\sigma_2}^{(2)}(x)\, \varphi_{\sigma_3}^{(3)}(x) \quad + \text{h.c.}$$

where ( ) denote the usual Wigner 3j symbols and φ_σ are (2j+1)-component fields transforming according to the (j,0) representations of SL(2,C). The fields φ_σ appearing here may be either of zero or of nonzero mass. We suppose that the particle '1' is massless. The covariant propagator for spin j and nonzero mass m is

$$S_{\sigma\sigma'}^{(j)}(x;m)=\frac{(-)^{2j}}{(2\pi)^4 i}\int\frac{d^4q D_{\sigma\sigma'}^{(j)}(\underset{\sim}{q})}{q^2+m^2-i\epsilon}\exp(iqx), \quad (1)$$

where $\underset{\sim}{q}$ is a 2X2 matrix constructed from the four-vector q as $\underset{\sim}{q}=q^{\mu}\sigma_{\mu}$ and $D_{\sigma\sigma'}^{(J)}$ is a finite-dimensional (j,o) representation matrices of SL(2,C). Here σ_0 is the unit matrix and σ_k, k=1,2,3, are the familiar Pauli matrices. The propagator for m=0 can be obtained by taking the limit $m\to 0$ of (1).

To understand the basic motivation of our technique let us consider the propagator (1). After integration over the angles we find that

$$S_{\sigma\sigma'}^{(j)}(x;m)=D_{\sigma\sigma'}^{(j)}(\underset{\sim}{\hat{x}})\frac{(-i)^{2j+1}}{4\pi^2|x|}\int_0^{\infty}\frac{dqq^{2+2j}}{q^2+m^2}J_{2j+1}(q|x|),$$

where $\hat{x}$ is the 2X2 matrix constructed from the unit vector $\hat{x}=x/x$ and J_{2j+1} is the well-known Bessel function. (We performed a Wick rotation.) The last integral has a meaning only if $-1\ \mathrm{Re}j\ 1/4$. In this region it is an analytic function of j which can be evaluated explicitly (see [4], p.96)

$$S_{\sigma\sigma}^{(j)}(x;m)=D_{\sigma\sigma'}^{(j)}(\underset{\sim}{\hat{x}})\frac{(-im)^{2j+1}}{4\pi^2|x|}K_{2j+1}(m|x|).$$

The expression for $S^{(j)}$ at other values of j must be obtaned by analytical continuation.

The idea, therefore, is the following. The divergent Feynman integrals may be defined in such a way that it is an analytic function of j which can be evaluated explicitly. And one can continue the resulting expressions analytically to the physical value j: the original ultraviolet divergences then reassert themselves as poles at the physical value j. Subtraction of these poles at the end of the calculation yields the desired finite portion of the integral.

In order to illustrate the new technique let us consider the second-order mass operator $M^{(j_3)}$ for particle '3'. After summation over all dummy indices and integration over angles it is written as

$$M^{(j_3)}(p) = 2^{2j_1-2}\pi^{-2}m_2^{2j_2+1}(2j_1+1)|p|^{-1}(-)^{j_1+j_2+j_3+1}$$

$$\times(-)^{\sigma_3-\sigma_3'} D_{-\sigma_3,-\sigma_3'}^{(j_3)}(\hat{\underset{\sim}{p}})\int_0^\infty dx x^{-1-2j_1} K_{2j_2+1}(xm_2) J_{2j_3+1}(|p|x).$$

(We work in Euclidean space.) Since remaining x integral is well defined analytic function of j, it can be readily computed using the formula 7.7 (31) of[4], so that

$$M^{(j_3)}(p) = (4\ )^{-2} m_2^{2j_1+2j_2-2j_3}(-)^{\sigma_3-\sigma_3'} D_{-\sigma_3,-\sigma_3'}^{(j_3)}(\underset{\sim}{p})$$

$$\times(-)^{j_1+j_2+j_3+1}\,\Gamma(2j_1+1)\,\Gamma(1-j_1+j_2+j_3)\Gamma(-j_1-j_2+j_3)$$

$$\times\frac{1}{\Gamma(2j_3+1)}\,{}_2F_1(1-j_1+j_2+j_3,-j_1-j_2+j_3;\ 2j_3+2;-\frac{p^2}{m_2^2}), \qquad (2)$$

where ${}_2F_1$ is the hypergeometric function. It is evident from (2) that the original ultraviolet infinities manifest themselves as poles of the Γ-function at the physical value j.

What can be done in more complicated cases? Since multiple-loop diagrams do not yield easily to straigtforward analytical or numerical methods, say, through Feynman parameters, one has to search for more subtle ideas.

One of the way of attacking multiple-loop diagrams is to employ a specific spliting formulae such as (A.2) and (A.3). The general scheme of the method is as follows:
(i) Write the Feynman integral formally in x-space.
(ii) Expand the obtained expression in the Bessel functions by means of eqs. (A.1), (A.2) and (A.3) and perform angular integrations by means of the Wigner formula (A.4).
(iii) Consider parameters j's complex and evaluate the remaining radial integrals within their convergency domain.

Appendix

Here we present formulae for evaluating multiple-loop Feynman integrals.[1]

[1] The formulae (A.2) and (A.3) are proven in[2].

$$\exp(iqx) = 2(|q||x|)^{-1} \sum_j i^{2j}(2j+1) J_{2j+1}(|p||x|)$$

$$\times \mathrm{Tr}\, D^{(j)}(\hat{\tilde{q}}^{-1}\hat{\tilde{x}}) \quad \text{(Tr means trace)}, \qquad (A.1)$$

$$|x_1-x_2|^{-1} D_{\sigma\sigma'}^{(j)}(\widetilde{\widehat{x_1-x_2}}) K_{2j+1}(m|x_1-x_2|) = 2(2j+1)^{-1}(m|x_1|)^{-1}$$

$$\times |x_2|^{-1} \sum_{j_1,j_2} \sum_{\sigma_1,\sigma_1'} \sum_{\sigma_2,\sigma_2'} (2j_1+1)(2j_2+1)\langle j_1\sigma_1, j_2\sigma_2 | j\sigma\rangle$$

$$\times \langle j_1\sigma_1', j_2\sigma_2' \; j\sigma'\rangle D_{\sigma_1\sigma_1'}^{(j_1)}(\hat{\tilde{x}}_1) D_{\sigma_2\sigma_2'}^{(j_2)}(\hat{\tilde{x}}_2) I_{2j_1+1}(m|x_1|)$$

$$\times K_{2j+1}(m|x_2|), \text{ if } |x_2| > |x_1| , \qquad (A.2)$$

$$\Gamma(2j+1) \; x_1-x_2^{-2-2j} D_{\sigma\sigma'}^{(j)}(\widetilde{\widehat{x_1-x_2}}) = (2j+1)^{-1}(|x_1||x_2|)^{-1}$$

$$\times \sum_{j_1} \sum_{\sigma_1\sigma_1'} \sum_{\sigma_2\sigma_2'} \frac{\Gamma(2j_1+2j+2)}{\Gamma(2j_1+1)} \langle j_1\sigma_1, j+j_1\sigma_2 | j\sigma\rangle$$

$$\times \langle j_1\sigma_1', j+j_1\sigma_2' | j\sigma'\rangle \; x^{2j_1+1} x^{-2j_1-2j-1} D_{\sigma_1\sigma_1'}^{(j_1)}(\hat{\tilde{x}}_1)$$

$$\times D_{\sigma_2\sigma_2'}^{(j+j_1)}(\hat{\tilde{x}}_2), \text{ if } |x_2| > |x_1| , \qquad (A.3)$$

$$\int D_{\sigma_1\sigma_1'}^{(j_1)}(\hat{\tilde{x}}) D_{\sigma_2\sigma_2'}^{(j_2)}(\hat{\tilde{x}}) D_{\sigma\sigma'}^{(j)*}(\hat{\tilde{x}}) d\hat{x}$$

$$= 8\pi^2(2j+1)^{-1} \langle j_1\sigma_1, j_2\sigma_2 | j\sigma\rangle \; \langle j_1\sigma_1', j_2\sigma_2' | j\sigma'\rangle \; . \qquad (A.4)$$

References

1. G.A.Kerimov and Yi.A.Verdiyev,Lett.Math.Phys.4,29(1980)
2. Yi.A.Verdiyev and G.A.Kerimov, In Proceedings of the II International Seminar on High Energy Physics and Quantum Field Theory,Protvino,1979,p.495 (in Russian).
3. S.Weinberg, Phys.Rev.133,B1318(1964);134,B882 (1964).
4. Higher Transcendental Functions, Vol.2, edited by A.Erdelyi, Mc. Graw-Hill, New York, 1953.

EXPLICIT REALIZATION OF E_8

Mehmet KOCA

Çukurova University
Dept. of Physics and Space Sciences P.O. Box 171-Adana
TURKEY

ABSTRACT

Generators of E_8 has been constructed using bilinear fermion fields in the bases of some interesting subgroups. Few comments are made on its possible use in physics.

1. INTRODUCTION

E_8 is the last member of the E series of the Cartan classification of simple Lie groups and the largest exceptional group of rank 8. First five members of E series are isomorphic to successfull gauge groups: $E_1 \simeq SU(2)_W$, $E_2 \simeq SU(2)_L x SU(2)_R$, $E_3 \simeq SU(3)_C x SU(2)_W$, $E_4 \simeq SU(5)$ and $E_5 \simeq SO(10)$. The remaining members are the exceptional groups E_6, E_7, and E_8. E_4[1], E_5[2], and E_6[3] have scored remarkable success concerning the unification of the strong and the electroweak interactions of a single family of fermions. N = 8 extended supergravity admits a global invariance under a non-compact version of E_7 which has SU(8) as a maximal subgroup[4].

E_8 has many interesting subgroups like SU(5)xSU(5), SO(10)xSU(4), SO(16), E_6xSU(3), E_7xSU(2), SU(9) and $F_4 x G_2$. There may be several ways of use of E_8 as a physical model of fundamental interactions; the first three subgroups can be regarded as the family extentions of SU(5), SO(10), and E_6 respectively, which are all unified in E_8[5]. If E_6 and SO(10) are rejected by nature we are still fully satisfied with E_8 which recognizes only 5 SU(5) families with their conjugates and super heavy ones. Symmetry breaking can always be arranged such that SO(10) and E_6 may not appear in the intermediate stages. Since E_8 involves SU(8) either via SU(9) $\supset$ SU(8)xU(1), SO(16) $\supset$ SU(8)xU(1) or E_7xSU(2) $\supset$ SU(8)xSU(2) it may serve as a superunification group.

The SO(16) ⊃ SO(8)xSO(8) subgroup may also play some role concerning supergravity, where the octonionic structure of E_8 could be emphasized.

Physicists are now familiar with the octonionic structures of the exceptional groups G_2, F_4, E_6, E_7, and E_8[6]. The smallest exceptional group G_2 is the automorphism group of the octonion algebra. Similarly the Tits' subgroups of the exceptional groups $SO(3)xG_2 \subset F_4$, $SU(3)xG_2 \subset E_6$, $SP(6)xG_2 \subset E_7$, and $F_4 x G_2 \subset E_8$ display the octonionic structures of the exceptional groups where E_8 incorporates two sets of octonions. The automorphism group $SU(3) \subset G_2$ of the Malcev's algebra of six octonionic units is identified as the color group by Gürsey[6]. Thus the reductions of the exceptional groups with respect to the color group $SU(3)_c$, $F_4 \supset SU(3)_f x SU(3)_c$, $E_6 \supset SU(3)_L x SU(3)_R x SU(3)_c$, $E_7 \supset SU(6)_f x SU(3)_c$, and $E_8 \supset E_6 x SU(3)_c$ lead to the identification of the respective flavour groups. Since E_8 uses octonions twice it contains two color subgroups, say, $SU(3)_c x SU(3)_H$ (H:hypercolor) in accord with the philosophy suggested by Gürsey. All these speculative arguments suggest a detailed investigation of E_8[7].

In this work I shall briefly illustrate how E_8 generators based on the SU(9) subgroup can be constructed as bilinear fermion fields. Using this result the generators in the basis of the subgroup SO(8)xSO(8) can be obtained. Representations of the generators in the bases of the other maximal subgroups can be found in ref.(7).

2. GENERATORS IN THE SO(8)xSO(8) BASIS

In ref.(7) we have obtained the generators in the SU(9) basis where

$$\underline{248} = \underline{80} + \underline{84} + \underline{84}^* \tag{1}$$

can be represented by

$$\underline{80}: T^a_b,\ T^a_a\ 0;\ \underline{84}: Q_{abc};\ \underline{84}^*: Q^{abc}\quad (a,b,c=1,2,\ldots,9) \tag{2}$$

The fundamental and the next lowest dimensional representation of E_8 in the SO(8)xSO(8) basis read

$$\underline{248} = \left< \begin{array}{l} \underline{120} = (\underline{28},\underline{1}) + (\underline{1},\underline{28}) + (\underline{8_s},\underline{8_v}) \\ \underline{128} = (\underline{8_s},\underline{8_v}) + (\underline{8_c},\underline{8_v}) \end{array} \right. \tag{3a}$$

$$
\underline{3875} = \left\langle \begin{array}{l} \underline{135} = (\underline{1},\underline{1}) + (\underline{1},\underline{35_s}) + (\underline{35_v},\underline{1}) + (\underline{8_v},\underline{8_s}) \\ \underline{1820} = (\underline{1},\underline{35_c}) + (\underline{1},\underline{35_v}) + (\underline{35_c},\underline{1}) + (\underline{35_s},\underline{1}) + \\ \qquad + (\underline{28},\underline{28}) + (\underline{56_v},\underline{8_s}) + (\underline{8_v},\underline{56_s}) \\ \underline{1920} = (\underline{8_s},\underline{56_c}) + (\underline{8_c},\underline{56_v}) + (\underline{56_c},\underline{8_v}) + (\underline{56_s},\underline{8_c}) + \\ \qquad + (\underline{8_s},\underline{8_c}) \quad (\underline{8_c},\underline{8_c}) \end{array} \right. \tag{3b}
$$

Decomposing the generators in SU(9) basis as SU(4)xU(1)xSU(4)xU(1) and then repgrouping each SU(4)xU(1) in SO(8) we obtain

$$(\underline{28},\underline{1}) \;:\; T^a_b,\ Q_{ab5},\ Q^{ab5} \tag{4a}$$

$$(\underline{1},\underline{28}) \;:\; T^a_a + T^5_5,\ T^{5+i}_{5+j},\ Q_{5(5+i)(5+j)}, Q^{5(5+i)(5+j)} \tag{4b}$$

$$(\underline{8_v},\underline{8_s}) \;:\; T^a_5,\ T^5_a,\ Q_{abc},\ Q^{abc},\ Q_{a(5+i)(5+j)},\ Q^{a(5+i)(5+j)} \tag{4c}$$

$$(\underline{8_c},\underline{8_v}) \;:\; T^a_{(5+i)},\ T^{(5+i)}_a,\ Q_{a5(5+i)},\ Q^{a5(5+i)} \tag{4d}$$

$$(\underline{8_s},\underline{8_v}) \;:\; T^5_{(5+i)},\ T^{(5+i)}_5,\ Q_{ab(5+i)},\ Q^{ab(5+i)}$$

$$\frac{1}{6}\,\varepsilon^{ijk\ell}\ Q_{(5+j)(5+k)(5+\ell)},\ \frac{1}{6}\,\varepsilon_{ijk\ell}\ Q^{(5+j)(5+k)(5+\ell)} \tag{4e}$$

$$(a,b,c = 1,2,3,4\ ;\ i,j,k = 1,2,3,4)$$

The branching rule for $\underline{3875}$ under SU(8)XSU(1) and SU(8)xSU(2) can be found:

$$E_8 \supset SU(8)xU(1)$$

$$
\begin{aligned}
\underline{3875} = {} & \underline{1} + \underline{63} + \underline{8} + \underline{8}^* + \underline{28} + \underline{36} + \underline{8} + \underline{168} + \underline{28}^* + \underline{36}^* \\
& + \underline{8}^* + \underline{168}^* + \underline{420} + \underline{56}^* + \underline{504} + \underline{70} + \underline{420}^* + \underline{56} \\
& + \underline{504}^* + \underline{70} + \underline{216}^* + \underline{63} + \underline{216} + \underline{720}
\end{aligned} \tag{5a}
$$

$$E_8 \supset SU(8)xSU(2)$$

$$
\begin{aligned}
\underline{3875} = {} & (\underline{1},\underline{1}) + (\underline{28}+\underline{28}^*,\underline{2}) + (\underline{63}+\underline{70},\underline{3}) \\
& + (\underline{63}+\underline{378}+\underline{378}^*+\underline{720},\underline{1}) + (\underline{36}+\underline{36}^*+\underline{420}+\underline{420}^*,\underline{2})
\end{aligned} \tag{5b}
$$

We notice that the fermionic multiplets of EGZ[8)] model appear in SU(8)xU(1) decomposition of E_8 whereas SU(8)xSU(2) decomposition involves the bosonic multiplets of the same model.

REFERENCES

1. H.Georgi and S.L.Glashow, Phys. Rev. Lett. 32, 438 (1974).

2. H.Georgi, in Particles and Fields-1974, Proceedings of the Williamsburg Meeting of the Division of Particles and Fields of the Americal Physical Society, edited by C.E.Carlson (AIP, New York, 1974) p.575; H.Fritzsch and P.Minkowski, Ann. Phys. (N.Y.) 93, 193 (1975).

3. F.Gürsey, P.Ramond and P.Sikivie, Phys. Lett. 60B, 177 (1976); for a review and further references see F.Gürsey and M.Serdaroğlu Nuovo Cimento 65A, 337 (1981).

4. E.Cremmer and B.Julia, Phys. Lett. 80B, 48 (1978), E.Cremmer and B.Julia, Nucl. Phys. B159, 141 (1979).

5. I.Bars and M.Günaydın, Phys. Rev. Lett. 45, 859 (1980); M.Koca, Phys. Lett. 107B, 73 (1981).

6. H.Freudenthal, Advances in Mathematics. 1, 145 (1965); B.A.Rozenfeld, Proc. Collog. Utrecht, P.135 (1962); J. Tits, Proc. Collog. Utrecht, P.175 (1962); M.Günaydın and F.Gürsey, J. Math. Phys. 14, 1651 (1973); F.Gürsey, in Group Theoretical Methods in Physics, Proceedings of the Fifth International Colloquium, edited by R.T. Sharp (Academic Press, New York, 1977) P.213.

7. M.Koca, Phys. Rev. D24, 2636 and 2645 (1981).

8. J.Ellis, M.K.Gaillard and B.Zumino, Phys. Lett. 94B, 343 (1980).

Dynamical Symmetry Breaking in SU(2) x U(1)
in Weak Coupling Limit

K. T. Mahanthappa
University of Colorado
Boulder, Colorado 80309

ABSTRACT

Possibility of dynamical symmetry breaking (DSB) in scalarless SU(2) x U(1) is studied in weak coupling limit using an effective potential approach. In the linearized approximation we find that the model with a quark doublet can have DSB whereas the model with a lepton doublet cannot. When non-linear effects, including those due to the non-abelian triple and quartic vector couplings, are included DSB is not realized.

The mechanism of dynamical symmetry (DSB) is viewed as better alternative to spontaneous symmetry breaking due to fundamental Higgs scalers which come in association with many arbitrary parameters that may require fine tuning. There has been considerable effort to execute DSB in SU(2) x U(1) using technicolor groups which bring new types of strong interactions[1]. In the following we describe a field theoretic study of DSB in SU(2) x U(1) without introducing any new interaction. In our study we use an effective potential which is a functional of composite operators[2].

Our Lagrangian density $L(\Phi(x))$ is the standard one of the electroweak interactions involving triplet A^i_μ and singlet B_μ gauge fields and left-handed doublet, L, and right-handed singlet, R, of fermions. We work in Landau gauge. $\Phi(x)$ represents all fields. The relevant generating functional is defined by

$$W[K] = \exp\{i\ Z[K]\} = \int[d\Phi]\ \exp\{i\ S(\Phi) + i\int d^4x\ d^4y$$
$$\times [\tfrac{1}{2} A^i_\mu(x)\ K^{i\mu\nu}_A(x,y)\ A^i_{\nu(y)} + \tfrac{1}{2} B_\mu(x)\ K^{\mu\nu}_B(x,y)\ B_\nu(y)$$
$$+ B_\mu(x)\ K^{\mu\nu}_{AB}\ (x,y)\ A^3_\nu\ (y) - \bar{\psi}_a(x)\ K_{\psi a}(x,y)\ \psi_a(y)]\}\ ,$$

where Φ represents all the fields generically, and S is the classical action, $S(\Phi) = \int d^4x\ L(\Phi(x))$. The effective action is obtained by a Legendre transform of Z[K],

$$\Gamma[G, \Delta_{Ai}, \Delta_B, \Delta_{AB}] = Z[K] - \int d^4x\ d^4y\ \ Tr[K_{\psi a}(x,y)\ G_a(x,y)]$$
$$+ \tfrac{1}{2}\ \Delta^i_{A\mu\nu}(x,y)K^{i\mu\nu}_A(x,y) + \tfrac{1}{2}\ \Delta_{B\mu\nu}(x,y)\ K^{\mu\nu}_B(x,y)$$
$$+ \Delta_{AB\mu\nu}(x,y)\ K^{\mu\nu}_{AB}(x,y)\}$$

The variational derivatives of Z with respect to K's give the full propagator functions G's and Δ's for fermions and vector mesons. The effective potential V[G,Δ] is given by

$$\Gamma[G,\Delta] = -V[G,\Delta] \int d^4x$$

Minimization of V with respect to G's and Δ's yields equations for the propagators. For our case we obtain

$$V[G,\Delta] = -i \sum_a \int \frac{d^4k}{(2\pi)^4} \mathrm{Tr}[\ln G^a(k) - S^{-1}(k)\, G^a(k)]$$

$$+ \frac{i}{2} \int \frac{d^4k}{(2\pi)^4} \mathrm{Tr}[\ln \Delta_B(k) - D^{-1}(k)\, \Delta_B(k) + \ln \Delta_{AB}(k)]$$

$$+ \frac{i}{2} \sum_i \int \frac{d^4k}{(2\pi)^4} [\ln \Delta^i_A(k) - D^{-1}(k)\, \Delta^i_A(k)] + V_2[G,\Delta] + \text{const},$$

$$S^{-1}(k) = -i \not{k}, \quad D^{-1}(k) = i(k^2 g^{\mu\nu} - k^\mu k^\nu) \equiv i\, k^2 K^{\mu\nu} .$$

Here V_2 is computed from two-particle irreducible vacuum graphs occuring in the expansion of Z, with the propagators in the graphs replaced by the appropriate G's and Δ's. In the lowest order the relevant graphs are:

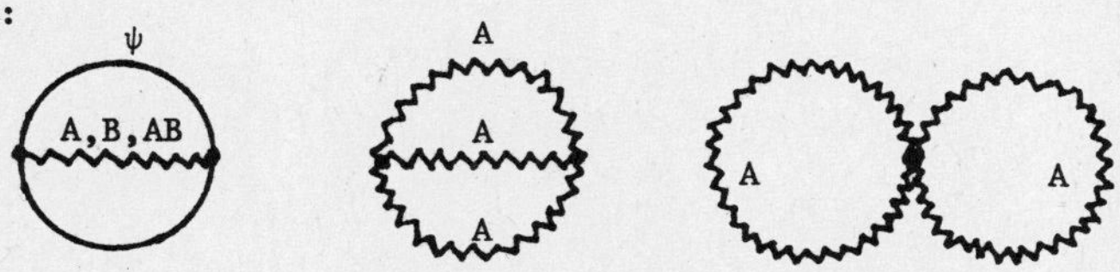

When contributions from these graphs are included in V, minimization of V with respect to G's and Δ's yields a set of coupled non-linear integral matrix equations for G's and Δ's. These can be solved in linearized approximation (LA) yielding for masses for fermions with $Y_{aL}Y_{aR} > 0$:

$$\mu_a(p) = m_a \left(\frac{-p^2}{m_a^2}\right)^{-r_a} \qquad \text{as } p \to \infty$$

with $r_a = (3/64\pi)^2 g_B^2 Y_{aL} Y_{aR}$. For fermions with $Y_{aL}Y_{aR} \leq 0$ there is no non-trivial solution and these fermions cannot acquire mass in LA. A similar expression for vector boson mass is obtained. Thus neutrino and d-quark do not acquire masses. All the nice features of the standard electroweak theory including the mixing and $M_Z^2 = M_W^2/\cos^2\theta$ are maintained in LA. Note that the trivial solution of the $m_a = 0$ exists. Which solution corresponds to the true vacuum? Effective potential enables us to answer this question and go beyond LA. In order to do this we adopt a variational technique[2]. We take the above functional form obtained for μ_a and a similar ones for vector meson masses and evaluate[3] $\Omega \equiv V(\text{broken}) - V(\text{sym}) = V(G,\Delta) - V(S,D)$. Minimization of Ω with respect to vector meson masses in LA yields same results as before. Using this (in LA) and evaluating Ω gives for $\sin^2\theta = 0.22$

$$g_B^2\Omega = +0.05\ m_f^4 \text{ (Leptons)}$$
$$= -9.8\ m_f^4 \text{ (quarks)}$$

Thus DSB occurs for the quark doublet not for the lepton doublet. LA means inclusion of only fermion loop. Now let us include non-abelian graphs. To what extent do the results of LA get altered? Again, consider the case of one fermion doublet with one fermion acquiring mass; we get an expression for Ω. Upon minimizing we find that all the attractive features of LA have been destroyed: (i) Vector meson masses are much smaller than fermion masses. (ii) $M_W^2/M_Z^2 \neq \cos^2\theta$ as now vector intermediate states have $I \neq \frac{1}{2}$. (iii) When Ω is evaluated, we find $\Omega = \omega m_f^4$ with $\omega > 0$ for $\sin^2\theta = 0.22$. (We do find $\omega < 0$ for $\sin^2\theta \approx 0$.) Thus, for this value of θ, symmetry does not break for either quarks or leptons. The non-abelian graphs tend to restore symmetry.

If there are two or more fermion doublets there will be residual Goldstone bosons if the symmetry breaks because the chiral symmetry is also broken; these are not included in our calculation of Ω. Consequently $\Omega > 0$ for non-zero masses does not necessarily mean the symmetry is not broken. But $\Omega < 0$ for non-zero masses is sufficient for DSB as Ω (true) $\leq \Omega$ (test). For example for one electron doublet + 3 color quark doublets + singlets Ω (test) > 0 and hence one cannot say whether DSB occurs; for 3 color quark doublets + singlets Ω (test) < 0 for $\sin^2\theta < 0.17$ and hence DSB occurs. Thus we can find cases in which SU(2) x U(1) breaks dynamically but none of them are realistic.

The main features manifested in our study are that the attractive features of LA do not survive the inclusion of the non-abelian effects, and even neglecting the non-abelian effects, the symmetry breaking solutions in LA do not always correspond to the true vacuum. DSB is strongly dependent on the number and quantum numbers of fermions.

This work was done in collaboration with James Randa. It was supported in part by the U. S. Department of Energy, Grant No. DE-AC02-81ER40025.

References

1. K. D. Lane and M. Perhkin, in Electroweak Interactions and Unified Theories, ed. J. Tran Thanh Van (Editions Frontieres, Dreux, France, 1980), Vol. II. p. 469 which contains references to previous work.
2. J. M. Cornwall, R. Jackiw and E. Tomboulis, Phys. Rev. D 10, 2428 (1974).
3. Details can be found in K. T. Mahanthappa and J. Randa, COLO-HEP-36 (1982) and (in preparation).

DYNAMICAL UNIFICATION OF FERMIONS AND GAUGE BOSONS FOR INTERNAL SYMMETRY AND GRAVITY

Heinrich Saller
Max-Planck-Institut für Physik und Astrophysik
Föhringer Ring 6, 8000 München 40
Fed.Rep.of Germany

ABSTRACT

Internal symmetry gauge bosons and the tetrad fields of gravity with their respective local transformation behaviour are constructed by more basic fermions. A physically sensible mass scale Λ^2 for a parametrization of the small distance behaviour is closely related to the gauge coupling constants (fine structure constant, Newton's constant). Local invariances determine uniquely the renormalizable fermion selfinteractions in a parameter free theory. The gauge boson couplings contain simultaneously the Higgs boson couplings which can lead to a symmetry breakdown by fermion pair condensation.

The effective "standard" theories, experimentally rather successful, are qualitatively unsatisfactory since they contain with the fermions, gauge and Higgs bosons three conceptually unrelated inputs. They are also quantitatively not determined, leaving open the values for the Yukawa couplings (fermion masses), for the Higgs potential masses (e.g. Fermi's constant) and for the gauge boson coupling constants.

There is a long history to replace the Higgs boson sector by fermion condensation[1] and to interpret the gauge bosons as fermionic bound states[2)3)4)]. The simplest ansatz for a bound state U(1) gauge boson $A_\mu(x) \sim \frac{1}{\Lambda^2}\bar{\chi}\gamma_\mu\chi(x)$ (χ fermion field, Λ^2 mass) fails-it cannot reproduce the characteristic gauge transformation behaviour $A_\mu \mapsto A_\mu + \partial_\mu\alpha$, requiring for the intrinsic dimension dim $A_\mu = 1$. A <u>dynamical incorporation of the dimension</u> leads to noncanonical fermions $\check{\chi}(x)$ with dimension $\dim\check{\chi}(x) = 1/2$ in contrast to the canonical value 3/2.

Now also composite gauge boson as the finite part bilinear product transform inhomogeneously

(1) $$\dim \check{\chi}(x) = 1/2 \;, \quad \langle \check{\chi}(\xi/2)\, \bar{\check{\chi}}(-\xi/2)\rangle \xrightarrow[\xi\to 0]{} -\frac{i}{8\pi^2}\frac{\gamma\cdot\xi}{\xi^2}$$

(2) $$\begin{cases} \check{\chi}(x) \longmapsto \exp[-i\alpha(x)]\,\check{\chi}(x) \\ \bar{\check{\chi}}(x_-)\gamma_\mu\check{\chi}(x_+) \longmapsto \bar{\check{\chi}}(x_-)\gamma_\mu\check{\chi}(x_+)\left[1 + i\xi^\nu\partial_\nu\alpha(x) + \dots\right] \;, \quad x_\pm = x \pm \xi/2 \\ i\xi^\nu \langle \bar{\check{\chi}}(x_-)\gamma_\mu\check{\chi}(x_+)\rangle \xrightarrow[\xi\to 0]{} \frac{1}{8\pi^2}\delta_\mu^\nu \\ A_\mu(x) = \frac{(4\pi)^2}{2} :\bar{\check{\chi}}\gamma_\mu\check{\chi}:(x) \longmapsto A_\mu(x) + \partial_\mu\alpha(x) \end{cases}$$

The canonical formulation for noncanonical fermion fields of dimension 1/2, $(i\partial\!\!\!/)^3\check{\chi} = 0$ is given by

(3) $$\begin{cases} \mathcal{L}(\check{\chi}) = -\left[\bar{\check{\chi}}\tfrac{i}{2}\overleftrightarrow{\partial\!\!\!/}\hat{\chi} + \bar{\hat{\chi}}\tfrac{i}{2}\overleftrightarrow{\partial\!\!\!/}\check{\chi} + \bar{\chi}\tfrac{i}{2}\overleftrightarrow{\partial\!\!\!/}\chi - (\bar{\chi}\hat{\chi} + \bar{\hat{\chi}}\chi)\right] \\ i\partial\!\!\!/\check{\chi} = \chi \;, \quad i\partial\!\!\!/\chi = \hat{\chi} \;, \quad i\partial\!\!\!/\hat{\chi} = 0 \\ \dim(\check{\chi},\chi,\hat{\chi}) = 3/2 + (-1,0,1) = (1/2, 3/2, 5/2) \end{cases}$$

The effective QED theory with basic fermion and gauge boson fields (χ, A_μ)

(4) $$\begin{cases} \mathcal{L}^{QED}_{eff} = \bar{\chi}(\tfrac{i}{2}\overleftrightarrow{D\!\!\!/} - M)\chi - \frac{\pi}{\alpha}(\partial_\nu A_\mu - \partial_\mu A_\nu)^2 \\ \tfrac{i}{2}\overleftrightarrow{D\!\!\!/} = \tfrac{i}{2}\overleftrightarrow{\partial\!\!\!/} - \gamma_\mu A^\mu \;, \quad 1/\alpha = \frac{4\pi}{e^2} \cong 137 \end{cases}$$

leads to the gauge boson field equations

(5) $$\begin{cases} \text{0. order: } (\eta_{\mu\nu} - \frac{p_\mu p_\nu}{p^2})\,\delta^\nu_\rho \frac{p^2}{e^2} A^\rho(p) = -J_\mu(p) = -\bar{\chi}\gamma_\mu\chi \\ \text{1. order: } (\eta_{\mu\nu} - \frac{p_\mu p_\nu}{p^2})(\delta^\nu_\rho \frac{p^2}{e^2} + \underbrace{J^\nu J_\rho}(p))A^\rho(p) = -J_\mu(p) \end{cases}$$

where the vacuum polarization $\underbrace{J^\nu J^\rho}(x-y) = \bar{\chi}\gamma^\nu\underbrace{\chi(x)\,\bar{\chi}}\gamma^\rho\chi(y)$ gives the 1. order correction.

A "gauge fermion" $(\check{\chi},\chi,\hat{\chi})$ - QED

(6) $$\mathcal{L}^{QED}_{bas} = -\left[\bar{\check{\chi}}(\tfrac{i}{2}\overleftrightarrow{D\!\!\!/} + M)\hat{\chi} + \bar{\hat{\chi}}(\tfrac{i}{2}\overleftrightarrow{D\!\!\!/} + M)\check{\chi} + \bar{\chi}(\tfrac{i}{2}\overleftrightarrow{D\!\!\!/} - M)\chi - (\bar{\chi}\hat{\chi} + \bar{\hat{\chi}}\chi) - \frac{\Lambda^2 - M^2}{2}(\bar{\chi}\check{\chi} + \bar{\check{\chi}}\chi)\right]$$

with the propagator $\underbrace{\check{\chi}\bar{\check{\chi}}}(p) = 1/(p\!\!\!/ - M)(\Lambda^2 - p^2)$ contains a composite U(1) gauge boson $A_\mu(x)$ (2). Its 0. order field equation[4)]

(7) $$(\eta_{\mu\nu} - \frac{p_\mu p_\nu}{p^2})(\delta^\nu_\rho + \underbrace{A^\nu J_\rho}(p))A^\rho(p) = -\underbrace{A_\mu A_\rho}(p)\,J^\rho(p)$$

is determined by the <u>0. order vacuum polarization</u> $\underbrace{A^{\nu}A^{\varrho}}(x-y)$, $\underbrace{A^{\nu}J^{\varrho}}(x-y)$ expressible by products of fermion propagators. The evaluation of the 0. order vacuum polarization leads to massless photons and determines the gauge coupling constant in terms of a mass ratio[4] $1/\alpha = 1/3 \;\log \Lambda^2/M^2$ $(M^2 \ll \Lambda^2)$.

The four fermion coupling of the gauge interaction (6) contains also <u>fermion pairs</u>, e.g. $\check{\bar{\chi}}\check{\chi}\ (x)$, whose <u>condensation</u> leads to a dynamical symmetry breakdown and associated gap equations which, in the end, may determine the relevant mass ratios.

In a similar procedure one can replace the <u>effective gravity</u> Lagrangian

$$\mathcal{L}_{eff}^{GRAV} = \sqrt{-g}\left[h_j^{\mu}(\bar{\chi}\gamma^j \tfrac{i}{2}\overleftrightarrow{D}_{\mu}\chi) - M\bar{\chi}\chi + \frac{\pi}{G(4\pi)^2}R\right] \tag{8}$$

with a basic tetrad field h_j^{μ}, $\frac{\pi}{G} = (2.1\ .\ 10^{19}\ \mathrm{GeV})^2$ and the 0. Order field equation

$$(9)\qquad \begin{aligned} &(\eta_{\mu\nu}p^2 - p_{\mu}p_{\nu})\, h_j^{\nu}(p) = (4\pi)^2 \frac{G}{\pi}\, \theta_{j\mu}(p) \\ &\theta_{j\mu}(x) = \bar{\chi}\gamma_j \tfrac{i}{2}\overleftrightarrow{\partial}_{\mu}\chi(x) \qquad \text{(translation currents)} \end{aligned}$$

by a <u>purely fermionic gravity</u>. The intrinsic dimension of the tetrad, $\dim h_j^{\mu} = 0$, and its local translation behaviour, $h_j^{\mu}(x) \mapsto h_j^{\mu}(x) - (\partial_{\nu}\epsilon^{\mu})\, h_j^{\nu}(x)$ for $x^{\mu} \to x^{\mu} + \epsilon^{\mu}(x)$, can be reproduced by using a subcanonical fermion $\check{\chi}(x)$ with $\dim \check{\chi}(x) = 3/2 - 2 = -1/2$.

$$h_{\mu}^{j}(x) = 3\,(4\pi)^2 :\bar{\check{\chi}}\gamma^j \tfrac{i}{2}\overleftrightarrow{\partial}_{\mu}\check{\chi}:(x) \tag{10}$$

The 0. order field equation for the composite tetrad h_{μ}^{j} (10)

$$\left[1 - \underbrace{h\,\theta}(p)\right] h(p) = \underbrace{h\,h}(p)\ \theta(p) \tag{11}$$

with the 0. order vacuum polarization $\underbrace{h\,\theta}(x-y)$, $\underbrace{h\,h}(x-y)$ ($\theta_{j\mu}(x)$ canonical translation current of the subcanonical theory) leads for a propagator $\underbrace{\check{\chi}\bar{\check{\chi}}}(p) = 1/(\not{p}-M)(p^2-\Lambda^2)$ with a physical mass M and a dilatation breakdown mass[5] Λ^2, $M^2 \ll \Lambda^2$, to Newtons constant $\frac{\pi}{G} = 2\Lambda^2$. For all those theories the mass Λ^2 (Planck mass, dilatation breakdown) determines the point where the <u>indefinite metric in the state space</u> becomes relevant.

References

1) J. Bardeen, L.N. Cooper, D.R.Schrieffer, Phys.Rev. 108, 1175 (1957); Y. Nambu, G. Jona-Lasinio, Phys.Rev. 122, 345 (1961); 124, 246 (1961).

2) H.P. Dürr, W. Heisenberg, H. Mitter, S. Schlieder, K. Yamazaki, Zs.f.Naturf. 14a, 441 (1959); J.D. Bjorken, Ann.Phys. 24, 174 (1963); A.D. Sakharov, Dokl.Akad.Nauk.SSSR 177, 70 (1967); Sov.Phys.JETP 12, 1040 (1968) .

3) H.P. Dürr, N.J. Winter, Nuov.Cim. 70A, 467 (1970); I.I. Bigi, H.P. Dürr, N.J. Winter, Nuov.Cim. 22A, 420 (1974).

4) H. Saller, Nuov.Cim. 4A, 404 (1971); 7A, 779 (1972); 12A, 349 (1972); 24A, 391 (1974); 34A, 99 (1976); 67A, 70 (1982); 68A, 324 (1982). Preprint MPI-PAE/PTh 17/82.

5) H. Saller, Preprint MPI-PAE/PTh 52/82.

$Q = I_3 + \frac{1}{2}Y$. WHY ?*

T. Schücker **

Département de Physique Théorique

Université de Genève

1211 Genève 4, Switzerland

ABSTRACT

A slightly different formulation of the standart $SU(2)\times U(1)$ theory is proposed, that makes some calculations more transparent without changing physical results.

The photon, corresponding to an exact symmetry in the standart $SU(2)\times U(1)$ theory [1], is massless. Still the mass matrix apparently mixes it with the Z-boson and the weak mixing angle θ_w appears analogous to the Cabbibo angle. This is merely a consequence of awkward definitions of generators and fields.

Before presenting a different formulation inspired by "Grand Unified" models let us review some generalities of gauge theories [2] : First there is the choice of a Lie algebra $\mathfrak{g}$ and a non-degenerate, bilinear, invariant, symmetric form (,) on $\mathfrak{g}$. Note that the choice of the bilinear form is unique up to a multiplicative constant - the coupling constant - if and only if $\mathfrak{g}$ is simple or one-dimensional [3]. Next we identify the generators corresponding to exact symmetries - in our case just the electric charge Q - as elements of $\mathfrak{g}$ and then we choose a basis of $\mathfrak{g}$ with the following properties :

1) It contains Q, the electric charge generator.
2) It is orthonormal with respect to the bilinear, invariant form.
3) It consists of eigenvectors of ad_Q (the commutation with Q).
4) If there is spontaneous symmetry breaking all basis vectors are also eigenvectors of the mass matrix.

3) and 4) can be fulfilled simultaneously since electric charge re-

---

* Dedicated to Youri Orlov

** Partially supported by the Swiss National Sciences Foundation

mains unbroken.

Finally to each basis vector we associate one physical gauge field. Condition 2) for instance assures that the kinetic energy term in the Lagrangian reads :

$$\frac{1}{4}(F_{\mu\nu},F_{\mu\nu}) = \frac{1}{4}F^1_{\mu\nu}F^1_{\mu\nu} + \frac{1}{4}F^2_{\mu\nu}F^2_{\mu\nu} + \ldots + \frac{1}{4}F^n_{\mu\nu}F^n_{\mu\nu}$$

where n is the dimension of $\mathfrak{g}$.

Now let

$$\mathfrak{g} = SU(2) \times U(1)$$

with basis

$$iI_k \quad k = 1,2,3 \quad \text{and} \quad iY \in U(1)$$

$$i^2 = -1, \; I_k = \text{Pauli matrices.}$$

A bilinear invariant form is defined by :

$$(I_k,I_\ell) := \frac{1}{2g^2}\mathrm{Tr}I_kI_\ell = \frac{1}{g^2}\delta_{k\ell}$$

$$(I_k,Y) := 0 \qquad (Y,Y) := \frac{1}{g'^2} \, .$$

The most general choice for the electric charge is :

$$Q := \sin\theta_w gI_3 + \cos\theta_w g'Y \, .$$

The other three basis vectors satisfying 2), 3) and 4) are determined uniquely up to a phase :

$$Z := \cos\theta_w gI_3 - \sin\theta_w g'Y$$

$$W^+ := g\sqrt{2}\begin{pmatrix}0 & 1\\ 0 & 0\end{pmatrix} \quad W^- := g\sqrt{2}\begin{pmatrix}0 & 0\\ 1 & 0\end{pmatrix} \, .$$

Note however that ad_Q can be diagonalized only in the complexified of $\mathfrak{g}$.

Let us list the advantages of this definition of the electric charge generator over the usual one :

$$Q = I_3 + \frac{1}{2}Y$$

which does not preserve the bilinear invariant form :

- The relation between the electromagnetic coupling e and $\sin\theta_w$ can be read off immediately.
- No mass matrix diagonalization is necessary to find the Z-mass : In the particular case of $SU(2) \times U(1)$ condition 4) is automatically satisfied since all broken generators have different charge.
- The calculation of $\sin^2\theta_w$ [4] after embedding of $SU(2) \times U(1)$ togegher with SU(3) of colour in a simple algebra becomes transparent.
- In more general theories with additional neutral currents θ_w, not being a mass mixing angle, now plays a distinct role from those additional angles that mix the massive neutral bosons [5].

All physically relevant quantities, of course, remain unchanged in the new formulation.

REFERENCES

[1] For a pedagogical presentation see for instance : J. Iliopoulos : An Introduction to Gauge Theories, CERN 76-11 (1976); J.C. Taylor : Gauge Theories and Weak Interactions, Cambridge University Press 1976.

[2] S. Weinberg : Phys. Rev. D7, 1068 (1973).

[3] T. Schücker : Z. Phys. C12, 81 (1982).

[4] H. Georgi, H.R. Quinn and S. Weinberg : Phys. Rev. Lett. 33, 451 (1974).

[5] P. Binétruy and T. Schücker : Z. Phys. C7, 35 (1980).

EXAMPLES OF GROUP CONTRACTION

A. Bohm with R. R. Aldinger
Center for Particle Theory, The University of Texas at Austin
Austin, Texas 78712, U.S.A.

ABSTRACT

Two limiting processes are applied to a model which may describe the relativistic quantum mechanical rotator. The first limiting process is defined by the contraction of the Poincaré group representation to the representation of the extended Galilei group (non-relativistic limit), and the second by the contraction of the deSitter group representation to the representation of the Poincaré group (elementary limit). In the elementary limit the model describes a relativistic elementary particle and in the non-relativistic limit it describes a non-relativistic rotator.

The concept of group contractions is now thirty years old. Though it has never been a very popular subject, it remains one of the most profound concepts for working with group-theoretical methods in the discovery of new physical theories. Group contractions were introduced in 1953 by Inönü and Wigner.[1] A more general limiting process between non-isomorphic groups had already been suggested in 1951 by J. Segal.[2] It is in honor of E. Inönü, who has devoted his life to the development of science in his country, that we have chosen the subject of group contractions for the group theory conference in Istanbul, Turkey.

In this short talk, I cannot describe the general theory and give a review of its applications. Therefore, I shall restrict myself to two examples. The first is the limiting process:

$$\text{Poincaré group } \mathcal{P} \longrightarrow \text{Galilei group } \mathcal{G} \tag{1}$$

for which the concept of a group contraction was originally conceived by Inönü and Wigner. The second example, which has interested me for many years, is the contraction:

$$\text{deSitter group } SO(4,1) \longrightarrow \text{Poincaré group } \mathcal{P} \tag{2}$$

The only new results that I shall discuss concern the interrelationships of these two limiting processes. I shall use these interrelationships to justify the definition of a model which I have called the "relativistic rotator."

The relativistic rotator has many ingredients which are based on concepts with a long history. However, none of these concepts justifies the name. The relativistic quantum mechanical rotator is a new entity that can be defined only by correspondence to models which are already well-known. This correspondence will be given by the use of the two contraction processes mentioned above.

This method of justifying new theories is a special case of the general approach in the conjecture of new theories: A new theory describing a new domain of physics is conjectured from an old known theory in such a way that the new theory, in a certain sense, corresponds to the old theory when the new domain of applicability is restricted to the old familiar domain. An example of such a correspondence is the limiting process in the passage from the relativistic domain to the non-relativistic domain when the velocity of light is taken to be infinite: $c \to \infty$ or $1/c \to 0$. The relativistic space-time is restricted to the Galilean space-time and their symmetry groups are related by the contraction:

$$\mathcal{P}_{P_\mu J_{\mu\nu}} \longrightarrow \mathcal{G}_{P_i H M G_i J_i} \qquad \begin{matrix} \mu,\nu = 0,1,2,3 \\ i,j = 1,2,3 \end{matrix} \tag{3}$$

The irreducible representations of the Poincaré group describe the relativistic system "elementary particle" which is characterized by (cm,s), while the irreducible representations of the Galilei group describe the non-relativistic "mass point" characterized by the eigenvalues of the three Casimir operators of the extended (quantum mechanical) Galilei group:

$$M = \text{central element} = m \; ; \quad U = H - \frac{1}{2M}\vec{P}^2 = 0 + \text{const.} \; ;$$
$$\vec{S}^2 = \left(\vec{J} - \vec{G} \times \frac{\vec{P}}{M}\right)^2 = s(s+1)$$

Therefore, this describes the non-relativistic mass point as the "contraction" limit of the relativistic elementary particle. The details of the contraction process are the following:

$\mathcal{P}$	$\mathcal{G}$	
P_i	P_i	
$cP_0 - Mc^2 = H^{(c)}$	$H^{(c)} \to H$	(4a)
$J_k = \frac{1}{2}\varepsilon_k{}^{ij} J_{ij}$	J_i	(4b)
$\frac{1}{c} K_j = \frac{1}{c} J_{j0} = G_j^{(c)}$	$G_i^{(c)} \to G_i = MQ_i$	(4c)

$$[P_\mu, P_\nu] = 0 \longrightarrow [H, P_i] = 0 \quad [P_i, P_j] = 0 \tag{4d}$$

$$[J_i, J_j] = i\varepsilon_{ijk} J_k \longrightarrow [J_i, J_j] = i\varepsilon_{ijk} J_k \tag{4e}$$

$$[J_i, P_j] = i\varepsilon_{ijk} P_k \longrightarrow [J_i, P_j] = i\varepsilon_{ijk} P_k \tag{4f}$$

$$[J_i, P_0] = 0 \longrightarrow [J_i, H] = 0 \tag{4g}$$

$$[K_i, K_j] = -i\varepsilon_{ijk} J_k \longrightarrow [G_i^{(c)}, G_j^{(c)}] = -i \frac{1}{c^2} \varepsilon_{ijk} J_k \to 0 \tag{4h}$$

$$[J_i, K_j] = i\varepsilon_{ijk} K_k \longrightarrow [J_i, G_j] = i\varepsilon_{ijk} G_k \tag{4i}$$

$$[K_i, P_j] = i\delta_{ji} P_0 \longrightarrow [G_i^{(c)}, P_j] = i\delta_{ij} (\frac{1}{c^2} H^{(c)} + M) \to i\delta_{ij} M \tag{4j}$$

$$[K_i, P_0] = iP_i \longrightarrow [G_i^{(c)}, H^{(c)}] = iP_i \tag{4k}$$

<u>Casimir Operators</u> — <u>Casimir Operators</u>

Define: $\hat{w}^\mu = \frac{1}{2} \varepsilon^{\mu\nu\rho\sigma} \hat{P}_\nu J_{\rho\sigma}$

(where: $\hat{P}_\nu = P_\nu \frac{1}{Mc}$)

$$\longrightarrow \hat{w}^0 = \vec{\hat{P}} \cdot \vec{J} = \frac{1}{Mc} \vec{P} \cdot \vec{J} \to 0 \tag{4ℓ}$$

$$\longrightarrow \vec{\hat{w}} = (\frac{H^{(c)}}{Mc^2} + \frac{Mc}{Mc}) \vec{J} - c\vec{G}^{(c)} \times \frac{\vec{P}}{Mc} \to$$

$$\to \vec{J} - \vec{G} \times \frac{\vec{P}}{M} = \vec{S} \tag{4m}$$

$$P_\mu P^\mu = M^2 c^2 \ \underline{\underline{\text{irrep}}}\ m^2 c^2 \longrightarrow (\frac{1}{c^2})(M^2 c^2) \longrightarrow M^2 \ \underline{\underline{\text{irrep}}}\ m^2 \tag{4n}$$

$$\hat{W} = -\hat{w}_\mu \hat{w}^\mu \ \underline{\underline{\text{irrep}}}\ s(s+1) \longrightarrow \hat{W} \to (\vec{J} - \frac{\vec{G} \times \vec{P}}{M})^2 = \vec{S}^2 \ \underline{\underline{\text{irrep}}}\ s(s+1) \tag{4o}$$

$$U = H^{(c)} - \frac{1}{2M} \vec{P}^2 \ \underline{\underline{\text{irrep}}}\ 0 \tag{4p}$$

(since $H^{(c)} \to H = \frac{\vec{P}^2}{2M}$; see argument preceding Eq. (5))

In the above, $\underline{\underline{\text{irrep}}}$ means equal to in an irreducible representation.

It is important to note that the contraction limit applied to the representation of the group is not only $1/c \to 0$. For in this case, $G_i^{(c)} \to 0$, and one would not obtain a faithful representation. Therefore, one must increase K_i along with c in such a way that $\frac{1}{c} K_i$ remains finite. This is accomplished by going through a sequence of representations (cm,s) of $\mathcal{P}$ i.e. $cm \to \infty$ when $1/c \to 0$ (note that we have written the eigenvalue of the invariant operator of $\mathcal{P}$ as $(cm)^2$). Then $(1/c)(cm) \to$ finite value = m and the operator P_0 "increases to infinity." But the operator

$$H^{(c)} = c(P_0 - mc) = c(\sqrt{\vec{P}^2 + m^2c^2} - mc)$$

$$= mc^2(\frac{\vec{P}^2}{2m^2c^2} + \cdots) = \frac{\vec{P}^2}{2m} + 0(1/c)$$

remains finite since it is the difference of two terms that grow to infinity:

$$H^{(c)} \rightarrow H = \frac{\vec{P}^2}{2m} \tag{5}$$

The above example is the physical system whose contraction limit is the "mass point." We now want to consider the physical system that corresponds by this contraction limit to the non-relativistic rotator.

For the non-relativistic quantum rotator the energy operator is not given by (5) but by

$$H = \frac{\vec{P}^2}{2M} + \frac{1}{2I_B}\vec{S}^2 \tag{6}$$

which corresponds to the case in which the operator U in (4p) is given by

$$U = \frac{1}{2I_B}\vec{S}^2 \quad .$$

The non-relativistic rotator is an "extended" object that can perform translational as well as rotational motion. In its simplest form, it can be visualized classically as a dumbbell with moment of inertia $I_B = \mu R^2$ (where μ is the reduced mass and R is the extension of the dumbbell), for which the angular momentum $\vec{S}$ is always perpendicular to the figure axis $\vec{D}$, which points from the center of mass (CM) to the center of charge (CQ):

$$D_i \cdot S_i = 0$$

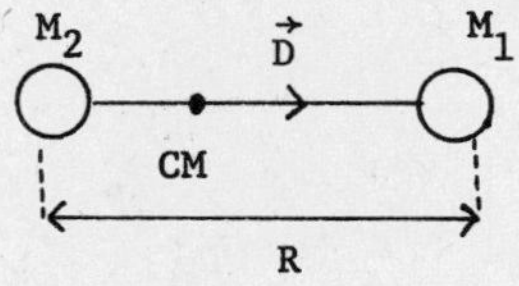

The first term in (6) is the kinetic energy of the CM, while the second term comes from rotations around the CM. Translations are generated by the P_i, time translations by H, proper Galilei transformations by $G_i = MQ_i$ (where Q_i = operator of position of CM) and rotations by $J_i = \varepsilon_{ijk}Q_jP_k + S_i$. In the limit of large R (extension of the interparticle distance) $I_B = \mu R^2 \rightarrow \infty$, the rotational degrees of freedom become frozen and we have a structureless "mass point" with translational energy only. In this sense the "mass point" is the elementary limit of the non-relativistic rotator.

If the motion of the translational degrees of the CM is ignored, one has a pure rotator where the observables S_i generate the "intrinsic" rotations of the extended object and D_i represents the unit vector in the direction of the symmetry axis which gives the position of the center of charge (CQ) relative to the CM. Therefore, it represents something like an intrinsic position operator or electric dipole operator. Therefore, one may conjecture that the D_i commute. Since they are obviously vector operators (with respect to J_i and S_i) one has the following c.r. for the algebra of the S_i and D_i:

$$[S_i,S_j] = i\varepsilon_{ijk}S_k \quad [S_i,D_k] = i\varepsilon_{ik\ell}D_\ell \quad [D_i,D_k] = 0 \tag{$\tilde{7}$}$$

This is the algebra of the three dimensional Euclidean group $E(3)_{D_iS_i}$, which is the spectrum generating group for the rotating dumbbell molecule.[3] The D_i transform between the different energy levels of the rotator describing radiative dipole transitions $M^* \to M + \gamma$

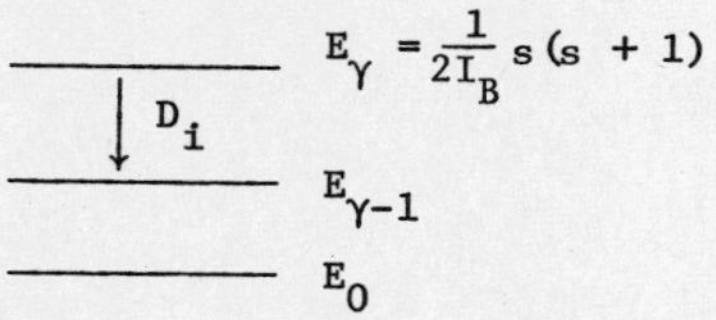

To obtain, in the non-relativistic contraction limit, the non-relativistic rotator (spectrum of (6)) and not the "mass point," the contraction process must result in the relation (6) between the two Casimir operators U and $\vec{S}^2$, up to an additive constant of G. It is natural to define the relativistic quantum mechanical rotator as the relativistic physical system which in the non-relativistic contraction limit $1/c \to 0$ goes into the non-relativistic rotator.

The relativistic rotator is -- like the non-relativistic rotator -- not an elementary system. (As in the non-relativistic case, elementary means a system whose internal structure is not resolved.) But in the elementary limit it should go into an elementary physical system i.e. an elementary particle which is described by an irrep. (cm,s) of the Poincaré group. Therefore, we have two requirements that the relativistic rotator model must satisfy: In the elementary limit it should go into an elementary particle and in the non-relativistic limit it should go into the non-relativistic rotator.[3a]

Relativistic Rotator ——elementary limit——→ Elem. Part.

↓ non-rel. limit

Non-relativistic Rotator

For the non-relativistic limit we will use the contraction process $P \rightarrow G$. The elementary limit is also described by a group contraction the result of which must be --because of the above condition --the Poincaré group. There are many groups that contract into the Poincaré group and it may be difficult to motivate the choice. The simplest are the deSitter group SO(4,1) and the anti-deSitter group SO(3,2). We will use SO(4,1) and see whether it goes into the non-relativistic rotator in the non-relativistic contraction limit.

There exists a remarkable connection between the Poincaré group and the deSitter group: If P_μ and $J_{\mu\nu}$ are the generators of P then

$$B_\mu = P_\mu + \frac{\lambda}{2}\{\hat{P}^\sigma, J_{\mu\sigma}\} \quad \text{where } (\hat{P}_\mu = \frac{P_\mu}{Mc}) \text{ and } J_{\mu\nu} \tag{7}$$

are the generators of SO(4,1)[7] where $R = 1/\lambda$ is the radius of the deSitter space in which SO(4,1) acts as the group of motion. It is obvious that the group $SO(4,1)_{B_\mu J_{\mu\nu}}$, contracts in the elementary limit $\lambda \rightarrow 0$ or $R \rightarrow \infty$ into the group $P_{P_\mu J_{\mu\nu}}$. The details of the contraction process are the following:

SO(4,1)	$\lambda = \frac{1}{R} \rightarrow 0$	P	
$J_{\mu\nu}$		$J_{\mu\nu}$	
$B_\mu^{(\lambda)} = P_\mu + \frac{\lambda}{2}\{\hat{P}_\nu, \hat{J}^\nu{}_\sigma\}$	$\longrightarrow$	P_μ	(8a)

$$[J_{\mu\nu}, J_{\rho\sigma}] = -i(g_{\mu\rho}J_{\nu\sigma} + g_{\nu\sigma}J_{\mu\rho} - g_{\mu\sigma}J_{\nu\rho} - g_{\nu\rho}J_{\mu\sigma}) \tag{8b}$$

$$[J_{\mu\nu}, B_\rho] = i(g_{\nu\rho}B_\mu - g_{\mu\rho}B_\nu) \longrightarrow [J_{\mu\nu}, P_\rho] = i(g_{\nu\rho}P_\mu - g_{\mu\rho}P_\nu) \tag{8c}$$

$$[B_\mu, B_\nu] = i\lambda^2 J_{\mu\nu} \longrightarrow [P_\mu, P_\nu] = 0 \tag{8d}$$

<u>Casimir Operators</u> (Using Principle series rep.) <u>Casimir Operators</u>

$$\text{Define: } \omega^\mu = \frac{1}{2}\varepsilon^{\mu\nu\rho\sigma}B_\nu^{(\lambda)}J_{\rho\sigma} \longrightarrow (Mc)\hat{w}^\mu = \frac{1}{2}\varepsilon^{\mu\nu\rho\sigma}P_\nu J_{\rho\sigma} \tag{8e}$$

$$\lambda^2 C_{(4)} = \lambda^2(\vec{K}\cdot\vec{J})^2 - \omega_\mu\omega^\mu \;\underline{\underline{\text{irrep}}} \tag{8f}$$

$$s(s+1)\lambda^2\alpha^2 + \lambda^2(s-1)s(s+1)(s+2) \rightarrow -(cM)^2\hat{w}_\mu\hat{w}^\mu = (cM)^2\hat{W} \;\underline{\underline{\text{irrep}}}\; (cm)^2 s(s+1) \tag{8g}$$

$$\lambda^2 C = B_\mu B^\mu - \frac{1}{2}J_{\mu\nu}J^{\mu\nu} \;\underline{\underline{\text{irrep}}}\; \lambda^2\alpha^2 \longrightarrow P_\mu P^\mu \;\underline{\underline{\text{irrep}}}\; m^2c^2 \tag{8h}$$

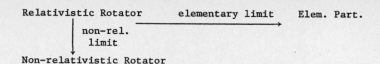

which (by using def. of B_μ, eq. 7) equals:

$$= P_\mu P^\mu + \frac{9}{4} \lambda^2 - \lambda^2 \hat{W} \tag{8i}$$

B_μ are the generators of translation along the deSitter sphere; $\vec{B}_\mu = \frac{1}{\lambda C} B_\mu$ is the dimensionless generator of the corresponding deSitter rotation.

As can be seen from (8h), the Casimir operator of SO(4,1), $\lambda^2 C$ constitutes a relation between the two Casimir operators $P_\mu P^\mu$ and $\hat{W}$ of the Poincaré group. If we impose the constraint relation:

$$\lambda^2 C = \lambda^2 \alpha^2 = \text{constant} \tag{9}$$

we obtain (by using the eigenvalues of (8g)) the mass vs. spin trajectory relation:

$$c^2 m^2 = \lambda^2 \alpha^2 - \frac{9}{4} \lambda^2 + \lambda^2 s(s+1) \tag{10}$$

The physical system that is described by a representation of SO(4,1) with the eigenvalue α^2 of the second order Casimir operator is a relativistic physical system. In the elementary limit, $\lambda \to 0$ $\alpha \to \infty$ such that $\lambda\alpha$ = arbitrary constant = cm, spin and mass become decoupled and with a given value of s any value of m is possible. For a finite value of $\frac{1}{R} = \lambda$ the mass is determined by the spin and the mass differences are of the same order as λ. Because of the $s(s+1)$ dependence of the mass square (10), we are tempted to call the physical system that is characterized by the eigenvalue α^2 of the above group SO(4,1) a relativistic rotator. In order to show that this physical system is actually a relativistic generalization of the rotator, we will now take the non-relativistic limit, $1/c \to 0$, and show that we indeed obtain the non-relativistic rotator as a limiting case.

The operators B_μ go to infinity in the non-relativistic limit. Therefore, we consider the dimensionless generators:

$$\overline{B}_\mu = \frac{1}{\lambda c} B_\mu = \frac{1}{\lambda c} P_\mu + \frac{1}{\lambda c} \frac{\lambda}{2Mc} (\{M_{\nu\mu}, P^\nu\} + \{S_{\nu\mu}, P^\nu\}) \tag{11}$$

Here, the splitting of the physical Lorentz generators $J_{\mu\nu}$ into an orbital part $M_{\mu\nu} = (Q_\mu P_\nu - Q_\nu P_\mu)$ and spin part $S_{\mu\nu}$ (which is <u>not</u> the spin)

$$J_{\mu\nu} = M_{\mu\nu} + S_{\mu\nu} \tag{12}$$

was used. This prepares us to introduce the operator:

$$d_\mu = S_{\mu\nu} \hat{P}^\nu \frac{1}{cM} = \frac{-1}{2(Mc)^2} \{S_{\nu\mu}, P^\nu\} \tag{13}$$

In the classical relativistic case, d_μ is the vector from the position Q_μ (position of charge) to the center of mass Y_μ (CM position). For the classical relativistic rotator, Y_μ describes a straight world line in the direction of the momentum P_μ, and d_μ is the vector from the (charge) position Q_μ to the CM position Y_μ. ($\vec{d}$) rotates around a straight world line giving rise to the spiral motion of Q around the straight world line of the CM position Y_μ, the so-called Zitterbewegung.[8] (See diagram on following page, where $\dot{\vec{d}}$ is the derivative of $\vec{d}$ with respect to τ, the parameter labelling the events along the world line.)

With (12) and (13) we can rewrite $\overline{B}_\mu$ in the following way:

$$\overline{B}_\mu = \frac{1}{\lambda c} P_\mu + \frac{\lambda}{2Mc} \frac{1}{\lambda c} \{(Q_\nu P_\mu - Q_\mu P_\nu), P^\nu\} - Md_\mu \tag{14}$$

For the space part, this can be written, after some calculation, in the following form:

$$\overline{B}_m = \frac{1}{\lambda c} P_m + \frac{1}{2Mc^2} \{Q_\nu P_m, P^\nu\} - \frac{1}{c} \frac{1}{2Mc^2} P_m - \frac{1}{2Mc^2} \{M^2 c^2, Q_m\} - Md_m$$

As

$$\frac{1}{\lambda c} P_m \rightarrow 0 , \quad \frac{1}{2Mc^2} \{Q_\nu P_m, P^\nu\} \rightarrow 0 \quad \text{and} \quad \frac{1}{c} \frac{1}{2Mc^2} P_m \rightarrow 0$$

we see that:

$$\overline{B}_m \rightarrow -MQ_m - Md_m^{(\infty)} = -MY_m \qquad \text{for} \quad c \rightarrow \infty \tag{15}$$

where

$$d_m \xrightarrow{c \rightarrow \infty} d_m^{(\infty)} \neq 0$$

exists and is $\neq 0$. Thus, $-\overline{B}_m$ contracts in the non-relativistic limit into something like the CM position multiplied by the mass. For the time part of $\overline{B}_\mu$ one sees that:

$$\overline{B}_0 \rightarrow -\frac{1}{\lambda} M = -RM \tag{16}$$

where M is the central element of the Galilei group into which $1/c^2(M^2c^2) = 1/c^2 \, P_\mu P^\nu$ goes in the non-relativistic contraction limit. It has an arbitrary value which we choose to be m = constant. (15) then shows that the space components of the deSitter boosts contract into the CM positions. The deSitter boosts are therefore in a certain sense the relativistic generalizations of the CM position.

The details of the contraction process are the following:

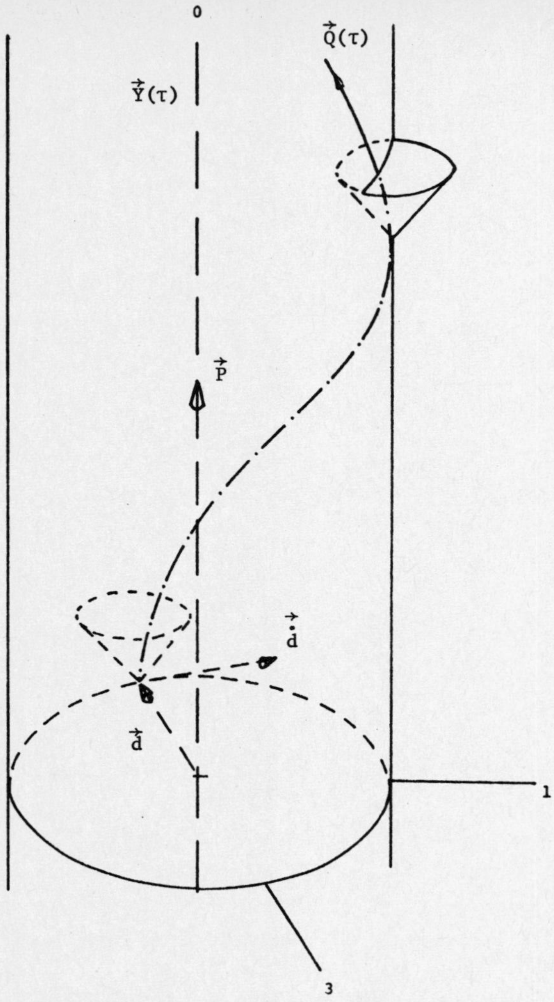

relativistic rotator SO(4,1)	non-relativistic rotator	
$SO(3,1)_{J_{\mu\nu}}$	$\rightarrow E(3)_{G_i J_i}$	
$[J_i,\bar{B}_j] = i\varepsilon_{ijk}\bar{B}_k$	$\rightarrow [M_i + S_i, Q_j + d_j^{(\infty)}] = i\varepsilon_{ijk}(Q + d)_k$	(17a)
$\frac{1}{c} K_i = \frac{1}{c} M_{i0} + \frac{1}{c} S_{i0}$	$\rightarrow G_i = Q_i M + g_i$	(17b)
$d_i = S_{i0}(\frac{1}{c} H + Mc)\frac{1}{c^2M^2} + S_{i\ell}P^{\ell} \frac{1}{c^2M^2}$	$\rightarrow g_i \frac{1}{M} = d_i^{(\infty)}$	(17c)
$[K_i,\bar{B}_j] = i\delta_{ij}\bar{B}_0$	$\rightarrow [G_i, Q_j + d_j^{(\infty)}] = 0$	(17d)
$[\bar{B}_i,\bar{B}_j] = \frac{1}{\lambda^2 c^2} i\lambda^2 J_{ij}$	$\rightarrow [Q_i + d_j^{(\infty)}, Q_j + d_j^{(\infty)}] = 0$	(17e)
$[\bar{B}_\mu, P_\nu] = i \frac{\lambda}{\lambda c} Mc(g_{\mu\nu} - \frac{P_\mu P_\nu}{M^2c^2})$	$\rightarrow [-MQ_m - Md_m, P_n] = iMg_{mn}$	(17f)
	$[Q_m, P_n] = i\delta_{mn}$	(17g)

i.e.

$$SO(4)_{\overline{B}_j J_i} \xrightarrow{1/c \to 0} E(3)_{G_i S_i}$$

Thus the two subgroups $SO(3,1)_{J_{\mu\nu}}$ and $SO(4)_{\overline{B}_i J_i}$ contract into the same group $E(3)_{G_i J_i} \subset G$.

In the $1/c \to 0$ limit we also rediscover in $E(3)_{d_i^{(\infty)}, S_i}$ the spectrum generating group $(\tilde{7})$ of the (translation free) non-relativistic rotator. It is interesting to notice that the dipole operators in the non-relativistic limit $D_i = d_i^{(\infty)}$ were obtained from non-commuting relativistic counterparts d_μ of (13).

According to (15), (17) and (4)

$$\frac{1}{c\lambda} \vec{B} = \vec{\overline{B}} \to - \vec{G} \qquad \frac{1}{c} \vec{K} \to \vec{G} \tag{18}$$

and consequently one obtains for the Casimir operator of SO(4,1) in the non-relativistic contraction limit:

$$\frac{\lambda^2}{c^2} C = \lambda^2 \overline{B}_0^2 - \lambda^2 \vec{\overline{B}}^2 + \frac{\lambda^2}{c} \vec{K}^2 - \frac{1}{2} \frac{\lambda^2}{c^2} L_{ij} L^{ij} \xrightarrow{1/c \to 0} M^2 \tag{19}$$

i.e. the Casimir operator of $SO(4,1)_{B_\mu L_{\mu\nu}}$ goes into the mass of the Galilei group.

In order that the K_i and $\overline{B}_i$ do not go into the zero operator in the contraction limit $1/c \to 0$

one must again go through a sequence of representations

$$\alpha^2 \to \infty \; (\lambda^2 \alpha^2 \to \infty) \text{ such that } \frac{1}{c^2} (\lambda^2 \alpha^2) \to m^2 = \text{finite eigenvalue of } M^2. \tag{20}$$

In this contraction process the difference $\lambda^2 \alpha^2 - c^2 m^2 \to 0$ or any finite value, for instance 'a'.

The mass formula (10) was obtained from the matrix elements of the SO(4,1) Casimir operator given in (8i):

$$\lambda^2 C = P_\mu P^\mu + \frac{9}{4} \lambda^2 - \lambda^2 \hat{W} \tag{8i}$$

with the constraint relation (9). In order to take the $1/c \to 0$ limit of this relation, we now write it out in detail.

$$\lambda^2 C = \frac{1}{c^2} H^2 + 2HM + (Mc)^2 - \vec{P}^2 + \lambda^2 \hat{w}^{\circ 2} - \lambda^2 \vec{\hat{w}}^2 + \frac{9}{4} \lambda^2 \tag{21}$$

According to (19):

$$\lambda^2 C - (Mc)^2 \to 0$$

and according to (4):

$$\frac{1}{c^2} H^2 \to 0 \qquad \overset{\wedge\circ}{w}{}^2 \to 0 \qquad \vec{w}^{\,2} \to \vec{S}^2$$

we obtain as the $1/c \to 0$ contraction limit of equation (20):

$$0 = 2HM - \vec{P}^2 - \lambda^2\vec{S}^2 + \frac{9}{4}\lambda^2$$

or

$$H = \frac{\vec{P}^2}{2M} + \lambda^2 \frac{\vec{S}^2}{2M} \left(-\frac{9}{4}\lambda^2\right) .$$

This is the energy operator (6) for the non-relativistic rotator.

Thus, the model given by SO(4,1), (and specified by the value α^2 of its second order Casimir operator) has the desired property of the model for a relativistic extended rotating object.

relativistic rotator $\xrightarrow{1/R \to 0}$ relativistic elementary particle (cm,s)

relativistic rotator $\xrightarrow{1/c \to 0}$ non-relativistic rotator $H = \frac{\vec{P}^2}{2M} + \frac{\lambda^2}{2M}\vec{S}^2$

Therefore we have justified the name relativistic rotator for the physical system described by this model. Whether this model is realized in nature, i.e. if such a physical system really exists, is an entirely different question which can only be answered by experiment, and is not the subject of this talk. Here it must suffice to mention that the resonance masses are more satisfactorily described by the mass-spin relation (10) than by linearly rising Regge trajectories and the value of $R = 1/\lambda$, determined from the experimental data, is $\approx\frac{1}{3} \times 10^{-13}$ cm. The theoretical support in favor of the relativistic rotator comes from the correspondence given by the Inönü-Wigner contraction process. Since the non-relativistic contraction limit gives the usual non-relativistic rotator which is realized by numerous quantum physical systems in molecular and nuclear physics, there is no reason why its relativistic generalization should not be realized in relativistic physics.

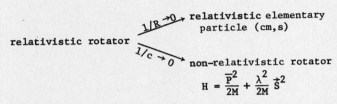

References

1. I. E. Segal, Duke Math. J. 18, 221 (1951).
2. E. Inönü, E. P. Wigner, Proc. N. A. S. 39, 510 (1953).

3a. There is a third correspondence, which I shall not discuss here, but which is very helpful in constructing the model of the relativistic quantum rotator. This is the classical limit establishing the connection to the classical relativistic rotator models of Takabayashi[4]; Mukunda, Biedenharn, van Dam[5]; Regge, Hanson.[6]

3. A. Bohm, Quantum Mechanics, Ch. III, Springer-Verlag, New York (1979).
4. T. Takabayashi, Prog. Theor. Phys. Suppl. 67, 1 (1979), and references therein.
5. N. Mukunda, H. van Dam and L. C. Biedenharn, Phys. Rev. D28, 1938 (1980).
6. A. J. Hanson, T. Regge, Annals of Phys. 87, 498 (1974).
7. F. Rohrlich, Nucl. Phys. B112, 177 (1978);
   H. S. Green, Aust. J. Phys. 29, 483 (1976);
   L. P. Staunton, Phys. Rev. D13, 3269 (1976);
   A. Bohm, Phys. Rev. 175, 1767 (1968);
   H. Bacry, J. Math. Phys. 5, 109 (1964);
   R. J. Finkelstein, Phys. Rev. 75, 1079 (1949);
   H. S. Snyder, Phys. Rev. 71, 38 (1947).
8. H. C. Corben, "Classical and Quantum Theories of Spinning Particles," Ch. II.8, Holden Day Inc., 1968.

Acknowledgement: Support from the U.S. DOE and the Alexander von Humboldt Foundation is gratefully acknowledged .

SYMMETRY BREAKING IN THE SPECTRUM GENERATING GROUP AND ITS EXPERIMENTAL TESTS

P.Kielanowski* and P.Magnollay

Center for Particle Theory
The University of Texas at Austin
Austin, Texas 78712
U.S.A.

ABSTRACT

The theoretical assumptions of the Spectrum Generating Groups (SGG) are explained in detail and then the theory is applied to the hyperon semileptonic decays and the hyperon magnetic moments. It is shown that all experimental data can be explained in the SGG frame but there are some serious discrepancies in the standard Cabibbo model. It is explained how the mass splitting in the octet is taken consistently into account by the SGG.

I- THEORETICAL INTRODUCTION

The idea of using the symmetries of a system in order to study its properties has been vastly employed in physics. In atomic physics, the symmetry group has been used to derive the selection rules and degeneracies of the energy levels. In particle physics the discovery of the SU(2) and SU(3) groups gave deep insight into the spectrum and many other properties of particles. However the SU(3) group is not a symmetry group of the system; the particle masses are not strictly constant within each multiplet, and therefore the four-momentum P cannot commute with the generators of SU(3). Nevertheless it still seems possible to use the irreducible representations of SU(3) in order to describe the internal quantum numbers if we reinterpret the meaning of the SU(3) group in particle physics. This new interpretation assumes that SU(3) is a spectrum generating group (SGG)[1]. A SGG model is a relativistic quantum mechanics based on the following general assumptions:

1- Invariance with respect to the homogeneous Lorentz group.

*On leave of absence from the University of Warsaw, Poland and Centro de Investigacion y de Estudios Avenzados del IPN, Mexico.

analysis of the magnetic moments and semileptonic hyperon decay data.

II- DERIVATION OF THE PHYSICAL QUANTITIES FROM THE THEORY

IIa) Hyperon Semileptonic Decays

Since $\hat{P}_\mu$ commutes with the SU(3) generator, our eigenvectors are going to be labeled by $\hat{p}_\mu$ in the SGG approach.

The semileptonic decay of a baryon B will be described by the following transition matrix element:

$$<\hat{p}'\alpha';p_\ell \ell;p_\nu \nu|T|\hat{p}\alpha> = \frac{G}{\sqrt{2}}\,\bar{u}(p_\ell)\gamma^\mu(1-\gamma^5)u(p_\nu)<\hat{p}'\alpha'|J_\mu|\hat{p}\alpha> \qquad (1)$$

where $\alpha = (I, I_3, Y)$ are the SU(3) quantum numbers, ℓ can be an electron or a muon.

J_μ is the Cabibbo current[3]:

$$J_\mu = \cos\theta(V_\mu^{\pm 1} - A_\mu^{\pm 1}) + \sin\theta(V_\mu^{\pm 2} - A_\mu^{\pm 2}) \qquad (2)$$

The $\cos\theta$ term is the strangeness conserving part and the $\sin\theta$ term is the strangeness changing part.

Since we have assumed that we have a complete set of commuting observables, the eigenvectors $|\hat{p}\alpha>$ are well defined octet basis vectors (which is not the case in the Cabibbo theory using the eigenvectors $|p\alpha>$). We can therefore simplify the equation (1) by use of the Wigner-Eckart theorem[4].

$$<\hat{p}'\alpha';p_\ell \ell;p_\nu \nu|T|\hat{p}\alpha> = \frac{G}{\sqrt{2}}\bar{u}(p_\ell)\gamma^\mu(1-\gamma^5)u(p_\nu)\left\{\begin{array}{l}\cos\theta \text{ for } \Delta S = 0\\ \sin\theta \text{ for } \Delta S = 1\end{array}\right\} \times$$

$$\times \sum_{\gamma=1,2} C(\gamma,\alpha\beta\alpha')<\hat{p}'\|V_\mu - A_\mu\|\hat{p}>_\gamma \qquad (3)$$

$C(\gamma,\alpha\beta\alpha')$ are the SU(3) Clebsch-Gordan coefficients

$\beta = +1, -1, +2$ or -2 and $\gamma = F$(antisymmetric) or D(symmetric)

2- The states are classified according to the SU(3) group. $|...II_3Y>$ span an octet representation.

3- Observables have adequate transformation properties under the Poincaré group extended by parity P, time reversal T and charge conjugation C.

4- There are eight Lorentz octet current operators V_μ^α and A_μ^α ($\alpha=\pm1,\pm2,\pm3,0,8$) which have definite transformation properties under C, P, T, the Lorentz group and SU(3).

The indices +1(-1), +2(-2), +3(-3) refer to the raising (lowering) currents of the third component of I-spin, V-spin, and U-spin, respectively. The following assumptions:

$$A_T V_\mu^\beta A_T = \varepsilon(\mu) V_\mu^\beta \qquad A_T A_\mu^\beta A_T = \varepsilon(\mu) A_\mu^\beta \qquad \varepsilon(\mu) = \begin{cases} +1 & \mu=0 \\ -1 & \mu=1,2,3 \end{cases}$$

and

$$V_\mu^{\beta+} = V_\mu^{-\beta} \qquad A_\mu^{\beta+} = A_\mu^{-\beta}$$

lead to "first class" currents.

5- CVC tells us that the weak vector current and the electromagnetic current are constructed from the same octet operator. The interaction Hamiltonian has the usual product form: $(\text{Hadronic})_\mu(\text{leptonic})^\mu$.

6- The generators E_α, H_i commute with the four velocity operator[2]

$$\hat{P}_\mu = \frac{P_\mu}{M} \qquad [\hat{P}_\mu, SU(3)] = 0.$$

SU(3) would be a symmetry group if the assumption $[\hat{P}_\mu, SU(3)]=0$ was replaced by $[P_\mu, SU(3)] = 0$ or $[P_\mu P^\mu = M^2, SU(3)] = 0$ which is obviously not the case.

Assumptions 1 to 6 form a set of axioms that are sufficient to calculate the one particle matrix elements of a current operator. This theory then gives a number of predictions for the static properties of mesons and baryons. Applied to the baryons, the theory allows an

$$<\hat{p}'\|V_\mu - A_\mu\|\hat{p}>_\gamma = (m_B m_{B'})^{3/2}\bar{u}(\hat{p}')\left[F_1^\gamma(\hat{q}^2)\gamma_\mu + F_2^\gamma(\hat{q}^2) i\sigma_{\mu\nu}\hat{q}^2 + F_3^\gamma(\hat{q}^2)\hat{q}_\mu + G_1^\gamma(\hat{q}^2)\gamma_\mu\gamma_5 + G_2^\gamma(\hat{q}^2)\, i\sigma_{\mu\nu}\hat{q}^2\gamma_5 + G_3^\gamma(\hat{q}^2)\hat{q}_\mu\gamma_5\right]u(\hat{p})$$

where

$$\hat{q}_\mu = \frac{P'_\mu}{m_{B'}} - \frac{P_\mu}{m_B}\ .$$

The "first class" condition tells us that $F_3^\gamma = G_2^\gamma = 0$. Hence

$$<\hat{p}'\|V_\mu - A_\mu\|\hat{p}>_\gamma = (m_B m_{B'})^{3/2}\bar{u}(\hat{p}')\left[F_1^\gamma(q^2)\gamma_\mu + F_2^\gamma(q^2) i\sigma_{\mu\nu}\hat{q}^2 + G_1^\gamma(\hat{q}^2)\gamma_\mu\gamma_5 + G_3^\gamma(\hat{q}^2)\hat{q}_\mu\gamma_5\right]u(\hat{p}) \qquad (4)$$

It is important to notice that the terms containing $\hat{q}_\mu$ can be easily expressed in terms of γ_μ, $i\sigma_{\mu\nu}q_\nu$ and q_μ. From the resulting expression it is easy to see that the SGG introduces "induced" second class current in the Cabibbo theory and explicitly shows the mass dependence of the standard form factor $f_i^\gamma(q^2)$ and $g_i^\gamma(q^2)$. We have for instance[5]:

$$g_1^\gamma(q^2) = G_1^\gamma(\hat{q}^2) - \frac{(m_B - m_{B'})^2}{2m_B m_{B'}}\, G_3^\gamma(\hat{q}^2) \qquad (5)$$

$$g_2^\gamma(q^2) = -\frac{(m_B - m_{B'})}{2m_B m_{B'}}\, G_3^\gamma(\hat{q}^2)$$

The striking result and prediction of the SGG is to introduce new terms proportional to the mass differences. Also we should notice that the correction to $g_1^\gamma(q^2)$ is much more important than the fact that $g_2^\gamma(q^2) \neq 0$. Actually we discovered that the correction to $g_1^\gamma(q^2)$, namely $-\frac{(m_B - m_{B'})^2}{2m_B m_{B'}}\, G_3^\gamma(\hat{q}^2)$, is large enough to change the sign of α_e for $\Sigma^- \to n\, e_\nu$.

Using assumption 5, the vector form factors can be determined from the charges and magnetic moments of the neutron and proton.

$$F_1^{\gamma=1} = \sqrt{6} \quad , \quad F_1^{\gamma=2} = 0$$

$$F_2^{\gamma=1} = (\frac{\mu_p - 1}{2} + \frac{\mu_n}{4}) \sqrt{6} = 1.02 \qquad (6)$$

$$F_2^{\gamma=2} = \mu_n \frac{\sqrt{30}}{4} = -2.61.$$

The value $F_1^{\gamma=1} = \sqrt{6}$ depends on our normalization of the Clebsch-Gordan coefficient[8]. Because of the shrinking of experimental errors corrections have to be included. In order to reduce the high value of χ^2, it seems that we are forced to introduce higher correction terms in our theory, such as: the higher multiplet expansion in the axial current[7,8]. From the reduction {8} x {8} we can see that we have none-zero contributions for {10}, {10*} and {27}-plet currents. We have introduced higher multiplet (10,10*,27) terms in the current as well as q^2-dependence of the form factors and radiative corrections.

Fits of the Cabibbo model and of the SGG model are presented in Table I. The SGG model shows no discrepancy and gives a fit with a total χ^2 of 12.10 for 15 degrees of freedom, which corresponds to a confidence level of 67 %.

IIb) Magnetic Moments of Baryons

The electromagnetic current is constructed from the vector currents V_μ^0 and V_μ^8 and takes the following U-spin scalar form (Gell-Mann-Nishijima current):

$$J_\mu^{el} = \frac{1}{\sqrt{2}} (V_\mu^0 + \frac{1}{\sqrt{3}} V_\mu^8) \qquad (1)$$

in our normalization.

We are interested in calculating the matrix elements of this current between two eigenvectors of the same baryon B. Again, using assumptions 1 to 6, we get:

$$\langle \hat{p}'_\alpha | J_\mu^{el} | \hat{p}\alpha \rangle = \sum_{\gamma=1,2} C(\gamma,\alpha\ el\alpha) \langle \hat{p}' \| J_\mu \| \hat{p} \rangle_\gamma \qquad (2)$$

where

$$\langle \hat{p}' \| J_\mu \| \hat{p} \rangle_\gamma = (m_B m_{B'})^{3/2} \frac{1}{\sqrt{2}} \bar{u}(\hat{p}') [F_1^\gamma(\hat{q}^2)\gamma_\mu + F_2^\gamma(\hat{q}^2) i\sigma_{\mu\nu}\hat{q}^2] u(\hat{p}) \qquad (3)$$

and

$$C(\gamma,\alpha\ el\alpha) = C(\gamma,\alpha\ 0\alpha) + \frac{1}{\sqrt{3}} C(\gamma,\alpha\ 8\alpha) \tag{4}$$

The magnetic moments are given by a linear combination of the form factors $F_1^\gamma(0)$ and $F_2^\gamma(0)$. In the SGG approach, the magnetic moment of the baryon B (described by the quantum number α) is given by:

$$\mu_B = \frac{m_P}{m_B} \sum_{\gamma=1,2} C(\gamma,\alpha\ el\alpha) \left[\frac{1}{\sqrt{2}} F_1^\gamma(0) + \frac{2}{\sqrt{2}} F_2^\gamma(0)\right] \tag{5}$$

where m_p is the mass of the proton and μ_B is in nuclear magnetons.

The magnetic moments in intrinsic magnetons, $m_B\mu_B$, are given by the group transformation property, i.e. they can be expressed in terms of products of C.G. coefficients and reduced matrix elements. Thus the assumption of a well-defined tensor character for the electromagnetic current (1) leads to a well-defined tensor character of $m_B\mu_B$ in the SGG approach.

The values of $F^\gamma_{1,2}(0)$ are determined by the charges and the magnetic moments of the nucleons (see equation (6) of part IIa). Therefore free parameters do not have to be put in the theory and the magnetic moments can be calculated from the group properties.

Unfortunately, the present data on magnetic moments are not compatible with equation IIb(5)[9]. In order to obtain acceptable agreement with experiment, one has to add correction terms to the octet current. Namely, one has to extend the Gell-Mann-Nishijima current to higher multiplets as well, since the electromagnetic current is a U-spin scalar, we take, for higher multiplets, a combination of currents that will be U-spin scalars. The generalized ansatz reads:

$$J_\mu^{el} = \frac{1}{\sqrt{2}} \left[V^0 + \frac{1}{\sqrt{3}} V_\mu^8 + V_\mu^{singlet}(U=0) + V_\mu^{27\text{-}plet}(U=0)\right] \tag{6}$$

where

$$V_\mu^{27\text{-}plet}(U=0) = \frac{5}{\sqrt{15}} V_\mu^{27\text{-}plet}(I=2,I_3=0,Y=0) + V_\mu^{27\text{-}plet}(I=1,I_3=0,Y=0)$$
$$- \frac{1}{\sqrt{3}} V_\mu^{27\text{-}plet}(I=0,I_3=0,Y=0) \tag{7}$$

The charges are already given by the octet current and therefore the higher multiplet form factors multiplying γ_μ for the baryon states must be zero. So all the F form factors vanish, except $F_1^{\gamma=1}$ which is equal to $\sqrt{6}$ in our normalization. The F_2 form factors are numerically determined by the fits. A fit of the magnetic moments is given in Table II.

III. CURRENT EXPERIMENTAL SITUATION

In the last two years, new experiments have been performed. Hyperon magnetic moments were measured to a very high accuracy and the semileptonic decay data were significantly improved. Because of this we are trying to test the theory to a high accuracy and to discover new effects, such as the fine structure effects of baryons. Such fine structure effects come from (1) the mass differences in the multiplets and (2) the admixture of other multiplets with the usual octet current.

The values of the magnetic moments are now available for all the hyperons of the octet, except Σ^o. For the nucleons the relative error is $\sim 6 \cdot 10^{-7}$ and no theory exists which can predict these values with such an accuracy. Therefore we use a theoretical error of 2 % for the magnetic moments of the nucleons, in order to avoid huge contributions to the χ^2.

The most significant measurement for the decays is α_e for $\Sigma^- \to ne\nu$. Since this value gives some important information concerning the sign of g_1/f_1 for this decay, it is important to notice that the mass difference $(m_{\Sigma^-} - m_n)$ is the largest one among all known hyperon decays. The new value for the $\Sigma^- \to \Lambda e\nu$ rate is also significant, because it gives, for the first time, an indication of the presence of symmetry breaking in the hyperon semileptonic decays.

IV- DISCREPANCIES IN THE HYPERON MAGNETIC MOMENTS AND SEMILEPTONIC DECAYS

Remarkably, the most serious discrepancy between the Cabibbo model and experiment is in α_e for $\Sigma^- \to ne\nu$. The predicted value in the Cabibbo model is -0.65 whereas the current experimental value is positive (0.26 ± 0.19). The leading term in the expression of α_e is proportional to $g_1 . f_1$ with the positive sign. The Cabibbo model predicts the value of -0.35 for $g_1 . f_1$ and therefore α_e is expected to be negative in this model. The SGG value of $g_1.f_1$ is 0.53 and forces the value of α_e to be consistent with the experimental value. The advantage of the SGG, as we have already seen, is that the sign change is explained very naturally in the SGG approach by the mass difference $(m_{\Sigma^-} - m_n)$.

The two fits of the magnetic moments, respectively, give a χ^2 of 196.9 and 9.7 for the Cabibbo model and the SGG model. Even with 2% error on the nuclear magnetic moments we get a χ^2 of 120.8 for the neutron. The SGG model does not show any discrepancies.

The largest contribution comes from the octet operator, so the Gell-Mann-Nishijima current is a good first approximation up to 20 % accuracy. The corrections of higher multiplets are about 1/10 of the octet terms.

Finally one can draw the following conclusions from the fits:

1- The fits of the SGG model always have a much better χ^2 than the standard Cabibbo model.

2- The SGG model very well explains and reproduces the experimental value for α_e in $\Sigma^- \to ne\nu$. The Cabibbo model predicts the opposite sign.

3- The baryon magnetic moments should be taken in intrinsic rather than nuclear magnetons, for both theoretical and phenomenological reasons.

4- In intrinsic magnetons, the magnetic moments are roughly described by the old SU(3) assumption with the Gell-Mann-Nishijima form for the electromagnetic current.

5- The SGG model takes the mass differences of hyperons consistently into account.

V- OUTLOOK

The new interpretation of the SU(3) group can predict one particle matrix elements of operators with definite transformation properties with respect to the SU(3) group. It therefore can be applied to processes where such matrix elements are present, e.g., static properties of particles or deep inelastic scattering. We also think that the SGG interpretation could be the basis of the unification of the mesons and baryons in the sector of the semileptonic decays.

The outstanding problem is the extension of our interpretation to the charmed particles. This extension, theoretically very simple and natural, cannot be compared with present experimental data. The measurements of magnetic moments and semileptonic decays of charmed particles will give new arguments in favor of one of the two interpretations of SU(4).

ACKNOWLEDGMENTS

The authors are indebted to A. Bohm for valuable discussions and considerable help on this paper. The work was supported in part by the U.S. Department of Energy.

TABLE CAPTIONS

Table I- Fits of hyperon semileptonic decays with correlation coefficients and asymmetries included. Radiative corrections and q^2-dependence of form factors have been taken into account. Decay rates are in 10^6 sec^{-1} except for neutron decay which is in 10^{-3} sec^{-1}.

Table II- Magnetic moments. We have 9 data and 4 parameters ($F_2^{\gamma=1,2}F^{27}$ and F^S) so that the number of the degrees of freedom n_D is 4. The errors of the experimental values are given by $\Delta\mu = \sqrt{(\Delta\mu_{exp})^2 + (0.02\ \mu_B)^2}$, where $\Delta\mu_{exp}$ is the experimental error and 0.02 μ_B is the theoretical error.

TABLE 1

Process	Experimental Value	Cabibbo		SGG	
		Predicted Value	Contribution to χ^2	Predicted Value	Contribution to χ^2
$n \to pe\nu$ (rate)	1.091 ± 0.017	1.083	0.221	1.072	1.282
$\Sigma^+ \to \Lambda e\nu$ (rate)	0.253 ± 0.059	0.229	0.171	0.232	0.126
$\Sigma^- \to \Lambda e\nu$ (rate)	0.378 ± 0.018	0.380	0.007	0.380	0.013
$\Lambda \to pe\nu$ (rate)	3.165 ± 0.053	3.164	0.001	3.152	0.061
$\Sigma^- \to ne\nu$ (rate)	7.085 ± 0.194	7.079	0.001	7.092	0.001
$\Xi^- \to \Lambda e\nu$ (rate)	3.244 ± 0.218	3.206	0.030	3.193	0.054
$\Xi^- \to \Sigma^o e\nu$ (rate)	0.524 ± 0.122	0.609	0.481	0.654	1.127
$\Lambda \to p\mu\nu$ (rate)	0.597 ± 0.133	0.601	0.001	0.555	0.098
$\Sigma^- \to n\mu\nu$ (rate)	3.086 ± 0.271	3.248	0.356	3.257	0.400
$\Xi^- \to \Lambda\mu\nu$ (rate)	1.580 ± 1.580	0.975	0.147	0.884	0.194
$n \to pe\nu$ ($\alpha_{e\nu}$)	−0.074 ± 0.004	−0.076	0.206	−0.077	0.481
$n \to pe\nu$ (α_e)	−0.084 ± 0.003	−0.084	0.007	−0.085	0.237
$n \to pe\nu$ (α_ν)	1.001 ± 0.038	0.988	0.110	0.988	0.116
$\Sigma^- \to \Lambda e\nu$ ($\alpha_{e\nu}$)	−0.412 ± 0.062	−0.412	0.000	−0.482	1.276
$\Sigma^+ \to \Lambda e\nu$ ($\alpha_{e\nu}$)	−0.4 ± 0.18	−0.404	0.001	−0.461	1.276
$\Sigma^- \to ne\nu$ ($\alpha_{e\nu}$)	0.279 ± 0.032	0.306	0.691	0.291	0.135
$\Sigma^- \to ne\nu$ (α_e)	0.26 ± 0.19	−0.647	22.767	0.076	0.939
$\Lambda \to pe\nu$ (α_e)	0.125 ± 0.066	0.018	2.635	0.064	0.866
$\Lambda \to pe\nu$ ($\alpha_{e\nu}$)	−0.009 ± 0.019	−0.004	0.083	−0.013	0.088
$\Lambda \to pe\nu$ (α_ν)	0.821 ± 0.06	0.973	6.405	0.943	4.165
$\Lambda \to pe\nu$ (α_p)	−0.508 ± 0.065	−0.580	1.239	−0.542	0.271
$\Xi \to \Lambda e\nu$ (A)	0.604 ± 0.120	0.599	0.002	0.577	0.052
$\Sigma^- \to \Lambda e\nu$ (A)	0.065 ± 0.072	0.054	0.023	0.043	0.093
$\Sigma^- \to \Lambda e\nu$ (B)	0.853 ± 0.070	0.898	(0.421)	0.911	(0.675)
TOTAL χ^2/n_D			35.59/15		12.19/15

TABLE II

Particle	Experimental Value of β_B	Cabibbo Model		SGG Model	
		Predicted Value	χ^2	Predicted Value	χ^2
p	2.792 ± 0.056	2.757	0.40	2.807	0.07
n	-1.913 ± 0.038	-1.492	120.80	-1.836	4.04
Λ	-0.614 ± 0.013	-0.612	0.01	-0.613	0.00
Σ	2.330 ± 0.138	2.546	2.45	2.215	0.69
Σ^o	(not measured)	1.147	0.00	1.494	0.00
Σ^-	-0.890 ± 0.141	-0.834	0.16	-0.839	0.13
Ξ^o	-1.250 ± 0.029	-1.492	71.61	-1.312	4.68
Ξ^-	-0.750 ± 0.062	-0.760	0.03	-0.76	0.03
$\Sigma\Lambda$	1.820 ± 0.253	1.524	1.37	1.756	0.06
χ^2/n_D			196.9/4		9.71/4

REFERENCES

1- A.Bohm, Nuovo Cim. Lett. 11, 13 (1974);
A.Bohm and J.Werle, Nucl. Phys. B106, 165 (1976);
A.Bohm, Phys. Rev. D13, 2110 (1976).

2- This relation has been suggested by J.Werle, "On a symmetry scheme described by a non-Lie algebra", ICPT preprint, Trieste, 1965.

3- N.Cabibbo, Phys. Rev. Lett. 10, 531 (1963).

4- For momentum eigenstates the Wigner theorem cannot be applied.

5- A.Bohm and R.B.Teese, J. Math. Phys. 18, 1434 (1977);
A.Bohm *et al*., Phys. Rev. D15, 689 (1977).

6- J.G.Kuriyan *et al*., J. Math. Phys. 6, 722 (1965).

7- S.Pakvasa, A.McDonald and S.P.Rosen, Phys. Rev. 181, 1948 (1969).

8- A.Garcia and P.Kielanowski, Phys. Rev., to be published;
A.Bohm *et al*., Phys. Rev., to be published.

9- A.Bohm and R.Teese, Phys. Rev. D26, 1103 (1982).

CONSTRUCTION OF THE DYNAMICAL SYMMETRY GROUP OF THE RELATIVISTIC HARMONIC OSCILLATOR BY THE INFELD-HULL FACTORIZATION METHOD

N.M.Atakishiyev
Institute of Physics
Azerbaijan Academy of Sciences
Baku - 370143, USSR

ABSTRACT

For the model of the harmonic oscillator in the relativistic configurational $\vec{r}$ -representation the quantum number ℓ raising and lowering operators are found and the dynamical symmetry group is constructed by the Infeld-Hull factorization method.

1. Introduction

The factorization method based on the replacement of a given second order differential equation with the boundary conditions by the equivalent pair of the first-order differential-difference equations is often applied to solve the quantum-mechanical eigenvalue problems[1]. Originally the factorization method was formulated to simplify the problem of finding the eigenvalues and successive calculation of the normalized eigenfunctions. However, from the group theory point of view it has deeper meaning and therefore in studying many quantum-mechanical systems there can be used the general algebraic approach, generalizing the well-known description of the nonrelativistic oscillator with the help of creation and annihilation operators[2].

In[3,4] an investigation has been made of the quasipotential equation which gives the relativistic generalization of the harmonic oscillator problem. The proposed model of oscillator is formulated in the relativistic configurational $\vec{r}$ -representation[5], in which the Hamiltonian is a finite-difference operator, having $U(3)$ -symmetry. This paper shows that the application of the factorization method even in this nontrivial case enables us to find the orbital quantum number ℓ raising and lowering operators and to construct the dynamical symmetry group.

The theory of dynamical symmetries is based on the fact that the operators, corresponding to the physical system under consideration, can be used to construct different Lie algebras, including not only the Hamiltonian and symmetry operators, but also other raising and lowering operators, which do not commute with the Hamiltonian[6,7]. In consequence of this, the independent eigenfunctions of the Hamiltonian form a basis of the irreducible representation of the dynamical Lie algebra, defined in this way.

2. Relativistic harmonic oscillator

Let us consider the quasipotential model of the relativistic harmonic oscillator with the radial part of the Hamiltonian of the form[3,4]:

$$\tilde{H}_\ell(r) = \frac{1}{2}\left\{ e^{-i\nabla_r} + (1+\omega^2 r^{(2)})\left[1+\frac{\ell(\ell+1)}{r^{(2)}}\right] e^{i\nabla_r} \right\} \tag{1}$$

It is shown in [4] that the Hamiltonian $\tilde{H}_\ell(r)$ is factorized in terms of the operators $b_\ell(r)$ and $b_\ell^+(r)$, i.e.

$$\tilde{H}_\ell(r) = \omega\left[b_\ell^+(r) b_\ell(r) + \gamma + \ell + 3/2 \right], \tag{2}$$

$$b_\ell(r) = \frac{i}{\sqrt{2\omega}}\left\{ e^{-\frac{i}{2}\nabla_r} - \omega(\gamma+ir) e^{\frac{i}{2}\nabla_r}\left(1+\frac{\ell+1}{ir}\right) \right\}, \quad b_\ell^+(r) = \frac{1}{i\sqrt{2\omega}}\left\{ e^{-\frac{i}{2}\nabla_r} - \omega\left(1-\frac{\ell+1}{ir}\right) e^{\frac{i}{2}\nabla_r}(\gamma - ir) \right\}$$

As $c \to \infty$ $\tilde{H}_\ell(r)$ coincides with the nonrelativistic harmonic oscillator Hamiltonian and when $\ell=0$ it corresponds to the exactly solvable relativistic model of the linear oscillator[8]. The calculation of the commutator of $b_\ell(r)$ and $b_\ell^+(r)$ leads to the expression

$$\left[b_\ell(r), b_\ell^+(r) \right] = \frac{1}{2}\left\{ 1 + \frac{(\ell+1)(\gamma+1/2)}{r^2+1/4} + \omega\alpha_\ell(r) e^{i\nabla_r} \right\},$$

$$\alpha_\ell(r) = \gamma + \ell + \frac{5}{4} - (\ell+1)\left(\gamma+\frac{1}{2}\right)\left[\frac{\ell(\gamma-1/2)}{r^{(2)}} - \frac{(\ell+1)(\gamma+1/2)}{(r+i/2)^{(2)}} \right]. \tag{3}$$

Using (3) it is now possible to define via $b_\ell(r)$ and $b_\ell^+(r)$ both the integrals of motion and the orbital quantum number ℓ raising and lowering operators. Indeed, it is easy to verify that the expression

$$B_\ell(r)=\sqrt{\frac{\omega}{2}}\,r\left\{e^{-\frac{i}{2}\nabla_r}b_\ell(r)-\frac{i}{\sqrt{2\omega}}\left[\widetilde{H}_\ell(r)-\omega\left(\gamma+\ell+\frac{3}{2}\right)\right]\right\}-\frac{1}{4\omega}\left\{\widetilde{H}_\ell(r)\left[\widetilde{H}_\ell(r)-\omega\right]-\omega^2\left(\gamma+\ell+\frac{1}{2}\right)\left(\gamma+\ell+\frac{3}{2}\right)\right\} \qquad (4)$$

is the integral of motion, i.e.

$$\left[B_\ell(r),\ \widetilde{H}_\ell(r)\right]=2\omega\,B_\ell(r). \qquad (5)$$

The second integral of motion $B_\ell^+(r)$ is obtained by the Hermitian conjugation of (4).

Having defined the generalized momentum operator $\mathcal{P}_\ell(r)$ through the commutator

$$[r,\widetilde{H}_\ell(r)]=i\mathcal{P}_\ell(r),$$

$$\mathcal{P}_\ell(r)=-\operatorname{sh}(i\nabla_r)-\frac{1}{2}\left\{\omega^2\left[\ell(\ell+1)+r^{(2)}\right]+\frac{\ell(\ell+1)}{r^{(2)}}\right\}e^{i\nabla_r}, \qquad (6)$$

it is possible to represent (4) in more compact form:

$$B_\ell(r)=\frac{1}{2}\left\{r\left[\omega r+i\mathcal{P}_\ell(r)\right]-\frac{1}{2\omega}\left[\widetilde{H}_\ell(r)\left(\widetilde{H}_\ell(r)-\omega\right)-\omega^2\ell(\ell+1)-1\right]\right\} \qquad (7)$$

After simple transformations the integral of motion $B_\ell(r)$ takes the final form:

$$B_\ell(r)=\frac{1}{4\omega}\left\{\left[\omega r+i\mathcal{P}_\ell(r)\right]^2-\frac{\ell(\ell+1)}{r^2+1}\right\}. \qquad (8)$$

The orbital quantum number ℓ raising operator $L_\ell(r)$, which is defined by the formula

$$L_\ell(r)\widetilde{H}_\ell(r)-\widetilde{H}_{\ell+1}(r)L_\ell(r)=\omega L_\ell(r), \qquad (9)$$

is expressed via $b_\ell(r)$ in the following manner:

$$L_\ell(r)=\omega e^{\frac{i}{2}\nabla_r}(\gamma-ir)b_\ell(r)+\frac{i}{\sqrt{2\omega}}\left[\widetilde{H}_\ell(r)-\omega(\gamma+\ell+3/2)\right]. \qquad (10)$$

To obtain a closed algebra of the operators $B_\ell(r)$, $B_\ell^+(r)$ and $\widetilde{H}_\ell(r)$ besides (5) it is necessary to calculate the commutator

$$\left[B_\ell(r),B_\ell^+(r)\right]=\frac{\omega}{4}\widetilde{H}_\ell(r)\left\{1+\frac{2}{\omega^2}\left[\widetilde{H}_\ell^2(r)-\omega^2\ell(\ell+1)-1\right]\right\}. \qquad (11)$$

As it follows from (11) to construct the dynamical algebra for the model under consideration it is sufficient to introduce the new operator

$$\tilde{B}_\ell(r) = B_\ell(r)\left\{\left[\tilde{H}_\ell(r)+\omega(\gamma-\ell-\tfrac{3}{2})\right]\left[\tilde{H}_\ell(r)+\omega(\ell-\gamma-\tfrac{1}{2})\right]\right\}^{-1/2} \qquad (12)$$

Now it is easy to check that the operators

$$M_+ = 2\tilde{B}_\ell^+(r), \quad M_- = -2\tilde{B}_\ell(r), \quad M_3 = \frac{1}{2\omega}\tilde{H}_\ell(r) \qquad (13)$$

satisfy the commutation relations

$$[M_3, M_\pm] = \pm M_\pm, \quad [M_+, M_-] = 2M_3 \qquad (14)$$

of the $SU(1,1)$ algebra (or isomorphic Lie algebras $SO(2,1)$, $s\ell(2,R)$).

Direct calculation shows that the Casimir operator in this case is equal to

$$M^2 \equiv M_3(M_3-1) + M_+M_- = \frac{1}{4}(\gamma+\ell+\tfrac{3}{2})(\gamma+\ell-\tfrac{1}{2}). \qquad (15)$$

As for the eigenvalue of M^2 the notation $s(s+1)$ is used and a representation of the group $SU(1,1)$ is characterized by the number s , then from (15) it follows that can take two values: $s_1 = \frac{1}{2}(\gamma+\ell-1/2)$ and $s_2 = -\frac{1}{2}(\gamma+\ell+3/2)$. The first value is discarded, because the corresponding representation is nonunitary. The second one determines the representation $D^+(-\frac{\gamma+\ell}{2}-3/4)$ with the eigenvalues of the operator $M_3 = \frac{1}{2\omega}\tilde{H}_\ell(r)$, which are bounded below and are equal to $-s_2+n = n+\frac{1}{2}(\gamma+\ell+3/2)$, $n=0,1,2,\ldots$. Thus the three-dimensional relativistic oscillator (1) is described by another representation of the same dynamical symmetry group of the nonrelativistic problem.

3. References

1. L.Infeld and T.E.Hull. Rev.Mod.Phys., 23, 21 (1951)
2. R.Hermann. Journ.Math.Phys., 22, 1163 (1981).
3. A.D.Donkov, V.G.Kadyshevsky, M.D.Mateev and R.M.Mir-Kasimov. Theoreticheskaya i mathematicheskaya physica, 8, 61 (1971).
4. N.M.Atakishiyev, R.M.Mir-Kasimov and Sh.M.Nagiyev. Proceedings of the IV International seminar on high energy physics and quantum field theory, vol.2. Protvino, 1981.
5. V.G.Kadyshevsky, R.M.Mir-Kasimov and N.B.Skachkov. Nuovo Cimento, 55A, 233 (1968).
6. A.O.Barut. Dynamical Groups and Generalized Symmetries in Quantum Theory. Christchurch, New Zealand, 1971.
7. I.A.Malkin, V.I.Man'ko. Dynamical symmetries and coherent states of quantum systems, "Nauka", Moscow, 1979.
8. N.M.Atakishiyev, R.M.Mir-Kasimov, Sh.M.Nagiyev. JINR preprint E2-12367, Dubna; Theoreticheskaya i mathematicheskaya physica, 44, 47 (1980).

DYNAMICAL SEMIGROUPS FOR RESONANCES IN RIGGED HILBERT SPACES

by

M.GADELLA

Facultad de Ciencias. Av. los Castros s/n. Santander.
Spain

We here present a recently developed formalism for resonances and virtual scattering states. To a resonance we shall associate a pair of (normalizable) vector states. One of them corresponds to the formation of the resonance and the other to its decay. They are called the Gamow vectors. The corresponding resonant poles of S(E) respectively are the eigenvalues of the total Hamiltonian H with these eigenvectors. To a virtual state we shall associate a pair of (non-normalizable) vector states. Both of them describe the virtual state equally well. The value of the corresponding pole of S(E) is an (generalized) eigenvalue of H, being the vectors for the virtual state being their (generalized) eigenvectors.

We assume that the total Hamiltonian H has an absolutely continuous spectrum only and $\sigma(H) = R^+$, also that S(E) on the second sheet does not grow faster than a polynomial and have any other singularities than simple poles (for the sake of simplicity we here assume that there exists a pair of resonant poles only)

We first have to construct a Gelfand triple $\phi \subset \mathcal{H} \subset \Phi^x$, where $\mathcal{H} = L^2(R^+)$. $\Phi = \Phi_+ + \Phi_-$ where: Φ_- is the set of all restrictions to R^+ of all entire functions which are Hardy-class on the lower half plane (second sheet) and Schwartz on R. Φ_+ is the set of all complex conjugates of Φ_-. Both Φ_+ and Φ_- are dense in $\mathcal{H}$, so Φ is dense in $\mathcal{H}$ too.

In the representation which diagonalizes H, $\Omega^+ \phi^{in} = \phi^+ \varepsilon \Phi_-$ and $\Omega^- \phi^{out} = \phi^- \varepsilon \Phi_+$. We obtain by using complex variable methods that

$$\phi^{+} = \text{background} + a|z_R^{-}> + b|-E_v^{-}>$$

$$\phi^{-} = \text{background} + c|z_R^{*+}> + d|-E_v^{+}>$$

where a,b,c,d are constants. $H|z_R^{-}> = z_R|z_R^{-}>$, $H|z_R^{*+}> = z_R|z_R^{*+}>$ and $H|-E_v> = -E_v|-E_v>$

ϕ^{+} is to be taken as the decaying state. It is the sum of the purely decaying state $|z_R^{-}>$ plus the virtual state $|-E_v^{-}>$ plus an always existing background in decaying processes. Both, $|z_R^{-}>$ and $|-E_v^{-}>$ are functionals on Φ_{-}. The second formula is an analogous decomposition for the growing vector ϕ^{-}.

Take now the dynamical group $U(t) = e^{-itH}$. We can prove that:

$$e^{itH}\ \Phi_{-} \subset \Phi_{-} \text{ if } t \geq 0;\ e^{itH}\ \Phi_{-} \not\subset \Phi_{-} \text{ if } t < 0$$

$$e^{itH}\ \Phi_{+} \subset \Phi_{+} \text{ if } t \leq 0;\ e^{itH}\ \Phi_{+} \not\subset \Phi_{+} \text{ if } t > 0.$$

Then using $<U^*(t)f|\phi> = <f|U(t)\phi>$ where $f \in \Phi\pm$ and ϕ to its dual:

$$U(t)|z_R^{-}> = e^{-itH}|z_R^{-}> = e^{-itE_R}\, e^{-\Gamma/2}|z_R^{-}>$$

and $U(t)|-E_v^{-}> = e^{itE_v}|-E_v^{-}>$ $\qquad z_R = E_R + i\Gamma/2$

for $t \geq 0$ but not for $t < 0$. This process corresponds to the decay of a quasistationary state. For $t \leq 0$ we have analogous formulas corresponding to the formation process.

Therefore, the total dynamical group has split in two semigroups: $U(t)$ with $t \geq 0$ and $U(t)$ with $t \leq 0$, respectively corresponding to the decay and formation of a quasistationary state.

REFERENCES

A.Bohm. J.Math.Phys. 22, 2813 (1981)

M.Gadella, "A Rigged Hilbert Space of Hardy Class Functions: Applications to Resonances", J.Math.Phys. (in press).

M.Gadella, "A Description of Virtual Scattering States in the Rigged Hilbert Space Formulation Of Quantum Mechanics".

EXTREMA OF LANDAU AND HIGGS POLYNOMIALS AND ZEROS OF RENORMALIZATION-GROUP EQUATIONS

Marko V. Jarić

Institut des Hautes Etudes Scientifiques, 91440 Bures-sur-Yvette, France and Freie Universität Berlin, Institut für Theoretische Physik, D-1000 Berlin 33 West-Germany

ABSTRACT

An algebraic technique designed to extract all the symmetry information contained in (non-linear) equations for the zeros of a G-covariant vector field is formulated.

1. Introduction

For a given linear group action $g \in G$ on a carrier space $\mathbb{R}^n$ a G-covariant vector field $f : \mathbb{R}^n \rightarrow \mathbb{R}^n$ satisfies

$$f(gx) = gf(x), \quad \forall g \in G \text{ and } \forall x \in \mathbb{R}^n. \tag{1}$$

Zeros of such fields determine, for example, broken symmetries in Landau and Higgs mechanisms or fixed points in renormalization-group theory of phase transitions. Therefore, there is great interest in developing techniques for determining zeros of G-covariant vector fields. Another paper at this conference [1] deals with the same problem in an invariant formulation (orbit space approach). Here, a carrier space approach is employed. Since this approach was presented with more details elsewhere [2], only its essential features will be outlined in the following section.

2. Carrier space approach

An action of the group G on the carrier space $\mathbb{R}^n$ defines, as is well known [3], certain geometric (algebraic) structures in the space: orbits Ω and strata Σ. These are G-invariant structures. Strata, in particular, are in one to one correspondence with the classes $[L]$ of isotropy subgroups L of G. An isotropy group defines also a geometric (algebraic) structure in the carrier space. This is a linear subspace FixL (since G-action is linear) of points fixed under the action of $L \leq G$. These linear subspaces are a natural generalization of rotation axes and reflection planes. They are simply related to the strata: topological closure of a stratum $\Sigma[L]$, $\cup_{[L] \leq [L']} \Sigma[L']$, is the union of the corresponding linear subspaces over the class $[L]$,

$$\bigcup_{[L'] \geq [L]} \Sigma[L'] = \bigcup_{L \in [L]} \mathrm{Fix} L. \qquad (2)$$

Just as a vector field at a rotation axis or at a reflection plane must be parallel to the axis or the plane, a G-covariant vector field at FixL must be parallel to FixL. This is equivalent to the fact that a G-covariant vector field at a stratum is tangential to the stratum [4].

If we denote by P_L a projector on FixL then the linear equations

$$(1-P_L)x = 0 \tag{3}$$

give n - i(L) independent equations of the plane FixL, where i(L) = dim FixL. The remaining i(L) equations for f(x) = 0 are

$$P_L f(x) = 0 \tag{4}$$

However, since

$$L_s > L \text{ iff } \quad \text{Fix}L_s < \text{Fix}L, \tag{5}$$

the solutions from $\text{Fix}L_s$ are also contained in Eq. (4). If $i(L_S) = i(L)-1$ these solutions may be factored out via

$$(1-P_{L_s})P_L f(x) = g_s(x)(1-P_{L_s})P_L x \tag{6}$$

leading to a system of equations of lower degree

$$g_s(x) = 0, \quad i(L_s) = i(L)-1 \ . \tag{7}$$

Since the group G acts in FixL via the normalizer N(L) of L in G the forms $g_s(x)$ for each class $[L_s]$ transform as the permutation representation of $N(L) \cap N(L_s)$ in N(L).

Finally, when the number m of different L_s is greater than i(L), the factors associated with s = 1,..., i(L) are removed first and then the projection technique, now for the action of N(L) on $g_s(x)$, s = 1, ... i(L), is used to remove solutions associated with L_s, s = i(L)+1,...,m.

In this fashion all the symmetry content is extracted from the equation f(x) = 0.

This work was supported in part by an Alexander von Humboldt fellowship and the Deutsche Forschungsgemeinschaft, Sonderforschungsbereich 161.

* Current address: Department of Physics, Montana State University, Bozeman, Montana 59717, USA

References

1 M.V. Jarić, L. Michel and R.T. Sharp, the same proceedings, p.

2 Marko V. Jarić, Phys. Rev. Lett. 48, 1641 (1982).

3 See for example, L. Michel, Rev. Mod. Phys. 52, 617 (1980).

4 L. Michel, Lecture Notes in Physics 6, 36 (1970).

GROUP CONTRACTION AND MACROSCOPIC QUANTUM SYSTEMS

Giuseppe Vitiello
Istituto di Fisica, Università di Salerno, Italia

The physical significance of group contraction in quantum field theories with spontaneous breakdown of symmetry is discussed. Low-energy theorems and macroscopic ordered state symmetry patterns and structures are observable manifestations of group contraction.

In this note we will not insist on formal details which can be found in earlier works; we would like instead to stress the physical significance of group contraction in spontaneously broken symmetry theories. In quantum field theory the mapping between the Heisenberg fields and the in- (out-) fields (quasi-fields in many body terminology) is a fundamental problem. The dynamics is given in terms of Heisenberg fields equations while the asymptotic in- (out-)fields are relevant to the description of the observable properties of the physical system. Thus the problem is the one of relating the dynamical level to the phenomenological level. From a mathematical point of view this is a linearization problem since we have to write Heisenberg fields $\psi(x)$ which satisfy non-linear dynamical equations, in terms of asymptotic fields $\phi(x)$ which satisfy linear free-field equations:

$$\langle a|\psi(x)|b(x)\rangle = \langle a|F[\phi(x)]|b\rangle \tag{1}$$

Of course the functional F in the "dynamical map" eq.(1) is determined by the dynamics specified by the non-linear Heisenberg equations. $|a\rangle$ and $|b\rangle$ are wave-packet states describing physical states and thus constructed in terms of in-fields ϕ. They are represented by vectors in certain Hilbert space H. The choice of H is not unique in general, since we are in the frame work of the quantum field theory which studies systems with infinite degrees of freedom. A particular choice of H is equivalent to a particular boundary condition for the dynamical field equations. Thus such a choice is not only a mathematical requirement but is also the physical input by which the system under study is characterized. Due to the non-linearity of the

Heisenberg field equations, the mapping (1) is expected to be non-linear.

Suppose the Heisenberg equations are invariant under a certain symmetry group G, i.e.

$$\Lambda(\partial)\psi(x) = J[\psi(x)] \tag{2}$$

are form invariant under

$$\psi(x) \to \psi'(x) = g(\psi(x)) \qquad g\in G \tag{3}$$

As a consequence of the non-linearity of the mapping (1) and of the particular choice of H, when ψ undergoes a G-transformation is not surprising that the field ϕ transforms under some group G^{in} which in general can be different from G:

$$\phi(x) \to \phi'(x) = h(\phi(x)) \qquad h\in G^{in} \tag{4}$$

such that

$$\langle a|g(\psi(x))|b\rangle = \langle a|F[h(\phi(x))]|b\rangle \tag{5}$$

Since the original invariance of the theory cannot disappear at the level of asymptotic fields, G^{in} must be the symmetry group for the free-field equations and for the S-matrix. When $G^{in}\neq G$ we are in the presence of the dynamical rearrangement of the symmetry[1]: the invariance of the theory manifests itself in different symmetry structures at the dynamical and at phenomenological level. In this way it is possible that the same dynamics creates different observable symmetry patterns (ordered states) and structures (as extended objects with topological quantum numbers) under different boundary conditions (different choices of H). It can be shown that the dynamical rearrangement of symmetry leads to group contraction under quite general conditions.[2]

Spontaneous breakdown of symmetry presents two strictly related features: disappearance of original symmetry and creation of macroscopic observable symmetry structures. From the above discussion we see that the first feature is related to the possibility of the choice of H with the ground state (vacuum) <u>not</u> symmetric under the original symmetry; the second feature is related to the possibility of building up those macroscopic symmetry structures dynamically. Let us clarify this last point. Homogeneous ordered states or localized structures (as dislocations, vortices, etc.) need of long range correlation forces, i.e. of massless quanta, to be built up. These are the Goldstone bosons which are always present in a spontaneously broken symmetry theory[3] (if not present as elementary fields, they are dynamically created as bound states). Macroscopic symmetry patterns and structures are created by the condensation of these Goldstone bosons B(x) induced by the "translation" transformation[2,4]

$$B(x) \to B'(x) = B(x) + \text{const.} \tag{6}$$

Since B is a massless field the transformation (6) is a symmetry transformation for the free-field equations and thus it belongs to G^{in}. Explicit computations have been carried out in examples of physical interest (relativistic SU(2) and chiral SU(2)xSU(2) models, cristal, ferromagnet, superconductivity, T-t Jahn-Teller systems, etc.[5-9]; it has been shown how an Abelian "translation" transformation comes into theory even if one starts with an original non-Abelian group only. In ref.2 the dynamical rearrangement of symmetry is proved to be a group contraction in the general case of SU(n) and SO(n). See also ref.10 for a discussion on the conditions allowing group contraction. Results of ref.11 support same conclusion in the case of non-Abelian gauge theories. The conformal group contraction in pure SU(n) Yang-Mills theories has been considered in ref.12. Infrared Goldstone bosons play a crucial role in the rearrangement of symmmetry since any macroscopic observation is a collection of local observations and in each of these one misses effects of order of 1/V with the volume $V \to \infty$. These missing effects give a finite contribution when integrated on the whole system. In other words, we actually measure matrix elements between wave-packet states which are insensitive to locally infinitesimal effects and this is the origin of group contraction[5,6]. Finally we note that many low-energy theorems, as Adler theorem in high energy physics and Dyson theorem in ferromagnetism, which follow from the S-matrix invariance, appear as observable manifestation of group contraction. Note also that these low-energy theorems express stability under "soft" perturbations. The central role played by the group contraction in the passage to macroscopic physics is thus understood: the original invariance group is dynamically rearranged to contracted group at the observational level; in this way Abelian boson transformations are introduced which regulate boson condensation. When a large number of bosons is condensed macroscopic symmetry patterns and structures are created, quantum fluctuations are very small and the system behaves as a classical one. A possible conclusion is then in support of the conjecture[13] that the passage from quantum to classical physics involves some group contraction phenomenon, in accordance also with the suggestions of the developments of the deformation theory[14].

REFERENCES

1- H.Umezawa, Renormalization and invariance in QFT, ed.E.R.Caianiello, Plenum Press 1974, p.275.

2- C.De Concini and G.Vitiello, Nucl.Phys.B116 (1976) 141; Phys.Lett. 70B(1977) 355.

3- J.Goldstone, NuovoCimento 19 (1961) 154; J.Goldstone, A.Salam and S.Weinberg, Phys. Rev. 127 (1962) 965.

4- H.Matsumoto and H.Umezawa, Symmetries in Science, eds. B.Gruber and R.S.Millman, Plenum Press, 1980.

5- M.N.Shah, H.Umezawa and G.Vitiello, Phys. Rev. B10 (1974) 4724.

6- H.Matsumoto, H.Umezawa, G.Vitiello and J.K.Wyly, Phys. Rev. D9 (1974) 2806; E.Weimar, Acta Phys.Austriaca 48(1978) 201.

7- H.Matsumoto, N.J.Papastamatiou, H.Umezawa, Nucl. Phys. B82 (1974) 45.

8- G.Vitiello, Phys. Lett. 58A (1976) 293.

9- H.Matsumoto, N.J.Papastamatiou, H.Umezawa, Phys.Rev. D13 (1976) 1054.

10- M.Hongoh, H.Matsumoto, H.Umezawa, Prog.Theor.Phys. 65 (1981), 315.

11- T.Kugo, I.Ojima, Progr. Theor. Phys. 61 (1979) 294.

12- P.Tataru-Mihai, G.Vitiello, Lett.Math.Phys., in print.

13- I.E.Segal, Duke Math.J. 18 (1951) 221; E.İnönü and E.P.Wigner, Proc.Nat.Acad.Sci. US 39 (1953) 510; R.Herman, Lie groups for physicists, Benjamin,N.Y. 1966 ch.11.

14- F.Bayen, M.Flato, C.Fronsdal, A.Lichnerowicz, D.Sternheimer, Ann. of Phys. 111 (1978) 61-151.

THE INTERACTING BOSON MODEL AND ITS CONNECTION WITH GROUP THEORY

Bruce R. Barrett*
Department of Physics and Arizona Research Laboratories, Bldg. 81
University of Arizona, Tucson, Arizona 85721, USA

ABSTRACT

The Interacting Boson Model for only one kind of s and d bosons is described, and its connection with the group U(6) of unitary transformations in six dimensions is demonstrated. It is then shown how the subgroup chains of U(6) facilitate the solution of the eigenvalue problem for the s and d bosons in certain limiting cases.

1. Introduction

One of the most exciting developments in nuclear structure theory of the last few years has been the Interacting Boson Model (IBM) of Arima and Iachello.[1)] The original, phenomenological IBM of Arima and Iachello has been extremely successful in describing the collective properties of medium-to-heavy-mass even-even nuclei.[1-3)] Since its creation, the IBM has been extended to include high-spin states,[4)] configuration mixing[5)] and α clustering,[6)] and even an Interacting Boson-Fermion Model has been developed for treating odd-A nuclei.[7)] A particularly intriguing feature of the IBM is that it appears to be directly related to the underlying fermion structure in terms of nuclear shell-model theory,[8)] using the generalized seniority scheme of Talmi.[9)] Since this is a Colloquium on Group Theoretical Methods in Physics, I will only comment on the fascinating group theory properties of the original IBM. Anyone interested in further information on the development of the IBM, its extension, its application, and its microscopic interpretation is referred to the Ph.D. theses and general review articles listed in References 2, 3, and 10-14.

---

*Supported in part by NSF Grant No. PHY-8100141

2. Description of the IBM

The original IBM of Arima and Iachello[1)] did not distinguish between proton and neutron bosons; this model is referred to as the IBM-1. Like the shell model, the IBM-1 is a truncation scheme for restricting the nuclear wave function to a few important degrees of freedom. In the case of medium-to-heavy-mass nuclei, the low-lying collective properties cannot be described in terms of a few shell-model configurations but would require millions or billions of configurations. The concept of the IBM-1 is that the bosons represent collective configurations, which contain the important degrees of freedom of these low-lying properties.

As in the shell model, one starts with a number of valence particles outside closed major shells and assumes that the structure of the low-lying levels is dominated by excitations among these particles. Secondly, one assumes that the important particle configurations for the low-lying levels of even-even nuclei are those for identical particles paired together in states with total angular momentum J=0 and J=2. The final assumption is that these pairs can be treated as bosons. Hence, the number of bosons is equal to the number of pairs of particles outside the closed shells and is a strictly conserved quantity. Proton (neutron) bosons with angular momentum J=0 are denoted by $s_\pi(s_\nu)$, while those with angular momentum J=2 are denoted by $d_\pi(d_\nu)$. In order to take into account the particle-hole conjugation in the particle space, the number of proton, N_π, and/or neutron, N_ν, bosons is taken as the number of hole pairs, if more than half of the shell is full. For example, ${}^{130}_{56}Ba_{74}$ has 6 protons outside the 50 closed shell or 3 proton-particle bosons and 8 neutron holes in the 82 closed shell or 4 neutron-hole bosons. In the IBM-1 one does not distinguish between proton and neutron bosons, so I will drop the subscripts π and ν.

In order to write down the appropriate operators for the bosons, it is convenient to use the second quantized formalism, introducing the creation $(s^\dagger, d^\dagger_\mu)$ and annihilation (s, d_μ) operators, for s and d bosons, respectively, where μ=0, ±1, ±2. These operators satisfy the standard Bose commutation relations

$$[s, s^\dagger] = 1,\ [s, s] = 0,\ [s^\dagger, s^\dagger] = 0\ ,$$

$$[d_\mu, d^\dagger_{\mu'}] = \delta_{\mu\mu'},\ [d_\mu, d_{\mu'}] = 0,\ [d^\dagger_\mu, d^\dagger_{\mu'}] = 0\ , \tag{1}$$

$$[s, d^\dagger_\mu] = 0,\ [s^\dagger, d^\dagger_\mu] = 0,\ [s, d_\mu] = 0,\ [s^\dagger, d_\mu] = 0\ .$$

The most general Hamiltonian for a system of s and d bosons, which conserves the total number of bosons and which contains only one- and two-body terms, can be written in the form[1)]

$$\begin{aligned} H = {} & \varepsilon_s(s^\dagger\cdot s) + \varepsilon_d(d^\dagger\cdot\tilde{d}) \\ & + \sum_{L=0,2,4} \tfrac{1}{2}(2L+1)^{1/2}\, C_L[[d^\dagger x d^\dagger]^{(L)} \times [\tilde{d}x\tilde{d}]^{(L)}]^{(0)} \\ & + \tfrac{1}{\sqrt{2}}\, \tilde{V}_2[[d^\dagger x d^\dagger]^{(2)} \times [\tilde{d}xs]^{(2)} + [d^\dagger x s^\dagger]^{(2)} \times [\tilde{d}x\tilde{d}]^{(2)}]^{(0)} \\ & + \tfrac{1}{2}\, \tilde{V}_0[[d^\dagger x d^\dagger]^{(0)} \times [sxs]^{(0)} + [s^\dagger x s^\dagger]^{(0)} \times [\tilde{d}x\tilde{d}]^{(0)}]^{(0)} \\ & + U_2[[d^\dagger x s^\dagger]^{(2)} \times [\tilde{d}xs]^{(2)}]^{(0)} + \tfrac{1}{2}U_0[[s^\dagger x s^\dagger]^{(0)} \times [sxs]^{(0)}]^{(0)}\ , \end{aligned} \tag{2}$$

where $\tilde{d}_\mu=(-1)^{2+\mu}d_{-\mu}=(-1)^\mu d_{-\mu}$, which is a spherical tensor under rotations, while $d^\dagger_\mu$ is already a spherical tensor. The [ ] denote the tensor product of two tensor operators, e.g., $[d^\dagger x d^\dagger]^{(L)}_M = \sum_{\mu_1\mu_2} \langle 2\mu_1 2\mu_2|LM\rangle\, d^\dagger_{\mu_1}d^\dagger_{\mu_2}$, while the ( ) denote the scalar product of two tensor operators, e.g.,
$(d^\dagger\cdot\tilde{d})=(-1)^2\sqrt{5}\,[d^\dagger x\tilde{d}]^{(0)}_0=\sqrt{5}\sum_{\mu_1\mu_2}\langle 2\mu_1 2\mu_2|00\rangle d^\dagger_{\mu_1}\tilde{d}_{\mu_2} = \sum_{\mu_1} d^\dagger_{\mu_1}d_{\mu_1} = n_d$= the number operator for d bosons. The above Hamiltonian contains nine parameters: the two single-bosons energies ε_s and ε_d and the seven two-body terms $C_L(L=0,2,4)$, $\tilde{V}_L(L=0,2)$ and $U_L(L=0,2)$. Since the total number of bosons is conserved ($N = n_s + n_d$), the Hamiltonian can be rewritten so that the <u>excitation energies</u> are independent of n_s and depend upon only six parameters.[2,11)]

The Hamiltonian in Eq. (2) can be directly applied to describe the low-lying spectra of medium and heavy mass nuclei by diagonalizing it in an appropriate basis. Such studies have been carried out.[1-3)] It is important to emphasize that the Hamiltonian in Eq. (2) is <u>completely</u>

general and can be used in a systematic manner to study the collective properties of any appropriate nucleus.

3. The U(6) Algebra and Subalgebras

There are a number of nuclei which exhibit special collective properties, which have already been explained and given specific names within the geometrical picture, such as vibrational properties and rotational properties.[15] One of the very interesting features of the IBM-1 is that these geometrical cases follow directly as limiting cases of the IBM-1, which can be solved analytically in terms of the group structure of Eq. (2). This is true since the second quantized operators of the IBM-1 (i.e., the $s^\dagger$, $d^\dagger_\mu$ and the s, $\tilde{d}_\mu$) can be combined pairwise (i.e., a creation operator with an annihilation operator) to form a set of 36 operators

$$G^{(k)}_\kappa(\ell\ell') = [b^\dagger_\ell \times \tilde{b}_{\ell'}]^{(k)}_\kappa = \sum_{\mu_1\mu_2'} \langle \ell\mu_1 \ell'\mu_2' | k\kappa\rangle b^\dagger_{\ell\mu_1} (-1)^{\mu_2'} b_{\ell',-\mu_2'} \tag{3}$$

where $\ell,\ell' = 0,2 \equiv s,d$ and

$$[b_{\ell\mu}, b^\dagger_{\ell'\mu'}] = \delta_{\ell\ell'}\, \delta_{\mu\mu'} \,. \tag{4}$$

The operators (3) satisfy the commutation relations

$$\begin{aligned}[G^{(k)}_\kappa(\ell\ell'), G^{(k')}_{\kappa'}(\ell''\ell''')] &= \sum_{k''\kappa''} (2k+1)^{1/2}(2k'+1)^{1/2} \langle k\kappa k'\kappa' | k''\kappa''\rangle \\ &\times \; (-1)^{k-k'} [(-1)^{k+k'+k''} \begin{Bmatrix} k & k' & k'' \\ \ell''' & \ell & \ell' \end{Bmatrix} \delta_{\ell'\ell''}\, G^{(k'')}_{\kappa''}(\ell\ell''') \\ &- \begin{Bmatrix} k & k' & k'' \\ \ell'' & \ell' & \ell \end{Bmatrix} \delta_{\ell\ell'''}\, G^{(k'')}_{\kappa''}(\ell''\ell')] \,. \end{aligned} \tag{5}$$

Operators X satisfying relations of the type

$$[X_a, X_b] = \sum_c C^c_{ab}\, X_c \tag{6}$$

form a Lie algebra with structure constants C^c_{ab}. Equation (5) defines the Lie algebra of the group U(6) of unitary transformations in six dimensions. The operators $G^{(k)}_\kappa(\ell\ell')$ are called the group generators. There are a total of $6^2 = 36$ generators, which are given by

$$G_0^{(0)}(ss) = [s^\dagger x s]_0^{(0)} \qquad 1$$

$$G_0^{(0)}(dd) = [d^\dagger x \tilde{d}]_0^{(0)} \qquad 1$$

$$G_\kappa^{(1)}(dd) = [d^\dagger x \tilde{d}]_\kappa^{(1)} \qquad 3$$

$$G_\kappa^{(2)}(dd) = [d^\dagger x \tilde{d}]_\kappa^{(2)} \qquad 5 \qquad (7)$$

$$G_\kappa^{(3)}(dd) = [d^\dagger x \tilde{d}]_\kappa^{(3)} \qquad 7$$

$$G_\kappa^{(4)}(dd) = [d^\dagger x \tilde{d}]_\kappa^{(4)} \qquad 9$$

$$G_\kappa^{(2)}(ds) = [d^\dagger x s]_\kappa^{(2)} \qquad 5$$

$$G_\kappa^{(2)}(sd) = [s^\dagger x \tilde{d}]_\kappa^{(2)} \qquad 5$$

$$\overline{36 = 6^2}$$

The next step is that of identifying all possible subalgebras of the full algebra. A subalgebra is a subset of generators which is closed under commutation.

Subalgebras I

If we delete the 11 operators which contain $s^\dagger$ and/or s, we find that the remaining 25 operators close under the algebra U(5), the group of unitary transformations in five dimensions.

A) U(5)

$$G_0^{(0)}(dd) = [d^\dagger x \tilde{d}]_0^{(0)} \qquad 1$$

$$G_\kappa^{(1)}(dd) = [d^\dagger x \tilde{d}]_\kappa^{(1)} \qquad 3$$

$$G_\kappa^{(2)}(dd) = [d^\dagger x \tilde{d}]_\kappa^{(2)} \qquad 5 \qquad (8)$$

$$G_\kappa^{(3)}(dd) = [d^\dagger x \tilde{d}]_\kappa^{(3)} \qquad 7$$

$$G_\kappa^{(4)}(dd) = [d^\dagger x \tilde{d}]_\kappa^{(4)} \qquad 9$$

$$\overline{25 = 5^2}$$

If we now delete from these 25 operators $G_0^{(0)}(dd)$, $G_\kappa^{(2)}(dd)$, and $G_\kappa^{(4)}(dd)$, we discover that the remaining 10 operators close under the algebra of O(5), the orthogonal group in five dimensions.

B) O(5)

$$G_\kappa^{(1)}(dd) = [d^\dagger x\tilde{d}]_\kappa^{(1)} \qquad 3$$

$$G_\kappa^{(3)}(dd) = [d^\dagger x\tilde{d}]_\kappa^{(3)} \qquad 7 \qquad (9)$$

$$10 = \frac{5x4}{2}$$

Deleting the 7 operators $G_\kappa^{(3)}(dd)$, we obtain the three operators which close under the algebra of O(3), the ordinary rotation group.

C) O(3)

$$G_\kappa^{(3)}(dd) = [d^\dagger x\tilde{d}]_\kappa^{(1)} \qquad 3 \qquad (10)$$

Finally, if we delete from these three operators the two operators $G_{\pm 1}^{(1)}(dd)$, we obtain the single generator of O(2), the group of rotations around the z axis.

D) O(2)

$$G_0^{(1)}(dd) = [d^\dagger x\tilde{d}]_0^{(1)} \qquad 1 \qquad (11)$$

Hence, one possible chain of subalgebras is

$$U(6)\supset U(5)\supset O(5)\supset O(3)\supset O(2)\ . \quad (I) \qquad (12)$$

<u>Subalgebras II</u>

A) U(3). Let us consider the operators

$$G_0^{(0)}(ss) + \sqrt{5}G_0^{(0)}(dd) = [s^\dagger xs]_0^{(0)} + \sqrt{5}\,[d^\dagger x\tilde{d}]_0^{(0)} \qquad 1$$

$$G_\kappa^{(1)}(dd) = [d^\dagger x\tilde{d}]_\kappa^{(1)} \qquad 3 \qquad (13)$$

$$G_\kappa^{(2)}(ds)+G_\kappa^{(2)}(sd)-\frac{\sqrt{7}}{2}\,G_\kappa^{(2)}(dd) = [d^\dagger xs+s^\dagger x\tilde{d}]_\kappa^{(2)} - \frac{\sqrt{7}}{2}\,[d^\dagger x\tilde{d}]_\kappa^{(2)} \qquad 5$$

$$9 = 3^2$$

These operators close under commutation and form the algebra of U(3). The remaining subalgebras are now obvious, i.e.,

B) O(3)

$$G_\kappa^{(1)}(dd) = [d^\dagger x\tilde{d}]_\kappa^{(1)} \qquad 3 \qquad (14)$$

and

C) O(2)

$$G_0^{(1)}(dd) = [d^\dagger x\tilde{d}]_0^{(1)} \qquad 1 \qquad (15)$$

Thus a second possible chain of subalgebras is

$$U(6) \supset U(3) \supset O(3) \supset O(2) \qquad (II) \qquad (16)$$

Subalgebras III

A) O(6). Let us now consider the operators

$$\begin{aligned} G_\kappa^{(1)}(dd) &= [d^\dagger x\tilde{d}]_\kappa^{(1)} && 3 \\ G_\kappa^{(3)}(dd) &= [d^\dagger x\tilde{d}]_\kappa^{(3)} && 7 \\ G_\kappa^{(2)}(ds) + G_\kappa^{(2)}(sd) &= [d^\dagger xs + s^\dagger x\tilde{d}]_\kappa^{(2)} && 5 \\ & && \overline{15 = \tfrac{6x5}{2}} \end{aligned} \qquad (17)$$

These operators close under commutation, yielding the Lie algebra of O(6), the orthogonal group in six dimensions. Once again the subalgebras are now obvious.

B) O(5)

$$\begin{aligned} G_\kappa^{(1)}(dd) &= [d^\dagger x\tilde{d}]_\kappa^{(1)} && 3 \\ G_\kappa^{(3)}(dd) &= [d^\dagger x\tilde{d}]_\kappa^{(3)} && 7 \\ & && \overline{10} \end{aligned} \qquad (18)$$

C) O(3)

$$G_\kappa^{(1)}(dd) = [d^\dagger x\tilde{d}]_\kappa^{(1)} \qquad 3 \qquad (19)$$

D) O(2)

$$G_0^{(1)}(dd) = [d^\dagger x\tilde{d}]_0^{(1)} \qquad 1 \qquad (20)$$

So a third possible chain of subalgebras is

$$U(6) \supset O(6) \supset O(5) \supset O(3) \supset O(2) \qquad \text{(III)} \qquad (21)$$

One can show that these are the only possible chains of subalgebras, if one insists that the angular momentum L be a good quantum number, i.e., that O(3) be contained in the chain. In fact, starting with U(6) we have considered U(5), U(3) and O(6). But O(6) is locally isomorphic to SU(4). Thus, we have considered all possible subgroups of U(6). To summarize, we have found that there are three and only three possible subgroup chains:

$$U(6) \begin{cases} \nearrow U(5) \supset O(5) \supset O(3) \supset O(2) & \text{(I)} \\ \rightarrow U(3) \supset O(3) \supset O(2) & \text{(II)} \\ \searrow O(6) \supset O(5) \supset O(3) \supset O(2) & \text{(III)} \end{cases} \qquad (22)$$

4. Classification of States

Since the Hamiltonian in Eq. (2) is built out of the operators $G_\kappa^{(k)}(\ell\ell')$, it has the group structure of U(6). Thus, once a group chain has been identified, one can use it to construct a basis in which the Hamiltonian (Eq. (2)) can be diagonalized. To do this, we must know the labels which characterize the irreducible representations of the various groups which appear in the chain. The general procedure by which one can determine the quantum numbers characterizing the irreducible representations of the various groups of interest to us is given in Ref. 12 and will not be discussed here. Only the results will be given, although it is worth noting that the representations of O(3) and O(2) are characterized by the quantum numbers L (angular momentum) and M (z projection of L), respectively, as is well-known.

Group chain I

The complete classification of group chain I is

U(6) [N]
U(5) (n_d)
O(5) v
O(3) L
O(2) M

The representations of U(6), U(5) and O(5) are the totally symmetric ones, since we are dealing with systems of bosons. In the above classification N is the total number of bosons, n_d is the number of d-bosons and v is called the seniority (i.e., the number of d-bosons not coupled pairwise to zero angular momentum). The step from O(5) to O(3) is not fully decomposable. We need an additional quantum number, which we call n_Δ and which can be identified with the number of d-boson triplets coupled to zero angular momentum.

The algorithm for finding the values of L contained in each representation n_d of U(5) is: First, partition n_d as

$$n_d = 2n_\beta + 3n_\Delta + \lambda, \tag{23}$$

where

$$n_\beta = (n_d-v)/2;\; n_\beta = 0,\, 1,\, ...,\, n_d/2 \text{ or } (n_d-1)/2. \tag{24}$$

Then,

$$L = \lambda,\, \lambda+1,\, \lambda+2,\, ...,\, 2\lambda-2,\, 2\lambda \text{ (Note that } 2\lambda-1 \text{ is missing).} \tag{25}$$

Hence, the complete classification scheme for chain I is

$$|[N]\;(n_d)\;v\;n_\Delta\;L\;M\rangle.$$

<u>Group chain II</u>

The labels needed to classify the states in chain II are

U(6) [N]
SU(3) (λ,μ)
O(3) L
O(2) M

Two quantum numbers (λ,μ) are needed to characterize the representations of SU(3), used here instead of U(3). The values of (λ,μ) contained in each [N] are given by

$$[N] = (2N,0)\oplus(2N-4,2)\oplus\;...\;\oplus\begin{Bmatrix}(0,N), \text{ N even}\\(2,N-1), \text{ N odd}\end{Bmatrix}$$

$$\oplus(2N-6,0)\oplus(2N-10,2)\oplus\;...\;\oplus\begin{Bmatrix}(0,N-3), \text{ N-3 even}\\(2,N-4), \text{ N-3 odd}\end{Bmatrix} \tag{26}$$

$$\oplus\;...$$

The step from SU(3) to O(3) is not fully decomposable. The simplest choice of the additional quantum number needed to classify the states is due to Elliott. The corresponding number is called K. The values of L contained in each (λ,μ) in the Elliott basis are given by the algorithm

$$L = K,\, K+1,\, K+2,\, ...,\, K+\max(\lambda,\mu) \tag{27}$$

where

$$K = \text{integer} = \min(\lambda,\mu),\ \min(\lambda,\mu)-2,\ \ldots,\ 1 \text{ or } 0 \tag{28}$$

with the exception of K = 0 for which

$$L = \max(\lambda,\mu),\ \max(\lambda,\mu)-2,\ \ldots,\ 1 \text{ or } 0. \tag{29}$$

So the complete classification for chain II is

$$|[N],\ (\lambda,\mu)\ K\ L\ M\rangle.$$

It is worth noting that the Elliott basis is not orthogonal. For this reason one oftens introduces another basis, called the Vergados basis, which is orthogonal. The procedure for doing this is described in Ref. 12.

Group chain III

For this chain the labels needed to classify the states are

U(6) [N]

O(6) (σ)

O(5) τ

O(3) L

O(2) M

The values of σ contained in each [N] are given by

$$\sigma = N,\ N-2,\ \ldots,\ 0 \text{ or } 1 \text{ for } N \text{ even or odd.} \tag{30}$$

The values of τ contained in each σ are given by

$$\tau = \sigma,\ \sigma-1,\ \ldots,\ 0 \tag{31}$$

Once again the step from O(5) to O(3) is not fully reducible, and one needs a further quantum number, called ν_Δ. The values of L contained in each τ are found by partitioning τ as

$$\tau = 3\nu_\Delta + \lambda, \qquad \nu_\Delta = 0,\ 1,\ \ldots \tag{32}$$

and taking

$$L = 2\lambda,\ 2\lambda-2,\ \ldots,\ \lambda+1,\ \lambda \text{ (again } 2\lambda-1 \text{ is missing).} \tag{33}$$

Thus the complete classification scheme for chain III is

$$|[N]\ \sigma\ \tau\ \nu_\Delta\ L\ M\rangle.$$

5. Solution of the Eigenvalue Problem: Dynamical Symmetries

Having constructed a classification scheme of the states, we are now in a position to diagonalize the Hamiltonian (Eq. (2)). After some manipulations, this Hamiltonian can be rewritten in terms of the Casimir operators of the groups U(6), U(5), O(5), O(3), SU(3), and O(6). [The group O(2) does not play any role unless the nucleus is placed in an external magnetic field, so we will neglect it.] For fixed boson number N, we can disregard the operators of U(6), since they contribute only to the binding energy. Since the Hamiltonian H is at most two body, our expression for H will involve at most quadratic operators. Only U(5) has a linear Casimir operator. Hence, the most general Hamiltonian H can be written as

$$H = \varepsilon C_{1U5} + \alpha C_{2U5} + \beta C_{2O5} + \gamma C_{2O3} + \delta C_{2SU3} + \eta C_{2O6}, \qquad (34)$$

which contains only 6 independent parameters, as we noted earlier. Here, C_{1U5} denotes the linear Casimir operator of U(5), C_{2U5} the quadratic Casimir operator of U(5), etc. The expressions for these Casimir operators in terms of $(s,\tilde{d}_\mu)$ and $(s^\dagger,d^\dagger_\mu)$ are given in Ref. 11. The Hamiltonian (34) is not diagonal in any of the group chains I, II or III. For example, C_{2U5} is not diagonal in chain II, etc. In order to find its eigenvalues one must diagonalize H numerically. This diagonalization can be done in any of the three group chains I, II or III, since all three form a complete basis.

In certain special cases, however, the eigenvalue problem for H can be solved in closed, analytic form. These special cases occur whenever the Hamiltonian can be written only in terms of Casimir operators of a group chain, in which case one says that H has a dynamical symmetry. It is clear from Eq. (34) that these dynamical symmetries correspond to the vanishing of some of the coefficients in H. In the IBM, since there are three possible group chains, there are also three possible dynamical symmetries.

Dynamical symmetry I

The group chain I is

$$U(6) \supset U(5) \supset O(5) \supset O(3) \supset O(2). \qquad (35)$$

This dynamical symmetry corresponds to the vanishing of δ and η in (34). The corresponding Hamiltonian is

$$H^I = \varepsilon C_{1U5} + \alpha C_{2U5} + \beta C_{2O5} + \gamma C_{2O3}. \qquad (36)$$

To find the expectation value of H^I in the representation $|[N]n_d\ v\ n_\Delta\ L\ M\rangle$ one needs to know the expectation values of the Casimir operators appearing in (36). These are given by standard group theoretical techniques.[12] The resulting expression is

$$\langle H^I\rangle = \varepsilon n_d + \alpha n_d(n_d+4) + 2\beta v(v+3) + 2\gamma L(L+1)\ . \qquad (37)$$

The structure of the corresponding spectrum is shown in Fig. 1 for ε, α, β and $\gamma > 0$ and is similar to that referred to in the geometrical model as corresponding to a spherical vibrator.[1]

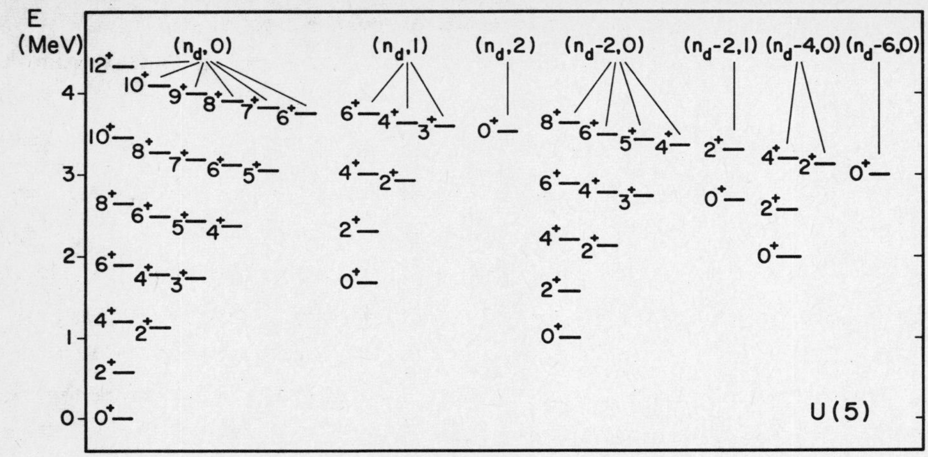

Fig. 1. A typical spectrum with U(5) symmetry and N=6. In parentheses are the values of v and n_Δ.

Dynamical symmetry II

The group chain II is

$$U(6) \supset SU(3) \supset O(3) \supset O(2). \tag{38}$$

The corresponding dynamical symmetry is obtained when ε, α, β and η vanish in Eq. (34), leading to the Hamiltonian

$$H^{II} = \delta C_{2SU3} + \gamma C_{203} \,. \tag{39}$$

The expectation value of H^{II} in the representation $|[N](\lambda,\mu)\ K\ L\ M\rangle$ is given by

$$\langle H^{II}\rangle = \delta\ \frac{6}{9}\ [\lambda^2+\mu^2+\lambda\mu+3(\lambda+\mu)] + \gamma 2L(L+1). \tag{40}$$

The structure of the corresponding spectrum for $\delta < 0$ and $\gamma > 0$ is illustrated in Fig. 2 and is similar to that referred to in the geometrical model as corresponding to a deformed rotor.[1)]

Dynamical symmetry III

The group chain III is

$$U(6) \supset O(6) \supset O(5) \supset O(3) \supset O(2). \tag{41}$$

The vanishing of ε, α and δ in Eq. (34) produces the corresponding dynamical symmetry, given by the Hamiltonian

$$H^{III} = \beta C_{205} + \gamma C_{203} + \eta C_{206}. \tag{42}$$

The expectation value of H^{III} in the representation $|[N]\ \sigma\ \tau\ \nu_\Delta\ L\ M\rangle$ is

$$\langle H^{III} \rangle = 2\beta\tau(\tau+3) + 2\gamma L(L+1) + 2\eta\sigma(\sigma+4). \quad (43)$$

Figure 3 shows an example of the structure of the corresponding spectrum for this case. In the geometrical picture a spectrum of this structure is referred as corresponding to a γ-unstable nucleus.[1)]

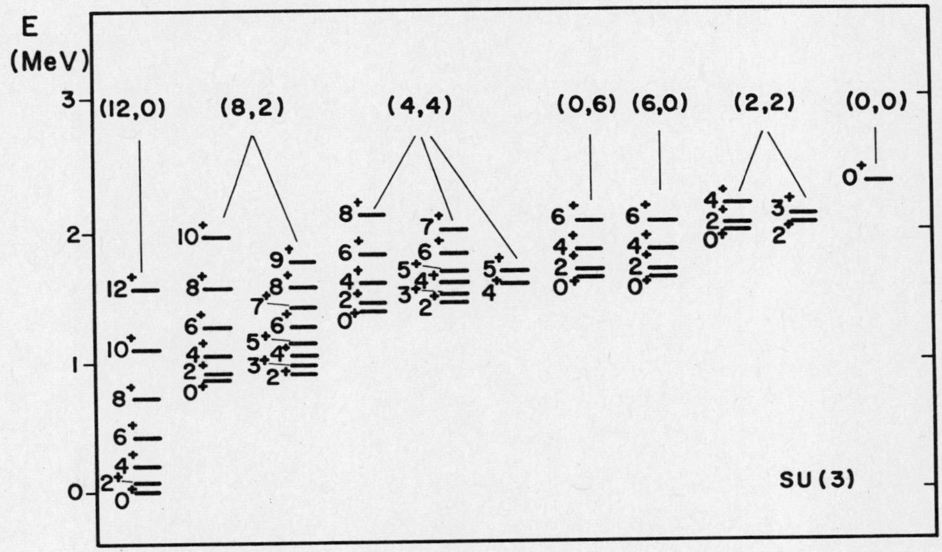

Fig. 2. A typical spectrum with SU(3) symmetry and N=6. In parentheses are the values of λ and μ.

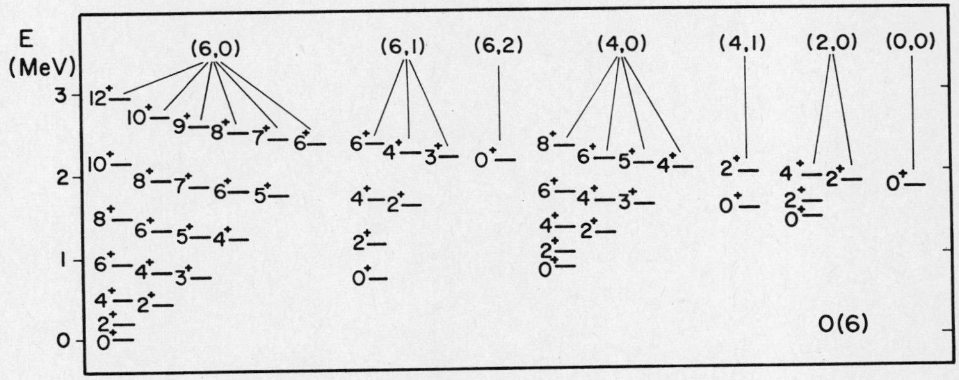

Fig. 3. A typical spectrum with O(6) symmetry and N=6. In parentheses are the values of σ and ν_Δ.

As stated above, there appear to be several nuclei whose spectra can be well described by one of these three limiting cases. Specific examples can be found in Refs. 1-3 and 11-14.

The s and d boson formalism can also be used to construct transition operators, quadrupole moments, magnetic moments and other physical operators. As with the Hamiltonian, the group structure of the IBM can be used in the limiting cases of the three group chains (22) to evaluate the matrix elements of these operators and/or to determine relevant selection rules. Details for doing this are given in Refs. 1-3 and 11-14.

6. Conclusions

The IBM-1 of only one kind of s and d bosons has been shown to be related to the group U(6) of unitary transformations in six dimensions. The subalgebras of U(6), containing O(3), lead to three subgroup chains, which can be utilized for solving analytically the eigenvalue problem for the boson Hamiltonian in the appropriate limiting cases. These limiting cases appear to be related to the structure of certain physically observed nuclei. Consequently, the group theory structure of the IBM-1 provides an extremely powerful tool for studying the properties of these nuclei and also of nuclei transitional in structure between the limiting cases (Refs. 1-3 and 11-14).

References

1. A. Arima and F. Iachello, Ann. Phys. (N.Y.) 99 (1976) 253; 111 (1978) 201; 123 (1979) 468.
2. O. Scholten, in "The Interacting Boson Approximation Model and Applications," Ph.D. thesis, University of Groningen, 1980.
3. Interacting Bosons in Nuclear Physics, ed. F. Iachello (Plenum Press, New York, 1979).
4. A Gelberg and A. Zemel, Phys. Rev. C22 (1980) 169.
5. P. D. Duval and B. R. Barrett, Phys. Lett. 100B (1981) 223; Nucl. Phys. A376 (1982) 213.
6. A. D. Jackson and F. Iachello, Phys. Lett. 108B (1982) 151.
7. Interacting Bose-Fermi Systems in Nuclei, ed. F. Iachello (Plenum Press, New York, 1981).
8. T. Otsuka, A. Arima, F. Iachello and I. Talmi, Phys. Lett. 76B (1978) 139.
9. I. Talmi, Nucl. Phys. A172 (1971) 1; in Frontier Research in Nuclear Physics, ed. D. H. Feng et al. (Plenum Press, New York, 1981) and Ref. 3, p. 79.
10. T. Otsuka, in "Boson Model of Medium-Heavy Nuclei," Ph.D. thesis, University of Tokyo, 1979.
11. F. Iachello, in "Proc. Int. Summer School on Nuclear Structure, Dronten, The Netherlands," ed. K. Abrahams et al. (Plenum Press, New York, 1981), p. 53.

12. F. Iachello, "Group Theory and Nuclear Spectroscopy," in Lecture Notes in Physics, Vol. 119, Nuclear Spectroscopy (Springer-Verlag, Berlin, 1980), p. 140.
13. B. R. Barrett, Rev. Mex. Fis. 27 (1981) 533.
14. A. Arima and F. Iachello, Ann. Rev. Nucl. Part. Sci., 31 (1981) 75.
15. A. Bohr and B. Mottelson, Nuclear Structure, Vol. II (Benjamin, Reading, 1975).

BOSON MAPPINGS IN NUCLEAR PHYSICS. A BRIEF AND PREJUDICED SURVEY.

Abraham Klein
DEPT. OF PHYSICS, U. OF PENNSYLVANIA
Philadelphia, PA. 19104
USA

ABSTRACT

The physical problems which stimulated the study of boson mappings are reviewed. The various mathematical techniques available for this problem are outlined. It is emphasized that the first mapping technique found in the literature, that of Holstein-Primakoff (HP), has been insufficiently exploited. A modified HP technique is illustrated. The unresolved problems in connection with the derivation of the interacting boson model are illustrated by means of a simplified model, and a proposal for a solution is made.

1. Physical Origin and Interest of the Problem.

The importance of the formal study of boson mappings for nuclear physics was first pointed out in a paper by Beliaev and Zelevinsky[1]. Consider, for instance, a magic nucleus approximated in its ground state by a Slater determinant. Let $a_p^+(a_{p'})$ be particle creation (annihilation) operators, $a_h^+(a_{h'})$ the corresponding hole operators. Collective excited states, e.g. giant multipole resonances, are created, approximately, by coherent particle-hole excitation operators

$$A_\lambda^+ = \sum_{ph} c_{ph}^\lambda a_p^+ a_h^+ \tag{1}$$

where

$$\sum_{ph} (c_{ph}^\lambda)^* c_{ph}^{\lambda'} = \delta_{\lambda\lambda'} \tag{2}$$

We have

$$[A_\lambda, A_{\lambda'}^+] = \delta_{\lambda\lambda'} - \sum_{h,h'} c_{ph}^\lambda (c_{ph'}^{\lambda'})^* a_{h'}^+ a_h - \sum_{p,p'} c_{ph}^\lambda (c_{p'h}^{\lambda'})^* a_{p'}^+ a_p . \tag{3}$$

For a collective state, we expect that $c_{ph} = O(N^{-\frac{1}{2}})$, where N is the number of terms in (1).

It follows from the assumed independent particle structure near the ground state that the first term on the right hand side of (3) is the dominant one, and we thus encounter boson commutation relations, approximately. With corresponding assumptions, the dynamics (linearized equations of motion or RPA) yields the harmonic spectrum of a set of uncoupled harmonic oscillators.

In reality N is never large enough nor are the non-linear effects small enough to permit us to ignore the fact that we are dealing in (1) with elements of a Lie Algebra. If the boson which corresponds to A_λ^+ in the linear approximation is denoted by a_λ^+, the problem suggested by the previous considerations is: Can we find expressions of the form

$$A_\lambda^+ = a_\lambda^+ + \text{"higher order terms"} \qquad (4)$$

so that (3) and the remaining commutators of the Algebra are satisfied.

The same type of mathematical problem arises in a different guise when, instead of a Slater determinant the ground state is superconducting. In the previous problem the basic boson is a particle-hole boson. In the latter it can be either a quasi-particle pair (increasing the number of quasiparticles by two) or an ordinary shell-model pair, depending on the formalism utilized.

The study of bosons in a superconducting formalism is related technically to the problem which has provided the strongest stimulus, until recently, for the investigation of boson mappings in nuclear physics, namely the problem of deriving the Bohr-Mottelson model (BMM) from the shell model. More recently the éclat associated with its very interesting variant, the interacting boson model (IBM) has brought fresh stimulation to the study of boson mappings. Both problems have led to major efforts, in the former case, of long standing[2)], in the latter case of recent vintage[3)], to derive these models from the shell model.

A succinct definition of the BMM problem can be given: Let b_μ^+ ($\mu = -2,-1,\ldots,2$) be a boson-valued tensor, conserving fermion number and of rank two under three-dimensional rotations. It is a reasonable inference from experiment that many low-lying bands of even nuclei not too near closed shell and for medium to heavy atomic weights can be modeled as states in the Hilbert space generated by the b_μ^+ - or more reasonably in a finite dimensional subspace of this Hilbert space. The problem is to understand how this description can be derived starting from the shell model defined by a

Hamiltonian, H_F, which is generally a polynomial in the generators of a Lie Algebra.

A derivaton of the BMM is understood to require two steps which can be considered in either order:(i) A mapping of H_F to a boson Hamiltonian,

$$H_F \rightarrow H_B \tag{5}$$

is carried out. The transform H_B may contain a larger space of boson variables than the b^+_μ alone and may also still contain fermion degrees of freedom. This mapping is in general a purely kinematical problem, a problem in Lie Algebras. (ii) The first step must be followed by a projection onto the collective subspace

$$H_B \rightarrow P_c H_B P_c, \tag{6}$$

$$P_c = \Sigma|n_\mu)(n_\mu| \;, \tag{7}$$

$$|n_\mu) = \prod_{n_\mu} (n_\mu!)^{-\frac{1}{2}}(b^+_\mu)^{n_\mu}|0), \tag{8}$$

$$\sum_\mu n_\mu \leq N_{max}. \tag{9}$$

This part of the problem is dynamical and requires procedures for the maximal possible decoupling of the collective from the non-collective degrees of freedom.

Several variants of the above sequence are preferred by active workers studying both the BMM and IBM problems. The favorite variant is to carry out the decoupling of the collective degree of freedom in the fermion space first and to map second. Still another possibility (preferred by the writer) involves defining a tractable subspace of the fermion space, large enough to contain the collective degrees of freedom, the mapping of this subspace (kinematics), followed by study of the dynamics in this subspace of the boson space.

We record here the opinion that the best available solutions of the BMM problem have been developed by Kishimoto and Tamura[2)] for vibrational and transitional nuclei and by Kumar[4)] for rotational and transitional nuclei. There remains room for improvement in both solutions.

A corresponding succint characterization of IBM (actually IBM 1) is the statement that the <u>same</u> states that can be modeled by BMM can also be modeled as states belonging to one of the fully symmetric representations of SU(6) spanned by the basis

$$|n_\mu, n_s\rangle = \prod_\mu (n_\mu!)^{-\frac{1}{2}} (d_\mu^+)^{n_\mu} (n_s!)^{-\frac{1}{2}} (s^+)^{n_s} |0), \tag{10}$$

$$\Sigma n_\mu + n_s = N = \text{constant}, \tag{11}$$

where s^+ and d_μ^+ are scalar and tensor of rank two, respectively under SO(3) and <u>increase</u> the fermion number by two. Many authors have recently discussed the relationship between BMM and IBM 1 (see [5] for a reasonably complete list of references). In particular, it has become clear that if we identify N_{max} of (7) with N in (11), then the models are isometric. In a somewhat more restricted form, this equivalence was obvious from the inception of IBM 1[6].

The derivation of IBM from the shell model is, from the technical side, in quite rudimentary form. In what follows, we hope to explain why this is so.

2. Exact Mappings of the Shell Model Algebra. Present Status.

We shall consider the problem of adding pairs of fermions rather than creating particle-hole excitations. Because of space limitations we consider only even systems. Let a_α^+ (a_β) be fermion single orbit creation (annihilation) operators. These are apportioned among levels of spin j_i; thus $\Lambda = \sum_i (2j_i + 1)$ is the total number of orbits. The Lie Algebra generated under commutation by the operators $a_\alpha^+ a_\beta^+$, $a_\beta a_\alpha$, and $[a_\beta^+ a_\alpha - \frac{1}{2}\delta_{\alpha\beta}]$ is that of the orthogonal group SO(2). The simplest basis for the antisymmetric representation of this algebra is the set of Slater determinants

$$|\alpha_1\beta_1, \alpha_2\beta_2 .. \alpha_p\beta_p\rangle = \prod_{i=1}^{p} a_{\alpha_i}^+ a_{\beta_i}^+ |0\rangle \quad , \tag{12}$$

$$a_\alpha |0\rangle = 0 \quad ,$$

where all indices must be distinct, by the Pauli principle. To obtain a formally exact mapping of this algebra, we introduce, for each pair $(a_\alpha^+ a_\beta^+)$, a boson $b_{\alpha\beta}^+ = -b_{\beta\alpha}^+$, antisymmetric in its indices and satisfying

$$[b_{\alpha\beta}, b_{\delta\gamma}^+] = \delta_{\alpha\delta}\delta_{\beta\gamma} - \delta_{\alpha\gamma}\delta_{\beta\delta} \; . \tag{13}$$

The generators a^+a^+, etc., can be given as functions of the bosons (13) in various forms, exact and approximate. At least three different forms are known: (i) The expressions

$$(a_\alpha a_\beta)_B = b_{\alpha\beta} \tag{14}$$

$$(a_\alpha^+ a_\beta^+)_B = b_{\alpha\beta}^+ - \sum_{\gamma\delta} b_{\alpha\gamma}^+ b_{\beta\delta}^+ b_{\gamma\delta} \quad , \tag{15}$$

$$(a_\alpha^+ a_\beta)_B = \sum_\gamma b_{\alpha\gamma}^+ b_{\beta\gamma} \; , \tag{16}$$

constructed by analogy with the famous SU(2) mapping of Dyson[8] satisfy the algebra SO(2Λ), but as usually interpreted to act in the space of boson states obtained by mapping

$$|0\rangle \rightarrow |0\rangle_B \equiv 0), \tag{17}$$

and

$$|\alpha_i\beta_i\ldots\alpha_p\beta_p\rangle \rightarrow |\alpha_1\beta_1\ldots\alpha_p\beta_p) = \prod_1^p (a^+_{\alpha_i} a^+_{\beta_i})_B |0), \tag{18}$$

(14) - (16) constitute a non-isometric mapping onto states which are mutually orthogonal but not normalized.

(ii) The scaling factors which normalize the states can be introduced in a formally exact manner, but the modified operator mapping which replaces (14) - (16) contains square roots of operators[7], in analogy with the Holstein-Primakoff (HP) mapping[9] for SU(2). Whereas the mapping (14) - (16) is well defined on the entire boson space generated by the $b^+_{\alpha\beta}$, the HP mapping is well defined only on the antisymmetric subspace (physical subspace).

(iii) By introducing into the mapping (ii) projection operators onto the antisymmetric subspace, the resulting mapping becomes well-defined in the entire boson space and matrix elements within the unphysical space and those which connect physical and unphysical space vanish. In practice these statements can, at best, be satisfied approximately because calculations can be carried out only if the irrational functions of the boson operators which represent the generators are expanded in series. These series were historically the first form of this mapping discovered[10].

Unfortunately these series do not converge, and it thus appears that the isometric forms of the mapping under discussion are not useful. The Dyson form has proved of some use for the practical calculation of spectra near closed shells[11]. But this use is far removed from our goal, the derivation of a Hamiltonian, Hermitian with respect to scalar products in the space (10) (and agreeing with experiment!). Toward such an end, the Dyson mapping can at best serve as an intermediate calculational tool[12,13].

Even more important to the ultimate success of the enterprise than the methodology, however, is the choice of a starting basis to map, one chosen according to the physical requirements of the problem. Before developing this most important point of our paper, we include a digression on methodology.

3. Methods of Boson Mapping.

In the following, a given method may produce either an isometric or non-isometric mapping so that we blur that distinction for the time being. In most of the standard literature, it is taken for granted that there are two techniques for carrying out boson mappings. In fact, we shall mention six methods below, the conventionally understood ones being the BZ[1] and MYT[10] methods.

(i) Holstein-Primakoff method[9]. As we reminded the nuclear physics community[14], the first boson mapping did not involve either of these methods. Using the original example of the SU(2) algebra, starting from the well-known matrices

$$\langle j,m|J_z|j,m\rangle = m, \quad -j \leq m < j \;, \tag{19}$$

$$\langle j,m+1|J_+|j,m\rangle = [(j-m)(j+m+1)]^{\frac{1}{2}} \;, \tag{20}$$

and introducing a non negative integer, n,

$$m = -j + n, \tag{21}$$

we have

$$\langle n|J_z|n\rangle = -j + n \;, \quad 0 \leq n \leq 2j, \tag{22}$$

$$\langle n+1|J_+|n\rangle = [(n+1)(2j-n)]^{\frac{1}{2}} \;. \tag{23}$$

One recognizes immediately that (22), (23) are the matrix elements of the boson operators

$$(J_Z)_B = -j + b^+b \tag{24}$$

$$(J_+)_B = (J_-^+)_B = b^+(2j - b^+b)^{\frac{1}{2}}, \tag{25}$$

acting on the boson basis

$$|n) = (n!)^{-\frac{1}{2}}(b^+)^n|0). \tag{26}$$

In fact this method can be generalized to any compact Lie Algebra: Given a matrix representation of the generators, these will depend on a fixed set of eigenvalues of the Casimir invariants (j...), and they can be written so as to depend furthermore on a set of non-negative integers which distinguish individual basis states. Introducing a boson for each of these integers, we obtain a mapping. This method has been used by several authors[15-17].

(ii),(iii). The BZ and MYT Methods. These techniques have been discussed at sufficient length in the literature[7,11].

(iv) Coherent State Method. By applying the SU(2) algebra to the generator coordinate state (z complex)

$$|\Psi_f\rangle = \int dz f(z) \exp(zJ_+)|0\rangle \qquad (27)$$

and integrating by parts, we obtain a representation in the space of functions f(z), namely

$$J_+ \to b^+ \quad , \qquad (28)$$

$$J_- \to (2j - b^+b)b \quad , \qquad (29)$$

$$J_Z \to -j + b^+b \quad , \qquad (30)$$

where $b^+ = -(d/dz)$ and $b = z$. This is a form of the Dyson mapping. Actually this method may be traced back to the paper of Jancovici and Schiff[18] and has recently been developed into a systematic and powerful calculus[19]. It remains to be seen what physically useful results can be obtained by this technique.

(v) Li's Method. C.T. Li has recently developed a powerful new method[13] for using the Dyson mapping as an intermediary in obtaining isometric mappings in physically interesting cases. It constitutes a possible viable alternative to the method emphasized below.

(vi) Modified HP (or BZ) Method.

Again consider the SU(2) algebra. We start with the boson basis (26). We take as "obvious" that the eigenvalues of J_Z are reproduced by (24). We ask for the solution (25) assuming the structure $(\hat{n} = b^+b)$

$$J_+ = b^+f(\hat{n}), \qquad (31)$$

which follows from the selection rule on n. Applying the commutation relation $[J_+,J_-] = 2J_Z$, we obtain the difference equation

$$(n + 1)f^2(n) - nf^2(n) = 2\,(j - n), \qquad (32)$$

with the solution

$$f(n) = (2j - n)^{\frac{1}{2}} \qquad (33)$$

If this example appears too trivial, the structure of the method we propose becomes clearer with a second example. Consider two levels, j_i, i = 1,2 and let us take $j_1 = j_2 = j$ and $2\Omega = 2j + 1$. The six operators

$$J_{+i} = \sum_{m>0} (-1)^{j_i - m} a_{j_i,m}{}^{+} a_{j_i,m}{}^{+} = (J_{-i})^{+}, \tag{34}$$

$$J_{zi} = -\tfrac{1}{2}\Omega + \sum_m a_{j_i,m}{}^{+} a_{j_i,m} , \tag{35}$$

generate the algebra SU(2) x SU(2). As is well-known, the linear combinations of "quasi spins"

$$\vec{J} = \vec{J}_1 + \vec{J}_2 , \vec{K} = \vec{J}_1 - \vec{J}_2, \tag{36}$$

generate the algebra SO(4) with the commutators.

$$[J_+,J_-] = [K_+, K_-] = 2J_Z, \tag{37a}$$

$$[J_+, K_-] = [K_+, J_-] = 2K_Z, \tag{37b}$$

$$[K_+, J_Z] = [J_+, K_Z] = 2K_+, \tag{37c}$$

$$[J_+, J_Z] = [K_+, K_Z] = 2J_+. \tag{37d}$$

A basis for the representation of the SO(4) algebra of interest to us, the one containing the vacuum state, can be obtained by orthonormalizing the set of vectors

$$(J_+)^{n_b}(K_+)^{n_c}|0), \tag{38}$$

under the restriction

$$0 \leq n_B + 2n_c \leq 2\Omega. \tag{39}$$

(The origin of this restriction will be evident below.) At this point the only information necessary to proceed is contained in (38) which informs us to look for a mapping onto a boson basis

$$(n_b!n_c!)^{-\frac{1}{2}}(b^+)^{n_b}(c^+)^{n_c}|0). \tag{40}$$

We build on the knowledge of SU(2) and take (40) to be the basis for the group chain SO(4) $\supset$ SU(2) $\supset$ SU(1), where Ω is associated with SO(4), n_c defines the pseudo-spin of SU(2) and n_b the magnetic quantum number. The SU(2) is that associated with the vector $\vec{J}$.

To obtain the mapping, we start with a "trivial" generalization of (24), (25),

$$J_Z = -\Omega + \hat{n}_c + \hat{n}_b, \tag{41}$$

$$J_+ = b^+[2\Omega - 2\hat{n}_c - \hat{n}_b]^{\frac{1}{2}} \equiv b^+ r(\hat{n}_b + 2\hat{n}_c). \tag{42}$$

These formulas show that for each value of n_c, the basis states (40) form, for varying n_b, a representation of SU(2) with pseudospin $J = \Omega - n_c$. It follows that the boson operator c^+ decreases the pseudospin by unity and increases its z component by the same unit.

Making use of the commutation relations (37) and the known selection rules, we can derive closed forms for the remaining three operators $\vec{K}$, which constitute a vector under the pseudospin defined by $\vec{J}$. The method has been described in detail for an SO(5) algebra[20,21].

We find[22]

$$K_+ = (K_-)^+ = c^+ r(\hat{n}_b + 2\hat{n}_c) r(\hat{n}_b + 2\hat{n}_c - 1) \emptyset(\hat{n}_c) - b^+ b^+ c\, \emptyset\, (\hat{n}_c - 1)\ , \tag{43}$$

$$K_Z = c^+ b r(\hat{n}_b + 2\hat{n}_c)\ \emptyset\ (\hat{n}_c) + \emptyset\ (\hat{n}_c) r(\hat{n}_b + 2\hat{n}_c) b^+ c, \tag{44}$$

where

$$\emptyset(n) = \left[\frac{2\Omega - n + 1}{(2\Omega - 2n - 1)(2\Omega - 2n + 1)}\right]^{\frac{1}{2}} . \tag{45}$$

For the application we have in mind, it is essential to remark that starting from (38) and (40) there is another mapping of interest. We note that $K_{\pm}$, J_Z also constitute an SU(2) subalgebra, and in consequence, we obtain upon examination of (37) a mapping simply by interchanging $K_{\pm} \leftrightarrow J_{\pm}$ and $b \leftrightarrow c$ in the formulas (41) - (45).

4. Application of the Previous Results to a Schematic Model.

The utility of the previous results may be illustrated by means of the Hamiltonian

$$H = -GJ_+J_- \ -F\, K_+K_- . \tag{46}$$

By analogy with the single j shell model we shall call the first term the pairing interaction and the second term the quadrupole interaction. For F = 0, H has the eigenvalues

$$\begin{aligned} H(F = 0) &= -G\, n_b(2\Omega - 2n_c - n_b + 1) \\ &= -G(n - n_c)(2\Omega - n - n_c + 1) \end{aligned} \tag{47}$$

where $n = n_b + n_c$ is half the number of fermions. For fixed n, we may interpret n_c as a vibrational quantum number. For low-lying states $(n_c/\Omega) \ll 1$. We expect this condition to hold even if $F \neq 0$, and thus as long as we are in the phase dominated by the first term of (43), we also expect that an accurate version of the mapping could be obtained from an expansion in powers of (n_c/Ω). The mapping in question may be called the seniority mapping and the approximate mapping the approximate seniority mapping.

By the same token, in the limit $G \to 0$, using the alternative mapping, we have

$$H(G = 0) = -F(n - n_b)(2\Omega \ - n - n_b + 1). \tag{48}$$

which interchanges the roles of n_b and n_c, as previously stated.

Here, the low-lying states have $(n_b/\Omega) \ll 1$. Approximations incorporating this condition would then be valid for the low-lying states in the deformed phase $G \ll F$. We may consider this discussion as a lesson of the two mappings.

5. Application of the Previous Considerations to the Problem of Deriving the IBM.

To illustrate the problem, we shall deal with a special case of the algebra $SO(2\Lambda)$ defined in Sec. 2, that of the single j shell, $\Lambda = 2j + 1$. The generators are conveniently taken as spherical tensors,

$$A_{JM}^{+} = 2^{-\frac{1}{2}} \sum_{mm'} (jm\ jm'|JM) a_{jm}^{+} a_{jm'}^{+} \tag{49}$$

$$B_{JM} = \sum_{mm'} (jm\ j\text{-}m'|JM)(-1)^{j-m'} a_{jm}^{+} a_{jm'}, \tag{50}$$

satisfying well-known commutation relations which we shall not reproduce[23]. For a conventional Hamiltonian

$$H = H_p + H_Q$$
$$= -GA_{\infty}^{+}A_{\infty} - F \sum_{M} B_{2M}B_{2-M}(-1)^{M} \quad , \tag{51}$$

in analogy with our remarks about the model with the symmetry of SO(4), we want at least two mappings of the algebra SO(2(2j+1). In contrast to the SO(4) case, where the availability of two mappings was a luxury, here it is a necessity because no mappings are known which satisfy the twin criteria of exactness and usefulness. We seek one mapping which diagonalizes the pairing interaction, H_p, and has a small expansion parameter in the pairing dominant phase (seniority mapping). We seek a second mapping which diagonalizes (at least approximately) the quadrupole interaction H_Q and has a small expansion parameter in the deformed region.

The importance of the seniority mapping was first noted by us long ago[23] and has since been reinvented a number of times, most recently in relation to the IBM problem in an improved form by Otsuka et al[24]. It has, unfortunately, been applied also to deformed nuclei, beyond its range of applicability.

We describe briefly our method of obtaining this mapping using the modified HP approach. We start with the non-orthonormal basis

$$(A_{\infty}^{+})^{n_s} \prod_{m} (A_{2m}^{+})^{n_m} |0\rangle , \tag{52}$$

utilyzing only pairs with angular momentum zero and two. For

$j \geq 3/2$ and for suitably restricted values of the integers n_s and n_m, these are a set of linearly independent vectors. We wish to map this vector space onto the space (10). Of course to obtain the seniority mapping we should replace the rudimentary basis (10) by the basis classified according to the group chain relevant to vibrational nuclei,

$$SU(6) \supset SU(5) \supset O(5) \supset O(3) \supset O(2). \qquad (53)$$

For our purposes in fact (10) will do because each state of this basis is an eigenstate of S^2 and S_z, where

$$S_z = 1/2(N - \Omega) = 1/2(2\hat{n}_s + 2\hat{n}_d - \Omega), \qquad (54)$$

$$S^2 = A_\infty^+ A_\infty + S_z^2 - S_z, \qquad (55)$$

and

$$A^+ = s^+ r^{\frac{1}{2}}, \; r = 1 - \Omega^{-1}[\hat{n}_s + 2\hat{n}_d] \; . \qquad (56)$$

The remaining operators of interest to us, namely A_{2m}^+ and B_{2m} are found by satisfying the relevant commutators as <u>well</u> <u>as</u> <u>possible</u> within the space (10) under constraints (54-56) and permitting an expansion only in the number of d-bosons[25]. We find, for example, the expressions

$$\begin{aligned} B_{2m} &= (2/\Omega)^{\frac{1}{2}} d_m^+ \, s \, r^{-\frac{1}{2}} f_1(\hat{n}_d) \\ &+ (2/\Omega)^{\frac{1}{2}} s^+ d_m (r + \Omega^{-1})^{\frac{1}{2}} f_1(\hat{n}_d - 1) \\ &- 10 \left\{ {222 \atop jjj} \right\} [d^+ x \tilde{d}]_m^{(2)} (r - s^+ s \Omega^{-1}) [1 - (2n_d/\Omega)]^{-1} + \ldots, \end{aligned} \qquad (57)$$

$$\begin{aligned} A_{2m}^+ &= d_m^+ [r(r - \Omega^{-1})]^{\frac{1}{2}} f_1(\hat{n}_d) \\ &\quad - \Omega^{-1} f_1(\hat{n}_d) s^+ s^+ \tilde{d}_m \\ &+ \frac{20}{(2\Omega)^{\frac{1}{2}}} \left\{ {222 \atop jjj} \right\} s^+ [d^+ x \tilde{d}]_m^{(2)} r^{-\frac{1}{2}} [1 - (2\hat{n}_d/\Omega)]^{-1} + \ldots, \end{aligned} \qquad (58)$$

where

$$f_1(n_d) = \left[\frac{\bar{r} + \Omega^{-1}(1 + 1/3 \, n_d)}{(\bar{r} - \Omega^{-1})(\bar{r} + \Omega^{-1})} \right], \qquad (59)$$

$$\bar{r} = 1 - (2\hat{n}_d/\Omega). \qquad (60)$$

Actually we have obtained the next term in (58) and corresponding expansions for the multi-level case. The dynamical method to be applied in conjunction with these results has also been described elsewhere[25]. Applications are underway which should complement results which are beginning to appear in the literature[3,26,27].

There remains the problem of what to do for deformed nuclei. We have only very recently girded our loins to try to attack this problem. One point of view is that we are looking for a mapping to a basis in the space (10) classified according to the group chain

$$SU(6) \supset SU(3) \supset O(3) \supset O(2). \qquad (61)$$

In the case represented by (53), we succeeded in expressing the operators associated with SU(6) and SU(5), $\hat{n} = \hat{n}_s + \hat{n}_d$ and $\hat{n}_d$ respectively, exactly in terms of generators belonging, in a restricted sense, to the fermion subspace (52). In the present instance we need corresponding expressions for the pair of quantum numbers which characterize the representations of SU(3). Another way of framing the same problem is to consider the generators of SU(3) in the boson basis, namely

$$(d^+ \times d)^{(1)}_m \qquad (62)$$

and

$$s^+ d_m + d_m^+ s - (\sqrt{7}/2)(d^+ \times d)^{(2)}_m \qquad (63)$$

and ask for those functions of the generators which map into these expressions. For (61) it is B_{1m}, up to a scale factor. By the same reasoning, comparison of the commutation relations, there can be no proportionality to B_{2m}. Our current efforts are directed toward resolving this dilemma.[28]

REFERENCES

1. S.T. Beliaev and V.G. Zelevinsky, Nucl. Phys. 39 (1962) 582. We refer to the method of this paper as BZ.
2. The literature may be traced from K.J. Weeks and T. Tamura, Phys. Rev. C22 (1980) 1323 and from T. Kishimoto and T. Tamura, Phys. Rev. C, to be published.
3. The literature may be traced from S. Pittel, P.P. Duval and B.R. Barrett, to be published in Annals of Physics (N.Y.).
4. The literature may be traced from K. Kumar, Erice Lectures on Dynamic Deformation Theory (1982), to be published in Prog. in Particle and Nuclear Physics, ed. by D. Wilkinson.
5. A. Klein, C.T. Li, and M. Vallieres, Phys. Rev. C25 (1982) 2733.
6. See the discussion by V. Paar in Interacting Bosons in Nuclear Physics, ed. by F. Iachello (Plenum, N.Y. 1979) p. 163.

7. For a good review, see E.R. Marshalek, Nucl. Phys. A347 (1980) 253.
8. F.J. Dyson, Phys. Rev. 102 (1956) 1217.
9. T. Holstein and H. Primakoff, Phys. Rev. 58 (1940) 1098.
10. T. Marumori, M. Yamamura, and A. Tokunoga, Prog. Theor. Phys. 31 (1964) 1009. We refer to the method of this paper as MYT.
11. For a review, see P. Ring and P. Shuck, Nuclear Many Body Problem (Springer, N.Y. 1980) Chap. 9.
12. H.B. Geyer and F.J.W. Hahne, Nucl. Phys. A363 (1981), 45; F.J.W. Hahne, Phys. Rev. C23 (1981) 2305.
13. C.T. Li, U. of Pennsylvania preprint UPR-0197-T (1982).
14. S.C. Pang, A. Klein, and R.M. Dreizler, Ann. Phys. (N.Y.) 49 (1968) 477.
15. S.Y. Li, A. Klein, and R.M. Dreizler, J. Math. Phys. 11 (1970) 975.
16. J.A. Evans and N. Kraus, Phys. Lett. B37 (1972) 455.
17. A. Arima, N. Yoshida, and J.N. Ginocchio, Phys. Lett. 101B, (1981) 209.
18. B. Jancovici and D.H. Schiff, Nucl. Phys. 58 (1964) 678.
19. J. Dobaczewski, Nucl. Phys. A369 (1981) 213; A369 (1981) 237.
20. A. Klein, H. Rafelski, and J. Rabelski, Nucl. Phys. A355 (1981) 189.
21. A. Klein, T.D. Cohen, and C.T. Li, Ann. Phys. 142 (1982) in press.
22. This mapping was first derived by K. Matsuyanagi, private communication.
23. For example, see S.Y. Li, R.M. Dreizler and A. Klein, Phys. Rev. C4 (1971) 1571.
24. T. Otsuka, A. Arima, and F. Iachello, Nucl. Phys. A309 (1978) 1.
25. Some details of this mapping have been given by A. Klein, C.T. Li, T.D. Cohen, and M.Vallieres, U. of Pennsylvania preprint UPR-0194T, to be published in "Collective Bands in Nuclei", D. Wilkinson, ed., (Pergamon, N.Y., 1982).
26. P.D. Duval and B. R. Barrett, Phys. Rev. C24 (1981) 1272.
27. Y.K. Gambhir, P. Ring, and P. Schuck, Phys. Rev. C25, (1982) 2858.

28. The route to a solution of this problem is to look for an inverse mapping from the boson basis characterized by (61) back onto the shell model basis. The difficulty is that there is no SU(3) subalgebra of the shell model Lie algebra. The resolution of this dilemma is to construct <u>non-linear</u> functions of the generators of the shell-model algebra which satisfy (approximately) the SU(3) algebra and to consrtuct irreducible tensors by a line procedure.

DOES ACCIDENTAL DEGENERACY IMPLY A SYMMETRY GROUP?

M.Moshinsky*
Instituto de Fisica, UNAM
Apdo. Postal 20-364, 01000 México, D.F. México

C.Quesne
Université Libre de Bruxelles
Physique Théorique et Mathématique - CP229
Brussels, Belgium

ABSTRACT

The object of this paper is to probe the relation, which is usually taken for granted, between the appearance of accidental degeneracy in the energy levels of a given Hamiltonian and its symmetry group. This is done by analyzing the very simple problem of an oscillator to which a particular spin-orbit and centrifugal force are added. We find the operators that connect all the states of given energy as well as their corresponding observables in the classical limit. The Poisson bracket relations between these observables leads to a Lie Algebra U(3)xSU(2), but it <u>does not</u> translate into a Lie Algebra for the commutators of the corresponding operators, as some matrix elements of commutators, corresponding to Poisson brackets that are zero, do not vanish. Thus while accidental degeneracy in the quantum problem may lead to a larger group in the classical limit, it is not always given by the dimensions of the irreducible representations of this group.

The full paper will appear in Journal of Mathematical Physics (1982).

* Member of the Instituto Nacional de Investigaciones Nucleares and El Colegio Nacional.

"A HIDDEN SYMMETRY IN COLLECTIVE EXCITATIONS OF MANY-BODY SYSTEMS"

M.Moshinsky
Instituto de Fısica, UNAM

Q.Castanos and A. Frank
Centro de Estudios Nucleares, UNAM
Circuito Exterior, C.U.
Delegacion Coyoacan
04510 México, D.F.

ABSTRACT

A collective Hamiltonian derived from an A-nucleon system is analyzed in the large A-limit and its hidden symmetry group identified.

1- INTRODUCTION

In several recent investigations[1-3] a collective Hamiltonian is projected out from the A-body Hamiltonian by restricting it to a definite irreducible representation (I.R.) of the orthogonal group O(n) associated to the n=A-1 Jacobi vectors. Studying an A-nucleon system interacting through harmonic oscillator forces, Deenen and Quesne[4] were able to prove that the symmetry group associated to the O(n)-scalar collective Hamiltonian is a U(6) group, which is very suggestive, given the phenomenological success of the Interacting Boson Model[5]. One should be aware, however, that the O(n)-scalar collective Hamiltonian is in general inconsistent with the Pauli Principle, and a more realistic problem is that of projecting the many - nucleon problem onto a definite I.R. of O(n) consistent with the latter. In this paper we present such an analysis for an A-nucleon system interacting through harmonic-oscillator forces in the limit when A is very large.

2- THE Sp(6) COLLECTIVE GENERATORS AND THE Z.D. TRANSFORMATION

The translationally invariant Hamiltonian for a system of A

particles interacting through harmonic-oscillator forces is given by

$$H = \frac{1}{2} \sum_{s=1}^{n} \sum_{i=1}^{3} (P_{is}^2 + X_{is}^2) \quad , \tag{2.1}$$

where X_{is} are the Jacobi coordinates, P_{is} their corresponding momenta and we chose units in which $\hbar$, the mass of the nucleon and an appropriate frequency are 1.

A possible decomposition of the symmetry group U(3n) is

$$\begin{array}{lllllll} & U(n) & \supset & O(n) & \supset & S_{n+1} & \\ & [h_1,h_2,h_3] & & (w_1,w_2,w_3) & & \{f\} & \\ U(3n) \supset & X & & & & & \\ [N] & U(3) & \supset & SU(3) & \supset \theta(3) \supset & \theta(2) & , \\ & [h_1,h_2,h_3] & & (\lambda,\mu) & L & M & \end{array} \tag{2.2}$$

where underneath each group we write down the quantum numbers characterizing its I.R. (We denote by S_{n+1} the symmetric group characterized by the partition $\{f\}$). The collective part of the Hamiltonian (2.1) is obtained by projecting on a definite I.R. (w_1,w_2,w_3) of O(n), which is obtained by constructing the set of "compact states"[6], i.e., the lowest possible energy states that satisfy the restrictions imposed by the Pauli Principle, then take linear combinations of these Slater determinants to build states characterized by the number of quanta N and the I.R. (λ_m, μ_m) corresponding to the maximal eigenvalue of the quadratic Casimir operator of the SU(3) group[7]. By construction, this ground state is such that

$$(w_1 w_2 w_3) = (h_1^o h_2^o h_3^o), \text{ where } \lambda_m = h_1^o - h_2^o \text{ and } \mu_m = h_2^o - h_3^o$$

The Sp(6) group plays the role of a dynamical group for the collective excitations of the system, due to the complementarity of Sp(6), O(n) in Sp(6n), (the last of which is a dynamical group for U(3n)) i.e.,

$$Sp(6n) \supset Sp(6) \times O(n) \tag{2.3}$$

We now write down the Sp(6) generators in Jacobi coordinates and momenta as

$$\hat{C}_{ij} = \frac{1}{2} \{\hat{q}_{ij} + \hat{T}_{ij} + i\, \hat{L}_{ij}\}, \tag{2.4a}$$

$$\hat{B}^{+}_{ij} = \frac{1}{2} \{\hat{q}_{ij} - \hat{T}_{ij} - n\, \delta_{ij} - i\, \hat{S}_{ij}\}, \tag{2.4b}$$

$$\hat{B}_{ij} = \frac{1}{2} \{\hat{q}_{ij} - \hat{T}_{ij} + n\, \delta_{ij} + i\, S_{ij}\}, \tag{2.4c}$$

where

$$\hat{q}_{ij} = \sum_{s=1}^{n} X_{is} X_{js} \,, \quad \hat{S}_{ij} = \sum_{s=1}^{n} (X_{is} P_{js} + X_{js} P_{is}),$$

$$T_{ij} = \sum_{s=1}^{n} P_{is} P_{js} \,, \quad \hat{L}_{ij} = \sum_{s=1}^{n} (X_{is} P_{js} - X_{js} P_{is}). \tag{2.6}$$

The next step is to carry out the Zickendraht-Dzublik (Z.D.) transformation[8], given by

$$X_{is} = \sum_{k=1}^{3} \rho_k \, D^{1_3}_{ki} (\theta_j) \, D^{1_n}_{n-3+k,s} (\phi), \tag{2.6}$$

where $D^{1_3}_{ki}$, $D^{1_n}_{n-3+k,s}$ are the fundamental matrix representations of the $\theta(3)$ and $O(n)$ groups respectively. This change of coordinates permits a separation of the collective and non-collective parts of the Hamiltonian[1,6].

Writting down the Sp(6) generators (2.4) in terms of the Z.D. coordinates is a rather difficult business[6,9] and we omit the details here. In the next section we consider a further transformation to make contact with variables appearing in phenomenological collective models of the nuclei and then consider the limiting situation when A is very large.

3- CONTRACTION OF THE Sp(6) GENERATORS AND IDENTIFICATION OF THE SYMMETRY GROUP

After expressing the operators (2.4) in terms of the new variables (2.6) it is convenient to carry out the additional coordinate transformations[6,10]:

$$\rho_k^2 = \frac{\rho^2}{3}\{1 + 2b\cos(\gamma - 2\pi k/3)\} \tag{3.1a}$$

$$1 + 2\frac{\beta^2}{\sigma^2} = (1-b^2)\,(1-3b^2 \mp 2b^3\cos 3\gamma)^{-1} \tag{3.1b}$$

$$\bar{\alpha} = \sqrt{2}\,\sigma\,\ell n\,(\rho/\sigma) \tag{3.1c}$$

where the parameter $\sigma^2 \equiv w_1 + w_2 + w_3 + \frac{3}{2}(n-4)$ is related to energy of the ground state and, by construction[6,10], behaves to highest order in A as $\sigma^2 \approx A^{4/3}$ so $A >> 1$ implies $\sigma >> 1$.

In terms of $\bar{\alpha}$, β, γ the collective Hamiltonian, defined as the matrix of (2.1) with respect to the Gelfand and Zetlin[11,12] states of O(n), is to highest order[6] in σ

$$\boldsymbol{H}_{coll} = \left(-\frac{\partial^2}{\partial\bar{\alpha}^2} + \bar{\alpha}^2\right).\boldsymbol{I}$$

$$+ \left(-\frac{1}{\beta^4}\frac{\partial}{\partial\beta}\beta^4\frac{\partial}{\partial\beta} - \frac{1}{\beta^2\sin 3\gamma}\frac{\partial}{\partial\gamma}\sin 3\gamma\frac{\partial}{\partial\gamma} + \beta^2\right)\boldsymbol{I} + \sum_{k=1}^{3}\frac{\boldsymbol{J}_k'^2}{4\beta^2\sin^2(\gamma - 2\pi k/s)} \tag{3.2}$$

where $\boldsymbol{I}$ is the unit matrix and

$$\boldsymbol{J}_{k'} = L_k'\,\boldsymbol{I} + \boldsymbol{L}_k', \quad k = 1, 2, 3, \tag{3.3a}$$

where L_k' are the components of the (real) angular momentum in the frame of reference fixed in the body, while $\boldsymbol{L}_k'$ are certain intrinsic operators, whose matrix elements are those of O(3) generators in a Unitary Gelfand and Zetlin basis $\boldsymbol{U}(3) \supset \boldsymbol{U}(2) \supset \boldsymbol{U}(1)$[6,9]. If we denote the $\boldsymbol{U}(3)$ generators by $\boldsymbol{C}_{ij}$, then

$$\begin{aligned} \boldsymbol{L}_1' &= i(\boldsymbol{C}_{23} - \boldsymbol{C}_{32}), \\ \boldsymbol{L}_2' &= i(\boldsymbol{C}_{31} - \boldsymbol{C}_{13}), \\ \boldsymbol{L}_3' &= i(\boldsymbol{C}_{12} - \boldsymbol{C}_{21}), \end{aligned} \tag{3.3b}$$

The Hamiltonian (3.2) has the structure of a one-dimensional oscillator in the variable $\bar{\alpha}$ plus a five dimensional oscillator of the

Bohr-Mottelson type. However, instead of the usual L'_k angular momentum components one encounters a "total angular momentum" $\boldsymbol{J}'_k$ composed of the latter and the intrinsic operators $\boldsymbol{L}'_k$ which behave as an additional spin for the system. Clearly $\boldsymbol{H}_{coll}$ has not in general a U(6) symmetry group, but has an additional degeneracy brought about by the presence of $\boldsymbol{L}'_k$.

To investigate the symmetry group associated to the Hamiltonian (3.2) we analize the behavior of the collective Sp(6) generators, defined analogously to $\boldsymbol{H}_{coll}$, in the same large A situation. To do so, we expand these operators in a power series in the parameter σ and then proceed to the limiting case, which in effect is a contraction procedure[13]. We find that the operators (whose explicit expressions we omit here)

$$\underset{\sim}{b}^{\dagger}_{\ell m} \equiv \lim_{\sigma\to\infty} \left(-\frac{\sqrt{3}}{2\sigma} \underset{\sim}{B}^{+}_{\ell}\right), \quad \ell = 0,\, 2 \quad , \tag{3.4a}$$

$$\underset{\sim}{b}^{\ell m} \equiv \lim_{\sigma\to\infty} \left(-\frac{\sqrt{3}}{2\sigma} \underset{\sim}{B}^{\ell m}\right), \quad \ell = 0,\, 2 \quad , \tag{3.4b}$$

$$\underset{\sim}{c}_{\ell m} \equiv \lim_{\sigma\to\infty} \left(\underset{\sim}{C}_{\ell m} + \frac{\sigma^2}{\sqrt{3}} \delta_{\ell o} \boldsymbol{I}\right), \quad \ell = 0,\, 1,\, 2 \quad , \tag{3.4c}$$

where $B^{+}_{\ell m}$, $B^{\ell m}$, $C_{\ell m}$ are the Sp(6) generators (2.4) written as spherical tensors. The $b^{+}_{\ell m}$, $b^{\ell m}$, $c_{\ell m}$ are the generators of a $W(6)\wedge U(3)$ Lie Algebra, where $\wedge$ indicates a semi-direct product, W(6) is a Weyl group[14] in 6 dimensions and U(3) a unitary group in three dimensions. We thus conclude that $W(6) \wedge U(3)$ constitutes the dynamical group for the A >> 1 collective Hamiltonian (3.2). In fact, on can verify by a direct calculation that

$$\boldsymbol{H}_{coll} = 2 \sum_{\ell m} \underset{\sim}{b}^{+}_{\ell m} \underset{\sim}{b}^{\ell m} + 6\boldsymbol{I} \quad . \tag{3.5}$$

What is the symmetry group of $\boldsymbol{H}_{coll}$? It is quite obvious from (3.5) that the set of operators

$$\underset{\sim}{\xi}^{\ell' m'}_{\ell m} = \underset{\sim}{b}^{+}_{\ell m} \underset{\sim}{b}^{\ell' m'} \quad \ell, \ell' = 0,\, 2 \quad , \tag{3.6}$$

commute with it, and generate a U(6) group. In regards to the U(3) group generated by the $C_{\ell m}$, eq. (3.4c)), we readily find

$$\underset{\sim}{C}_{oo} = - \frac{1}{\sqrt{3}} \boldsymbol{H}_{coll} = - \frac{1}{\sqrt{3}} \{2 \sum_{\ell m} \underset{\sim}{\xi}_{\ell m}^{\ell m} + 6\boldsymbol{I} \} \quad , \tag{3.7}$$

and thus $\underset{\sim}{C}_{oo}$ is already contained in the U(6) algebra generated by the operators (3.6). The SU(3) subalgebra generated by $\underset{\sim}{C}_{1m}$ and $\underset{\sim}{C}_{2m}$ is not, however, contained in the U(6) algebra. To show this point, we explicitly construct the generators of the SU(3) subgroup of U(6), which we denote by SU*(3)

$$\boldsymbol{T}_{1m} = \sqrt{6} \sum_{\mu,\mu'} < 2\mu\ 1m | 2\mu' > \underset{\sim}{b}^{+}_{2\mu'}\ \underset{\sim}{b}^{2\mu}, \tag{3.8a}$$

$$\boldsymbol{Q}_{2m} = \sqrt{\frac{7}{3}} [\underset{\sim}{b}^{+}_{2} \times \underset{\sim}{b}_{2}]^{2}_{m} + \sqrt{\frac{4}{3}} (\underset{\sim}{b}^{+}_{oo} \underset{\sim}{b}_{2m} + \underset{\sim}{b}^{+}_{2m} \underset{\sim}{b}_{oo}) \quad , \tag{3.8b}$$

and compare the resulting expressions with (3.4c) for $\ell = 1, 2$. The results are as follows:

$$\underset{\sim}{C}_{1m} = - \frac{1}{\sqrt{2}} (\boldsymbol{T}_{1m} + \boldsymbol{O}_{1m}) \quad , \tag{3.9a}$$

$$\underset{\sim}{C}_{2m} = - (\boldsymbol{Q}_{2m} - \boldsymbol{S}_{2m}) \quad , \tag{3.9b}$$

where

$$\boldsymbol{O}_{1m} = D'_{om} \boldsymbol{L}'_{3} + \frac{1}{\sqrt{2}} (D'_{-1m} - D'_{1m}) \boldsymbol{L}'_{1} - \frac{i}{\sqrt{2}} (D'_{-1m} + D'_{1m}) \boldsymbol{L}'_{2} \quad , \tag{3.10a}$$

$$\begin{aligned} \boldsymbol{S}_{2m} = & -i\ (D^{2}_{1m} - D^{2}_{-1m})\ (\boldsymbol{C}_{23} + \boldsymbol{C}_{32}) \quad , \\ & + (D^{2}_{1m} - D^{2}_{-1m}) (\boldsymbol{C}_{31} + \boldsymbol{C}_{13}) + i\ (D^{2}_{2m} - D^{2}_{-2m}) (\boldsymbol{C}_{12} + \boldsymbol{C}_{21}) \quad , \end{aligned} \tag{3.10b}$$

with $\boldsymbol{L}'_{k}$, $\boldsymbol{C}_{ij}$ defined in (3.3). Thus it is clear that the two groups SU(3) and SU*(3) are distinct. Furthermore, since $[\boldsymbol{H}_{coll}, \underset{\sim}{C}_{\ell m}] = -\sqrt{3}\ [\underset{\sim}{C}_{oo}, \underset{\sim}{C}_{\ell m}] = 0$, $\ell = 1, 2$ SU(3) is an additional symmetry group for $\boldsymbol{H}_{coll}$. Finally, computing the commutators between the U(6) and SU(3) generators we conclude that the symmetry group for $\boldsymbol{H}_{coll}$ is the semi-direct product group.

$$G = U(6) \wedge SU(3) \tag{3.11}$$

From (3.9) it is clear that for the particular case where one takes either the scalar or the closed Shell ((w w w)) I.R. of O(n), the SU(3) and $SU^*(3)$ groups are identical, since the matrix elements of the U(3) generators, $\mathcal{C}_{ij}$ $i \neq j$ vanish. Thus, for these cases G=U(6), in accordance to previous investigations[4,15].

REFERENCES

1- V.Vanagas, "The Microscopic Nuclear Theory", Lecture Notes in Physics (University of Toronto, Toronto, 1977); "The Microscopic Theory of the Collective Motion in Nuclei", p.220, in "Group Theory and Its Applications in Physics-1980" Ed. by T.H. Seligman (AIP, New York, 1980).

2- G.F.Filippov, V.I. Ovcharenko,Yu, F.Smirnov, "Microscopic Theory of Collective Excitations in Nuclei" (in Russian) (Navkova Dumka, Kiev, 1981).

3- E.Chacon, M.Moshinsky, V.Vanagas, J. Math. Phys. 22 605 (1981).

4- J.Deenen, C.Quesne, J. Math. Phys. (in press); J.Deenen, C.Quesne, private communication.

5- A.Arima, F.Iachello, Ann. of Phys. 99 (1976) 253; 123 (1977) 468; 111 (1978), 201.

6- O.Castanos, A.Frank, E.Chacon, P.O.Hess, M.Moshinsky, J.Math.Phys. (in press).

7- L.Sabaliauskas, Liet. fiz. rink. 19, 5 (1979) (in Russian).

8- W.Zickendraht, J.Math.Phys. 10, 30 (1969); 12, 1663 (1971); A. Ya Dzublik, V.I. Ovcharenko, A.I.Steshenko, G.F.Filippov, Yad.Fiz. 15, 869 (1972) Sov.J.Nucl.Phys. 15, 487 (1972).

9- O.Castanos, A.Frank, to be published.

10- O.Castanos, A.Frank, E.Chacon, P.O.Hess, M.Moshinsky, Phys.Rev. C25, (1982) 1611.

11- I.M.Gelfand, M.L.Zetlin, Dokl. Akad Navk. USSR 71, 1017 (1950).

12- S.C.Pang, K.T.Hecht, J.Math.Phys. 8, 1233 (1967).

13- R.Gilmore, "Lie Groups, Lie Algebras, and some of their Applications" (John Wiley, N.Y., 1974).

14- B.G.Wybourne, "Classical Groups for Physicists", (John Wiley, N.Y. 1973).

15- P.Kramer, preprint.

THE U(6) SYMMETRY IN THE MICROSCOPIC COLLECTIVE MODEL

J. Deenen and C. Quesne*

Service de Physique Théorique et Mathématique CP 229,
Université Libre de Bruxelles, Bd du Triomphe, B 1050 Brussels,
BELGIUM

ABSTRACT

For any number of particles, the collective part of the harmonic oscillator Hamiltonian is shown to reduce to a six-dimensional harmonic oscillator. There is therefore a hidden U(6) symmetry in the microscopic collective model.

In the microscopic collective model of the nucleus proposed by Vanagas [1)] and Filippov [2)], the collective subspace of the A nucleon space is assumed to be spanned by the translationally invariant A nucleon states characterized by a definite irreducible representation (irrep) $(\lambda_1\ \lambda_2\ \lambda_3)$ of the orthogonal group O(n) associated with the n = A-1 Jacobi vectors. The collective part of the Hamiltonian of the A nucleon system is obtained by projecting the latter on the irrep $(\lambda_1\ \lambda_2\ \lambda_3)$. As it is often the case in nuclear physics, the simple Hamiltonian corresponding to harmonic oscillator interactions between the nucleons turns out to have some interesting properties. It was shown that in the limit $A \to \infty$, it provides us with a complete set of states for the A nucleon system in appropriate coordinates including the collective ones [3)]. In the present contribution, we wish to show that for any A its collective part reduces to the Hamiltonian of a six-dimensional oscillator and therefore has a U(6) symmetry group.

---

* Maître de recherches F.N.R.S.

A nucleon states can be classified according to the following chains of groups [4]

$$Sp(6n,R) \supset Sp(6,R) \times O(n) \quad , \quad Sp(6,R) \supset U(3) \supset SO(3). \quad (1)$$

The $Sp(6n,R)$ group is generated by the bilinear operators in the boson creation and annihilation operators η_{is} and ξ_{is}, i = 1, 2, 3, s = 1, ..., n, associated with the Jacobi coordinates and momenta, while $Sp(6,R)$ is generated by those bilinear operators which are invariant under O(n). Collective states characterized by the irrep $(\lambda_1 \lambda_2 \lambda_3)$ of O(n) belong to a single irrep of $Sp(6,R)$, specified by its minimum weight $\langle \frac{n}{2}+\lambda_3, \frac{n}{2}+\lambda_2, \frac{n}{2}+\lambda_1 \rangle$. The $Sp(6,R)$ group is therefore the dynamical group of collective states.

Let us consider the representation of collective states in a Bargmann space of analytic functions in 3n complex variables z_{is}, i = 1, 2, 3, s = 1, ..., n. When restricted to the subspace of Bargmann space, characterized by the irrep $(\lambda_1 \lambda_2 \lambda_3)$ of O(n), the $Sp(6,R)$ generators assume a specific form, that we are now going to derive. The corresponding group will be denoted by $Sp_c(6,R)$. The derivation proceeds in two steps : first we shall consider the case where $(\lambda_1 \lambda_2 \lambda_3)$ is the scalar representation, i.e., $\lambda_1 = \lambda_2 = \lambda_3 = 0$; second we shall treat the general case by making use of the results previously obtained.

When collective states are assumed to be invariant under O(n), their representations in Bargmann space are analytic functions in the six variables $w_{ij} = \sum_s z_{is} z_{js} = w_{ji}$ invariant under O(n). In analogy with what is done in Schrödinger representation [1,2,3], the w_{ij} are called collective variables. The generators of $Sp_s(6,R)$ (when the s index reminds us that they are restricted to scalar states) can be written in terms of w_{ij} and $\Delta_{ij} = (1+\delta_{ij})\partial/\partial w_{ij}$, as

$$\underset{\sim}{\mathcal{D}}^{s+} = \underset{\sim}{w} \quad , \quad \underset{\sim}{\mathcal{D}}^{s} = \underset{\sim}{\Delta}\left[\underset{\sim}{\mathcal{C}}^{s} + (n-4)\underset{\sim}{I}\right] \quad , \quad \underset{\sim}{\mathcal{E}}^{s} \doteq \underset{\sim}{\mathcal{C}}^{s} + \frac{n}{2}\underset{\sim}{I} \quad , \quad (2)$$

where we use a matrix notation, and $\underset{\sim}{\mathcal{C}}^{s} = \underset{\sim}{w}\,\underset{\sim}{\Delta}$ denotes the 3x3 matrix of the $U_s(3)$ subgroup generators.

The analytic functions in the six variables w_{ij} can be considered as the representation in Bargmann space of boson states built from six independent boson operators $a^+_{ij} = a^+_{ji}$, i, j = 1, 2, 3, [or, equivalently, unnormalized boson operators defined by $\bar{a}^+_{ij} = (1+\delta_{ij})^{1/2} a^+_{ij}$]. One can establish [4] that there is a one-to-one correspondence between scalar collective states classified according to the chain $\mathcal{S}p_s(6,R) \supset \mathcal{U}_s(3)$ and boson states classified according to the chain $\mathcal{U}(6) \supset \mathcal{U}(3)$. In such a mapping, the generators of both U(3) groups correspond to one another, i.e., $\underset{\sim}{\mathcal{C}}^s = \underset{\sim}{\mathfrak{C}} \equiv \underset{\sim}{\bar{a}}^+ \underset{\sim}{\bar{a}}$. In this way, the scalar part $\mathcal{H}^s_{osc}$ of the harmonic oscillator Hamiltonian $\mathcal{H}_{osc}$, which is just the first-order Casimir operator of $\mathcal{U}_s(3)$, is mapped upon $2\,\Omega_{osc} = 2 \sum_{i \leq j} a^+_{ij} a_{ij}$, which is a six-dimensional oscillator whose frequency is the double of that of $\mathcal{H}_{osc}$. By studying the Dyson and Holstein-Primakoff representations of $\mathcal{S}p_s(6,R)$, and by inverting the latter in order to express the boson creation and annihilation operators in terms of the $\mathcal{S}p_s(6,R)$ generators, one gets [4] the unitary representation in quantum mechanics of the classical canonical transformation relating $\mathcal{H}^s_{osc}$ and Ω_{osc}:

$$\underset{\sim}{\bar{a}}^+ = \left[\underset{\sim}{\mathcal{C}}^s + (n-4)\underset{\sim}{I} \right]^{-\frac{1}{2}} \underset{\sim}{\mathcal{D}}^{s+} , \qquad \underset{\sim}{\bar{a}} = \underset{\sim}{\mathcal{D}}^s \left[\underset{\sim}{\mathcal{C}}^s + (n-4)\underset{\sim}{I} \right]^{-\frac{1}{2}} . \tag{3}$$

The generators $a^+_{ij} a_{kl}$ of the $\mathcal{U}(6)$ symmetry group of $\mathcal{H}^s_{osc}$ can then be expressed in terms of the $\mathcal{S}p_s(6,R)$ generators.

Turning now to the case of a general irrep $(\lambda_1 \lambda_2 \lambda_3)$, it can be shown [5] that in Bargmann space the highest weight state of any such irrep is represented by a function depending upon twelve variables instead of the whole set of 3n variables z_{is} : the six collective variables w_{ij}, already encountered in the scalar case, and six internal variables $\bar{w}_{ij}$, $i \leq j = 1, 2, 3$. When acting upon these functions of w_{ij} and $\bar{w}_{ij}$, the generators of $\mathcal{S}p_c(6,R)$ can be written as

$$\underset{\sim}{\mathcal{D}}^{c+} = \underset{\sim}{\mathcal{D}}^{s+} , \quad \underset{\sim}{\mathcal{D}}^c = \underset{\sim}{\mathcal{D}}^s + \underset{\sim}{\tilde{\bar{\mathcal{C}}}} \underset{\sim}{\Delta} + \underset{\sim}{\Delta} \underset{\sim}{\bar{\mathcal{C}}} , \quad \underset{\sim}{\mathcal{E}}^c = \underset{\sim}{\mathcal{C}}^c + \frac{n}{2} \underset{\sim}{I} , \tag{4}$$

where

$$\underset{\sim}{\mathcal{C}}^c = \underset{\sim}{\mathcal{C}}^s + \underset{\sim}{\bar{\mathcal{C}}} . \tag{5}$$

Here $\underset{\sim}{\mathcal{D}}^{s+}$, $\underset{\sim}{\mathcal{D}}^s$, and $\underset{\sim}{\mathcal{C}}^s$ only depend upon w_{ij} and $\partial/\partial w_{ij}$ and are given by Eq. (2), while $\bar{\mathcal{E}}$ only depends upon $\bar{w}_{ij}$ and $\partial/\partial\bar{w}_{ij}$. The $\bar{\mathcal{E}}_{ij}$ operators are the generators of a U(3) group, that we denote by $\bar{\mathcal{U}}(3)$. This group is not the most general U(3) group : it only has a single irrep, characterized by $[\lambda_1 \lambda_2 \lambda_3]$. Since the operators $\mathcal{E}^s_{ij}$ and $\bar{\mathcal{E}}_{ij}$ commute, $\mathcal{U}_c(3)$ is a subgroup of $\mathcal{U}_s(3) \times \bar{\mathcal{U}}(3)$.

The collective part $\mathcal{H}^c_{osc}$ of the harmonic oscillator Hamiltonian $\mathcal{H}_{osc}$, which is the first-order Casimir operator of $\mathcal{U}_c(3)$, only differs from $\mathcal{H}^s_{osc}$ by the constant $\lambda_1+\lambda_2+\lambda_3$. It therefore reduces to a six-dimensional oscillator and has a $\mathcal{U}(6)$ symmetry group. However, each boson state is now associated with a set of collective states, whose number is equal to the dimension of the irrep $[\lambda_1 \lambda_2 \lambda_3]$ of $\bar{\mathcal{U}}(3)$. The collective Hamiltonian $\mathcal{H}^c_{osc}$ has therefore an additional $U(3)$ symmetry group.

In conclusion, we have shown that for any A there is a hidden $U(6)$ symmetry in the microscopic collective model. In the limit $A\to\infty$ such a result was also obtained by Castaños and Frank [6] in the Schrödinger representation. This U(6) symmetry, which is exact for the harmonic oscillator Hamiltonian, will in general be a broken symmetry for more realistic Hamiltonians.

References

1. V. Vanagas, "The Microscopic Nuclear Theory Within the Framework of the Restricted Dynamics", Lecture Notes in Physics (University of Toronto, Toronto, 1977).
2. G.F. Filippov, V.I. Ovcharenko, and Yu. F. Smirnov, "Microscopic Theory of Collective Excitations in Nuclei" (in Russian) (Naukova Dumka, Kiev, 1981).
3. O. Castaños, A. Frank, E. Chacón, P.O. Hess, and M. Moshinsky, Phys. Rev. C25, 1611 (1982), and to be published.
4. J. Deenen and C. Quesne, J. Math. Phys. 23, 878 (1982), and to be published in J. Math. Phys.
5. J. Deenen and C. Quesne, to be published.
6. O. Castaños and A. Frank, private communication.

SO(2n+1) IN AN $[SU(2)]^n$ BASIS : SYMMETRIC REPRESENTATIONS

H. De Meyer(*), G. Vanden Berghe and P. De Wilde(°)
Seminarie voor Wiskundige Natuurkunde, R.U.G.
Krijgslaan 281-S9, B-9000 Gent, BELGIUM

The number of internal labels of symmetric irreducible representations (SIR's) of SO(2n+1) equals 2n-1, as can be verified from a simple counting rule of Seligman and Sharp [1]. One possible way to define such labels is to reduce the SIR's of SO(2n+1) into IR's of its maximal subalgebra $SO(2n-3)\otimes SU(2)\otimes SU(2)$ and to verify that no degeneracies are left. In order to derive the branching rule (restricted to SIR's) various methods are at ones disposal. Recently, the present authors used one of these to obtain the expression

$$SO(2n+1) \rightarrow SU(2)\otimes SU(2)\otimes SO(2n-3) :$$

$$[\underbrace{v,0,\dots,0}_{\substack{n-1\\ \text{zeros}}}] \rightarrow \sum_{\mu=0}^{v}\sum_{\lambda=0}^{[v-\mu/2]} (\mu/2\,,\,\mu/2\,,[\,v-\mu-2\lambda,\underbrace{0,\dots,0}_{\substack{n-3\\ \text{zeros}}}]) \tag{1}$$

From (1) it is obvious that no labels are missing in the reduction.

In the following we are concerned with the application of (1) in the case n=3. The main reason is that the reduction of SIR's of SO(7), which are closely related to the nuclear octupole phonon states, is a relevant step for solving the octupole phonon state labelling problem. In particular, one is interested in the construction of a basis of 'good' angular momentum states, i.e. states characterized by the representation label ℓ and internal label m associated to the principal SO(3) subalgebra of SO(7). The 21 generators of SO(7) divide into the generators ℓ_0, $\ell_\pm$ of that SO(3), a seven-dimensional SO(3) tensor representation q_μ $(-3\leq\mu\leq 3)$, and an eleven-dimensional SO(3) tensor representation p_ν $(-5\leq\nu\leq 5)$. On the other hand, in the chain

(*) Research associate N.F.W.O. Belgium)
(°) Research assistant I.W.O.N.L. (Belgium)

$SO(7) \supset [SU(2)]^3$ the adjoint representation reduces according to

$$[1,1,0] \to (1,0,0) + (0,1,0) + (0,0,1) + (1/2,1/2,1) \ , \qquad (2)$$

clearly showing that the 21 generators can be divided into three sets of SU(2) generators, respectively denoted by $s_0, s_\pm$; $t_0, t_\pm$; and $u_0, u_\pm$, and a twelve-dimensional bispinor-vector representation $T^{[1/2\ 1/2\ 1]}$ with respect to $[SU(2)]^3$. The explicit relationship between the two operator bases has been established elsewhere [2]. Let us mention here the relation

$$\ell_0 = s_0 + 3t_0 + 3u_0 \qquad (3)$$

which will play a role in projecting good angular momentum states out of $[SU(2)]^3$ states.

According to (1) the states belonging to SIR's of SO(7) can unambiguously be written as $|v;s,t,u;s_0,t_0,u_0\rangle$ whereby $s=t=0,1/2,\ldots,v/2$ and $u=v-2s,v-2s-2,\ldots,1$ or 0. One also wants to find the (reduced) matrix elements of the algebra generators in this basis of states. Only the calculation of the matrix elements of the bispinor-vector is non trivial and can proceed as in the case of $SO(5) \supset [SU(2)]^2$. Hence, three methods can be used : a recursion method [3], a polynomial basis method [4] or the shift operator technique [5]. The result reads :

$$\langle v;s+1/2,s+1/2,u\pm1 \| T^{[1/2\ 1/2\ 1]} \| v;s,s,u\rangle$$

$$= \tfrac{1}{2}\{(s+1)(2s+1)(2u+1\pm1)[v-u+2\mp2(s+1)][v+u+3\pm2(s+1)]\}^{1/2}. \qquad (4)$$

---

1) T. SELIGMAN and R. SHARP, Orsay (France) preprint IPNO/TH82-17 (1982)
2) H. DE MEYER, P. DE WILDE and G. VANDEN BERGHE, *J. Phys. A* (1982) (in press)
3) N. KEMMER, D. PURSEY and S. WILLIAMS, *J. Math. Phys.* 9 (1968) 1224
4) R. SHARP and S. PIEPER, *J. Math. Phys.* 9 (1968) 664
5) J. HUGHES, *J. Math. Phys.* (in press)

A FORMALISM FOR THE MICROSCOPIC INTERACTING BOSON MODEL WITH NON-DEGENERATE ORBITS

A. Frank and P. Van Isacker
Centro de Estudios Nucleares, UNAM
Circuito Exterior, C. U.
Delegación Coyoacán
04510 México, D. F.

ABSTRACT

We discuss a formalism for the microscopic theory of the interacting boson model with non-degenerate shell model orbits, which is concise and simple for numerical calculations. We briefly indicate two of the possible applications of this formalism.

In a recent paper of Pittel et al.[1)] the connection between the interacting boson model (IBM) and the shell model was established for the general case, i.e., when the protons and neutrons are distributed over an arbitrary number of non-degenerate shell model orbits. This work is based on the same ansatz used in ref. 1, namely a binomial expansion of the correlated S-pair, but by means of a unifying approach and through algebraic manipulations, it is possible to simplify the problem significantly and arrive to compact expressions for all matrix elements of interest. Here we sketch the main features of the procedure.

In order to study the connection between the IBM and the shell model, we have taken the following steps:

1) The matrix elements of the shell model Hamiltonian (which is assumed as input) between correlated S- and D-pair states, are written in terms of similar matrix elements of one- and two-body proton or neutron operators. This is explained in detail in ref. 1, for the proton-neutron interaction.

2) We express the matrix elements of one- and two-body fermion operators between correlated S- and D-pairs, in terms of certain norms,

denoted by $\eta_m^{(r)}(j_1,j_2,\ldots j_r)$, where m represents the total number of particles and r the number of particles outside the correlated S-pairs. The index r typically increases with the complexity of the matrix elements. For instance, a one-body operator between states of generalized seniority zero (number of D-pairs zero) involves norms $\eta_m^{(r)}(j_1\ldots j_r)$ with $r \leq 1$; a one-body operator between states of generalized seniority two involves $\eta_m^{(r)}(j_1\ldots j_r)$ with $r \leq 3$ etc.

3) Simple recurrence relations are derived for the norms $\eta_m^{(r)}(j_1\ldots j_r)$ which give rise to great simplification in the evaluation of the matrix elements of interest.

We are currently applying this formalism to the following problems:

1) The determination of the structure of the S- and D-pairs in vibrational nuclei, by minimizing the expectation value of the fermion Hamiltonian in the $|0_1^+\rangle$ state and the $|2_1^+\rangle$ state.

2) Assessment of the importance of cubic terms in the IBM Hamiltonian. This study requires the calculation of the matrix elements of a one-body operator between states of generalized seniority four and thus leads to norms $\eta_m^{(5)}(j_1j_2j_3j_4j_5)$. The bigger problem to deal with however, is the recoupling of five angular momenta, which inevitably requires the use of 12-j symbols.

1. S. Pittel, P.D. Duval and B.R. Barrett, to be published in Ann. of Phys.

SU(3) IN AN O(3) BASIS : EIGENVALUE SPECTRUM OF THE Q_ℓ^0 AND O_ℓ^0 OPERATORS

G. Vanden Berghe, H. De Meyer(*) and J. Van der Jeugt(°)
Seminarie voor Wiskundige Natuurkunde, R.U.G.
Krijgslaan 281-S9, B-9000 Gent, BELGIUM

Some years ago Hughes [1] introduced for the SU(3) ⊃ O(3) group-subgroup system two invariants which he denoted as O^0 and Q_ℓ^0. He developed a method based on shift operator techniques by which it was possible to derive general expressions in closed form for the eigenvalue spectra of both operators. The analysis was based upon relations between O(3) quadratic and triple scalar product operators of the O_ℓ^k ($k=0,\pm1,\pm2$) type. Recently we have improved Hughes' method by considering also the relations between the O(3) non-scalar quadratic product operators of the O_ℓ^k type [2]. Moreover it was obvious that these relations give rise to the O_ℓ^0 eigenvalues in a much easier way than before. However, it turns out that certain signs could only be fixed by making use of one of the triple product operators of Hughes, which are quite involved and difficult to work with.

The new approach which we want to comment on here, consists in constructing shift operators Q_ℓ^k $(-2\leqslant k\leqslant 2)$, which are quadratic in the generators q_μ $(|\mu|\leqslant 2)$, forming together with the three O(3) generators ℓ_0, $\ell_\pm$ the algebra SU(3). The scalar Q_ℓ^0 introduced previously by Hughes, belongs to that class. It can be proven that all Q_ℓ^s operators are expressible in terms of products of the type $O_{\ell+k}^j O_\ell^k$ $(j+k=s)$. Following results, which may be regarded as an extension of eq.(45) of Ref. [1], have been obtained :

$$Q_\ell^{-2} = \frac{1}{2\sqrt{6}(2\ell-1)}\,(O_\ell^{-2}O_\ell^0 - O_{\ell-2}^0 O_\ell^{-2}) \tag{1}$$

$$Q_\ell^{-1} = \frac{1}{2\sqrt{6}\ell}\,(O_\ell^{-1}O_\ell^0 - O_{\ell-1}^0 O_\ell^{-1}) \tag{2}$$

$$= \frac{1}{4\sqrt{6}}\,[O_{\ell+1}^{-2}O_\ell^{+1}/\ell(\ell+1)^2 - O_{\ell-2}^{+1}O_\ell^{-2}/\ell(\ell-1)^2] \tag{3}$$

(*) Research associate N.F.W.O. (Belgium)
(°) Research assistant N.F.W.O. (Belgium)

$$Q_\ell^0 = \frac{1}{(2\ell+1)} [O_{\ell-1}^{+1}O_\ell^{-1}/\ell^2 - O_{\ell+1}^{-1}O_\ell^{+1}/(\ell+1)^2] + 72\ell(\ell+1)I_2$$

$$-6\ell(\ell+1)(2\ell^2+2\ell+9) \quad (4)$$

$$= \frac{1}{4(2\ell+1)} [O_{\ell+2}^{-2}O_\ell^{+2}/(\ell+1)^2(\ell+2)^2 - O_{\ell-2}^{+2}O_\ell^{-2}/\ell^2(\ell-1)^2]$$

$$- 72(2\ell^2+2\ell+3)I_2 + 6\ell(\ell+1)(4\ell^2+4\ell+3) \ . \quad (5)$$

Here I_2 denotes the second order SU(3) Casimir operator.

It is now again possible to derive relations between O(3) scalar and non-scalar product operators of the type $Q_{\ell+k}^j Q_\ell^k$, $O_{\ell+k}^j Q_\ell^k$ and $Q_{\ell+k}^j O_\ell^k$. Also it can be shown that newly derived relations containing the mixed O-Q type operators are connected with the relations between triple product operators of the O-type, previously reported by Hughes [1]. It turns out that the O-Q relations which can replace the O-relations completely, allow the unambiguous eigenvalue determination of O_ℓ^0 and Q_ℓ^0 in a much easier and faster way. New results can also be obtained, e.g. the Q_ℓ^0 eigenvalues for the twofold degenerate $\ell = p-2$ state which are given by :

$$\langle Q^0 \rangle_{\ell=p-2} = -2[(2p^4-12p^3+121p^2-165p+78)-4q(p-q)(p+1)(2p+1)]$$

$$+(-1)^{\ell-1}24\sqrt{\gamma} \ , \ (i=1,2) \quad (6)$$

and where γ is defined by :

$$\gamma = p^2(4p^4-4p^3+5p^2-2p+1)-4q(p-q)(2p^4+5p^3-4p^2-2p+1)+4q^2(p-q)^2p^2. \quad (7)$$

---

1) J.W.B. Hughes, J. Phys. A6 (1973) 48, 281

2) H.E. De Meyer, G. Vanden Berghe and J.W.B. Hughes, J. Math. Phys. 22 (1981) 2360, 2366.

SU(4) IN AN SO(4) BASIS : SHIFT OPERATOR TECHNIQUE

J. Van der Jeugt(*), G. Vanden Berghe and H. De Meyer(°)
Seminarie voor Wiskundige Natuurkunde, R.U.G.
Krijgslaan 281-S9, B-9000 Gent, BELGIUM

The reduction of irreducible representations (IR's) (p p'p") of SU(4) into IR's (s,t) of SO(4) = SU(2) ⊗ SU(2) leads to a two missing label problem. This problem has been solved in principle by Moshinsky and Nagel [1)], who have constructed a pair of commuting Hermitian labelling operators (Ω,Φ). Some eigenvalues of Ω and Φ have been calculated numerically by Quesne [2)] and Partensky and Maguin [3)].

Within the shift operator formalism a basis of four functionally independent SO(4) scalars in the enveloping algebra of SU(4) can be constructed. The basis consists of the SO(4) scalar shift operator of lowest degree (which is proportional to Ω) and three linear combinations of quadratic shift operator products of scalar type. A vast amount of relations connecting these basis elements can be set up. With the aid of these, closed analytic formulae producing Ω- and Φ-eigenvalues are easily deduced, even for cases of degeneracy in the SO(4) labels.

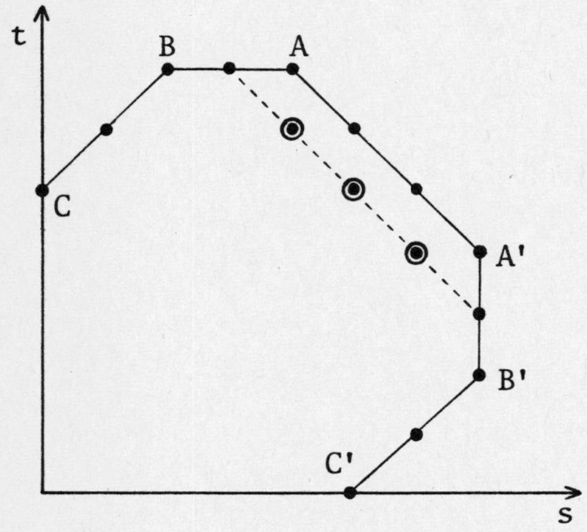

The adjoining figure shows some general features of an SU(4) spin-isospin degeneracy diagram. We summarize the original expressions obtained for ω and φ, respectively the eigenvalues of Ω and Φ, for all situations indicated in the diagram.

(*) Research assistant N.F.W.O. (Belgium)
(°) Research associate N.F.W.O. (Belgium)

(1) for the states $|s,t\rangle$ on the line AA' :

$$\omega = \frac{1}{2}(s+1)(t+1)p''$$

$$\varphi = -\frac{1}{4}\{(2st+3s+3t+4)p(1-p') - [(s+1)^2+(t+1)^2]p''^2 + (2s^2t^2-4st-s-t)\}$$

(2) for the states $|s,p\rangle$ on the line AB :

$$\omega = \frac{1}{2}\, p''(p+1)(p'+1)$$

$$\varphi = \frac{1}{4}\,\{p''^2[(p+1)^2+(p'+1)^2]+(p+1)^2[p'(p'+2)-s(s+1)]+3ps(s+1)-3p(p+1)\}$$

(3) for the states $|p''-k,p-k\rangle$ $(k=0,1,\ldots,[p''])$ on the line BC :

$$\omega = \frac{1}{2}(p-k+1)(p'+1)(p''-k)$$

$$\begin{aligned}\varphi = \frac{1}{4}\{&(p''-k)^2[(p'+1)^2+(p-2)]+(p''-k)[p(k+1)+k-2]\\ &+(p-k+1)^2[(p'+1)^2-p''-4]-(p-k+1)[k(p''+1)-3]\\ &-2(p''-k)(p-k+1)[k^2-k(p+p''+1)-p''-1]\}\end{aligned}$$

(4) for the degenerate $|s,t\rangle$ states satisfying $s+t = p+p'-1$:

$$\omega_{1,2} = \frac{1}{4}[(s+1)(t+2)+(t+1)(s+2)]p'' \pm \frac{1}{4}\sqrt{\Delta}$$

$$\varphi_{1,2} = -\omega_{1,2}+\delta(s,t) \ ,$$

where $\Delta = (s-t)^2p''^2+4(s+1)(t+1)[st-p(p'-1)]$,

and
$$\begin{aligned}\delta(s,t) = &-\frac{1}{12}[st-p(p'-1)-p''^2][(s+1)(2s+3)+(t+1)(2t+3)\\ &+6(s+2)(t+2)-3] - \frac{1}{8}[s(s+1)(2s+3)+t(t+1)(2t+3)\\ &-2(s+1)(t+1)(2s+2t+3)] + \frac{1}{24}[s(s+1)+t(t+1)]\\ &\times \{2[p(p+4)+p'(p'+2)+p''^2-s(s+1)-t(t+1)]-9\} \ .\end{aligned}$$

---

1) M. Moshinsky and J.G. Nagel, Phys. Lett. 5 (1963), 173.

2) C. Quesne, J. Math. Phys. 17 (1976), 1452.

3) A. Partensky and C. Maguin, J. Math. Phys. 19 (1978), 511.

The algebraic geometry of multimonopoles

Werner Nahm
MPI for mathematics, Bonn

Let G be the symmetry group of the laws of nature. The physical vacuum is not invariant under G, but only under a subgroup H. Thus any element of the coset space G/H yields a different possible local vacuum state. Now consider some accumulation of matter which is surrounded to a good approximation by vacuum. Taking a closed surface S in the surrounding space, one obtains a map $S \to G/H$, and in particular an element of $\pi_2(G/H)$. As $\pi_2(G)$ is trivial, $\pi_2(G/H)$ can be identified with the kernel of the imbedding map $\pi_1(H) \to \pi_1(G)$. Now H contains the electromagnetic group $U(1)_{em}$, such that $\pi_1(H)$ has a subgroup isomorphic to Z. On the other hand one believes, partly for esthetic reasons and partly to achieve renormalizability, that G is simple or at least semisimple. Thus $\pi_1(G)$ should be finite. Accordingly, $\pi_2(G/H)$ should have a subgroup Z_{mag} isomorphic to Z, like $\pi_1(H)$. The element of Z_{mag} is the magnetic charge of the matter accumulation.

As $U(1)_{em}$ is coupled to a long range field, magnetic charges have long range interactions, too. Configurations with non-zero magnetic charge are called magnetic monopoles. According to sign and magnitude of the charge one may distinguish anti-monopoles, multimonopoles etc.

Monopoles can be acted upon by the invariance group of the physical vacuum. They have a translational degree of freedom and also can be rotated, if one undoes the effect of the rotation on the map $S \to G/H$ by a suitable G-transformation. Less obvious is the effect of internal symmetry transformations $h \in H$. If the monopoles are far apart, such that the state of the space in between differs little from the physical vacuum, h has little effect on locally measurable quantities. However, consider loops C_i, $i=1,2$, which start from a fixed point in the vacuum and go through the center of the i-th monopole. Gauge fields integrated along these loops yield group elements g_i, and the combined loop C_1+C_2 yields the gauge invariant trace $tr(g_1 g_2)$. If the first monopole is transformed by h, the trace changes to $tr(h^{-1} g_1 h g_2)$, which is a large effect, independent of the monopole distance. If one quantizes this internal degree of freedom, one obtains monopoles with electric charges.

So far we have considered a model of the real world which has a good chance of being correct. Now let us consider a simple Lagrangian which shares the properties considered above, but is otherwise quite unrealistic. However, it will turn out to be very interesting from a mathematical point of view.

A state of our system will be given by a gauge potential A_μ, i.e. by a connection

$$A = A_\mu dx^\mu \in \Omega^1(L(G)) \tag{1}$$

and by a Higgs field

$$\varphi : R^4 \to L(G), \tag{2}$$

where L(G) is the Lie algebra of G. We shall work with G=SU(N), though generalization to orthogonal and symplectic groups is easy. We represent the elements of L(G) by the anti-hermitean matrices of the fundamental representation, without explicit notation.

Let F be the field strength 2-form, i.e. the curvature of the connection form A, and take the Lagrangian

$$L = -\frac{1}{2} \operatorname{tr} \int_{R^4} (F \wedge *F + D\varphi \wedge * D\varphi). \tag{3}$$

We shall only consider classical properties of the model, though the quantized version of its N=4 supersymmetric extension is at least as interesting.

We shall look for static configurations of finite energy, namely for extrema of the energy functional

$$E = -\frac{1}{2} \operatorname{tr} \int_{R^3} (F \wedge *F + D\varphi \wedge *D\varphi), \tag{4}$$

where $*$ is now the Hodge operation in R^3.

At the sphere at infinity S^2_{inf} the Higgs field is given by a map $S^2_{inf} \to G\varphi_o$, where $\varphi_o \in L(G)$ can be prescribed arbitrarily. Because of the Bianchi identity, the quantity

$$Q = \operatorname{tr} \int D\varphi \wedge F \tag{5}$$

only depends on this map at infinity and is not affected by variations in compact domains. Accordingly for positive Q the minima of

$$E = -\frac{1}{2} \operatorname{tr} \int ((*F + D\varphi) \wedge *(*F + D\varphi)) + Q \tag{6}$$

are the solutions of the Bogomolny equation

$$*F = -D\varphi \tag{7}$$

The sign of Q can be changed by a parity transformation, which also changes the minus sign in eq. (7) into a plus sign.

In addition to the minima, there exist further stationary points of the energy functional[1], but they have not yet been constructed explicitely.

Let us describe the behaviour of the Higgs field at infinity in more detail. One has

$$i\varphi(x) = \sum_{l=1}^{N} n_l(x) (z_l + k_l/2r) + O(1/r^2), \tag{8}$$

where the k_l are integers, the z_l form a non-decreasing sequence of real numbers, and the n_l are hermitean matrices which project onto the eigenvectors of $i\varphi$.

The Bogomolny equation can be written as the self-duality equation for the curvature of a connection in a euclidean space R^4, with the additional condition that the connection only depends on the three coordinates x^m, m=1,2,3, not on the dummy coordinate x^4. The Higgs field φ has to be identified with the fourth component of the potential.

This self-duality equation with three variable and one dummy can be solved by a transformation to a self-duality equation with one variable and three dummies: Consider an orthonormal basis $\psi_r(x,z)$, r=1,..,k(z) of the normalizable solutions of the Weyl equation

$$(\sigma^m D_m - i\varphi + z)\, \psi(x,z) = 0, \tag{9}$$

where z is a real parameter. The number k(z) is given by[2]

$$k(z) = \frac{1}{2\pi} \operatorname{tr} \sum_l \int_{S^2_{inf}} (dn^l \wedge dn^l)\, n^l \operatorname{sgn}(z_l - z) \tag{10}$$

This expression can be rewritten in the form[3)]

$$k(z) = \frac{-i\,tr_\Sigma}{4\pi e}\int (Fn_e)\,\mathrm{sgn}(z_e - z) = \sum_e k_e\,\theta(z - z_e). \tag{11}$$

Now form the matrices

$$T^m(z)_{rs} = -i\int \psi_r^+ x^m \psi_s\, d^3x . \tag{12}$$

They satisfy the self-duality equation for a connection

$$T = T^m dp_m + T^4 dz, \tag{13}$$

which only depends on z, not on the dummy variables p_m.
If one transforms T^4 to zero by a z-dependent gauge transformation, which corresponds to the condition

$$\int \psi_r^+ \frac{\partial}{\partial z}\psi_s\, d^3x = 0, \tag{14}$$

one finds

$$\frac{dT^k}{dz} = \frac{1}{2}\varepsilon_{klm}\left[T^l T^m\right]. \tag{15}$$

Now consider the Weyl equation for the connection T:

$$\left(-i\frac{d}{dz} + \sigma^m(T^m + ix^m)\right) v(z,x) = 0. \tag{16}$$

Let $v_m(z,x)$, $m=1,\ldots,N$ be an orthonormal basis of its normalizable solutions. Then

$$A_k(x)_{mn} = \int v_m^+ \frac{\partial}{\partial x^k} v_n\, dz \tag{17}$$

$$i\varphi(x)_{mn} = \int v_m^+ z\, v_n\, dz \tag{18}$$

gives back the solution of the Bogomolny equation.

The functions $v_{rn}(z,x)$ and $\psi_{nr}(x,z)$, $r=1,..,k(z)$, $n=1,..,N$, are related by the actions of the corresponding covariant Laplace operators:

$$\left(\mathcal{D}^2 + (\varphi + iz)^2\right) v^+(z,x) = 2(2\pi)^{1/2} \varepsilon \psi(x,z) \tag{19}$$

$$\left(-\frac{d^2}{dz^2} + (T^m + ix^m)(T_m + ix_m)^+\right)\psi(x,z) = 2(2\pi)^{-1/2}\varepsilon v(z,x). \tag{20}$$

Here $\varepsilon = i\sigma_2$ is the charge conjugation matrix for the σ_m.

If some of the k_l are zero, the transformation described above is incomplete. In this case one also has to take into account the solutions of the Laplace equation

$$\left(\mathcal{D}^2 + (\varphi + iz_l)^2\right)\sigma = 0, \tag{21}$$

which behave as a constant at infinity. The eqs. (15) and (16) aquire additional source terms proportional to $\delta(z-z_l)$ with coefficients related to those solutions.

All properties of the transformation can be verified by direct calculation, though they were originally derived from twistor methods. More details can be found elsewhere[4,5,6].

It is easy to see that all solutions of the eqs. (15) and (16) yield regular configurations (A,φ). In particular, eq. (18) yields

$$z_1 \leq i\varphi(x) \leq z_N \tag{22}$$

and from eq. (17) one obtains

$$*F_{mn} = 2i\,dx^k \int v_m^+ \sigma_k \left(-\frac{d^2}{dz^2} - (T+ix)^2\right)^{-1} v_n \, dz \tag{23}$$

and

$$-2\pi^{-2}(z_N - z_1)^2 \leq iF^k \leq 2\pi^{-2}(z_N - z_1)^2 \tag{24}$$

The ADHM instanton construction[7] may be described as a transformation from a self-duality equation with four variables and no dummy coordinates to a self-duality equation with zero variables and four dummies. The latter is somewhat obscured by the fact that one always has additional source terms. Now we have a transformation from three variables and one dummy to one variable and three dummies. This analogy was pointed out to me by E.Corrigan, who also conjectured that it should extend to two variables and two dummies. This might be helpful for the construction of vortex configurations.

The utility of our transformation for the explicit construction of multimonopoles depends on the possibility to solve the eqs. (15) and (16). Eq. (16) may be solved with the help of the adjoint equation

$$\left(-i\frac{d}{dz} - \sigma_m (T^m + ix^m)\right) w(z,x) = 0. \tag{25}$$

One has

$$\frac{\partial}{\partial z} v^+ w = 0, \tag{26}$$

such that v(z) is essentially the inverse of w(z). Note that we look for local solutions of the equation, the normalizable ones can be easily picked out later.

Eq. (25) can be solved by the ansatz

$$w(z,x) = \left(1 + \sigma_m u^m(x)\right) f(z,x), \tag{27}$$

where u is a fixed unit vector in R^3. This yields

$$\left(i\frac{d}{dz} + u_m (T^m(z) + ix^m)\right) f = 0 \tag{28}$$

and

$$y_m (T^m + ix^m) f = 0, \tag{29}$$

where y is a complex vector determined by

$$u \times y = -iy. \tag{30}$$

The compatibility equation for eqs. (28) and (29) is again eq. (15), so we have a kind of Lax pair.

We need not distinguish vectors y which differ by a multiplicative constant. Thus y will be considered as an element of the cone of CP^2

$$y \in \{(y_1, y_2, y_3) \in CP^2 \,/\, y^2 = 0\}. \tag{31}$$

This cone can be parametrized by a complex variable ζ, e.g.

$$y = \left(\frac{1-\zeta^2}{2}, \frac{1+\zeta^2}{2i}, \zeta\right). \tag{32}$$

Eq. (29) can be solved, if

$$P(\zeta, \eta) = \det(y_m T^m + \eta) = 0 \tag{33}$$

where

$$\eta = i y_m x^m. \tag{34}$$

The Lax pair yields

$$\frac{\partial}{\partial z} \det(y_m T^m + \eta) = 0, \tag{35}$$

as long as z does not cross one of the z_1. Thus each multi-monopole solution of the Bogomolny equation yields a finite number of algebraic curves of type (33), one less than the distinct eigenvalues of $i\varphi$. This family of curves is called the spectrum of the multimonopole.

What algebraic geometers call a curve in the two-dimensional complex space with variables ζ, η, function theorists call a compact Riemann surface, counting two real parameters instead of a single complex one. The curve may have isolated singular points. In similar contexts, mathematicians have shown, how to deal with this complication[8)].

Now we shall solve eq. (25) explicitly. Let $M(\zeta, \eta, z)$ be the matrix of subdeterminants of codimension 1 of $y(T+ix)$, such that

$$(yT + \eta) M = \det(yT + \eta) \cdot 1_k. \tag{36}$$

Let λ be any fixed k-dimensional vector, and write

$$f = M \lambda h \tag{37}$$

with a scalr function h. Because of

$$\frac{\partial M}{\partial z} = [u_m T^m, M] \tag{38}$$

one has

$$(\mu^T M \lambda) \frac{\partial h}{\partial z} = -(\mu^T M (uT + \gamma)\lambda)\, h \tag{39}$$

for any fixed vector μ, and h is given by an ordinary integral.

Now let us consider eq. (15). From eq. (35) one obtains k^2+2k integration constants, namely the coefficients of the polynomial p. As the equations are invariant under constant SU(k) transformations, this leaves $(k-1)^2$ integration constants to be determined.

Now the solutions f of eq. (29) yield a holomorphic line bundle over the curve (33). This line bundle may be characterized by the zeros of a cross section $M(\zeta,\eta,z)\lambda$.

How can one do this explicitly? Our curve has genus $(k-1)^2$, such that there are $(k-1)^2$ globally holomorphic differentials, which we denote collectively by ω. Now let $(\zeta_s,\eta_s)(z)$ be the zeros of $M\lambda$ on the curve. Form the $(k-1)^2$-dimensional vector

$$\Omega(z) = \sum_s \int_{(\zeta_0,\eta_0)}^{(\zeta_s,\eta_s)(z)} \omega\,, \tag{40}$$

where (ζ_0,η_0) is a fixed point of the curve. This vector is determined up to integrations around the $2(k-1)^2$ independent cycles of the surface and has to be regarded as a point on a torus. This torus is called the Jacobian manifold of the curve.

Now Abel's theorem states that $\Omega(z)$ does not depend on λ, but only on the line bundle. How does it vary with z? The vector $d\Omega/dz$ is a symmetric function of the (ζ_s,η_s), thus it is a rational function of the matrix elements of the $T^m(z)$. As the differentials are regular, it turns out that

$$\frac{d\Omega}{dz} = \text{const.}, \tag{41}$$

with constants given by the coefficients of $p(\zeta,\eta)$.

The line bundle can be reconstructed from the corresponding point of the Jacobian by Riemann theta-series, such that the matrix elements of the $T^m(z)$ become rational functions of Riemann theta-functions, with arguments replaced by linear functions of z. Thus we have at least a scheme for the solution of eq. (15), though its application in concrete cases needs some additional work.

In our treatment of eqs. (15) and (16) we used the adjoint Weyl equation in z-space, eq. (25). The success of our transformation suggests to consider the same equation in x-space:

$$(\sigma^m D_m + i\varphi - z')\chi(x) = 0. \tag{42}$$

Again we use the ansatz

$$\chi(x) = (1 + u^m\sigma_m)k(x), \tag{43}$$

which yields

$$(u^m D_m + i\varphi - z')k(x) = 0 \tag{44}$$

and

$$y^m D_m k(x) = 0, \tag{45}$$

with y given by eq. (30). The two equations (44) and (45) are compatible because of the Bogomolny equation (7). Eq. (26) is replaced by

$$\partial_m(\psi^+\sigma^m\chi) = 0. \tag{46}$$

Eq. (44) is an ordinary linear differential equation on a line with orientation u. It has been used by Hitchin in his treatment of multimonopoles[9)]. Oriented lines can be characterized by y and iyx, where x is any point of the line. Thus the oriented lines in R^3 form exactly the two-dimensional complex space considered above. Now Hitchin considers the subspace for which eq. (44) has square integrable solutions and finds algebraic curves.

If the asymptotic Higgs field only has two distinct eigenvalues, in particular in the case G=SU(2), which Hitchin studied in detail, his construction yields a single spectral curve. We shall see that this curve agrees with our spectrum given by eq. (33).

The general solution of eq. (44) is given by

$$k(x) = \int v^{+}(z,x)\exp(\alpha(z-z'))\sum_{\ell=1}^{N} c_\ell v_\ell(z,x_0)\,dz, \tag{47}$$

where x_o is a fixed point on the line and

$$\alpha = u_m (x_0 - x)^m. \tag{48}$$

The c_n are arbitrary constants. To achieve square integrability, the exponential factor in eq. (47) must be cancelled for large α. Now for $\alpha \to s'\infty$, $s'=\pm 1$, one has a basis

$$v_\ell(z,x) \sim$$

$$\sim (1 - s_\ell s' \sigma^m u_m)\exp(-s_\ell r(z_\ell - z))(z-z_\ell)^{(s_\ell k_\ell - 1)/2}\,\theta(s_\ell(z-z_\ell))\,a_\ell, \tag{49}$$

where

$$s_\ell = \operatorname{sgn} k_\ell . \tag{50}$$

The a_1 are constant vectors which depend algebraically on u. The $v_1(z,x)$ yield the asymptotic behaviour of φ given by eq. (8) when inserted into eq. (18). To obtain square integrability, one needs

$$a_\ell^{+}(1 - s''_\ell \sigma^m u_m)(z-z_\ell)^{(1-s_\ell k_\ell)/2}\sum_m c_m v_m(z,x_0)\Big|_{z\to z_\ell} = 0, \tag{51}$$

with

$$s''_\ell = \operatorname{sgn}(z'-z_\ell)\operatorname{sgn} k_\ell . \tag{52}$$

In general, these N conditions on the N constants c_1 only admit the trivial solution. However, if the z_1 only assume two distinct values z_a, z_b and z' lies between those values, some u yield nontrivial solutions. Indeed, if the oppositely oriented line belongs to the spectrum given by eq. (33), the N conditions are not independent. A linear combination of them is implied by the integral of eq. (26) between z_a and z_b, with w(z,x) given by eq. (27).

Thus the spectrum given by eq. (33) yields square integrable solutions of eq. (44). The opposite implication also is true, but a proof would need additional arguments. For Higgs fields with more than two asymptotic eigenvalues the relationship between Hitchin's and our spectral curves still has to be clarified.

References

1) C.Taubes, Comm.Math.Phys. 86 (1982) 257 and 299.
2) C.Callias, Comm.Math.Phys. 62 (1978) 213, with comments by R.Bott and R.Seeley, Comm.Math.Phys. 62 (1978) 235.
3) Hou Bo-Yu et al., Scientia Sinica 21 (1978) 446.
4) W.Nahm, All self-dual monopoles for arbitrary gauge groups, CERN TH-3172 (1981).
5) W.Nahm, in: Proceedings of the Symposium on Particle Physics, Z.Horváth et al. eds., Visegrád 1981.
6) W.Nahm, in: Monopoles in Quantum Field Theory, Proceedings, N.Craigie et al. eds., Trieste 1981.
7) M.Atiyah, N.Hitchin, V.Drinfeld, and Yu.Manin, Phys. Lett. 65A (1978) 185.
8) D.Mumford and P.v.Moerbeke, Acta Mathematica 143 (1979) 93.
9) N.Hitchin, Comm.Math.Phys. 83 (1982) 579.

Dimensional Reduction, Spinor Fields and Characteristic Classes

H. Römer

Fakultät für Physik, Universität Freiburg
Hermann-Herder-Str. 3
D-7800 Freiburg i.Br., Germany

Abstract:

The effect of dimensional reduction, both global and local, on natural vector bundles over higher dimensional manifolds and on characteristic classes is investigated. In particular spinor fields are considered and a new mechanism for introducing a generalized spin structure by dimensional reduction is proposed.

The idea of dimensional reduction has been considered for several decades. One conceives either formally or in reality that space-time has dimension $D > 4$. The apparent four-dimensionality of space-time is explained by a hypothesis of dimensional reduction: Space-time is shrunk or curled up in $D - 4$ spacelike dimensions or complete homogeneity in $D - 4$ directions is assumed. The $D - 4$ additional dimensions thus give rise to "external degrees of freedom". The rise of supergravity theories has provided new motivation to envisage mechanisms of dimensional reduction[1]. The purpose of this talk is to present some calculations of characteristic classes in arbitrary dimensions, and to give a general discussion [2] of the consequences of dimensional reduction for natural vector bundles over space-time and for the characteristic classes, which "measure" their degree of nontriviality. To be more definite, we assume space-time to be a $D(>4)$-dimensional oriented Riemannian manifold X. For all of our considerations we could as well assume X to be pseudo Riemannian except for the parts in which applications of the Atiyah-Singer index theorem [3] are presented. We chose X to be Riemannian for convenience, for the sake of such applications and because classical Riemannian field theory seems to be the right starting point for quantum field theory.

For us, dimensional reduction amounts to one of the following assumptions

(A) X is a smooth product of manifolds: $X = M \times Y$
(global reduction)

or, more generally,

(B) There is a fibration $\pi : X \to M$ (local reduction).

In either case M is assumed to be an oriented four-dimensional (pseudo) Riemannian manifold and identified with "observed" space-time.

The natural vector bundles over X to be discussed are the following ones

(1) The tangent bundle of X : TX .
The Riemannian structure of X allows the identification of TX with the cotangent bundle.

(2) The complexified tangent bundle $T_{\mathbb{C}} X = TX \otimes \mathbb{C}$

(3) The n-fold tensorial product of $T_{\mathbb{C}} X$ with itself: $(T_{\mathbb{C}} X)^{\otimes n}$

(4) The n-th exterior power of $T_{\mathbb{C}} X$: $\Lambda^n X$
(differential n-forms on X are sections of $\Lambda^n X$)

(5) The symmetric tensorial powers of $T_{\mathbb{C}}X$: $S^n X$

(6) The Clifford bundle of $T_{\mathbb{C}}X$: $CX = \bigoplus_{i=0}^{D} C^i X$
(there are vector bundle isomorphisms $C^i X \cong \Lambda^i X$.)

(7) The spinor bundles [4] $\Delta X = \Delta^+ X \oplus \Delta^- X$, which splits into the spinor bundles $\Delta^{\pm} X$ of even and odd chirality

(8) The generalized spinor bundles $S^n X \otimes \Delta^{\pm} X$, $\Lambda^n X \otimes \Delta^{\pm} X$, etc.

In addition it will be convenient to consider the generating functions

(9a) $$\Lambda_t X = \bigoplus_{n=0}^{D} t^n \Lambda^n X$$ and

(9b) $$S_t X = \bigoplus_{n=0}^{\infty} t^n S^n X$$.

The bundles (1) - (6) exist for every smooth manifold, whereas the existence of the bundles (7) and (8) requires that X have a so-called spin structure[4,5]. Let us briefly explain what this means. The bundle BX of oriented orthogonal frames of tangent vectors of X is a principle bundle over X with structural group SO(D). Let $\tilde{B}X$ be a principal bundle over X with structural group Spin (D), the two-fold covering of SO(D). A spin structure on X is said to exist, if there exists a commutative diagram

(10) $$\begin{array}{ccc} \tilde{B}X & \xrightarrow{f} & BX \\ & {\scriptstyle\tilde{\pi}}\searrow \quad \swarrow{\scriptstyle\pi} & \\ & X & \end{array}$$,

where $\tilde{\pi}$ and π are the projections of $\tilde{B}X$ and BX .
It is well-known that a spin structure exists on X if and only if the second Stiefel Whitney class w_2 of X vanishes

(11) $$w_2(X) = 0$$

and that, if a spin structure exists, it is in general not unique, the number of inequivalent strutures being given by $H^1(X, \mathbb{Z}_2)$, the number of elements in the first cohomology module of X with coefficients in $\mathbb{Z}_2$ (for the D-dimensional torus this number is 2^D).
The bundles (1) - (8) are the building blocks for all the fields appearing in supergravity theories. A field of given spin[6,7] is a section of a virtual bundle, a linear combination of bundles of type (1) - (8) with integer coefficients, where negative coefficients belong to ghost fields, which compensate nonpropagating degrees of freedom. The characteristic classes, whose behaviour under dimensional reduction will be described are for complex vector bundles over X:

(12) The Chern class $c(E) = \sum c_i(E)$ with
$c_i(E) \in H^{2i}(X,\mathbb{Z})$, the $2i$-th cohomology module of X with integer coefficients. The Chern class has the Whitney property
$c(E \oplus F) = c(E)\, c(F)$

(13) The Todd class $td(E) = \sum td_i(E)$ with
$td_i(E) \in H^{2i}(X,\mathbb{Z})$ and $td(E \oplus F) = tdE\ tdF$

(14) The Chern character $chE = \sum ch_i E$ with
$ch_i E \in H^{2i}(X,\mathbb{Z})$ and
$ch(E \oplus F) = chE + chF$
$ch(E \otimes F) = chE \cdot ch\, F$

and for real vector bundles over X :

(15) The Euler class $e(TX) \in H^D(X,\mathbb{Z})$

(16) The Pontryagin classes $p_i(E) = (-1)^i c_{2i}(E) \in H^{4i}(X,\mathbb{Z})$

(17) The Stiefel-Whitney class $w(E) = \sum w_i(E)$ with
$w_i(E) \in H^i(X,\mathbb{Z}_2)$
$w(E \oplus F) = w(E)\, w(F)$.

All characteristic classes behave naturally under the transition to induced bundles.[8)]
The precise definition of the classes (12) - (16) in terms of formal splittings will be given below, they can be represented by closed differential forms, which are well defined polynomials in the curvature quantitis associated to any connection in the vector bundles. For the bundles (1) - (8) natural connections are induced by the Riemannian connection on TX.
Now let X be a compact oriented Riemannian manifold without boundary of even dimension $D = 2\ell$. Let $\Sigma^{\pm}X$ be one of the bundles

(18) $$(T_c X)^{\otimes n} \otimes \Delta^{\pm} X\ , \quad \Lambda^n X \otimes \Delta^{\pm} X\ , \quad S^n X \otimes \Delta^{\pm} X\ .$$

The Dirac operator is an elliptic operator

(19) $$\mathcal{D} : \Gamma\Sigma^+ X \to \Gamma\Sigma^- X$$

from section of $\Sigma^+ X$ (spinor fields of even chirality) to sections of $\Sigma^- X$.

(20) $\text{index}\,\mathcal{D} = \dim \text{kernel}\,\mathcal{D} - \dim \text{kernel}\,\mathcal{D}^*$,

at first sight a purely analytical quantity, actually has a topological meaning. The index theorem of Atiyah and Singer[3] asserts that it is given by

(21) $$\text{index}\,\mathcal{D} = (-1)^{\frac{D}{2}(D+1)} \frac{(ch\,\Sigma^+X - ch\,\Sigma^-X)\; td\, T_c X}{e(TX)}\,[X]$$

i.e. by a well defined characteristic class, which has to be evaluated on the compact manifold X. This characteristic class, expressed as a polynomial in the Riemannian curvature is, up to a trivial numerical factor identical[9] with the anomaly of the axial current for the affiliated quantum field theory, even for noncompact X. How the curvature polynomial on the right hand side of eq.(21) has to be evaluated will be explained below.
At this place it is worth noticing that eq.(21) is independent on the spin structure and even on the precise form of the elliptic operator $\mathcal{D}$ and dependent only on the manifold X.
As a first illustration we mention the case of generalized gauge fields[10,11]. For arbitrary dimensions $D=2\ell$ they are antisymmetric tensor fields and sections of the (virtual) bundles

(22) $$G_p X = \bigoplus_{r=1}^{p} (-1)^r (r+1)\, \Lambda^{p-r} X$$

of fibre dimension

(23) $$\dim G_p X = \dim G_{D-p-2} X = \binom{D-2}{p} .$$

The fields in $G_p X$ and $G_{D-p-2} X$ are only locally equivalent but globally inequivalent, which reflects itself in the fact that their associated trace anomalies differ by a multiple of the Euler class[10,11].
Generalized spinorial gauge fields are sections of

(24) $$\Sigma_p^{\pm} X = \Delta^{\pm} X \otimes G_p X$$

Again there is local equivalence but global inequivalence of $\Sigma_p^{\pm} X$ and $\Sigma_{D-p-2}^{\pm} X$ (different trace anomalies). Using eq.(21) it can be shown[11] , however, that the axial anomalies coincide, a reassuring fact in view of the possible phenomenological significance of axial anomalies.

The calculation of characteristic classes is performed by formal splitting methods[5].

Let E be a complex vector bundle which is splittable into a sum of one-dimensional vector bundles:

(25) $$E = \bigoplus_i E_i \quad .$$

Denoting $c_1(E_i) = y_i$ one obtains, using eq.(12)

(26) $$c(E) = \prod_i (1 + y_i) \; .$$

Furthermore, by definition

(27a) $$\mathrm{ch}E = \sum_i e^{y_i}$$

(27b) $$\mathrm{td}E = \prod_i \frac{y_i}{1 - e^{-y_i}} \quad ,$$

and because of the additivity and multiplicativity of the Chern character

(28a) $$\mathrm{ch}E^{\otimes n} = \left(\sum_i e^{y_i}\right)^n$$

(28b) $$\mathrm{ch}\,\Lambda_t E = \prod_i (1 + te^{y_i})$$

(28c) $$\mathrm{ch}\,S_t E = \prod_i \frac{1}{1 - te^{y_i}}$$

Now, of course not every bundle is splittable into one-dimensional bundles but, according to the splitting principle every bundle can be induced from a splittable bundle such that the corresponding mapping for the cohomologies of the base spaces is injective. This implies that any relation between characteristic classes, which holds for splittable bundles is true in general.

The complexified tangent bundle $T_{\mathbb{C}}X$ has a special formal splitting of the form $(D = 2\ell)$

(29) $$T_{\mathbb{C}}X = \sum_{i=1}^{\ell} (E_i \oplus \bar{E}_i) \,,$$

where the bundles E_i and $\bar{E}_i$ are complex conjugates of oneanother such that

(30) $$c_1(E_i) = -c_1(\bar{E}_i) := x_i$$

and

(31) $$e(TX) = \prod_{i=1}^{\ell} x_i \quad .$$

Furthermore

(32) $p(X) = \prod_{i=1}^{\ell} (1 + x_i^2)$

(33) $\mathrm{ch}\,\Delta^+X - \mathrm{ch}\,\Delta^-X = \prod_{i=1}^{\ell} (e^{\frac{x_i}{2}} - e^{-\frac{x_i}{2}})$

(34a) $\mathrm{ch}(T_{\mathbb{C}}X)^{\otimes n} = \left[\sum_{i=1}^{\ell} 2\cosh x_i\right]^n$

(34b) $\mathrm{ch}\,\Lambda_t X = \prod_{i=1}^{\ell} (1 + t^2 + 2t\cosh x_i)$

(34c) $\mathrm{ch}\,S_t X = \dfrac{1}{(1 + t^2 - 2t\cosh x_i)}$

One notices that all the characteristic classes (33) - (34) are expressible as polynomials in the Pontryagin classes $p_i(X) := p_i$

For $D \leq 12$ contributions up to sixth order in the quantities x_i have to be kept. We here present the results for the bundles $(T_{\mathbb{C}}X)^{\otimes n}$, $\Lambda^n X$ and $S^n X$, which are vital, for instance, for the evaluation of axial anomalies in supergravity theories of arbitrary dimension.

(35a) $$\mathrm{ch}(T_{\mathbb{C}}X)^{\otimes n} = D^n + nD^{n-1}p_1 + \left\{\frac{n}{12}D^{n-1} + \binom{n}{2}D^{n-2}\right\}p_1^2$$

$$-\left\{\frac{n}{6}D^{n-1} + 2\binom{n}{2}D^{n-2}\right\}p_2 + \left\{\frac{n}{360}D^{n-1} + \binom{n}{3}D^{n-3}\right\}p_1^3$$

$$+\left\{-\frac{n}{120}D^{n-1} + \frac{1}{12}\binom{n}{2}D^{n-2} - 3\binom{n}{3}D^{n-3}\right\}p_1p_2$$

$$+\left\{\frac{n}{120}D^{n-1} - \frac{1}{4}\binom{n}{2}D^{n-2} + 3\binom{n}{3}D^{n-3}\right\}p_3 + \dots$$

(35b) $$\mathrm{ch}\,\Lambda^n X = a_n + b_n p_1 + \frac{1}{12}b_n p_1^2 + \left(c_n - \frac{1}{6}b_n\right)p_2$$

$$+\frac{1}{360}b_n p_1^3 + \frac{1}{12}\left(c_n - \frac{1}{10}b_n\right)p_1p_2$$

$$+\left(\frac{1}{120}b_n - \frac{1}{4}c_n + d_n\right)p_3 + \dots$$

with $a_n = \binom{D}{n}$, $b_n = \binom{D-2}{n-1}$, $c_n = \binom{D-4}{n-2}$, $d_n = \binom{D-6}{n-3}$,

(35c)

$$(-1)^n \mathrm{ch}\, S^n X = \hat{a}_n - \hat{b}_n p_1 + (\hat{c}_n - \tfrac{1}{12}\hat{b}_n) p_1^2 + (\tfrac{1}{6}\hat{b}_n - \hat{c}_n) p_2$$

$$- (\tfrac{1}{360}\hat{b}_n - \tfrac{1}{6}\hat{c}_n + \hat{d}_n) p_1^3 + (\tfrac{1}{120}\hat{b}_n - \tfrac{5}{12}\hat{c}_n + 2\hat{d}_n) p_1 p_2$$

$$- (\tfrac{1}{120}\hat{b}_n - \tfrac{1}{4}\hat{c}_n + \hat{d}_n) p_3 \quad + \ldots$$

with $\hat{a}_n = \binom{-D}{n}, \hat{b}_n = \binom{-D-2}{n-1}, \hat{c}_n = \binom{-D-4}{n-2}, \hat{d}_n = \binom{-D-6}{n-3}$.

For obtaining the right-hand side of eq (21) and, hence, the axial anomaly for the fields in $\Sigma^{\pm} X$ one has to multiply (35a) or (35b) or (35c) with the $\hat{A}$-class

(36)

$$\hat{A} = (-1)^{\frac{D}{2}(D-1)} \frac{\mathrm{ch}\,\Delta^+ X - \mathrm{ch}\,\Delta^- X}{e(TX)} \mathrm{td}\, T_{\mathbb{C}} X$$

$$= 1 - \frac{1}{24} p_1 + \frac{1}{5760}(-4p_2 + 7p_1^2)$$

$$- \frac{1}{967\,860}(16 p_3 - 44 p_1 p_2 + 31 p_1^3) + \ldots$$

Formulae (35) and (36) contain the axial anomalies for generalized spinor fields for dimension $D \leq 12$, if one expresses the Pontryagin classes in terms of the Riemannian curvature 2-form R :

$$p_1 = -\frac{1}{8\pi^2} \, tr\, R^2$$

(37) $$p_2 = -\frac{1}{128\pi^4}\{2\, tr\, R^4 - tr\, R^2\, tr\, R^2\}$$

$$p_3 = -\frac{1}{3072\pi^6}\{8\, tr\, R^6 - 6\, tr\, R^4\, tr\, R^2 + tr\, R^2\, tr\, R^2\, tr\, R^2\}$$

Notice that in spite of the big denominators in eqs. (35)- (37), the integral over the resulting index density eq.(21) has to be an integer for compact X without boundary, if X has a spinstructure. This illustrates how restrictive this latter requirement actually is.
By the way, from the characteristic classes (35) - (37) one immediately obtains the corresponding secondary Chern-Simons classes[12,4] for dimension D - 1 (in particular for dimension 11).
We already mentioned that the index of the Dirac operator and the axial anomaly are independent of the spin structure of X.
The g-index of the Dirac operator, which we are going to describe now is an example of a quantity which does depend on the spin structure.
Let a group Γ act on the manifold X by orientation preserving isometries. This action of Γ induces actions on the bundles eqs. (1) - (6) in an evident way. It is, however, not necessarily possible to "lift" the action of Γ to the spinor bundle ΔX. The problem here is not in lifting the action of an individual element $g \in \Gamma$ but in the preservation of the group structure and of continuity (if Γ is a continuous group).
A lifting can be arranged such that the action of Γ commutes with the Dirac operator. Hence, for compact X without bundary, kernel D and kernel D* are finite dimensional representation spaces of Γ, and the g-index of D:

(38) $$\mathrm{index}_g \mathcal{D} := tr\, g\,|_{\mathrm{kernel}\,\mathcal{D}} - tr\, g\,|_{\mathrm{kernel}\,\mathcal{D}^*}$$

can be defined.
The G-index theorem[13] now tells us that $(D = 4\ell)$

(39)
$$\mathrm{index}_g \mathcal{D} = \frac{(ch_g\, j^*\Sigma^+X - ch_g\, j^*\Sigma^-X)\, td\, T_c X^g}{ch_g(\Lambda_{-1} N^g \otimes \mathbb{C})\, e(TX^g)}\,[X^g] \quad .$$

Here X^g is the fixed point set of $g \in \Gamma$ (all of its components have even dimension), N^g is the normal bundle of X^g in TX , and j* means restriction to X^g , and ch_g is the so-called equivariant Chern character. Physically interesting fixed point sets are the "nuts" and "bolts", the fixed point sets of euclidean time translations in quantum gravity. The evaluation of eq. (39) proceeds by generalized splitting methods.[13] For simplicity we only exhibit the case that X^g consits of isolated points $(z_\alpha)_{\alpha=1,\dots,N}$ (nuts) so no integration has to be performed.
We find at z_α

$$\text{(40a)} \quad ch_g\, j^*(T_{\mathbb{C}}X)^{\otimes n} = 2^n \left[\sum_{j=1}^{2l} \cos\theta_j^\alpha \right]^n ,$$

$$\text{(40b)} \quad ch_g\, j^* \Lambda_t X = \prod_{j=1}^{2l} (1 + t^2 + 2t\cos\theta_j^\alpha) ,$$

$$\text{(40c)} \quad ch_g\, j^* S_t X = \prod_{j=1}^{2l} \frac{1}{1 + t^2 - 2t\cos\theta_j^\alpha} ,$$

where the angles θ_j^α are obtained by bringing the rotation of g about the fixed point z_α into diagonal form.
Furthermore at z_α :

$$\text{(41)} \quad \frac{ch_g\, j^*\Delta^+X - ch_g\, j^*\Delta^-X}{ch_g(\Lambda_{-1}N^g \otimes \mathbb{C})} = \varepsilon_\alpha \left(\frac{i}{2}\right)^{2l} \prod_{i=1}^{2l} \frac{1}{\sin^2 \frac{\theta_j^\alpha}{2}} ,$$

where $\varepsilon_\alpha = \pm 1$, depending on the spin structure and on the lift of the action.
$\text{Index}_g\, \mathcal{D}$ is now simply obtained by multiplying either (40a) or (40b) or (40c) with (41) and summing over all fixed points.
The dependence of $\text{index}_g\, \mathcal{D}$ on the spin structure is now clear:
Going over to the opposite lift of the action changes all the signs ε_α simultaneously. So, the spin structure reflects itself in the quantities $\varepsilon_\alpha/\varepsilon_\beta$.
Now we shall investigate the behaviour of the bundles and characteristic classes discussed so far under dimensional reduction. It suffices to discuss the case of local reduction $\pi : X \longrightarrow M$, where M is an oriented four-dimensional Riemannian manifold, because global reduction is just a special case.

Then there is in general only a local immersion $M \to X$ of four-dimensional space-time into the D-dimensional manifold X and, hence, opposite to the global case, there is some freedom in identifying M locally inside X.

From the very definition of a fibre bundle it is clear, that the tangent bundle TX splits (globally) into a horizontal and a vertical subbundle:

(42) $$TX = HTX \oplus VTX \; .$$

Here VTX consists just of those tangent vectors of X, which are annihilated by (the differential of) the projection π.

The horizontal subbundle HTX is uniquely determined only up to isomorphism (at least for genuinely local dimensional reduction).

One has

(43) $$HTX \cong \pi^* TM \quad ,$$

the horizontal part is isomorphic to the bundle induced from the tangent bundle TM of M by means of the projection π. The structure group SO(D) of TX is reduced to

(44) $$SO(D-4) \times SO(4) \subset SO(D) \quad .$$

To be more definite, we now introduce the additional hyothesis that the fibration $X \xrightarrow{\pi} M$ be a principal fibration[8,14] $P \xrightarrow{\pi} M$ with total space P, base M and structure group G. The theories thus arising should be called local Kaluza-Klein theories[15]. In this case, in addition to (42) - (44) one can state that

(45) $$VTP = \pi^* (P \times_G \mathfrak{g}) := \pi^* VM \qquad :$$

The vertical bundle is obtained by induction with the projection π from the bundle $P \times_G \mathfrak{g} := VM$, the vector bundle with fibre $\mathfrak{g}$ (Lie algebra of G), associated to the principal bundle $P \xrightarrow{\pi} M$ by means of the adjoint representation of G. In addition, there is a further reduction of the structural group of TP :

(46) $$G \times SO(4) \subset SO(D-4) \times SO(4) \subset SO(D) \quad .$$

The Riemannian connection along the locally immersed parts of M splits into a Riemannian connection and a Yang-Mills connection in $\pi^* VM$.

The splitting

(47) $$TP = \pi^* TM \oplus \pi^* VM$$

leads to the following splittings (in evident notation)

(48) $$T_{\mathbb{C}} P = \pi^* T_{\mathbb{C}} M \oplus \pi^* V_{\mathbb{C}} M$$

(49) $$\Lambda_t P = \pi^* \Lambda_t M \otimes \pi^* \Lambda_t V_{\mathbb{C}} M$$

(50) $$S_t P = \pi^* S_t M \otimes \pi^* S_t V_{\mathbb{C}} M$$

(51) $$CP = \pi^* CM \quad \pi^* CV_{\mathbb{C}} M \quad ,$$

and, if X and M have spinstructures

(52) $$\Delta P = \pi^* \Delta M \otimes \pi^* \Delta V_{\mathbb{C}} M$$

(53) $$\Delta^+ P - \Delta^- P = (\pi^* \Delta^+ M - \pi^* \Delta^- M) \otimes (\pi^* \Delta^+ V_{\mathbb{C}} M - \pi^* \Delta^- V_{\mathbb{C}} M)$$

These splittings, by Whitney's property and by the additivity and multiplicativity of the Chern character result in reduction formulae for the characteristic classes of natural bundles over P :

(54) $$e(TP) = \pi^* e(TM) \quad \pi^* e(VM)$$

(55) $$w(TP) = \pi^* w(TM) \quad \pi^* w(VM)$$

(56) $$c(T_{\mathbb{C}} P) = \pi^* c(T_{\mathbb{C}} M) \quad \pi^* c(V_{\mathbb{C}} M)$$

(57) $$p(P) = \pi^* p(M) \quad \pi^* p(VM)$$

(58a) $$\mathrm{ch} \Lambda_t P = \pi^* \mathrm{ch} \Lambda_t M \quad \pi^* \mathrm{ch} \Lambda_t V_{\mathbb{C}} M$$

(58b) $$\mathrm{ch}\, S_t P = \pi^* \mathrm{ch} S_t M \quad \pi^* \mathrm{ch}\, S_t V_{\mathbb{C}} M$$

(58c) $$\mathrm{ch}\, \Delta P = \pi^* \mathrm{ch} \Delta M \quad \pi^* \mathrm{ch} \Delta V_{\mathbb{C}} M$$

(58d) $$\mathrm{ch} \Delta^+ P - \mathrm{ch} \Delta^- P = (\pi^* \mathrm{ch} \Delta^+ M - \pi^* \mathrm{ch} \Delta^- M) \quad (\pi^* \mathrm{ch} \Delta^+ V_{\mathbb{C}} M - \pi^* \mathrm{ch}\, \Delta^- V_{\mathbb{C}} M) \text{ etc.}$$

As an example, let us have a closer look at eq. (57). p(VM) is nothing but the Chern class of the vector bundle $V_{\mathbb{C}} M$. So, the geometric Pontryagin classes $p_i(P)$ of P arise as combinations of pull-backs of Pontryagin classes $p_i(M)$ and of Chern classes $c_i(V_{\mathbb{C}} M)$.
Interesting compensations of characteristic classes (and anomalies) occur, when the Pontryagin classes of P vanish. Examples of this situation are

(A) The Hopf-fibration

(59) $P = S^5 \xrightarrow{\pi} \mathbb{C}P(2) = M$,

where S^5 is the five-dimensional sphere, whose Pontryagin classes vanish, because it is a hypersurface, and $\mathbb{C}P(2)$ is the complex projective space;

(B) A homogeneous space

(60) $P = G \xrightarrow{\pi} G/H = M$,

where $H \subset G$ are Lie groups. Here $p(G) = 0$ because G is parallelizable and therefore has a trivial tangent bundle.
Example (59) teaches us that P may have a unique spinstructure whereas M may be such that it does not admit any spin structure. This opens a possibility to define spinor fields with additional external degrees of freedom also for manifolds M without spin structure like $\mathbb{C}P(2)$. The construction envisaged here is related to but different from the construction describes by S.Hawking e.a.[16] The general feasibility of this construction, which may be of general use e.g. in quantum gravity is guaranteed by the existence of a universal principal bundle [8,14] $P_G \xrightarrow{\pi} M_G$, from which every G-principal bundle can be obtained by induction. For the universal bundle P_G is contractable and, thus has a unique spin structure. It would even suffice to take a 4-universal bundle[8]. More details will be presented in a separate publication[17].

References

1) E.Cremmer, B.Julia, J.Scherk: Nucl.Phys. 76B, 409 (1978)

2) H.Römer, unpublished

3) M.F.Atiyah, I.M.Singer: Bull.Ann.Math.Soc. 69, 422 (1964)
Ann.Math. 87, 484, 546 (1968)

4) See,e.g. T.Eguchi, A.J.Hanson: Physics Reports 66, 214 (1980)
and references therein

5) F.Hirzebruch: Topological Methods in Algebraic Geometry
Grundlehren der Math.Wissenschaften, Vol.131, Springer New York (1966)

6) S.M.Christensen, M.J.Duff: Phys.Lett. 76B, 571 (1978)
Nucl.Phys. B154, 301 (1979)

7) H.Römer: Phys.Lett. 83B, 172 (1979)

8) J.W.Milnor, J.D.Stasheff: Characteristic Classes, Annals of Math.
Studies 76, Princeton Univ.Press (1974)

9) See, e.g. N.K.Nielsen, H.Römer, B.Schroer: Nucl.Phys. B127, 493 (1977)
and references therein

10) M.J.Duff, P.van Nieuwenhuizen: Phys.Lett. 94B, 179 (1980)

11) H.Römer: Phys.Lett. 101B, 55 (1981)

12) S.S.Chern, J.Simons: Ann.Math. 99, 48 (1974)

13) M.F.Atiyah, J.B.Segal: Ann.Math. 87, 531 (1968)

14) N.Steenrod: The Topology of Fibre Bundles, Princeton Math.Series
No.14, Princeton University Press

15) T.Kaluza: Sitzungsbericht Preuss.Akademie Wiss. (1921), 966

16) S.W.Hawking, C.N.Pope: Phys.Lett. 73B, 42 (1978)

17) H.Römer: in preparation.

GRADED BUNDLES IN THE OGIEVETSKY-SOKATCHEV SUPERGRAVITY

Janusz Czyż
00-950 Warsaw 137
POLAND

Abstract. The trivial smooth graded bundles and the trivial and non-trivial analytic ones are presented. The applications of graded bundles in supergravity are discussed.

1. Why graded bundles ?

The duality of a bosonic and a fermionic behaviour of particles, which is perhaps a most profound feature of matter, implies the necessity of putting on an equal footing symmetric and anti-symmetric relations in mathematical theories describing the physical reality. The differential geometry fails this postulate because for vector fields on a manifold the anti-symmetric commutator of vector fields defined by means of the Leibnitz rule

$$X(fg) = X(f)g + fX(g) \tag{1}$$

makes sense but the anti-commutator does not. The search for a correct symmetric relation between vector fields led to several geometric models like a super-manifold of Konstant [6], where the manifold as a space remains unaltered but the local algebra of smoth functions $C^{\infty}_{loc}(M)$ is replaced by an algebra of local sections $\Gamma^{\infty}_{loc}(M,\Lambda E)$ of a Grassmann bundle ΛE and a G^{∞}- supermanifold of Alice Rogers [8], where coordinate charts are modelled on flat Grassmann moduli and coordinate functions have Taylor expansions in a Grassmann algebra. But these notions adopted only the category of smooth manifolds for needs of bosonic and fermionic fields. On the other hand it is natural to consider them being attached to fibres of a fibred space instead of being tangent to a usual manifold. Furthermore, the success of the ideas of Penrose made complex (analytic) structures on manifolds not less important than real ones.

Graded bundles seem to satisfy all these expectations. Their misery in the class of C^{∞}-functions they fairly compensate with their wealth in the domain of holomorphic maps.

2. Graded bundles: the notion.

As a vector (n-)bundle is, roughly speaking, a certain manifold which projects onto a base manifold M and locally looks like $U_\alpha \times \mathbb{C}^n$ or $U_\alpha \times \mathbb{R}^n$, where open sets U_α cover M, as a graded (n-)bundle is a manifold which projects onto a base M too and locally looks like $\overset{n}{\oplus}\Lambda E_{|U'_\alpha}$ where E is a vector bundle over M called the auxiliary bundle. Note that graded bundles admit richer algebraic structures than the vector ones: as fibres V_x, $x \in M$, of a vector bundle are vector spaces as fibres G_x of a graded bundle are Grassmann moduli over the Grassmann algebras being fibres of the auxiliary Grassmann bundle ΛE at the same points $x \in M$.

As each vector bundle is determined by a cocycle $g_{\alpha\beta}: U_{\alpha\beta} \to GL(n,\mathbb{C})$ which functions are sections of a product bundle $M \times GL(n,\mathbb{C})$ as each graded bundle is by a cocycle $G_{\alpha\beta}: U_{\alpha\beta} \longrightarrow GL(n,\Lambda E)$ but its values take place at fibres $GL(n,\Lambda E_x)$ of the bundle of groups $GL(n,\Lambda E) \to M$ of matrices with Grassmann algebra elements (almost all such matrices are invertible).

Similarly like in the case of vector bundles we may define equivalent graded bundles and identify the equivalence classes with elements of the first cohomology $H^1(M, GL(n,\Lambda E))$.

We will consider graded bundles with respect to various classes of mappings: C^∞, G^∞, analytic and algebraic.

Consider the subgroup $HL(n,\Lambda E_x) \subset GL(n,\Lambda E_x)$ consisting of matrices $[a_{ij}]$ such that the scalar part $[a_{ij}^{(o)}]$ is the unit matrix. Repeating the previous part for the groups $HL(\cdot)$ we obtain a class of HL-graded bundles. Any HL-graded n-bundle can be interpreted as a composition of a system of n mutually interferring identical objects represented by the auxiliary bundle.

For more information about graded bundles see [2],[3],[4].

3. Triviality and non-triviality.

A graded bundle is said to be trivial if it is equivalent to a bundle admitting a cocycle

$$u_{\alpha\beta} \ni x \longrightarrow G_{\alpha\beta}(x) \in GL(n,\mathbb{C}) \subset GL(n,\Lambda E_x) \qquad (2)$$

($\mathbb{C}$ may be replaced by any field). If the cocycle $G_{\alpha\beta}$ determines a vector n-bundle W then the graded bundle admits the form $W \otimes \Lambda E$. Any trivial HL-graded bundle is of the type $\oplus \Lambda E$

Theorem

1° Each smooth graded bundle over a paracompact base is trivial.

2° There are non-trivial G^{∞}-graded bundles

3° There are non-trivial holomorphic and algebraic graded bundles. The sufficient but not necessary condition for their existence is $H^{1}(M,E) \neq 0$.

For the proof see [3],[4].

Comments

α) The theory of analytic graded bundles includes the classical Cousin problems from complex analysis: additive, which we can see in the HL-bundles and multiplicative one. It also includes "distributive" problems connected with the equivalence relation $f'_{\alpha\beta} \sim s_{\alpha} \wedge s_{\beta} + f_{\alpha\beta}$ and other ones.

β) The theorem can be summarized: the non-triviality of graded bundles is an effect of analytic global structures of manifolds.

4. Graded bundles in supergravity.

The graded bundle seems to be a suitable global algebraic and geometric structure for spaces of parameters in supergravity. Then the rank n corresponds to the number of considered basic fermionic fields so that "proper" graded bundles, i.e. $n > 1$, can be applied in the extended supergravity which will be abbreviated by e.s., see [10].

The triviality of each smooth graded bundle causes that all the models of e.s. using the ordinary real smooth structure for the space-time look like a gauge theory in a trivial flat bundle. It means that all the basic fermion fields can be separated each from another one by a suitable choice of a "gauge". Hence the classical idea "mutual interactions of the fermion fields are measured by a curvature" seems to be not adequate in this case . Perhaps that is why T.Ross in [10] could not generalize his geometric approach to supergravity onto the case of e.s.

The model of supergravity with a complex (holomorphic) manifold as a base space had been proposed by Ogievetsky-Sokatchev [7] and was elaborated by A.S.Schwarz in his paper [9] about a "space-field democracy". In this model the space of parameters ξ is provided with the structure of a complex flat superspace generated by the variables $(x, \theta, \bar{\theta})$, where $x \in \mathbb{C}^{4}$, θ (resp. $\bar{\theta}$) is a left (resp.right) 2-spinor (the variables Θ, Θ form a sector of anticommuting variables).

The above method can be extended onto the case of a complex 4-manifold M instead of $\mathbb{C}^4$. Then we have to assume that a real structure $\sigma : M \longrightarrow M$ is given ($\sigma^2 = \mathrm{id}$ and σ generated an anti-linear map of the tangent bundle), fix-points of σ form an oriented, real 4-manifold F_M and certain real integrable 4-manifolds H_α (we shall mention them below) are orientable too. Then real subspaces of tangent spaces give rise uniquely to spaces of spinors.

Under these assumptions the Grassmann bundle ΛTM seems to be a primeval structure of the parameter space for a simple supergravity ($n = 1$) and a HL-graded n-bundle having TM as the auxiliary bundle is that of an e.s.

The dynamics and field equations are determined by a family of spinor bundles over surfaces $H_\alpha \subset M$. These bundles are certain smooth transformations of the spinor bundle over F_M and the family H_α fulfills infinitesimaly equations 4.1 - 4.3 in [9]. The family H_α corresponds to the fundamental superfield $H(\xi, \Theta, \Theta), \xi \in F_M$ in [7] up to super-symmetric transformations.

If $n = 1$ then the Ogievetsky-Sokatchev model is equvalent locally to a model with a real and smoth manifold of space-time. The theorem about triviality of graded bundles makes impossible to H_α prolongate this equivalence onto all the theory. Namely in the case of a global e.s. when $H^1(M, TM) = 0$ then there exist different models associated with non-trivial HL-graded bundles which are equivalent in the sense of the smooth base space-time manifold.

Examples:

a) $H^1(P^4\mathbb{C}, TP^4\mathbb{C}) = 0$ (it follows from the Euler exact sequence).

b) $H^1(Q_4, TQ_4) = 0$, where Q_4 is the non-degenerate quadric in $P^5\mathbb{C}$. If follows from the standard exact sequences of sheaves over $P^5\mathbb{C}$ and $Q_4 \subset P^5\mathbb{C}$.

c) The case where the first cohomology does not vanish takes place if $M = R_1 \times R_2 \times R_3 \times R_4$, where R_i are Riemann surfaces given by equations with real coefficients (then we have real structure on M and the orientability condition holds) and $g(R_1) > 0$ (g is genus). Then

$$\dim H^1(R, TR) = \begin{cases} 0 & g(R) = 0 \\ 1 & g(R) = 1 \\ 3(g-1) & g(R) > 1 \end{cases} \tag{3}$$

makes $H^1(M, TM) \neq 0$

References

1. Atiyah, M.F., R.S. Ward, Comm. Math. Phys 55 (1977) 177
2. Czyż, J., Lecture Notes in Math. 838 (1981) 110
3. Czyż, J., preprint Luminy-Marseille, CNRS, CPT 81/PE1328.
4. Czyż, J. "On first cohomology methods at super-spaces" in preparation
5. Gawędzki K., Ann. Inst. Henri Poincarée 27 (1977) 355
6. Kostant, B., Lecture Notes in Math. 570 (1977) 170
7. Ogievetsky, V., E.Sokatchev Phys. Lett. 79B (1978) 222
8. Rogers, A., J.Math. Phys. 21 (1980) 1952
9. Schwarz, A.S., Nucl. Phys. 171B (1981) 154
10. "Supergravity", Proceedings of the Stonybrook Symposium, 1979, P. van Nieuvenhuizen ed.

ANALYSIS AND COMPARISON OF DIFFERENT WAYS OF IDENTIFICATION OF SPIN FUNCTIONS VARIABLES

Z.Marić and M.Božić
Institute of Physics, P.O.Bob.57, Beograd, Yugoslavia

I. Spinor components variables in the nonrelativistic theory of spin

It is well known that Pauli have introduced into physics a two-component quantity $\binom{u}{v}$ which, by definition, belongs to the two-dimensional representation of the rotation group and which we call today, in accordance with Cartan[1], spinor. Such a definition leaves open the question: "What are the variables of functions u and v which change under rotation and consequently induce the mentioned transformation property?"

It quantum mechanics of many electron systems Pauli introduced the spin variable σ, which takes two values $\pm 1/2$, and the functions $u = C_{1/2}(\sigma) = \delta_{1/2,\sigma}$, $v = C_{-1/2}(\sigma) = \delta_{-1/2,\sigma}$. This choice turned out to be useful and suitable in the construction of many-electron antisymmetric functions but it does not give the answer to the above question. Another remark associated with the fact that spinor components are probability amplitudes may be stated in connection with this choice. Namely, the argument of the function $C_m(\sigma)$ is the same quantity as the index of the function, and consequently probability amplitude takes two values only: 0 and 1.

From the work of Cartan[1] and Kramers[2] it follows that for the spinor components one may take one of the two pairs (χ_1, χ_2) or (χ_3, χ_4) from the following set of functions

$$\chi_1(\alpha,\beta,\gamma) = D^{1/2}_{1/2,1/2} = (8\pi^2)^{-1/2}\cos(\beta/2).\exp(i(\alpha+\gamma)/2)$$

$$\chi_2(\alpha,\beta,\gamma) = D^{1/2}_{-1/2,1/2} = (8\pi^2)^{-1/2}\sin(\beta/2).\exp(+i(-\alpha+\gamma)/2)$$

$$\chi_3(\alpha,\beta,\gamma) = D^{1/2}_{1/2,-1/2} = (8\pi^2)^{-1/2}\sin(\beta/2).\exp(i(\alpha-\gamma)/2)$$

$$\chi_4(\alpha,\beta,\gamma) = D^{1/2}_{-1/2,-1/2} = (8\pi^2)^{-1/2}\cos(\beta/2).\exp(-i(\alpha+\gamma)/2)$$

$$0 \le \alpha \le 4\pi, \quad 0 \le \beta \le \pi, \quad 0 \le \gamma \le 2\pi \tag{1}$$

where α, β, and γ are the Euler angles of the three-dimensional coordinate system associated with the particle. In this choice of spinor components the Euler angles play the role of the spin functions variables.

Bopp and Haag[3] found the differential operator $\hat{s}$ such that $\hat{s}_x$, $\hat{s}_y$, $\hat{s}_z$ form the Lie algebra of the rotation group whereas $\hat{s}^2$ have eigenvalues s(s+1), s taking integer and half-integer values. They also assumed that the coordinate system is attached to the particle. The orientation of the system with respect to the fixed coordinate system is determined either by the Euler angles α,β,γ or with the aid of two vectors $\vec{r}_1$ and $\vec{r}_2$. Consequently two equivalent forms for the differential operator $\hat{s}$ are obtained. They are:

$$\hat{s} = -i\hbar(\vec{r}_1 \times \frac{\partial}{\partial\vec{r}_1} + \vec{r}_2 \times \frac{\partial}{\partial\vec{r}_2}) \tag{2}$$

and

$$\hat{s}_x = i\hbar(\sin\alpha \frac{\partial}{\partial\beta} + \cot\beta.\cos\alpha \frac{\partial}{\partial\alpha} - \frac{\cos\alpha}{\sin\beta} \frac{\partial}{\partial\gamma})$$

$$\hat{s}_y = i\hbar(-\cos\alpha \frac{\partial}{\partial\beta} + \cot\beta.\sin\alpha \frac{\partial}{\partial\alpha} - \frac{\sin\alpha}{\sin\beta} \frac{\partial}{\partial\gamma})$$

$$\hat{s}_z = -i\hbar \frac{\partial}{\partial\alpha}$$

$$\hat{s}^2 = -\hbar^2\left[\frac{1}{\sin\beta} \frac{\partial}{\partial\beta} (\sin\beta \frac{\partial}{\partial\beta}) + \frac{1}{\sin^2\beta} (\frac{\partial^2}{\partial\alpha^2} + \frac{\partial^2}{\partial\gamma^2}) - \frac{2\cos\beta}{\sin^2\beta} \frac{\partial^2}{\partial\alpha\partial\gamma}\right] \tag{3}$$

The common eigenfunctions of the operators $\hat{s}^2$ and $\hat{s}_z$ have the form

$$\phi_{\mu\nu s}(\alpha,\beta,\gamma) = \exp[i.(\mu\alpha+\nu\gamma)].\phi_s(\beta) \tag{4}$$

where $\phi_s(\beta)$ is the solution of the hypergeometric differential equation s is an integer and half integer and

$$|(\mu+\nu)/2| + |(\mu-\nu)/2| - s = 0, -1, -2 \ldots$$

It happens that the functions (1) are special cases of the functions (4) corresponding to the case s = 1/2.

Combining this result with the correspondance principle $-i\hbar\nabla \to \hat{p}$ and with the standard definition of angular and magnetic momentum vector of a mass and charge distribution the conclusion has been derived that the spin may be associated with a charged body in which the ratio of the distribution of charge and mass depends on coordinate.

II. Spinor components variables in the relativistic theory of spin

Dirac relativistic theory of electron has been credited for a long time to be the theory in which spin appears in a most natural way. That this attitude is changing is illustrated by Leblond[4] with the aid of quotations from literature. In Dirac theory spinor components variables are space-time coordinates of the particle.

Recently Barut and Bracken[5] have re-examined the Dirac equation and Schrödinger work on Zitterbewegung. Those authors argue that Dirac equation describes an extended particle for which the center of charge and the center of mass are not at the same point. According to Barut and Bracken $\vec{r}$ in Dirac equation and associated solutions should be interpreted to be the center of charge and not the center of mass. The relative momentum and microscopic coordinate in the rest frame of center of mass are introduced. Spin appears as the orbital angular momentum in this system.

Dahl[6] have contributed to the relativistic theory of spin in the direction which makes this theory to be more closer to the non-relativistic one. By incorporating the results of Bopp and Haag into general principles of the relativistic quantum mechanics Dahl found that Dirac particle and quantum mechanical relativistic rotor are identical dynamical systems. Dahl reached this conslusion by showing that Dirac equation may be regarded as the matrix representation of the Schrödinger equation with Hamiltonian

$$\hat{H} = \frac{2}{\hbar} m_o c^2 \hat{\zeta}_3 + \frac{4c}{\hbar^2} \hat{\zeta}_1 . (\hat{s}.\hat{p}), \qquad (5)$$

$\hat{s}$ is the operator (3) constructed by Bopp and Haag, $\hat{\zeta}_i = \hat{s} \cdot \vec{e}_i$, $\hat{p} = -i\hbar\nabla$. The Hamiltonian[5] operates in the space of wave functions

$\psi(\vec{r},\alpha,\beta,\gamma,t)$ of particle coordinates $\vec{r}$ and internal variables α,β,γ. This function is represented by the general expansion

$$\psi(\vec{r},\alpha,\beta,\gamma,t) = \sum_{i=1}^{4} u_i(\vec{r},t)\chi_i(\alpha,\beta,\gamma) \tag{6}$$

where χ_i are given in (1) and $u_i(\vec{r},t)$ satisfy Dirac equation. The important step in Dahl's proof is the finding that 4-dimensional matrices σ_i and ρ_i occuring in Dirac's paper are matrix representatives of operators $\hat{s}_i$ and $\hat{\varrho}_i$ in the space of functions χ_i, i=1...4, multiplied by 2/h.

Dahl solved the Schrödinger equation with the Hamiltonian (5). The solutions associated with positive energy eigenvalues tend in nonrelativistic limit to the products of plane waves and spinor components χ_1 and χ_2.

III. Verification of coherent spinor rotation of fermions

Recently, the transformation properties of spinor components under rotations around z-axis have been verified[7,8] experimentally in a crystal interferometer shown in the Fig.1. Unpolarized neutrons with a wavelength λ=1.82 were used. They are coherently split into beams I and II and exposed to magnetic field along their path. The intensities of two beams emerging from the interferometer are measured as functions of the difference of the magnetic field action on beam I and II [$\Delta\int\vec{B}d\vec{s} = \int B_z ds$ (path I)$-\int B_z ds$(path II)]. The ideal experiment in which one of the beams is exposed to magnetic field along its path serves to describe the experiment theoretically.

The part of the evolution operator which operates on the spinor is

$$T = \exp\,(i\mu\hat{\sigma}\int\vec{B}dt/\hbar) \tag{7}$$

where

$$\mu = \frac{ge}{2mc}\frac{\hbar}{2}$$

This operator has the form of the operator of rotations in the space of spinors

$$R(\vec{\phi}) = \exp(-i\vec{\phi}.\hat{\sigma}/2) \tag{8}$$

and the vector $\mu\int Bdt/\hbar$ plays the role of the vector $-\vec{\phi}/2$. If the magnetic vector is in the z-direction, the rotation operator takes a particularly simple form so that spinor components $\chi_1(\alpha,\beta,\gamma)$ and $\chi_2(\alpha,\beta,\gamma)$ transform into functions:

$$\chi_1'(\alpha,\beta,\gamma,\phi) = \{\sin(\beta/2).\exp[i(\alpha-\phi+\gamma)/2]\}/\sqrt{8\pi^2}$$
$$\chi_2'(\alpha,\beta,\gamma,\phi) = \{\cos(\beta/2).\exp[-i(\alpha-\phi-\gamma)/2]\}/\sqrt{8\pi^2} \tag{9}$$

where

$$\phi = -(2\mu/\hbar).\int B_z dt = -(2\mu m/\hbar^2 k).\int B_z ds$$

Calculation of the relative intensity of the beam in the forward direction (0) with a magnetic field in beam I yields

$$\frac{I_o(\phi)}{I_o(0)} = \frac{|\chi'_{1,I} + \chi'_{1,II}|^2 + |\chi'_{2,I} + \chi'_{2,II}|^2}{|\chi_{1,I} + \chi_{1,II}|^2 + |\chi_{2,I} + \chi_{2,II}|^2}$$

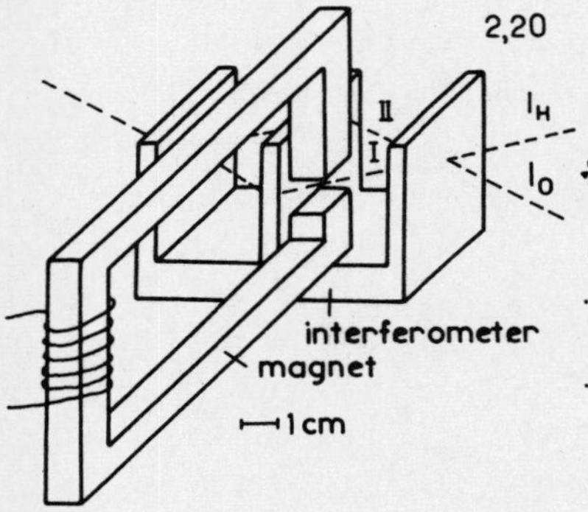

Fig.1- Sketch of the experimental setup.

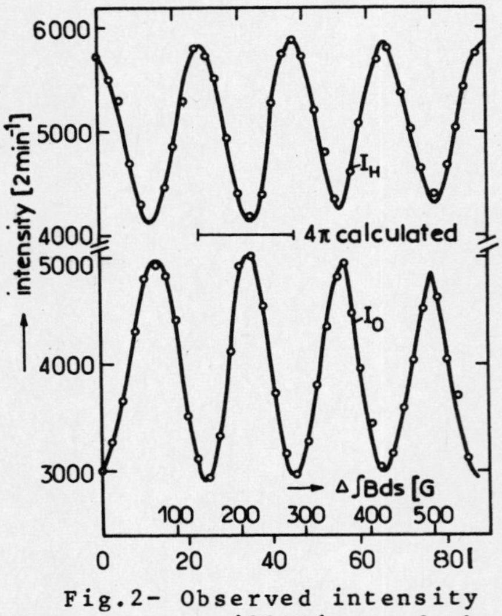

Fig.2- Observed intensity oscillations of the 0- and H- beam[7].

$I_o(0)$ and $I_o(\phi)$ are the intensities of the forward beam when the beam I is not/is subjected to the magnetic field.

REFERENCES

1- Cartan,M.Z. Lecons sur la Theorie des Spineurs, Hermann, Paris, (1938).

2- Kramers,H.A. Grundlagen der Quantheorie, Akad.Verlagsgeselschaft, Leipzig, (1938).

3- Bopp,F. and Haag,R. (1950) Z.Naturforschg. 5a, (1950) 644.

4- Levy-Leblond, Riv.Nuov.Cim. 4 (1974) 99.

5- Barut,A.O. Bracken,A.J. Phys.Rev. D.23 (1981) 2454.

6- Dahl,J.P. Det Kongelige Danske Videnskabenes Selskab Matematisk-fysiske Meddeleser 39 (1977) 12.

7- Rauch,H.Zeillinger, A.Badurker, A.Wilfing, G.Bauspiess, W.Phys. Lett. 54A (1975) 425.

8- Werner,S.A. Colella,A.W. Overhauser,A.W. Eagen,C.F. Phys.Rev.Lett. 35 (1975) 1053.

A GENERALIZED IMPRIMITIVITY THEOREM FOR A CLASS OF POV - MEASURES

S. Twareque Ali
Department of Mathematics
Concordia University
Montréal, P. Q.
Canada H4B 1R6

ABSTRACT

We discuss some recently obtained results on the problem of generalizing the imprimitivity theorem of Mackey, to the case where the underlying projection valued measure is replaced by a commutative positive operator valued measure. A possible extension of the group representation to a K-space is also discussed.

1. Introduction

Let X be a metrizable, locally compact topological space, B(X) the set of all Borel sets of X, G a metrizable locally compact group, $\mathcal{H}$ a separable Hilbert space and $L(\mathcal{H})^+$ the set of all bounded positive operators on $\mathcal{H}$. Let $E \mapsto a(E)$ be a normalized positive operator valued (POV)- measure[1)] on X, i. e., $E \in B(X)$ and $a(E) \in L(\mathcal{H})^+$. Assume that $\forall$ E, $F \in B(X)$, a(E) and a(F) commute. Let $g \mapsto U_g$, $g \in G$ be a strongly continuous unitary representation of G on $\mathcal{H}$. Let X be a transitive G-space, so that

$$X = G/H \tag{1}$$

for some subgroup H of G.

Definition 1

The pair $\{a, U\}$ is said to form a commutative system of covariance, if $\forall g \in G$, $E \in B(X)$,

$$U_g\, a(E)\, U_g^* = a(g[E]). \qquad (2)$$

A commutative normalized POV - measure can always be written[1] as a Choquet integral, with respect to a probability measure, over a set of extreme points. Thus, given a, there exists a unique probability measure ν, carried by the set $\mathfrak{G}$ of all normalized projection valued measures $E \mapsto P(E)$, $E \in B(X)$, $P(E) \in A(a)$, (the von Neumann algebra generated by a), such that, $\forall \phi, \psi \in \mathcal{H}$, $E \in B(X)$,

$$(\phi, a(E)\psi) = \int_{\mathfrak{G}} (\phi, P(E)\psi)\, d\nu(P). \qquad (3)$$

Definition 2

The commutative system of covariance $\{a, U\}$ is said to be transitive if the induced action of U_g on the spectrum of A(a) is transitive[2].

2. The main results.

The following proposition[3] achieves a generalization of the imprimitivity theorem[4] of Mackey, to transitive, commutative systems of covariance. The proof depends on the integral representation (3) and is given in Ref. 4.

Proposition

Let $g \mapsto U_g$, G, $\mathcal{H}$ and X be as above. Then there exists a normalized POV-measure a on B(X), for which $\{a, U\}$ is a transitive, commutative system of covariance, if and only if U_g is a representation which is induced from some subgroup M of G, and there exists a probability measure ν on B(X) which is invariant under M. Furthermore, given a,

ϑ is uniquely fixed and vice versa.

Finally, consider the union of Hilbert spaces

$$K = \bigvee_{y \in Y} \mathcal{H}_y \tag{4}$$

where $Y = G/M$, and, $\forall y$, $\mathcal{H}_y$ is isomorphic to some fixed Hilbert space $\mathcal{H}_o$. Such a space is called a K-space[5] in the literature. It is then possible[1], given any transitive, commutative system of covariance, to lift U_g to a unique, minimal K-representation $\tilde{U}_g$ of G on K. Details will be published elsewhere.

3. References

1) S.T. Ali, Lecture Notes in Mathematics, Vol. 905, Springer-Verlag, Berlin (1982), pp. 207-228.

2) M. Takesaki, Acta Math. 119 (1967), 273.

3) S.T. Ali, 'Commutative systems of covariance and a generalization of Mackey's imprimitivity theorem', to appear.

4) G.W. Mackey, Proc. Natl. Acad. Sci. U.S.A. 35 (1949), 537.

5) A. Giovannini, J. Math. Phys. 22 (1981), 2389.

INVERSIONS IN TWISTOR SPACE

P.M. van den Broek
Department of Applied Mathematics
Twente University of Technology
7500 AE Enschede, The Netherlands

Twistor space T is equal to $\mathbb{C}^4$ with Hermitian form

$$\langle L,L'\rangle = L_0\bar{L}'_2 + L_1\bar{L}'_3 + L_2\bar{L}'_0 + L_3\bar{L}'_1 \tag{1}$$

Elements of T are called twistors; the rays of T, called projective twistors, form projective twistor space PT. If $L \in T$ and $L \neq 0$ we denote by $\underline{L}$ the projective twistor with $L \in \underline{L}$. If $\langle L,L'\rangle = 0$ for each $L \in \underline{L}$ and each $L' \in \underline{L}'$ then $\underline{L}$ and $\underline{L}'$ are said to be orthogonal. Projective twistors correspond to congruences of null lines in compactified Minkowski space[1]. Let C be the conformal group; C has the coset decomposition

$$C = C_o + pC_o + tC_o + ptC_o \tag{2}$$

where C_o is the restricted conformal group, p denotes space inversion and t denotes time inversion. C is isomorphic to the group of transformations of PT which preserve orthogonality; the action of C on PT provides an explicit realisation of this isomorphism. To each transformation of PT which preserves orthogonality there corresponds a unique ray of semilinear transformations (ROST) of T; this follows from Uhlhorns version [2] of Wigners theorem, when modified for the Hermitian form of equation (1).

In turn, a ROST corresponds uniquely to an orthogonality preserving transformation of PT if its elements satisfy one of the equations

$$\langle UL,UL'\rangle = C \langle L,L'\rangle \qquad \forall L, L' \in T \tag{3}$$

$$\langle UL,UL'\rangle = C\,\overline{\langle L,L'\rangle} \qquad \forall L, L' \in T \tag{4}$$

where C is a real constant. ROSTs which satisfy equation (3) with $C > 0$ correspond to conformal transformations of C_o; the restriction $\det U = 1$ now gives the well-known 4 : 1 homomorphism of SU(2,2) and

C_o which is determined explicitly by Klotz[3). We see that the fact that this homomorphism is a 4 : 1 homomorphism stems from the fact that T is four-dimensional. ROSTs which satisfy equation (3) with $C < 0$ correspond to conformal transformations of the coset ptC_o; ROSTs satisfying equation (4) with $C > 0$ and $C < 0$ correspond to the cosets tC_o and pC_o respectively. The action of the inversions p and t on T may now be determined explicitly; details will be published elsewhere.

1) R. Penrose, Journ. Math. Phys. 8, 345 (1967)
2) U. Uhlhorn, Arkiv för Fysik 23, 307 (1963)
3) F.S. Klotz, Journ. Math. Phys. 15, 2242 (1974)

PHYSICAL GROUP THEORY AND EUCLIDEAN SPACE

Herwin Venables
Portsmouth Polytechnic, PO1 2EG
England

ABSTRACT

An outline is given of the considerations involved in using group elements in place of Euclidean space.

Newtonian mechanics gives rise to the philosophical discussion of the nature of space, as by Leibniz, Mach, Einstein and others. The concepts present are made greater under quantum conditions.

It is suggested that group elements might be used at least where the Euclidean metric function does not seem physically appropriate enough. Symmetry requires that $d(x,y) = d(y,x)$. For a pair of points this seems to assume an accurate measurement in the opposite direction between the same, seemingly undisturbed, points. The triangular condition $d(x,z) \leqslant d(x,y) + d(y,z)$ requires three measurements for sides that may be attended by quite different physical circumstances, and two of such measurements would be difficult.

In any neighbourhood there may be signals, stresses and interactions, and these might be better described by group elements as a preliminary to the adjunction of such spatial properties as are needed for propagation. Criticism of Quantum Mechanics is made of its usage of Euclidean space with its set of precise distances, which might imply the existence of a nondenumerable set of exact measurements. Surrounding an atom a union of neighbourhoods can be visualized, each with undetermined physical properties. From this union propagation might be launched with an evolving Euclidean grid of position markers.

The selection of group elements will depend on the parameters required to model the physical conditions. Quaternions, SU(2), D_4, Dirac group, SU(3), etc. are widely used. The parameters will necessarily have distributed values because of the infeasibility of full measurement.

A Cartan-Killing metric can be defined for each group element by taking the trace of their matrix representations. Each point, as in diffraction patterns may have several matrices associated with it for different directions, phases and wavelengths. Thus the trace will contain partial data of position mixed in with data of interation, mass, etc. When this trace does not yield the raw data to assemble a more

Euclidean type of metric function, it is desirable to speak of the separation of points, rather than their displacement. These separations inevitably contain interactive contributions. Until these contributions are weak enough or fortuitously cancelled, then position is not adequately defined.

Between the hydrogen nucleus and the orbitting electron, there are regions of separation, for which distance might be defined via the radius of curvature of a Bohr patch on a manifold. Such a radius would be a sizable fraction of the Bohr radius, a_o. For numerical values $r \ll a_o$, only impure separations exist, and $-Ze^2/r$ does not have a satisfactory meaning for such values.

This however will leave us with a heavy program of development for the hydrogen atom, or Kepler problem placed in a region of group elements from which an adquate amount of dynamics must be extracted.

The photons can be described by quaternions and spin matrices,

$$q = |q|e^{n\omega t} \quad \text{and } \sigma = |\sigma|e^{in\omega t},$$

where the axis of free propagation is represented by

$$n = j \cos\alpha + k \cos\beta + l \cos\gamma$$

When interation is included groups of higher order must be used, which gives more room for scattering. Now the Kepler problem, whilst solvable in R^2, can also be solved with quaternions and spin matrices and with groups of higher order. Each photon of the region makes a contribution to be summed, i.e. we arrive at a sum of quaternions or spin matrices to simulate the equations of ordinary Kepler motion. In passing it might be said there are interactions from all about the atom so that a more fully considered Bohr atom is not properly a two dimensional problem. Until the separations have become pure enough to permit the existence of the inverse square law of force between the nucleus and electron there exists more mixed interation.

The dynamics is best developed by the Hamilton-Jacobi method. Each photon has its own equation of motion and constants of motion such as frequency and axis of propagation. Also the Coulomb potential after the Bohr patch formation can be regarded as an averaged effect and, further, the constants of motion and quantization can be obtained. At greater distances the group elements may be left out but they can still be usefully retained.

REFERENCE

Gilmore, R. Lie Groups, Lie Algebras, and some of Their Applications, p.248. New York, John Wiley and Sons, 1974.

SOME ASPECTS OF RANDOM WALKS ON GROUPS

P.W. Kasteleyn
Instituut-Lorentz voor Theoretische Natuurkunde
Nieuwsteeg 18, 2311 SB Leiden
THE NETHERLANDS

ABSTRACT

This paper is concerned with random walks on countable groups. After the introduction of the main concepts three special subjects are treated: random walks on free groups with a finite number of generators, random walks on certain non-abelian extensions of the additive groups $\mathbb{Z}^d$ and random walks on groups with a stochastic colouring.

1. Introduction

In classical random walk theory a random walk is a Markov process with the d-dimensional Euclidean space $\mathbb{R}^d$ or the lattice of d-dimensional integers $\mathbb{Z}^d$ as state space and with a spatially homogeneous transition probability. More explicitly, a random walk on $\mathbb{R}^d$ is a Markov process with a transition probability density $\rho(x,y)$ which is invariant under all translations of $\mathbb{R}^d$, i.e. which satisfies $\rho(x+z, y+z) = \rho(x,y)$ for all $x,y,z \in \mathbb{R}^d$, and a random walk on $\mathbb{Z}^d$ is a Markov chain with a transition probability $p(\ell,m)$ which is invariant under all translations leaving $\mathbb{Z}^d$ invariant, i.e. which satisfies $p(\ell+n, m+n) = p(\ell,m)$ for all $\ell,m,n \in \mathbb{Z}^d$. Random walks on other lattices in $\mathbb{R}^d$ are defined analogously.

One calls the transitions described by $\rho(x,y)$ and $p(\ell,m)$ <u>steps</u> and one studies the stochastic properties of <u>walks</u> consisting of a number of consecutive steps.

Because of the translation invariance of the transition probability (density) a random walk on $\mathbb{R}^d$ or $\mathbb{Z}^d$ is completely determined by the <u>stepping probability (density)</u> $\rho(x) := \rho(0,x)$ or $p(\ell) := p(0,\ell)$, respectively.

We remark in passing that a random walk on $\mathbb{Z}^d$ can be considered as a special case of a random walk on $\mathbb{R}^d$, characterized by a stepping probability density concentrated on $\mathbb{Z}^d$: $\rho(x) = \sum_{\ell \in \mathbb{Z}^d} p(\ell)\delta(x-\ell)$.

A vast amount of literature, among which the famous monograph by Spitzer [1], has been devoted to classical random walks, in particular on $\mathbb{Z}^d$, and their applications; refs. 1 and 2 contain an extensive list of references to, respectively, the mathematical and the physical literature in this field. We mention a few of the many properties of random walks on $\mathbb{Z}^d$ that have been investigated: the mean square displacement after n steps, the average number of distinct sites visited in n steps, the probability $F(\ell)$ that a walk starting at 0 ever passes through the point ℓ and the average number of steps $\langle n(\ell)\rangle$ which this requires. Of particular interest is the probability of return to the origin $F := F(0)$. If $F=1$ the random walk is called <u>recurrent</u>, if $F<1$ it is <u>transient</u>.

Random walks on $\mathbb{R}^d$ and $\mathbb{Z}^d$ (and other lattices) have in common that the state space is an (additively written) abelian group and that the essential condition imposed on the transition probability is that of invariance under this group. It is evident that the concept of a random walk may in a natural way be generalized to a larger class of groups. It is the aim of this paper to discuss some aspects of these generalized random walks, especially random walks on countable groups, to be introduced in section 2.

Kesten [3] was the first to consider (symmetric) random walks on countable groups; he investigated in particular properties of the spectrum of the transition probability matrix. Random walks on groups with a finite number of generators were investigated a few years later by Dynkin and Malyutov [4].

Random walks on countable abelian groups were studied in the 1960's by Dudley [5], at one place in Spitzer's monograph (ref. 1, pp. 91-95), and, most extensively, by Kesten and Spitzer [6]. The emphasis was on recurrence vs. transience and on the existence and asymptotic properties of the so-called potential kernel associated with a random walk.

More recently, a number of mathematicians in France have developed a theory of random walks on locally compact groups [7,8]. They derived, among other results, criteria for recurrence and asymptotic properties of potential kernels.

2. Random walks on countable groups

Let G be a countable (finite or infinite) group. By a <u>random walk on G</u> we understand a Markov chain with state space G and with a transition probability p(g,h) which is <u>left-invariant</u> under G, i.e. which satisfies $p(fg,fh) = p(g,h)$ for all $f,g,h \in G$. It follows that for a given element $g \in G$ the probability of a step from an element g' to g'g is equal to p(e,g), where e is the identity element of G, and so is independent of g'. Hence the steps of the random walk consist in the right multiplication with an element of G. A random walk thus defined may be called a right random walk in order to distinguish it from a left random walk, proceeding by left multiplication; however, we shall consider only right random walks, and omit the word "right".

Since a random walk is completely determined by G and the <u>stepping probability</u> $p(g) := p(e,g)$, we may alternatively define a random walk on G as a pair (G,p), where p is a probability on G. Random walks on continuous groups are defined in a similar way, a general probability measure taking the role of p.

If G is a subgroup of a group G', a random walk on G may in a trivial way also be considered as a random walk on G'; to this end the probability p is extended from G to G' by putting $p(g)=0$ for all $g \in G' \setminus G$. Conversely, if G_p is the smallest subgroup of G such that $p(g)=0$ for $g \in G \setminus G_p$ (i.e. the subgroup generated by the support of p), (G,p) is essentially a random walk on G_p. If $G_p = G$ the random walk is called <u>aperiodic</u>.

If H is an invariant subgroup of G, a random walk (G,p) induces in a natural way a random walk on the factor group G/H. If for each coset A of H we define

$$p^H(A) := \sum_{g \in A} p(g) = \sum_{h \in H} p(ah), \tag{1}$$

where $a \in G$ is such that $A = aH$, then $(G/H, p^H)$ is a random walk on G/H.

Consider a random walk (G,p). Let $P_n(g)$, $g \in G$, be the probability of going from e to g in n steps. These probabilities obey the Chapman-Kolmogorov or convolution equation:

$$P_n(g) = \sum_{h \in G} P_{n-1}(gh^{-1})p(h), \quad n \geq 1. \tag{2}$$

Hence the generating function $P(g;z) := \sum_{n=0}^{\infty} P_n(g) z^n$, to be called the Green's function of the random walk, obeys the equation

$$P(g;z) - \delta_{ge} = z \sum_{h\in G} P(gh^{-1};z)p(h) . \tag{3}$$

Just as in the classical theory of random walks [9], many properties of a random walk are readily extracted from its Green's function. We give several examples: (a) the probability that a walk starting at e ever visits the element g (excluding the start as a visit to e) is given by

$$F(g) = \lim_{z\uparrow 1} \frac{P(g;z)-\delta_{ge}}{P(e;z)} ; \tag{4}$$

(b) the average number of steps made before such a visit to g is

$$\langle n(g)\rangle = \{F(g)\}^{-1} \left[\frac{d}{dz}\left(\frac{P(g;z)-\delta_{ge}}{P(e;z)}\right)\right]_{z=1} ; \tag{5}$$

(c) the generating function for the average number of distinct elements visited in n steps is

$$\sum_{n=0}^{\infty} \langle S_n\rangle z^n = (1-z)^{-2} \{P(e;z)\}^{-1} . \tag{6}$$

From eq. (4) we see in particular that the probability of return to the starting point is $F = F(e) = 1 - \{P(e;1)\}^{-1}$. If $P(e;1) < \infty$ we have $F<1$, i.e. the random walk is transient. It can be shown that in this case $P(g;1) < \infty$ for all g; the function $P(g;1)$ is called the potential of the random walk. If $P(e;1) = \infty$, the random walk is recurrent ($F=1$).

If H is an invariant subgroup of G, the Green's function of the random walk $(G/H, p^H)$ can be simply expressed in terms of that of (G,p). If we denote the probability of going in G/H from the identity element E(=H) of G/H in n steps to an element A by $P_n^H(A)$ and the corresponding Green's function by $P^H(A;z)$, then obviously $P_n^H(A) = \sum_{g\in A} P_n(g)$ and hence $P^H(A;z) = \sum_{g\in A} P(g;z)$.

Since $P_n(g) \geq 0$ for all $g\in H$ we have $P_n^H(E) \geq P_n(e)$ and hence $P^H(E;1) \geq P(e;1)$. Therefore, if (G,p) is recurrent, $(G/H,p^H)$ is also recurrent.

In sections 3 and 4 we shall discuss examples of random walks on countable non-abelian groups.

3. Random walks on free groups with a finite number of generators

To the author's knowledge the only case of a random walk on a non-abelian group for which $P(g;z)$ is known explicitly for at least some g is that where G is the free group with $k(>1)$ generators and $p(g)=(2k)^{-1}$ if either g or g^{-1} is a generator, $p(g)=0$ otherwise. For this walk (which in statistical physics would be called a simple random walk on a Cayley tree or Bethe lattice of valency $2k$) Kesten[3] has shown that

$$P(e;z) = \frac{\{k^2-(2k-1)z^2\}^{\frac{1}{2}}-(k-1)}{1-z^2} = \frac{2k-1}{k-1+\{k^2-(2k-1)z^2\}^{\frac{1}{2}}}\;; \qquad (7)$$

the second expression is more convenient if one wants to substitute $z=1$. The complete Green's function is easily found from eqs. (3) and (7):

$$P(g;z) = \left[\frac{z}{k+\{k^2-(2k-1)z^2\}^{\frac{1}{2}}}\right]^{\nu(g)} \frac{2k-1}{k-1+\{k^2-(2k-1)z^2\}^{\frac{1}{2}}}\;, \qquad (8)$$

where $\nu(g)$ is the length of the shortest word representing g (= the <u>distance</u> between e and g, i.e. the minimum number of steps required to go from e to g); $\nu(e)=0$.

Applying eq. (4) we obtain

$$F(g) = \left(\frac{1}{2k-1}\right)^{\nu(g)} \text{ for } g\neq e\;, \qquad F = F(e) = \frac{1}{2k-1}\;. \qquad (9)$$

Since $F<1$ the random walk is transient [4]. From eq. (5) we further find

$$\langle n(g)\rangle = \begin{cases} \nu(g)\,\dfrac{k}{k-1} & \text{for } g\neq e\;, \\ \dfrac{2k-1}{k-1} & \text{for } g=e\;. \end{cases} \qquad (10)$$

By eq. (6), the average number of distinct elements visited in n steps is the coefficient of z^n in $(2k-1)^{-1}(1-z)^{-2}[k-1+\{k^2-(2k-1)z^2\}^{\frac{1}{2}}]$. Its asymptotic behaviour for $n\to\infty$ is most easily derived from a well-known relation valid for all transient random walks: $\langle S_n\rangle \simeq (1-F)n$ for

$n\to\infty$. Thus $\langle S_n\rangle \simeq \{(2k-2)/(2k-1)\}n$ $(n\to\infty)$ in this case.

So far we have excluded k=1 because the free group with one generator is abelian; it is (isomorphic to) the additive group $\mathbb{Z}$. The (well-known) Green's function for this case is still given by eq. (8) with k=1, but now F= 1, so that the random walk is recurrent, and the expression for the asymptotic behaviour of $\langle S_n\rangle$ for $n\to\infty$ is different from that given above.

4. Random walks on the group $\mathbb{Z}^{d,i}$

After having discussed in the previous section random walks on groups with the maximum possible degree of non-commutativity let us now, as a counterpart, consider random walks on a class of non-abelian groups which differ only "slightly" from abelian groups. To this end we introduce the semi-direct product of $\mathbb{Z}^d$ and the group C_i consisting of two automorphisms of $\mathbb{Z}^d$, the identity map E and the inversion I (defined by $I\ell=-\ell$ for all $\ell\in\mathbb{Z}^d$). In other words, we consider the set $C_i\times\mathbb{Z}^d$, the elements of which we denote by $\{A|\ell\}$ with $A\in C_i$ and $\ell\in\mathbb{Z}^d$, equipped with the multiplication rule

$$\{A|\ell\}\{A'|\ell'\} = \{AA'|A\ell'+\ell\} \ ;$$

the identity element is $\{E|0\}$ and the inverse of $\{A|\ell\}$ is $\{A^{-1}|-A^{-1}\ell\}$ (see e.g. ref. 10, p. 106).

On this group, which we denote by $\mathbb{Z}^{d,i}$ (and which is in a trivial way isomorphic to a subgroup of the group of all automorphisms of $\mathbb{Z}^d$), we consider an arbitrary random walk. For convenience we write $p(\{A|\ell\}) =: p_A(\ell)$ and $P_n(\{A|\ell\}) =: P_{nA}(\ell)$. Eq. (2) can then be written in the form

$$P_{nE}(\ell) = \sum_{m\in\mathbb{Z}^d}\left\{P_{n-1,E}(\ell-m)p_E(m) + P_{n-1,I}(\ell-m)p_I(-m)\right\} ,$$

$$P_{nI}(\ell) = \sum_{m\in\mathbb{Z}^d}\left\{P_{n-1,E}(\ell-m)p_I(m) + P_{n-1,I}(\ell-m)p_E(-m)\right\} , \qquad (11)$$

or, in matrix notation,

$$\underset{\sim}{P}_n(\ell) = \sum_m \underset{\sim}{p}(m)\underset{\sim}{P}_{n-1}(\ell-m) , \qquad (12)$$

$$\underset{\sim}{p}(m):=\begin{pmatrix} p_E(m) & p_I(-m) \\ p_I(m) & p_E(-m)\end{pmatrix}, \quad \underset{\sim}{P}_n(\ell):=\begin{pmatrix} P_{nE}(\ell) \\ P_{nI}(\ell)\end{pmatrix}.$$

The eqs. (11) may be given the following interpretation. A random walker moves on a system consisting of two copies of the lattice $\mathbb{Z}^d$: the "E-lattice" and the "I-lattice", say. These lattices differ in that the stepping probability for points of the I-lattice is inverted ($p_A(\ell)\to p_A(-\ell)$) with respect to that for points of the E-lattice. Alternatively, one may think of the random walk as taking place on a single lattice $\mathbb{Z}^d$, but in such a way that at each step by a vector ℓ there is a probability $p_I(\ell)$ that "the wind shifts to the opposite direction", i.e. that from that step onward $p_A(\ell)$ is inverted. Or, equivalently, one may consider the walker to have the possibility of being in two different internal states, characterized by "opposite" stepping probabilities, with at each step a certain probability to change his state. A random walk of the latter type is a composite stochastic process as discussed by Van Kampen [11], and a generalization of a type of process known as a <u>two-state random walk</u> [12]. As such it has found various physical applications.

From eq. (12) it follows that the vector $\underset{\sim}{P}(\ell;z):=\sum_{n=0}^{\infty}\underset{\sim}{P}_n(\ell)z^n$ satisfies the equation $\underset{\sim}{P}(\ell;z)-\delta_{\ell 0}\binom{1}{0}=z\sum_m \underset{\sim}{p}(m)\underset{\sim}{P}(\ell-m;z)$. This equation can be solved by standard Fourier transform. For the Green's function at the identity element we find

$$P_E(0;z)=(2\pi)^{-d}\int_{-\pi}^{\pi}\!\!\dots\!\int_{-\pi}^{\pi} d\phi\,\frac{1-z\,\mathrm{Re}\,\hat{p}_E(\phi)}{1-2z\,\mathrm{Re}\,\hat{p}_E(\phi)+z^2\left(|\hat{p}_E(\phi)|^2-|\hat{p}_I(\phi)|^2\right)},\quad(13)$$

with $\hat{p}_A(\phi):=\sum_{\ell\in\mathbb{Z}^d} e^{i\ell\cdot\phi}\,p_A(\ell)$ for $A=E,I$ and $\phi\in\mathbb{R}^d$.

From now on we restrict ourselves in this section to aperiodic random walks with a transition probability which factorizes: $p_A(\ell)=p_A p(\ell)$ (with $p_E+p_I=1$, $\sum_\ell p(\ell)=1$) with $p(\ell)$ having finite first and second moments; observe that aperiodicity implies that $p_I>0$. In this case eq. (13) reduces to

$$P_E(0;z)=(2\pi)^{-d}\int_{-\pi}^{\pi}\!\!\dots\!\int_{-\pi}^{\pi} d\phi\,\frac{1-zp_E\,\mathrm{Re}\,\hat{p}(\phi)}{1-2zp_E\,\mathrm{Re}\,\hat{p}(\phi)+z^2(p_E^2-p_I^2)|\hat{p}(\phi)|^2},\quad(14)$$

where $\hat{p}(\phi):=\sum_\ell e^{i\ell\cdot\phi}p(\ell)\,(=\hat{p}_E(\phi)+\hat{p}_I(\phi))$.

To decide whether or not the random walk is recurrent we must study the zeroes of the denominator of the integrand in eq. (14) for z=1. This function of ϕ, to be denoted by $N(\phi)$, can be written in the form

$$N(\phi) = p_E^2 \left[\sum_\ell p(\ell) \{1 - \cos(\ell\cdot\phi)\}\right]^2 + p_E^2 \left[\sum_\ell p(\ell) \sin(\ell\cdot\phi)\right]^2$$

$$+ 2p_E p_I \sum_\ell p(\ell)\{1-\cos(\ell\cdot\phi)\} + p_I^2 \sum_{\ell,m} p(\ell)p(m)\{1-\cos\big((\ell-m)\cdot\phi\big)\},$$

and is thus ≥ 0. Obviously, $N(\phi)=0$ iff $(\ell\cdot\phi)=0$ for all ℓ for which $p(\ell)>0$. We may now apply the argument given by Spitzer for the case of random walks on $\mathbb{Z}^d$ (ref. 1, pp. 68-69) to conclude that this is only the case for $\phi=0$. To second order in ϕ, $N(\phi)$ is a (positive definite) quadratic form in ϕ. Since the numerator in eq. (14) is non-zero for $\phi=0$, $P_E(0;1)$ is infinite for $d\leq 2$ and finite for $d\geq 3$. Hence, all random walks of the type considered on $\mathbb{Z}^{d,i}$ are recurrent for $d\leq 2$ and transient for $d\geq 3$. This result, which is a special case of a more general result derived for random walks on the Euclidean group in d dimensions [8], is in contrast with the well-known results for random walks on $\mathbb{Z}^d$, where for $d\leq 2$ an aperiodic random walk with finite first and second moments is transient iff its first moment is non-zero, i.e. if there is a systematic drift. The reason for the difference is, of course, the random "shifting of the wind", which neutralizes the effect of this drift.

Since $(\mathbb{Z}^{d,i},p)$ is a Markov chain on $C_i \times \mathbb{Z}^d$ it may be interesting to consider its "projections" on C_i and $\mathbb{Z}^d$, i.e. the stochastic processes formed by the components in C_i and $\mathbb{Z}^d$ of the elements visited. The projection on C_i is itself a random walk on a group (the simplest non-trivial one); it is essentially the induced random walk on the factor group of $\mathbb{Z}^{d,i}$ modulo its invariant subgroup $\mathbb{Z}^d$ and describes the random inversion of the stepping probability.

The projection on $\mathbb{Z}^d$, however, is not a random walk; it is not even a Markov chain because the transition probability itself changes randomly. Let us calculate the average displacement during the nth step $\langle\Delta\ell_n\rangle$ and the average position after the nth step $\langle\ell_n\rangle$. If we define $P_{nA} := \sum_\ell P_{nA}(\ell)$ we can write $\langle\Delta\ell_n\rangle = \sum_\ell \ell\{P_{n-1,E}p(\ell)+P_{n-1,I}p(-\ell)\}$. It is easily verified that $P_{nE} = \frac{1}{2}\{1+(1-2p_I)^n\}$, $P_{nI} = \frac{1}{2}\{1-(1-2p_I)^n\}$, so that $\langle \Delta\ell_n \rangle = (1-2p_I)^{n-1}\mu$, where $\mu := \sum_\ell \ell p(\ell)$. Thus there is at each step a non-zero drift if $\mu \neq 0$, but it goes to zero for $n\to\infty$ (remember that $p_I > 0$). For $\langle \ell_n \rangle$ we find

$$\langle \ell_n \rangle = \sum_{m=1}^{n} \langle \Delta\ell_m \rangle = (2p_I)^{-1} \{1-(1-2p_I)^n\}\mu ,$$

which converges to $(2p_I)^{-1}\mu$ for $n\to\infty$. Another property of this projected walk which distinguishes it from a random walk is that the steps are correlated. It turns out that the displacements during the nth and the (n+k)th step (k>0) have the following covariance matrix:

$$\langle \Delta\ell_n^i \Delta\ell_{n+k}^j \rangle - \langle \Delta\ell_n^i \rangle \langle \Delta\ell_{n+k}^j \rangle = (1-2p_I)^k \{1-(1-2p_I)^{2n-2}\} \mu^i \mu^j ,$$

where i and j denote components (i,j=1,...,d).

The analysis given in this section can be extended in a straightforward way to random walks on, successively, arbitrary groups of automorphisms of $\mathbb{Z}^d$, automorphism groups of arbitrary lattices in $\mathbb{Z}^d$ and d-dimensional space groups. For a generalization to continuous groups see ref. 8.

5. Random walks on groups with a stochastic colouring

In the literature on random walks much attention has been paid to random walks on lattices containing special points, e.g. so-called absorbing points (traps) or partially absorbing points (imperfect traps); for references, see refs. 2 and 13. One is interested in the probability distributions for the number of steps required to reach a special point and for the number of steps made between two successive visits to special points, and in particular in the averages of these numbers.

The methods developed in this field can be immediately generalized to arbitrary countable groups. It is evident, however, that as soon as certain elements of a group are designated as being special the symmetry of the group is lost, or at best lowered. Still there is a way to divide the elements of a group into two classes and to investigate properties such as those referred to above while retaining the full symmetry of the group. The idea is simple: instead of selecting a fixed set of special elements we assume that the state of an element (special or regular) is a random variable and that the whole system is described by a joint probability distribution for all these random variables which is <u>invariant</u> under the group. In this section we work out this idea for a relatively simple case, viz. that of random walks which proceed independently of the nature of the elements, i.e. which have a stepping probability that is the same for special and for regular points. In doing so we generalize to arbitrary countable groups results obtained earlier by W.Th.F. den Hollander and the author for random walks on d-dimensional lattices [13,14]. Instead of the terms "special" and "regular" we use the more neutral terms "black" and "white".

We consider a countable group G of which the elements are coloured black and white according to a given joint probability distribution $\mathcal{P}$. More precisely, to each pair (U,V) of disjoint finite subsets of G we attribute a number $\mathcal{P}(U,V)$, the probability that all elements of U are black and all elements of V are white. If this is done in a consistent way these probabilities define a unique probability measure $\mathcal{P}$ on the set of all colourings of G (see ref. 15, p. 34). We assume that the probability $\mathcal{P}$ is <u>left-invariant</u> under G, or, equivalently, that $\mathcal{P}(gU,gV) = \mathcal{P}(U,V)$ for all $g \in G$.

Examples of left-invariant probability distributions are: (a) the <u>random distribution</u>, defined by $\mathcal{P}(U,V) = q^{|U|}(1-q)^{|V|}$ (i.e. the colours of the elements are independently and identically distributed); (b) <u>factor distributions</u>, obtained by choosing an invariant subgroup H of G such that $N := |G/H|$, the cardinality of G/H, is finite, and defining $\mathcal{P}(U;V) = N^{-1}$ if there is an element $g \in G$ such that $U \subset gH$, $V \subset G \backslash gH$, and $\mathcal{P}(U,V) = 0$ otherwise (i.e. H or one of its cosets is entirely black, each one with probability N^{-1}, while its complement is entirely white); (c) the <u>uniform distribution</u>, defined by $\mathcal{P}(U,\emptyset) = q$, $\mathcal{P}(\emptyset,V) = 1-q$, $\mathcal{P}(U,V) = 0$ otherwise (i.e. G is entirely black with probability q or entirely white with probability 1-q). Factor distributions form a generalization of so-called strictly periodic distributions for colourings of $\mathbb{Z}^d$ [13]. One simply convinces oneself that the left-invariant probability distributions $\mathcal{P}$ form a convex cone.

We next consider a random walk on G in which the steps are independent of the colouring of G. If we register the colours of the elements visited in succession we get a sequence such as WWWWBBBWBWWB ...

with B = black, W = white. Such a sequence is a realization of a stochastic process $(X_0,X_1,X_2,\dots) =: (X_n)_0^\infty$ where X_n, the colour of the element visited at step n, is a random variable taking the values B and W. This process is, of course, entirely determined by the probabilities $\mathcal{P}$ and p. It is in general not a Markov process.

The process $(X_n)_0^\infty$ has many interesting properties. In this section we consider several properties that are related to the successive visits to black elements (hits). We study in particular the stochastic behaviour of the numbers $n_0,n_1,n_2,\dots$, where n_0 is the number of steps made until the first hit, i.e. the lowest value of n such that X_n=B, $(n_0 \geq 0)$ and n_i (i=1,2,...) is the number of steps made between the ith and (i+1)th hit $(n_i \geq 1)$. Observe that the sequence $(n_0,n_1,\dots)$ may be finite because it may happen that after a certain hit, or even from the start on, only white elements are visited. If an (i+1)th hit takes place we say that the ith run is completed, and call n_i the length of that run. Averages such as $\langle n_i \rangle$ should always be understood as conditional averages given that the runs involved are completed.

Let $F_{n_0 n_1 \dots n_i}$ be the probability that at least the 0th run, of length n_0, and i more runs, of lengths $n_1,\dots,n_i$, are completed, and $P_{n;n_1\dots n_i}$ the probability that at step n a black element (not necessarily the first one) is visited and that subsequently i runs, of lengths $n_1,\dots,n_i$ are completed.

Since a visit to a black element is either the first hit or takes place, say, m steps after a preceding hit we have

$$P_{n;n_1\dots n_i} = F_{nn_1\dots n_i} + \sum_{m=1}^{n} P_{n-m;m\,n_1\dots n_i} \; ; \qquad (15)$$

for n=0 the sum in the right-hand side is empty.

Now for any walk the subwalk beginning with the (n+1)th step may be considered as a complete walk in itself. By the left-invariance of the probability $\mathcal{P}$, however, the probability distribution for the sequence $(X_n,X_{n+1},\dots)$ is identical to that for $(X_0,X_1,X_2,\dots)$. This means that the process $(X_n)_0^\infty$ is stationary; it implies that $P_{n;n_1\dots n_i} = P_{0;n_1\dots n_i} = F_{0n_1\dots n_i}$. Indeed, for i=1, e.g., we have

$$P_{n;n_1} = \sum_{g_0}\sum_{g_1} \dots \sum_{g_{n_1}} P_n(g_0)p(g_1)\dots p(g_{n_1})\mathcal{P}\Big(\{g_0,g_0g_1\dots g_{n_1}\};$$

$$\{g_0g_1,\dots,g_0g_1\dots g_{n_1-1}\}\Big)$$

$$= \sum_{g_0}\sum_{g_1} \dots \sum_{g_{n_1}} P_n(g_0)p(g_1)\dots p(g_{n_1})\mathcal{P}\Big(\{e,g_1\dots g_{n_1}\};$$

$$\{g_1,\dots,g_1\dots g_{n_1-1}\}\Big)$$

$$= \sum_{g_1} \dots \sum_{g_{n_1}} p(g_1)\dots p(g_{n_1})\mathcal{P}\Big(\{e,g_1\dots g_{n_1}\};\{g_1,\dots,g_1\dots g_{n_1-1}\}\Big)$$

$$= P_{0;n_1} = F_{n_1} ,$$

where we have used the invariance of $\mathcal{P}$ under left multiplication by g_0^{-1} and the fact that $\sum_{g_0} P_n(g_0) = 1$. For general i the proof is identical, but the formulae are more lengthy.

We now can write eq. (15) in the form

$$F_{nn_1\dots n_i} = F_{0n_1\dots n_i} - \sum_{m=1}^{n} F_{0mn_1\dots n_i} . \tag{16}$$

From this relation we can draw a number of conclusions of which we mention the following ones.

A. For i=0 eq. (16) reads $F_n = F_0 - \sum_{m=1}^{n} F_{0m}$. Subtracting this equation for two consecutive values of n we get

$$F_{n-1} - F_n = F_{0n} . \tag{17}$$

Since $F_{0n} \geq 0$, F_n is a monotonically decreasing function of n. Further, since $F^f := \sum_{n=1}^{\infty} F_n$ is the total probability that at least one hit occurs, and thus ≤ 1, F_n goes to zero for $n\to\infty$. If we now sum eq. (17) over all n we find that the total probability that the walk begins with a hit and that at least one more hit takes place is

$$F_0^f := \sum_{n=1}^{\infty} F_{0n} = F_0 = q , \tag{18}$$

where q denotes the probability that an element is black, which we assume to be >0. Hence the conditional probability F_0^f/F_0 that at least two hits occur given that the walk begins with a hit (i.e. starts from a black element) equals one. More generally, it can be shown that whenever a hit occurs then with probability one a next hit will occur. Hence, the probability that the sequence $(n_0,n_1,\ldots)$ is infinite equals the probability F^f that at least one hit occurs.

B. The average number of steps between the first and the second hit given that the walk starts with a hit can also be calculated with the aid of eq. (17). We find:

$$\langle n_1|B\rangle := \sum_{n=1}^{\infty} nF_{0n} \Big/ \sum_{n=1}^{\infty} F_{0n} = F_0^{-1} \sum_{n=1}^{\infty} n(F_{n-1} - F_n)$$

$$= F_0^{-1} \sum_{n=1}^{\infty} F_n = \frac{F^f}{q} \quad . \tag{19}$$

It can further be shown that $\langle n_i|B\rangle = \langle n_1|B\rangle$ for all $i>1$. [13)]

The probability F^f that at least one hit occurs depends on G, $\mathcal{P}$ and p. It can be shown that F^f equals the probability that at least one element of the subgroup G_p generated by the support of p is black; we intend to give the proof in a subsequent paper. For the random distribution we thus find $F^f = 1-(1-q)^{|G_p|}$; it reduces to 1 if G_p is infinite. For the factor distribution determined by an invariant subgroup H we have $F^f = N_p/N$, where $N = |G/H|$ and N_p is the number of cosets of H having elements in common with G_p; it equals one if the random walk is aperiodic $(G_p=G)$. For the uniform distribution $F^f = q$.

The results (18) and (19) are valid for arbitrary countable groups, for all left-invariant colour distributions and for all random walks (recurrent or transient, with finite and/or infinite moments). As such they are in sharp contrast with results that have been (or can easily be) derived if the condition of a hit at the start is dropped. For the averages $\langle n_i\rangle$, i=1,2,..., e.g. one can obtain wildly varying results by suitably choosing G, $\mathcal{P}$ and p.

Further consequences of eq. (16) for random walks on d-dimensional lattices have been discussed in ref. 13. They are equally valid for arbitrary countable groups.

Note added in proof: on p.10 $\mathcal{P}(\emptyset,\emptyset)$ must always be equal to 1.

References

1. F. Spitzer, Principles of Random Walk (Van Nostrand, Princeton, 1964).
2. G.H. Weiss and R.J. Rubin, Adv. in Chem. Phys. (to appear).
3. H. Kesten, Trans. Amer. Math. Soc. 92 (1959) 336-354.
4. E.B. Dynkin and M.B. Malyutov, Doklady Akad. Nauk SSSR 137 (1961) 1042-1045 [English translation: Soviet Math. Doklady 2 (1961) 399-402].
5. R.M. Dudley, Proc. Amer. Math. Soc. 13 (1962) 447-450.
6. H. Kesten and F. Spitzer, Acta Math. 114 (1965) 237-265.
7. Seminaire KGB sur les marches aléatoires, Astérisque 4 (1973), and references mentioned there.
8. P. Crepel, in: Théorie ergodique, Actes des Journées Ergodiques, Rennes 1973/1974, Lecture Notes in Mathematics 532 (Springer-Verlag, 1976), pp. 50-69.
9. E.W. Montroll, Proc. Symp. Appl. Math. 16 (1964) 193-220.
   E.W. Montroll and G.H. Weiss, J. Math. Phys. 6 (1965) 167-181.
10. T. Janssen, Crystallographic Groups (North-Holland Publ. Co., Amsterdam-London, 1973).
11. N.G. van Kampen, Physica 96A (1979) 435-453.
12. G.H. Weiss, J. Stat. Phys. 15 (1976) 157-165.
13. W.Th.F. den Hollander and P.W. Kasteleyn, Physica (to appear).
14. P.W. Kasteleyn and W.Th.F. den Hollander, J. Stat. Phys. (to appear).
15. C.J. Preston, Gibbs states on countable sets, Cambridge Tracts in Mathematics 68 (Cambridge University Press, London, 1974).

ISING MODEL ON FINITELY PRESENTED GROUPS

Mario Rasetti

Dipartimento di Fisica del Politecnico

Torino , Italy

In statistical mechanics the study of critical phenomena and phase transitions has long been,to a large extent,centered around lattice systems. Besides for their intrinsic interest as models of real physical situations,lattice systems are often attractive to theorists for the possibility of obtaining exact non-trivial solutions.

For reasons of this sort space-time lattices were recently introduced as a technical device to obtain cut-off field theories,whose solution would give some insight in field theories defined in continuum Minkowski space-time. Once a lattice field theory has been formulated,the problem can be formally reduced to one of statistical mechanics /1/. The general philosophy is that at the critical point the theory should loose memory of the lattice structure,and the continuous space time results be recovered. Lattice gauge theories are especially remarkable for their relation to the classical spin systems of statistical mechanics /2/.

Exactly solvable models are very few in number,and the Ising model still stands at the very frontiers of our present knowledge. So far no exact general solution has been found for it in more than two dimensions.

A great deal of interest has recently risen as well in the so called glassy states of solids. The latter are amorphous systems exhibiting a typical long range positional disorder of atoms. This results from the long range random deviation from the perfect form of densest lattice packing of a set of hard spheres. Typically such systems show a local (short range) order : five binding rings,originated from the prevalence of close-packed ordered sets of 5 tetrahedra /3/. Tetrahedra do not fill $\mathbf{R}^3$ regularly,but - at the espense of small elastic deformations - arrange themselves in small regular clusters. In fact a finite set of tetrahedra is known to tesselate the 3-dimensional sphere $S^{(3)} \subset \mathbf{R}^4$,on which they constitute a regular polytope. Since the volume of $S^{(3)}$ is finite,such a polytope can be mapped on a compact portion of $\mathbf{R}^3$:the mapping induces distortions of length and implies cuts in $S^{(3)}$,which are the images of the boundary of this compact region. Indeed it can be made one-to-one by the introduction of disclination lines.

One is therefore led to consider a manifold with curvature (typically negative) and hence non-Euclidean (Lobachevskii), even though diffeomorphic to $\mathbf{R}^3$,in order to be able to describe an amorphous solid as an ordered set of atoms.

A possible model-picture for a spin glass is thus that of a system of Ising spins on a regular lattice globally embedded in the non-Euclidean manifold by the action of any transitive discrete subgroup of its isometry group.

These the physical motivations of present work,which deals with solvable spin models in statistical mechanics,defined on lattices homogeneous under finitely presented groups.

There are several powerful avenues of attack of Ising models. One of the most promising and rich of structure is the so called Pfaffian method. It was recently restated /4/ in a form which - through the relabeling of the positional degrees of freedom in terms of anticommuting Grassmann variables associated with the group under which the lattice is homogeneous - reduces the calculation of the partition function (indeed a

combinatorial problem:a linear graph enumeration problem) to the evaluation of a finite number of determinants of matrices whose rank is connected to the dimension of the irreducible representations of the group itself.
Several instances of Ising models have been thus exactly solved,by restricting the attention to a class of lattices homogeneous under a special family of groups (the modular group and some of its arithmetic subgroups of finite index /5/) :

i) on a generalized Cayley tree /6//4/

ii) on a lattice embedded in a spherical surface,homogeneous under the icosahedral group /7/

iii) on a 3-dimensional lattice homogeneous under the triangle group $\mathbf{T}(2,3,7)$ /8/.

Here the theory is formulated in its global setting,whereby the possibility of utilizing it for general 3-dimensional structures is derived.

Let Γ be a graph homogeneous under a group G. Assume G is a finite,discrete group of order $|G|$. Let g be a generic element of G,and $\mathscr{Z}(G)$ denote the center of G.
Γ can be represented on a 2-dimensional orientable surface Σ, of genus γ.
Consider the complex $\mathbf{C}_p$,of dimension p,on Γ.
$\mathbf{C}_p$ is the set of formal sums (chains) mod 2

$$c_p = \sum_{\alpha} P_p^{(\alpha)} \varepsilon_\alpha \tag{1}$$

where $\varepsilon_\alpha \in \{0,1\}$ (mod 2), α is a multi-index of n elements ($n \geqslant p+1$),and

$P_0^{(i)}$; $i = 1,\dots,N_\Sigma$ are the vertices of Γ, $N_\Sigma = 2\nu$ denoting the total number of vertices of the graph (whis is assumed to be even)

$P_1^{(i,j)}$; $i \neq j$; $i,j = 1,\dots,N_\Sigma$ are the edges of Γ,connecting $P_0^{(i)}$ and $P_0^{(j)}$

$P_2^{(i_1,i_2,\ldots,i_n)}$; $n \geqslant 3$; $i_1 \neq i_2 \neq \cdots \neq i_n$; $i_j = 1,\ldots,N_\Sigma$ are the faces (n-agons) or plaquettes, of perimeter $P_1^{(i_1,i_2)}, P_1^{(i_2,i_3)}, \ldots, P_1^{(i_{n-1},i_n)}$, $P_1^{(i_n,i_1)}$ and vertices $P_0^{(i_1)}, \ldots, P_0^{(i_n)}$.

Let us define

$$\Gamma_p = \sum_\alpha P_p^{(\alpha)} \in \mathbf{C}_p \qquad (2)$$

There exists a boundary operation ∂ such that

$$\partial^2 = 0 \qquad (3)$$

whereby the above complex can be factorized as follows

$$\partial P_0^{(i)} = 0$$

$$\partial P_1^{(i,j)} = P_0^{(i)} + P_0^{(j)}$$

$$\partial P_2^{(i_1,\ldots,i_n)} = \sum_{q=1}^{n} P_1^{(i_q,i_{q+1})} \qquad (i_{n+1} \quad i_1)$$

all the sums being mod 2.

There exists a mapping $\mu(g), g \in G$; $\mu: \Gamma \to \Gamma$ such that

$$\mu(g)\ P_p^{(\alpha)} = P_p^{(g\alpha)} \qquad (4)$$

where $g\alpha$ is a multi-index denoting the collection of vertices obtained from those labeled by α by the action of g.

Define the sets:

$$\mathbf{B}_p = \partial \mathbf{C}_p \subset \mathbf{C}_{p-1} \qquad (5)$$

A generic element $b_p \in \mathbf{B}_p$ belongs to the boundary of some $c_p \in \mathbf{C}_p$.

$$\mathbf{Z}_p = \{ c_p \in \mathbf{C}_p \mid \partial c_p = 0 \} \qquad (6)$$

A generic element $z_p \in \mathbf{Z}_p$ is a closed chain (i.e. a chain with no boundary).

$$H_p = Z_p / B_p \tag{7}$$

The elements $h_p \in H_p$ are referred to as cycles.
Associate to each site $P_0^{(i)}$, $i = 1,\ldots,N_\Sigma$ an element $a_i \in \mathscr{A}(\Gamma)$ of a Grassmann algebra $\mathscr{A}$, characterized by the anticommutative product $\cap$:

$$a_i \cap a_j = - a_j \cap a_i \tag{8}$$

On Γ define then the oriented two-form

$$f = \sum_{i,j=1}^{N} x_{ij}\, a_i \cap a_j \tag{9}$$

where $x \equiv \{x_{ij}\}$ is an $N_\Sigma \times N_\Sigma$ generalized incidence matrix of Γ, defined in the following way:

i) $x_{ij} = - x_{ji}$; $x_{ii} = 0$; x is skew-symmetric
ii) if $\mathrm{sgn}(x_{ij}) = +1$, the bond $P_1^{(i,j)}$ is oriented with an arrow pointing from $P_0^{(i)}$ to $P_0^{(j)}$
iii) if $x_{ij} = 0$, there is no bond between $P_0^{(i)}$ and $P_0^{(j)}$
iv) for any plaquette $P_2^{(i_1,\ldots,i_n)} \subset \Gamma$, if the sequence $i_1, i_2, \ldots, i_n$ is anticlockwise with respect to the orientation of Σ,

$$\prod_{q=1}^{n} \mathrm{sgn}(x_{i_q i_{q+1}}) = -1 \tag{10}$$

The form f is said to be invariant under G if $x_{gi,gj} = x_{ij}$.
Let μ^* be the mapping induced by μ on $\mathscr{A}(\Gamma)$ according to

$$\mu^*(g)\, a_i = a_{gi} \tag{11}$$

A dimer configuration is a chain c_1 such that

$$\partial c_1 = \Gamma_0 \tag{12}$$

The dimer generating function for a given signature of Γ is

$$Z_s(x) = Pf(x) \tag{13}$$

where

$$2^{\nu} \nu! \, f^{\cap\nu} = Pf(x) \, a_1 \cap a_2 \cap \cdots \cap a_{N_\Sigma} \tag{14}$$

Let $\mathbf{D}(\Gamma)$ denote the set of all dimer configurations on

$$\mathbf{D}(\Gamma) \equiv \{ c \in \mathbf{C}_1 \; ; \; \partial c = \Gamma_0 \} \tag{15}$$

(whenever it can be done without ambiguity,we write c instead of c_1,etc.).

In general,for $c, c' \in \mathbf{D}(\Gamma)$,

$$c + c' \in \mathbf{Z}_1 \tag{16}$$

Two dimer configurations,say c and c',are defined to be equivalent,which is denoted $c \sim c'$,if

$$c + c' \in \mathbf{B}_1 \tag{17}$$

There exists a one-to-one correspondence between the dimer configurations and the monomials in the polynomial $Z_s(x)$. This can be checked in the following way. Due to its very definition,the matrix x is alternating.

For N_Σ even,Pf(x) is a polynomial function of the entries in x such that $Det(x) = \left(Pf(x) \right)^2$ (Cayley theorem /9/).

Denote by $x^{(i,j)}$ the alternating $(N_\Sigma -2) \times (N_\Sigma -2)$ matrix obtained from x by deleting the i-th and j-th row and the i-th and j-th column. Then,

$$Pf(x) = \sum_{j<i} (-)^{i+j-1} x_{ij} \, Pf(x^{(i,j)}) + \sum_{j>i} (-)^{i+j} x_{ij} \, Pf(x^{(i,j)}) \tag{18}$$

If one deletes the same $N_\Sigma -2s$ rows and columns from x,one gets an alternating $2s \times 2s$ matrix,whose Pfaffian is called a Pfaffian of x of order 2s. Denoting by $Pf_{2s}(x)$ the ideal generated by all such Pfaffians,iteration of (18) implies,for all $1 \leqslant s \leqslant \nu-1$,

$$Pf_{2s+2}(x) \subseteq Pf_{2s}(x) \tag{19}$$

from which the statement follows in a straightforward manner. For $c \in \mathbf{D}(\Gamma)$,

$$c = \sum_{q=1}^{\nu} P_1(i_{2q-1}, i_{2q}) \tag{20}$$

let $s(c)$ be the sign of the corresponding monomial in $Z_s(x)$,

$$s(c) = \eta(\mathscr{P}) \operatorname{sgn}(\prod_{q=1}^{\nu} x_{i_{2q-1} i_{2q}}) \tag{21}$$

where $\eta(\mathscr{P})$ is the parity of the permutation $\mathscr{P}$ $(i_1,\ldots,i_{N_\Sigma})$ of $(1,\ldots,N_\Sigma)$.

Let $\mathbf{C}_c$ be the module generated by the formal sums $y = \sum_i n_i c^{(i)}$, $n_i \in Z_2$ (the ring of integers mod 2),and set

$$Sgn(y) = \prod_i \left[s(c^{(i)}) \right]^{n_i} \tag{22}$$

$Sgn(y)$, $y \in \mathbf{C}_c$ corresponds to an element $z^1 \in \mathbf{C}^1(\Gamma, Z_2)$, the group of cochains mod 2,through the relation

$$Sgn(y) = (-)^{<z^1, y>} \tag{23}$$

where $<z^1,y>$ is a linear functional of y.

Define

$$\mathbf{Z}_c = \mathbf{C}_c \cap \mathbf{Z}_1 \tag{24}$$

and

$$\mathbf{B}_c = \mathbf{Z}_c \cap \mathbf{B}_1 \subset \mathbf{Z}_c \tag{25}$$

where $\cap$ denotes cycle intersection.
If $\sum_i n_i = 0$,then $y \in \mathbf{Z}_c$. When the functional $\langle z^1, y \rangle = 0$ for $y \in \mathbf{B}_c$, z^1 is said to be weakly closed.
We will confine henceforth our attention to graphs and signatures thereof such that z^1 is weakly closed.
Let now $c^{(o)}$ be a dimer configuration,referred to as standard, chosen in such a way that $s(c^{(o)}) = 1$.
We have the following lemma.
If z^1 is weakly closed,

$$s(c) = \sigma(\{c\}) \tag{26}$$

where $\{c\}$ denotes the homology class of $c + c^{(o)}$ in $\mathbf{H}_c = \mathbf{Z}_c / \mathbf{B}_c$;moreover $\sigma(\{c\})$ is multiplicative

$$\sigma(\{c\} + \{c'\}) = \sigma(\{c\})\, \sigma(\{c'\}) \tag{27}$$

The proof of (26) is based on the fact that $\{c\} = \{c'\}$ implies $c + c^{(o)} = c' + c^{(o)} + b$,where $b \in \mathbf{B}_c$;whence $c \sim c'$ (since $c = c' + b$). Then $\langle z^1, c+c' \rangle = \langle z^1, b \rangle = 0$ and $s(c') = s(c)$. Thus $s(c)$ is a function of $\{c\}$ alone,as stated.
Let now $\{c''\} = \{c\} + \{c'\}$,i.e. $c'' + c^{(o)} \sim c + c' + 2c^{(o)} = c + c'$,or $c + c' + c'' + c^{(o)} \sim 0$. Then

$$s(c)\, s(c')\, s(c'')\, s(c^{(o)}) = 1 \tag{28}$$

or,recalling that $s(c^{(o)}) = 1$,

$$s(c'') = s(c)\, s(c') \tag{29}$$

By (26) the latter is equivalent to (27).
Let $\{\Omega_i \mid i = 1,\dots,m\}$ be a base of the homology group

$\mathbf{H}_1(\Gamma, Z_2)$, and set $z^1 = \sum_{i=1}^{m} q_i\, \Omega_i$; $y = \sum_{i=1}^{m} n_i\, \Omega_i$; $q_i, n_i \in Z_2$, then

$$<z^1, y> = \sum_{i=1}^{m} q_i\, n_i \tag{30}$$

and

$$\sigma(\{y\}) = (-)^{\sum_{i=1}^{m} q_i n_i} \tag{31}$$

Of special interest is the case when

$$\mathbf{H}_c = \mathbf{Z}_c / \mathbf{B}_c \sim \mathbf{Z}_1 / \mathbf{B}_1 = \mathbf{H}_1 \tag{32}$$

One can write then

$$<z^1, y> = |h \cap \{y\}| \quad , \quad h \in \mathbf{H}_1 \tag{33}$$

In order to do this, define the incidence matrix $\Lambda = \{\lambda_{ij}\}$, where

$$\lambda_{ij} = \lambda_{ji} = \Omega_i \cap \Omega_j \in Z_2 \tag{34}$$

Upon setting $h = \sum_{i=1}^{m} t_i\, \Omega_i$, we have

$$|h \cap \{y\}| = \sum_{i,j=1}^{m} \lambda_{ij}\, n_i\, t_j \tag{35}$$

Choosing finally the t_j's in such a way that $\sum_{j=1}^{m} \lambda_{ij}\, t_j = q_i$ one recovers - by comparison with (31) -

$$\sigma(\{y\}) = (-)^{|h \cap y|} \tag{36}$$

One can now proceed to redefine the orientations on Γ in a way such that $\sigma(\{y\}) = 1$, namely Sgn(y) = 1 for all the sets of cycles $c \in \mathbf{D}(\Gamma)$. To this purpose let first realize h on the dual graph Γ^* of Γ , by a sum of plaquettes $P_2^{(\alpha)}$, $\alpha = 1, \ldots, q$, such that, if $P_1^{(\alpha)}$ denotes the edge common to

$P_2^{(\alpha)}$ and $P_2^{(\alpha+1)}$,

$$h = \sum_{\alpha,\mathrm{mod}\ q} P_1^{(\alpha)} \tag{37}$$

Consider a point A^α internal to $P_2^{(\alpha)}$. Let s_1^α be the line segment joining A^α to $A^{\alpha+1}$. The closed line $c = \sum_{\alpha,\mathrm{mod}\ q} s_1^\alpha$ is a closed chain on Γ^*.

Such a cycle on Σ corresponds to the homology element h,we will denote it c(h). Let y be any closed chain on Γ^*. The intersection $y \cap c(h)$ is a set of points

$$Q^\alpha = s_1^\alpha \cap P_1^{(\alpha)} \tag{38}$$

with $\alpha \in Q$, Q being a subset of $\{1,\ldots,q\}$.

Upon redefining the signature by reversal of the orientation of all the edges $P_1^{(\alpha)}$, $\alpha = 1,\ldots,q$ one gets a new functional

$$\mathrm{Sgn}'(y) = \mathrm{Sgn}(y)\ (-)^{|Q|} \tag{39}$$

where $|Q| = |y \cap c(h)|$ (mod 2) is the number of elements of Q. It follows that

$$\mathrm{Sgn}'(y) = 1 \qquad \forall\, y \tag{40}$$

The following three lemmas permit to check whether a given graph Γ has the proper signature for the solution of the Ising model (by the Pfaffian method,after the lattice has been decorated).

i) (Kasteleyn /10/) In any closed chain homologous to zero,the number of anticlockwise arrows has parity opposite to the number of internal points.

ii) Let's define a plaquette even (odd) if it has an even (odd) number of sides. In any closed circuit the number of sides L_{ext} has the same parity as the number p_o of odd plaquettes.

Let l_α denote the number of sides of $P_2^{(\alpha)}$. Obviously

$$\sum_\alpha l_\alpha = 2L_{int} + L_{ext} \tag{41}$$

where the sum is extended to all the edges of the surface $\subset \Sigma$ delimited by the circuit, and L_{int} is the number of internal edges. From (41), taken mod 2, it follows

$$L_{ext} = p_o \tag{42}$$

iii) p_o is even.

Indeed, any $b \in \mathcal{B}_c$, given by $b = \partial \sum_\alpha P_2^{(\alpha)}$ is of the form $b = \sum_i b^i$, where b^i are connected closed chains of even length. This is so because each b^i is the superposition of an even number of dimer configurations. According to lemma ii), $\sum_\alpha P_2^{(\alpha)}$ must therefore contain an even number of odd plaquettes.

In the class of graphs we are considering, the group $\mathcal{B}_c$ is constrained only by the latter condition. In other words it is the group generated by all possible combinations with p_o even and p_e (the total number of even plaquettes enclosed by b) arbitrary.

Such a condition selects the graphs for which the method is applicable. On these graphs the global signature is simply checked by requiring that the functional Sgn is equal to +1 for all the elementary cycles corresponding to $p_o = 2$ and $p_e = 1$.

Let G be finitely presented, generated by u generators w_i, $i = 1,\dots,u$ of which v have finite orders $k_1,\dots,k_v$ respectively ($k_i > 1 \ \forall i$), with presentation

$$< w_1,\dots,w_u; w_1^{k_1},\dots,w_v^{k_v}, r_{v+1},\dots,r_{v+t} > \tag{43}$$

r_s, $1 \leqslant s \leqslant v+t$ denote the relators of G; where for $1 \leqslant s \leqslant v$

$r_s = w_s^{k_s}$,whereas for $v+1 \leq s \leq v+t$ the relations involve more than one generator.

(43) means that $G = \mathcal{F}/\mathcal{N}$,where $\mathcal{F}$ is the free group on $\{w_1,\dots,w_u\}$ and $\mathcal{N}$ is the normal subgroup generated by $\mathcal{R} = \{r_1,\dots,r_{v+t}\}$.

Let φ be the homomorphism of group rings $\varphi : R[\mathcal{F}] \to R[G]$, which extends the natural map $\mathcal{F} \to \mathcal{F}/\mathcal{N}$ and let W_i denote the element φw_i of G. Let moreover ϱ be a representation of G on Aut(V),where V is a finite dimensional vector space over a field K. The first cohomology group $\mathbf{H}^1(G,V)$ is also a vector space over K.

An arbitrary map $J : \{W_1,\dots,W_u\} \to V$ extends to a 1-cocycle of G in V if and only if the 1-cocycle of $\mathcal{F}$ determined by $w_i \to J(W_i)$ vanishes on the relators.

More precisely the following sequence is exact :

$$0 \to \mathbf{Z}^1(G,V) \overset{J}{\to} V^{(u)} \overset{D}{\to} V_1 \oplus V_2 \oplus \cdots \oplus V_{v+t} \tag{44}$$

Here $\mathbf{Z}^1(G,V)$ is the space of 1-cocycles, $V^{(u)}$ is the direct sum of u copies of V and

$$D : (x_1,\dots,x_u) \to \Big(\sum_j \frac{\partial r_1}{\partial w_j} x_j, \dots, \sum_j \frac{\partial r_{v+t}}{\partial w_j} x_j \Big) \tag{45}$$

where $\dfrac{\partial r_i}{\partial w_j}$ is the Fox derivative /11/.

G is said residually finite if given $1\!\!1_G \neq g \in G$, there exists a finite quotient G' of G in which the image of g is $\neq 1$. A theorem by Mal'cev guarantees that all finitely generated linear groups over a field are residually finite.

A natural action of G on the group algebra $V = K[G']$ can be introduced as follows : if $g \in G$ and $v \in V$, gv is defined to be the product g'v in K[G'] ,where g' is the image of g in G'. Then $V^G = \{v \in V \mid gv = v, \forall g \in G\}$ is the one-dimensional subspace generated by

$$S = \sum_{g' \in G'} g' \qquad (46)$$

Let $G' = \{g'_1, \ldots, g'_d\}$ and π be the permutation of $(1, \ldots, d)$ such that $g' g'_i = g'_{\pi i}$. If g' has order k, then π is the product of d/k disjoint cycles : $\pi = (i_1, \ldots, i_k)(i_{k+1}, \ldots, i_{2k}) \ldots$ The fixed point space of an element $g \in G$, i.e. $\{v \in V \mid gv = v\}$ is then the d/k dimensional subspace of V generated by the elements

$$\sum_{j=1}^{k} g'_{i_j} \; , \; \sum_{j=k+1}^{2k} g'_{i_j} \; , \ldots \qquad (47)$$

Upon introducing the quantities

$$\xi = \sum_{i=1}^{v} |W_i|^{-1} \quad ; \quad \tau = \sum_{i=1}^{v} k_i^{-1} \qquad (48)$$

where $|W_i|$ is the order of the image of W_i in G', these properties follow:

i) the minimum number of generators of G is $\geqslant u - t - \tau$

ii) if the above lower bound is actually attained, then G is free, of this rank

iii) G is finite if $u - t - \tau \geqslant 1$

iv) $\xi \geqslant \tau$

v) $\dim H^1(G,V) \leqslant (u - \tau - 1)d + 1$

vi) if t is finite, $\dim H^1(G,V) \geqslant (u - t - \xi - 1)d + 1$

Every compact connected 2-dimensional manifold is homeomorphic to either $S^{(2)}$, a γ-fold torus $\mathcal{T}_\gamma$ or a surface $\mathcal{U}_h$ ($\mathcal{U}_1$ is the projective plane, $\mathcal{U}_2$ the Klein bottle, etc.).
We disregard for the moment the latter possibility.
In the case of $S^{(2)}$ the homology is trivial.
The fundamental group π_1 of $\mathcal{T}_\gamma$ is free abelian on 2γ generators. Its homology group is then itself a finitely generated module over the Noetherian ring constituted by 2γ copies of Z_2:

$$H_1(\mathscr{T}_\gamma) \sim Z_2^{2\gamma} \tag{49}$$

There are thus $2^{2\gamma}$ possible orientations of the classes of independent 1-cycles which freely generate H_1.
Let $X \subset \mathscr{Z}(G)$ be the subset of elements in the center of G corresponding to the handles of Σ ($\gamma \geq 1$) and $\overline{X}$ its complement in $\mathscr{Z}(G)$; $x^2 = \overline{x}^2 = \mathbb{1}_G$, $x \in X$, $\overline{x} \in \overline{X}$.
We consider the central extension of G such that $\overline{x}_k = -\mathbb{1}_G$ $\forall \overline{x}_k \in \overline{X}$, $x_j = z_j \mathbb{1}_G$, $z_j = \pm 1$, $\forall x_j \in X$.
$\mathscr{A}(\Gamma)$ is called centralized if one identifies $a_{x_j i} = z_j a_i$, $a_{\overline{x}_k i} = - a_i$, $i = 1,\ldots,N_\Sigma$.
Whenever no ambiguity arises, such a central extension will still be denoted as G.
Thus if Σ is a manifold homeomorphic to $\mathscr{T}_\gamma$ the global dimer generating function is

$$Z(x) = \frac{1}{2\gamma} \sum_s \zeta_s Z_s(x) \tag{50}$$

where s labels the inequivalent signatures and $\zeta_s = \pm 1$.
In what follows we want to show that it is possible to find a more general extension G¨ of G, whereby (50) can be reduced to the calculation of a single Pfaffian, successively to be expanded in terms of the irreducible representations of G¨ (and hence split in a set of smaller Pfaffians).
Let $\{\omega_i\}$ denote the set of simple, mutually disjoint, orientation preserving loops on Σ, each ω_i based at some point o_i. Let p_i be some path joining o_i to some fixed point $0 \in \Sigma$, and let λ_i be the element of $\pi_1(\Sigma, 0)$ corresponding to $p_i^{-1}\omega_i p_i$. Let $\mathscr{H}$ be the smallest normal subgroup of $\pi_1(\Sigma, 0)$ containing all of the elements λ_i. $\mathscr{H}$ is well defined at any base point, and depends only on the free homotopy classes of the ω_i.

The ζ_s's are multiplicative linear functionals of the characters of the representatives of certain elements of $\mathcal{H}$.
More precisely, let M be the mapping class group of Σ, namely the group of isotopy classes of orientation preserving self-diffeomorphisms of Σ : $M = \pi_0 \, \mathrm{Diff}^+(\Sigma)$.
Here it is sufficient to interpret the definitions relative to M in the sense of the piecewise-linear category.
M operates from the right on the simplexes of Γ, and it does it cellularly [i.e. if $m \in M$, $m : \Gamma \rightarrow \Gamma$ in such a way that for $P_p^{(\alpha)} \in \Gamma$, $m(P_p^{(\alpha)}) = P_p^{(\beta)} \in \Gamma$; moreover $m|\,\mathrm{int}\, P_p^{(\alpha)}$ is a fibration over the image of $m(P_p^{(\alpha)})$; where $\mathrm{int}\, P_p^{(\alpha)}$ denotes the open cell $(P_p^{(\alpha)} - \partial P_p^{(\alpha)})$]. M is further transitive on Γ_0.
Let us consider on Σ an unordered collection $\mathcal{C}$ of γ disjoint smoothly embedded circles $\{C_1, \ldots, C_\gamma\}$. $\mathcal{C}$ is a maximal non separating system of disjoint cycles, whose complement on Σ, $\overline{\mathcal{C}} = \Sigma - \{C_1, \ldots, C_\gamma\}$ is a γ-punctured sphere : the isotopy class of such systems $\langle C_1, \ldots, C_\gamma \rangle$ is called a cut system.
Let H denote the stabilizer subgroup of M, namely the group of diffeomorphisms of Σ which preserve the cut system $\langle C_1, \ldots, C_\gamma \rangle$ (elements of H can permute the C_i's and reverse their orientations). $\mathcal{H}$ is isomorphic to the subgroup of those elements of M/H consisting of one simple move : let C_i' be a circle on Σ intersecting C_i transversally in one point and disjoint from C_j for $j \neq i$; the operation of M/H whereby C_i is replaced in the cut system $\langle C_1, \ldots, C_\gamma \rangle$ by C_i' is called a simple move.
The ζ_s's are functions over M/H of the form $\zeta_s = e^{i\pi\varepsilon_s}$, $\varepsilon_s \in Z_2$: indeed they are representatives of the elements in the center of $\mathcal{H}$ which reverse the boundary of the (non-Euclidean) polygon Π obtained from $\overline{\mathcal{C}}$ by γ additional cuts.
There is an exact sequence [12]:

$$Z \to Z^{\gamma} \oplus B_{2\gamma-1} \to H \to \pm S_{\gamma} \to 0 \qquad (51)$$

where Z denotes the ring of integers, $B_{2\gamma-1}$ is the pure braid group on $2\gamma-1$ strands, and $\pm S_{\gamma}$ is the group of signed permutations of γ objects. This implies that H has a finite presentation.

Consider now the set of $3\gamma-1$ cycles $\{\alpha_i, \beta_i, i = 1,\ldots,\gamma ; \gamma_j, j = 1,\ldots,\gamma-1\}$ represented in Fig.1), and construct the element $\sigma \in M$ locally supported by the transformation shown in Fig.2). We have,

i) H and σ generate M

ii) elements m_j of M represented by words μ_j whose letters belong to $\{\sigma^k; k \in Z\} \cup H$, belong to H

iii) the words $\mu_j m_j^{-1}$ generate relations of M, namely generate as invariant subgroup the kernel of the natural morphism $H \circ Z \to M$ associated with σ.

It follows that since H is finitely presented, also M has finite presentation /13/. From the properties of the presentation obtained according to i), ii), iii) above it is possible to deduce that all relations follow from relations supported in subsurfaces of Σ, finite in number, of genus at most 2.

The natural representation of the braid group by permutations is, up to an inner automorphism, an intrinsic property of the abstract group; therefore the subgroup of all braids with identity permutation is a characteristic sybgroup : this guarantees that there exists a finite matrix representation of M.

Z(x) can then be considered as a function over the manifold $\mathscr{X} = \mathscr{X}(G'', 2\gamma)$, which is the set of pairs (g,i) with $g \in G''$ and $i \in \{1,\ldots,2\gamma\}$. $G'' \sim (M/H)/\mathscr{H}$ acts on $\mathscr{X}$ by

$$g(g',i) = (gg',i) \qquad \forall g,g' \in G'' \qquad (52)$$

Let $G_{2\gamma}$ denote the group of all G''-automorphisms of $\mathscr{X}$.

Let $\varkappa_i = (\Uparrow_G, i) \in \mathscr{X}$ for $i \in \{1,\ldots,2\gamma\}$. Then $\vartheta \in G_{2\gamma}$ is uniquely defined by

$$\vartheta(\varkappa_i) = \varsigma_i^{\vartheta} \circ \varkappa_{\pi(i)} \tag{53}$$

where $\varsigma_i^{\vartheta} \in G''$, and π is a permutation of $\{1,\ldots,2\gamma\}$.
$G_{2\gamma}$ is isomorphic to the wreath product of G'' and $S_{2\gamma}$, the permutation group of 2γ objects. The isomorphism is realized via the map $\vartheta \to (\Theta, \pi)$ where $\Theta : \{1,\ldots,2\gamma\} \to G''$ is given by $\Theta(i) = (\varsigma^{\vartheta}_{\pi^{-1}(i)})^{-1}$.
The map $p : G_{2\gamma} \to S_{2\gamma}$ given by $p(\vartheta) = \pi$ is an epimorphism with kernel K isomorphic to $G \otimes G \cdots \otimes G$ (2γ factors).
Let $\{\chi_j\}$ denote the ordinary simple characters of G''; then the simple characters of K are

$$\chi_{i_1,\ldots,i_{2\gamma}} = \chi_{i_1} \otimes \cdots \otimes \chi_{i_{2\gamma}} \tag{54}$$

Write $\lambda \infty 2\gamma$ if λ is a partition of 2γ. Let $\{\Phi^{\lambda} \mid \lambda \infty 2\gamma\}$ be a complete set of characters of $S_{2\gamma}$, and for each λ define a character Ψ^{λ} of $G_{2\gamma}$ by

$$\Psi^{\lambda}(z_{\pi}\,\vartheta) = \Phi^{\lambda}(\pi) \quad ; \quad z_{\pi} \in \Xi\,, \quad \vartheta \in K \tag{55}$$

where $\Xi = \{ z_{\pi} \in G_{2\gamma}\,,\ \pi \in S_{2\gamma} \mid z_{\pi}(\varkappa_i) = \varkappa_{\pi(i)}\ ;\ i \in \{1,\ldots,2\gamma\}\}$.
Finally let $\hat{\chi} : G_{2\gamma} \to \mathbb{C}$ be the function defined in the following way. If $B : G'' \to GL(d, \mathbb{C})$, with $d = \chi_j(\Uparrow_G)$, is a representation affording the character χ_j, then

$$A : K \to GL(d^{2\gamma}, \mathbb{C}) \ ;\ A(\vartheta) = B(\varsigma_1^{\vartheta}) \otimes \cdots \otimes B(\varsigma_{2\gamma}^{\vartheta}) \tag{56}$$

is a $d^{2\gamma} \times d^{2\gamma}$ matrix. Define another $d^{2\gamma} \times d^{2\gamma}$ matrix F_{π} by

the equations $F_{\pi}(B_1 \otimes \cdots \otimes B_{2\gamma})\, F_{\pi}^{-1} = B_{\pi^{-1}(i)} \otimes \cdots \otimes B_{\pi^{-1}(2\gamma)}$.

From the property $F_\pi F_{\pi'} = F_{\pi\pi'}$, it follows that

$$\hat{\chi}(z_\pi \vartheta) = \mathrm{Tr}\,\{F_\pi \circ A(\vartheta)\} \qquad z_\pi \in \Xi,\ \vartheta \in K \tag{57}$$

is a character of $G_{2\gamma}$ which extends χ.
Then the induced character $\chi^{G_{2\gamma}}$ can be decomposed into simple characters /14/

$$\chi^{G_{2\gamma}} = \sum_{\lambda \infty 2\gamma} c_\lambda\, \hat{\chi} \circ \varphi^\lambda \tag{58}$$

where $c_\lambda = \Phi^\lambda(\Uparrow_G)$ and the characters $\hat{\chi} \circ \varphi^\lambda$ $(\lambda \infty 2\gamma)$ are simple and distinct characters of $G_{2\gamma}$:

$$\hat{\chi} \circ \varphi^\lambda(z_\pi \vartheta) = \chi_j(C_1)^{\beta_{11}+\cdots+\beta_{2\gamma,1}} \cdots \chi_j(C_r)^{\beta_{1r}+\cdots+\beta_{2\gamma,r}}\, \Phi^\lambda(\pi) \tag{59}$$

where β_{ts}, $1 \leqslant t \leqslant 2\gamma$, $1 \leqslant s \leqslant r$, r denoting the number of coniugacy classes of $G^{\cdot\cdot}$, C_s, is the number of disjoint cycles of length t into which the permutation $p \in G_{2\gamma}$ can be decomposed within the given class.
The characters (59) furnish a basis for the expansion of $Z(x)$. Since the regular representation of a finite group is the direct sum of its irreducible representations,each contained as many times as its dimension

$$\varrho\,\{G_{2\gamma}\} = \bigoplus_J\, [J]\, D^{(J)}(G_{2\gamma}) \tag{60}$$

where $\varrho\{g\}$ denotes the regular representation of $g \in G_{2\gamma}$, $D^{(J)}$ is the J-th irreducible representation and $[J] = \dim D^{(J)}$, the eveluation of such an expansion can be reduced to a finite number of finite determinants.
Indeed,if the form f is invariant under $G_{2\gamma}$,we set

$$a_g = \sum_J \sum_{m,m'} D^{(J)*}_{m,m'}(g)\, \hat{a}^J_{m,m'} \tag{61}$$

where

$$\hat{a}^{J}_{m,m'} = \sum_{g \in G_{2\gamma}} D^{(J)}_{m,m'}(g)\ a_g \qquad (62)$$

J labeling the irreducible representation, m and m' the basis; a_g denoting the Grassmann variable associated with the point in Γ obtained from a standard vertex P_0 - arbitrarily associated to $\Uparrow_G$ - by μ^*.
Upon inserting (61) into (9) one gets

$$f = \sum_J f^{(J)} \qquad (63)$$

where the form $f^{(J)}$ contains only the Grassmann variables $\hat{a}^J$. Of course the sum over J now implicitly sums over all the signatures of the lattice consistent with the combinatorial requisites. From (63) it follows

$$Z(x) = \sum_J \left[\text{Det } M^{(J)} \right]^{\frac{1}{2}[J]} \qquad (64)$$

where $M^{(J)}$ is now a matrix of rank [J].

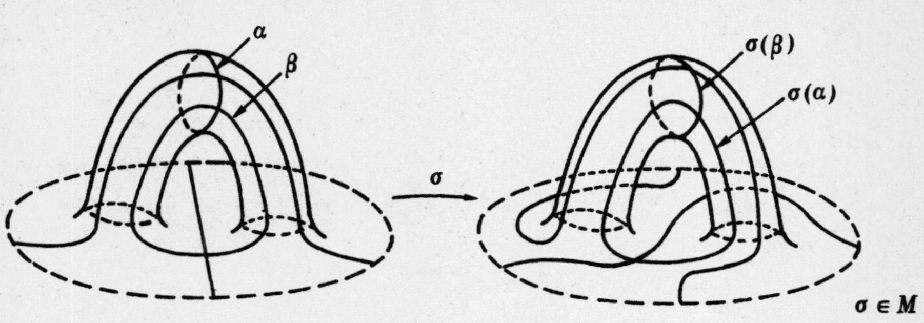

Fig. 1

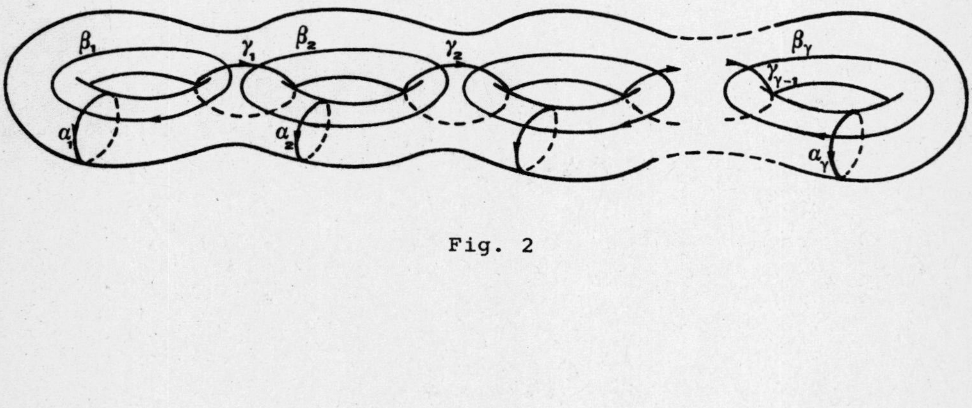

Fig. 2

References

/1/ M.Creutz and B.Freedman,Ann. Phys. (N.Y.) 132,427 (1981)

/2/ J.B.Kogut,Revs. Mod. Phys. 51,659 (1979)

/3/ J.F.Sadoc,J.Dixmier and A.Guinier,J. Non Cryst. Solids 12,46 (1973)

/4/ F.Lund,M.Rasetti and T.Regge,Commun. Math. Phys. 51,15(1977)
F.Lund,M.Rasetti and T.Regge,Teor. Mat. Fiz. 11,246 (1977)

/5/ H.S.M.Coxeter and W.O.Moser,"Generators and Relations for Discrete Groups",Springer-Verlag,Berlin,1965

/6/ A Cayley tree in which the branching number is 2,and the edges are replaced by hexagons.

/7/ M.Rasetti and T.Regge,Rivista Nuovo Cimento 4,1 (1981)
M.Rasetti and T.Regge,in "Symmetries and Broken Symmetries in Condensed Matter Physics",N.Boccara ed., IDSET,Paris,1981

/8/ M.Rasetti,in "Selected Topics in Statistical Mechanics", N.N.Bogolubov,jr. and V.N.Plechko eds.,J.I.N.R. Publ.,Dubna,1981
M.Rasetti,in "Non-perturbative Aspects of Quantum Field Theory",J.Julve and M.Ramòn Medrano eds.,World Scientific Publ. Co.,Singapore,1982

/9/ P.Heymans,Proc. London Math. Soc. 19 (1969)

/10/ P.W.Kasteleyn ,J. Math. Phys. 4,287 (1963)

/11/ R.H.Crowell and R.H.Fox,"Introduction to Knot Theory", Ginn and Co. Publ.,Boston,1963

/12/ A.Hatcher and W.Thurston,A Presentation for the Mapping Class Group of a Closed Orientable Surface, to be publ.

/13/ F.Laudenbach,Astérisque 66-67,267 (1979)

/14/ A.Kerber,"Representations of Permutation Groups I",Springer-V. Lecture Notes in Mathematics 240 (1971)

SELF-TRIALITY IN STATISTICAL MECHANICS AND FIELD THEORY

R. SHANKAR
J.W. GIBBS LABORATORY
YALE UNIVERSITY
NEW HAVEN, CT 06511, U.S.A.

ABSTRACT

I present here a system which exhibits self-triality, a natural extension of the Kramers-Wannier self-duality. There are three complete sets of variables ψ, R and L for describing the system. Any two are disorder variables with respect to the third and the hamiltonian has the same form when expressed in terms of ψ, R or L.

Forty years ago Kramers and Wannier[1] discovered, upon examining the Ising model in two dimensions, a phenomenon called self-duality which has since been encountered in other contexts[2]. Here I present an illustration of self-triality[3] which we will see is a natural extension of self-duality.

To refresh your memory on self-duality, let me begin with a simpler example than the Ising model: the oscillator hamiltonian

$$H(m,\omega) = \frac{P^2}{2m} + \frac{1}{2} m\omega^2 X^2 \tag{1}$$

with

$$[X,P] = i\hbar \tag{2}$$

Let us now define new variables

$$\hat{X} = -P \quad , \qquad \hat{P} = X \tag{3}$$

in terms of which

$$H(m,\omega) = (m\omega)^2 \left[\frac{\hat{P}^2}{2m} + \frac{1}{2} m\tilde{\omega}^2 \hat{X}^2\right] \tag{4}$$

where $\tilde{\omega} = 1/m\omega^2$. Since $\hat{X}$ and $\hat{P}$ are isomorphic to X, P, i.e. $[\hat{X},\hat{P}] = i\hbar$ also, Eq. (4) tells us

$$H(m,\omega) = (m\omega)^2 \, H(m, 1/m\omega^2) \tag{5}$$

This equation provides strong constraints on the dependence of the energy levels, wavefunctions, etc. on m and ω. This simple example illustrates one feature of self-duality: under an isomorphic change of variables, the hamiltonian preserves its form and only the parameters change (because the kinetic and potential terms get exchanged).

We now turn to the Ising model which illustrates yet another aspect of self-duality for which there is no analog in the oscillator case since it involves the notion of order and disorder variables. Now, the Ising model is a classical statistical mechanics problem on a 2-dimensional lattice, with nearest neighbour interactions between spins that can take values ±1. Through the transfer matrix we can map it to a quantum spin problem in one dimension. For a range of parameters, the problem is governed by the following hamiltonian[4] defined on a one-dimensional lattice whose points are labelled by an integer n:

$$H = -\sum_{n=-\infty}^{\infty} \sigma_3(n)\, \sigma_3(n+1) - \lambda \sum_{n=-\infty}^{\infty} \sigma_1(n) \tag{6}$$

Here $\sigma_i(n)$, i=1 or 3 are Pauli matrices which obey the following algebra:

$$[\sigma_i(n) \,,\; \sigma_j(m)] = 0 \qquad n \neq m \tag{7a}$$

$$\{\sigma_i(n) \,,\; \sigma_j(n)\} = 2\delta_{ij} \qquad i,j = 1 \text{ or } 3 \tag{7b}$$

and λ is a parameter related to the temperature T of the 2-dimensional model. For λ=0, (T=0) H has two vacua called $|+\rangle_0$ and $|-\rangle_0$ in which all spins are up or down respectively w·r·t σ_3. As we turn on λ, σ_1 will start flipping spins and destroying the order. But there will still be two vacua, call them $|\pm\rangle_\lambda$ with $\langle\sigma_3\rangle_\lambda = \pm c$ where c<1. Finally, at some λ_c, there will be a phase transition with $\langle\sigma_3\rangle=0$ thereafter. This corresponds to the Curie temperature of the 2-dimensional model since $\langle\sigma_3\rangle_{\text{vacuum}}$ may be identified with the magnetization[4-6].

Let us now introduce a <u>dual lattice</u> whose sites are located half-way between those of the old one and define variables $\mu_i(n)$ as follows:

$$\mu_1(n) = \sigma_3(n)\ \sigma_3(n+1) \tag{8a}$$

$$\mu_3(n) = \prod_{-\infty}^{n} \sigma_1(n') \tag{8b}$$

One may verify that the μ_i are isomorphic to the σ_i and that

$$H = -\sum_{-\infty}^{\infty} \mu_1(n) - \lambda \sum_{-\infty}^{\infty} \mu_3(n)\ \mu_3(n+1) \tag{9}$$

Thus, we get, as in Eq. (4),

$$H(\lambda) = \lambda H(1/\lambda) \tag{10}$$

Now at λ_c, not only does $\langle\sigma_3\rangle$ vanish, the ground state energy of H also becomes singular, since it corresponds to the free energy of the 2-dimensional model[4-6]. But Eq. (10) implies that the same must happen at $1/\lambda_c$. _If_ there is only one critical point, it must occur at $\lambda_c=1/\lambda_c=1$. Thus self-duality can help locate the critical temperature if it is unique. We call $\lambda_c=1$ the _self-dual point_ and note that $H\rightarrow H$ there under $\sigma\leftrightarrow\mu$.

In the ordered phase $\lambda<1$, $\langle\sigma_3\rangle\neq 0$; we can view μ_3 as the _disorder variable_ for the following reason. At $\lambda=0$, we have the vacua $|\pm\rangle_0$. When $\mu_3(n)$ acts on $|+\rangle_0$ say, it flips all the spins from $-\infty$ to n, i.e., it produces a "kink" configuration that interpolates between the two vacua as x goes from $-\infty$ to $+\infty$. While boundary conditions in the ordered state forbid a kink, a pair of kinks can produce, between them, an island of spins aligned opposite to the boundary spins. As $\lambda\rightarrow\lambda_c$, the mass of the pair becomes smaller and smaller, and order is gradually destroyed and finally wiped out at λ_c when the kinks become massless.

By the same logic, for $\lambda>1$, the vacuum is ordered in μ_3 and $\sigma_3(n) = \prod_{-\infty}^{n} \mu_1(n')$ plays the role of the disorder variable. _Thus the dual transformation exchanges order and disorder variables_. Note the symmetry between μ_3 and σ_3: either can be called the order parameter and the only absolute thing is that one disorders the other.

Now we turn to an example of self-triality. Here we will begin with a hamiltonian, find a non-vanishing order parameter and find the kink operators. We shall see that the kink operators form _two_ complete sets of variables and that H can be expressed in terms of either. Self-triality is then the result that H has the same form in all three versions. It will also be seen that any one variable can be identified with the order and that w·r·t it, the other two are disorder variables.

The theory in question is the O(8) Gross-Neveu model[7]. It is given by the following Lagrangian density in 1+1 dimensions:

$$L = \frac{1}{2} \sum_{a=1}^{8} \bar{\psi}_a i\partial\!\!\!/ \psi_a + g_o \left(\sum_{a=1}^{8} \bar{\psi}_a \psi_a \right)^2 \tag{11}$$

Here ψ is a Majorana spinor that transforms as an isovector under the O(8) symmetry. L is invariant under the discrete chiral symmetry $\psi \to \gamma^5 \psi$. We shall see that this symmetry gets spontaneously broken i.e. $\langle\bar{\psi}\psi\rangle \neq 0$ and there are "positive" and "negative" vacua with $\langle\bar{\psi}\psi\rangle = \pm 4M$, respectively, where M is a dynamically generated mass set equal to unity hereafter. It will turn out that there are 16 kinks interpolating between positive and negative vacua, created by operators R_a and L_a, a=1,...8. Self-triality will then emerge as the result that $L(\psi)$ can be reexpressed as a local field theory in terms of R or L., and furthermore $L(\psi) = L(R) = L(L)$! Since L is invariant, we are at the self-triality point (the analog of $\lambda=1$ where $H \to H$).

All these results are shown by bosonization, i.e., upon rewriting the theory in terms of bosonic fields. I present only the relevant details. First we form four Dirac fields (since only these can be bosonized), $\Psi_1 = (\psi_1 + i\psi_2)/\sqrt{2}$, $\Psi_2 = (\psi_3 + i\psi_4)/\sqrt{2}$, etc. The bosonization rule for Ψ_1, say, is

$$\bar{\Psi}_1 i\partial\!\!\!/ \Psi_1 = \frac{1}{2}(\bar{\psi}_1 \partial\!\!\!/ \psi_1 + \bar{\psi}_2 \partial\!\!\!/ \psi_2) = \frac{1}{2}(\partial_\mu \phi_1)^2 \tag{12a}$$

$$J_1{}^\mu = \bar{\Psi}_1 \gamma^\mu \Psi_1 = \frac{i}{2}[\bar{\psi}_1 \gamma^\mu \psi_2 - \bar{\psi}_2 \gamma^\mu \psi_1] = \pi^{-1/2} \varepsilon^{\mu\nu} \partial_\nu \phi_1 \tag{12b}$$

$$\bar{\Psi}_1 \Psi_1 = \frac{1}{2}(\bar{\psi}_1 \psi_1 + \bar{\psi}_2 \psi_2) = \frac{1}{2} \cos\sqrt{4\pi}\, \phi_1 \tag{12c}$$

(Normal ordering mass, symbols, etc., are omitted. See ref. 8 for details.) Similar equations hold for Ψ_2, Ψ_3 and Ψ_4 in terms of ϕ_2, ϕ_3 and ϕ_4. In terms of ϕ_i,

$$L = \frac{1}{2} \sum_{i=1}^{4} (\partial_\mu \phi_i)^2 + g_\circ \sum_{i \neq j=1}^{4} \sum^{4} \cos\sqrt{4\pi}\ \phi_i \cos\sqrt{4\pi}\ \phi_j \qquad (13)$$

(See refs. 8 and 9 for why $i \neq j$ above.)

Let us now explore the possible vacua of this theory by considering the minima of the potential energy term. Clearly we want the cosines to be all equal to +1 (positive vacua with $\bar{\psi}\psi > 0$) or all equal to -1. These obtain for $\phi_i/\sqrt{\pi} = n_i$ and $\phi_i/\sqrt{\pi} = n_i + 1/2$ respectively where n_i is any integer. In particular, the origin $\phi=0$ is a positive vacuum.

Consider now solitons that interpolate from one vacuum to another as x goes from $-\infty$ to $+\infty$. Without loss of generality, we may assume $\phi_i(-\infty)=0$ since if it weren't so it may be suitably translated without changing L. Let us consider the lightest solitons, interpolating as $x \to \infty$, to the positive or negative vacua nearest to the origin. Of the former there are 8, corresponding to unit shifts (in $\phi/\sqrt{\pi}$) up or down one of the 4 axes. <u>These are just the original Dirac fermions Ψ_i</u>. We arrive at this conclusion as follows. The group O(8) has four commuting generators which we choose to be H_1, H_2, H_3 and H_4 that generate rotations in the 1-2, 3-4, 5-6 and 7-8 planes respectively. Now $\Psi_1 = (\Psi_1 + i\Psi_2)/\sqrt{2}$ clearly has spin 1 under H_1 rotations and zero under other. A similar conclusion holds for Ψ_2, Ψ_3 and Ψ_4. Now look at Eq. (12b). It says

$$H_i = \int_{-\infty}^{\infty} J_1^0 \ dx = \frac{1}{\sqrt{\pi}} \int_{\infty}^{\infty} \partial_x \phi_i \ dx = \frac{\phi_i(\infty)}{\sqrt{\pi}} \qquad (14)$$

since $\phi_i(-\infty)=0$. Thus $\phi_i(\infty)/\sqrt{\pi}$ are the O(8) weights and a unit step up or down one of the axes (in $\phi/\sqrt{\pi}$) corresponds to Ψ_i.

Consider now the $2^4=16$ negative vacua nearest to the origin with coordinates $\phi_i/\sqrt{\pi} = (\pm 1/2, \pm 1/2, \pm 1/2, \pm 1/2)$. The solitons connecting the origin to these evidently have half-integral charges and correspond to isospinors. By this one means that if one constructed Dirac matrices γ^μ, $\mu=1\ldots, 8$, they will act on a 2^4 dimensional space of these isospinors. Since $\gamma^9 = \gamma^1 \ldots \gamma^8$ commutes with the generators of rotations ($\sigma^{\mu\nu}$), the spinor breaks up into two irreducible 8-dimensional components w·r·t O(8) rotations. We call these R or L according as $\gamma^9 = \pm 1$. A better way for us to tell them apart is that R(L) will have an even (odd) number of negative components in the weight vector $(\pm 1/2,\ldots,\pm 1/2)$. We shall reserve the word "kink" for R and L solitons since only these (and not Ψ) disorder the condensate in $\bar{\psi}\psi$, i.e. connect positive and negative vacua.

While an analysis of this kind is valid for any O(2N), O(8) is remarkable in that

(1) the lattice points corresponding to R and L are all at unit distance from the origin (in $\phi/\sqrt{\pi}$), just like the points corresponding to Ψ.

(2) just as the latter define an orthonormal basis, so do the points corresponding to R or L.

Given this symmetry it is natural to ask how the physics looks if we switch from the old basis (along which ϕ_i were measured) to a new one generated by R. If η_i are the coordinates in the new basis,

$$\begin{aligned}
\eta_1 &= \tfrac{1}{2}(\phi_1 + \phi_2 + \phi_3 + \phi_4)\\
\eta_2 &= \tfrac{1}{2}(- \phi_1 - \phi_2 + \phi_3 + \phi_4)\\
\eta_3 &= \tfrac{1}{2}(\phi_1 - \phi_2 - \phi_3 + \phi_4)\\
\eta_4 &= \tfrac{1}{2}(- \phi_1 + \phi_2 + \phi_3 - \phi_4)
\end{aligned} \tag{15}$$

In this basis we will find that (i) R kinks correspond to unit displacement (in $\eta/\sqrt{\pi}$) along the coordinate axes (ii) ψ and L have half-integral coordinates. More precisely, $\psi \leftrightarrow R$ under $\phi \leftrightarrow \eta$. To see how the theory looks when cast in terms of R, let us rewrite $L(\phi)$ in terms of η. The free field part is clearly invariant under this orthogonal transformation. More remarkably so is the potential part! Thus $L(\phi) = L(\eta)$! If we now run the bosonization routine backwards and write $L(\eta)$ in terms of the operators R that create the R-kinks, we will, of course, regain the Gross-Neveu model, i.e., $L(\psi) = L(R)$. Likewise, one can show that $L(\psi) = L(R) = L(L)$. This is self-triality. Since L is invariant, we are at the self-triality point: a more general theory away from this point may exist possessing self-triality but I am not aware of it.

Readers familiar with Dynkin diagrams will recall that O(8) has a special symmetry among the vector and the two spin representations and that there is no absolute basis for distinguishing one from the others. What we have here is a dynamical realization of this symmetry - what you call a particle and what you call a kink is not absolute, since starting with $L(\psi)$ or $L(R)$ or $L(L)$ we can get the other two particles as kinks. Other close connections between the general O(2N) case and the corresponding Dynkin diagrams may be found in ref. [9].

Finally note that although the relation between ϕ and η is simple (and local) ψ and R (or L) are related by some very complicated non-local transformation (which we never considered explicitly). I believe this to be a general feature: whenever some local hamiltonian goes into itself (miraculously) under a complicated nonlocal transformation, there must exist an intermediate basis in which this transformation is local and simple. In ref.[3] I show that this is so for the Ising model and also use this idea to construct and solve a spin model with self-triality. Ref. [10] establishes self-triality for the Ashkin-Teller model[11].

I thank the organizers of this conference for their kind invitation and the Department of Energy and A.P. Sloan Foundation for funding the travel.

References

1. H. A. Kramers and G.H. Wannier, Phys. Rev. 60, 252 (1941).
2. R. Savit, Rev. Mod. Phys. 52, 453 (1980).
3. R. Shankar, Phys. Rev. Lett. 46, 379 (1981).
4. E. Fradkin and L. Susskind, Phys. Rev. D17, 2637 (1978).
5. J.B. Kogut, Rev. Mod. Phys. 51, 659 (1979).
6. K.G. Wilson and J.B. Kogut, Phys. Rep. 12C, No2,(1974).
7. D.J. Gross and A. Neveu, Phys. Rev. D10, 3235 (1974).
8. E. Witten, Nucl. Phys. B142, 285 (1978).
9. R. Shankar, Phys. Lett., 92B, 333 (1980).
10. R. Shankar, Phys. Rev. Lett., 50, No. 11, 1983.
11. J. Ashkin and E. Teller, Phys. Rev. 64, 178, 1983.

SYMMETRIES OF FINITE HARD ROD SYSTEMS

P.Kasperkovitz and J.Reisenberger
Institut für theoretische Physik
Technische Universität Wien
A-1040 Vienna, Karlspl.13, Austria

We consider N particles of mass m and diameter d moving on a line and interacting via elastic collisions. The particles are assumed to be enclosed by a freely movable, massless box of length L. This device makes the system slightly different from that with periodic boundary conditions ; even more marked is the formal difference between our approach and previous ones[1].

The phase space of the system is $R^2 \times (R^{N-1} \times C)$ where R^2 is the phase space of the (free) motion of the center of mass. This is assumed to be at rest in the following. $R^{N-1} \times C$ is the phase space of the relative motion of the particles and C is an (N-1)-dimensional cell composed of N! simplices $S_1, ..S_{N!}$ each belonging to a certain order of the particles (e.g. $S_1 : x_1 < x_2 - d < ..x_N - (N-1)d < x_1 + L - Nd$). It can be shown that, as time evolves, the point representing the relative position of the particles moves inside one of the simplices, elastically reflected at its boundaries but otherwise free. Instead of following this complicated motion we generalize Hobson's method[2] and extend the configuration space C to a covering of R^{N-1} by simplices, generated from S_1 by iterated reflections. In the extended phase space R^{2N-2}, endowed with a periodic structure in R^{N-1}, the representative point moves like a free particle and its instantaneous position uniquely determines the relative velocities and positions of the particles. This can be expressed in closed form using suitably defined functions. It can also be shown that the evolution commutes with permutations of the particles and a cyclic symmetry group C_N originating from the massless box which makes the first and the last particle collide like next neighbours. If the motion of the center of mass is included the evolution also commutes with rigid motions (translations, inversion).

Since the order p and the set of velocities $\{v_i\}=\{v_1,..v_N\}$ are constants of motion the evolution cannot be mixing or ergodic on the whole energy surface F_E. But F_E decomposes into N! parts $F_{E,p}= K_E \times S_p$ (sphere × simplex) on which the evolution is mixing with respect to S_p: if k(measurable)$\subset K_E$ and $s \subset S_p$ then a constant distribution on k×s tends towards one on k×S_p. Each $F_{E,p}$ decomposes further into an overcountable set of parts $F_{\{v_i\},p}$ on which the evolution is ergodic, i.e. for a system starting in $F_{\{v_i\},p}$ the time average is is equal to the average over $F_{\{v_i\},p}$. Beside these smallest stationary ensembles (SSE) we also consider microcanonical (MCE) and canonical ensembles (CE). Their invariance groups (containing transformations which are not canonical) are $D_N \times S_N$ for SSE and $D_N \circledS S_N \times O_{N-1}$ for MCE and CE.

The symmetries listed above and those typical for the observable under consideration have been exploited in calculating expectation values. For all three kinds of ensembles (SSE,MCE,CE) and fixed parameters m,d,N,L we found closed expressions or series representations for the following quantities :

(1) The pressure (defined as time average of momentum transfer) ;

(2) the collision frequency ;

(3) the probability for a particle not to collide within a given time interval ;

(4) the probability of finding two particles at a given distance ; and

(5) the velocity autocorrelation function. This function is almost periodic for SSE ; for MCE it oscillates infinitely often finally decreasing as $t^{-N/2}$; and for CE it changes sign only once and decreases ultimately as $t^2 \exp\{-\gamma t^2\}$, $\gamma=\gamma(m,d,N,L,\beta)>0$.

The results (1)-(5) confirm and supplement similar investigations[1]. They also show that qualitative insight can be gained already from very small systems ($N \geq 3$). However convergence toward the thermodynamic limit may be slow in some cases and rather large systems ($N \leq 20$) are needed for reliable extrapolations.

1) D.W.Jepsen, J.Math.Phys. 6(1965)405
J.L-Lebowitz and J.K.Percus, Phys.Rev. 155(1967)122
J.W.Evans, Physica 95A(1979)225

2) A.Hobson, J.Math.Phys. 16(1975)2210

INFLUENCES OF LATTICE DIMENSION d AND CHARACTER OF SPIN-INTERACTIONS ON THE THERMODYNAMICS OF ISING MODELS

H.Nencka-Ficek

Institute of Molecular Physics, Polish Academy of Sciences, Poznan, Smoluchowskiege 17 / 19, POLAND

ABSTRACT

An exact solution is obtained for hypercubic d-dimensional Ising model with multi-spin interactions.

1- INTRODUCTION

It is well known that the eight-vertex model solved by Baxter[1] leads to the solution of an Ising model with two- and four-spin interactions. The properties of the Baxter's solutions differ from those of the ordinary Ising models. Hence there is an interest in studying other similar models e.g. Ising with multi-spin interactions.

In the present note we investigate the thermodynamics of a d-dimensional Ising model with 2^k-spin interactions in the hypercubic simple lattice, d=2,3,...; k=d or d-1. k is a dimension of hypersurface representing a single interaction. For example in two dimensions we consider the square lattice with 4-spin interactions, each unit square represents a single interaction, for d=3 we can consider either a cubic lattice with 4- or 8-spin interactions on each unit cube. The Hamiltonian is

$$-\beta H = K_w \sum_{<i_1 \ldots i_w>} \sigma_{i_1} \ldots \sigma_{i_w} \tag{1}$$

w being a number of interacting spins, $\sigma=\pm1$ Ising variable, $K_w=\beta J_w$ a coupling constant.

2- RESULTS

In the case when w=4, d=2 and the square lattice with free boundary conditions, the specific heat is

$$C_V = a\, k(J_4/kT)^2\, \mathrm{sech}^2(J_4/kT) \tag{2}$$

a is a number of squares in the lattice.

If w=4, d=2 but we consider some stripes formed of n squares in one direction and of m=1 or 2 squares in second directions, we have

$$C_V^{(1)} = C_V\{ch^3K_4 - \frac{t^n}{t^n+1}\left[2-sh^{-2}K_4(n\,\frac{t^n}{t^n+1} + n-1)\right]\} \tag{3a}$$

$$C_V^{(2)} = C_V\{ch^3K_4 - \frac{t^n}{1+2t^n + t^{2n}}\left[2(1+t^n) - sh^{-2}K_4chK_4\right.$$

$$\left. .(2n\,\frac{t^n(1+t)^2}{1+2t^n+t^{2n}} - (2n-1)t^n + 1 - n)\right]\} \tag{3b}$$

with $t = thK_4$.

If $w=2^d$, d>2 in a hypercubic lattice with free boundary conditions the specific heat is

$$C_V = a\;\; k(J_w/kT)^2\, \mathrm{sech}^2(J_2/kT) \tag{4}$$

If $w=2^{d-1}$, d>2 the partition function is

$$Z_{w,d} = \nu(Z_{1,2})^d \tag{5}$$

where ν is some real number, $Z_{1,2}$ is the partition function for the nearest neighbour Ising model on the square lattice.

3- CONCLUSIONS

One concludes that in the hypercubic simple lattice with free boundary conditions with $w=2^d$-spin interactions there is no phase transition. From (2) and (4) it is evident that these models are equivalent to an infinite Ising linear chain with w=2-spin interactions. Eq.5 convinces us that the models with $w=2^{d-1}$-spin interactions are uqivalent up to a constant factor in the thermodynamics functions to the Ising model on the square lattice with w=2-spin interactions. Let us remark that the models with 2^d-spin interactions have no closed hypercontours but there are some hyperloops for systems with $w=2^{d-1}$-spin interactions in d-dimensional hypercubic lattice[2]. Our conclusions agree with the Wood and Griffiths' criterion for phase transitions[3].

4- REFERENCES

1- Baxter,R.J., Phys. Lett. 26, 832, 1971.

2- Nencka-Ficek,H. Physica A to appear.

3- Wood,D.W. and Griffiths,H.P., J.Phys. C6, 2533, 1973.

GAUGE INVARIANCE IN THE STRONG COUPLING BCS-MODEL*

A. Rieckers
Institut für Theoretische Physik
Universität Tübingen, Germany

We want to describe here some results of an ab initio treatment of the strong coupling BCS-model by means of C^*-algebraic methods which covers also single electron excitations. Thus, we need the full electronic CAR-algebra and not only the pair-subalgebra as in the quasi-spin formulation of [1]. Let us denote by c_κ^* the creation operator for an electron in the state $\kappa = (k,\sigma)$, where k is the wave-vector and σ the spin. The pair operators are $b_k^* = c_\kappa^* c_{-\kappa}^*$, $b_k^z = 1 - c_\kappa^* c_\kappa - c_{-\kappa}^* c_{-\kappa}$, where $-\kappa = (-k,-\sigma)$. For every cube Λ_a with length a the pairing Hamiltonian in the strong coupling limit of a k-independent kinetic energy ε is

$$H_a = \sum_{k \in J_a} \varepsilon (1 - b_k^z) - \frac{g}{|J_a|} \sum_{k,k' \in J_a} b_k^* b_{k'} \tag{1}$$

where J_a is the set of wave vectors corresponding to Λ_a in a shell around the Fermi surface, and $|J_a|$ the cardinality of J_a. $\mathcal{A}_a$ is the CAR-algebra associated with J_a and $\mathcal{A} = \overline{\bigcup_a \mathcal{A}_a}$ the quasilocal CAR-algebra, where a varies in a countable set. The local Gibbs state is $\langle \omega_{a'}^\beta ; A \rangle = tr_{a'}[\exp(-\zeta_{a'} - \beta H_{a'})A]$, $A \in \mathcal{A}_{a'}$, $a \le a'$. The most fundamental result, and also the most difficult one to obtain, is the convergence to the limiting Gibbs state:

$$\lim_{a'} \langle \omega_{a'}^\beta ; A \rangle =: \langle \omega^\beta ; A \rangle = \int_0^{2\pi} \langle \omega^{\beta\theta} ; A \rangle \frac{d\theta}{2\pi} \tag{2}$$

where $\omega^{\beta\theta}$ is locally given by the density matrix

$$\rho_a^{\beta\theta} = \exp\{-\xi_a - \beta[\varepsilon(|J_a| - b_a^z) - 2gw(e^{-i\theta} b_a^* + e^{i\theta} b_a)]\}$$

with $b_a^z := \sum_{k \in J_a} b_k^z$, and so on.

---

*Contribution to the XIth International Colloquium on Group Theoretical Methods in Physics, 23-28 August 1982, Istanbul

The calculation of the parameter $w=w(\beta)$ requires a discussion of the free energy based on the full sum over states and goes beyond the gap equation. If $f_g^P(\beta,y)$, $y \in \mathbb{R}^2$, denotes the free energy of the quasi-spin formulation [1] and y^g its minimal point for the coupling constant g, the free energy of the full system is given by

$$f(\beta,z) = z f^P_{gz}(\beta, y^{gz}) + (1-z)\varepsilon - \frac{1}{\beta}(1-z)\log 2 \\ - \frac{1}{\beta}[-z\log z - (1-z)\log(1-z)], \quad z \in (0,1), \tag{3}$$

where the physical value is attained at the minimal point $z=\hat{z}(\beta)$. z gives the ratio of the number of paired to the total number of electrons. Its equilibrium value $\hat{z}(\beta)$ tends to unity for $\beta \to +\infty$. Equation (3) exhibits in addition to the free energy of the pairs (with reduced coupling constant) the kinetic energy and the entropy of the singles and a mixing entropy. Given $\hat{z}(\beta)$ one obtains the implicit equation

$$\hat{\beta}_c = \frac{1}{\varepsilon} \operatorname{artanh}\left(\frac{2\varepsilon}{g\hat{z}(\hat{\beta}_c)}\right) \tag{4}$$

for the critical temperature and, finally,

$$w(\beta) = \begin{cases} ((\hat{z}\hat{y}_1/2)^2 - (\varepsilon/g)^2)^{1/2}, & \hat{\beta}_c < \beta \\ 0, & 0 < \beta \leq \hat{\beta}_c \end{cases}, \tag{5}$$

where $y^{g\hat{z}} =: \hat{y} = (\hat{y}_1, \hat{y}_2)$.

The limiting Gibbs state ω^β being now completely determined we can reconstruct the quantum mechanical formalism via the GNS-representation and obtain by means of the spatial decomposition theory (2, Ch. 4.4)

$$(\pi^\beta, \mathcal{H}^\beta, \Omega^\beta) = \int_0^{2\pi}{}^{\oplus} (\pi^{\beta\theta}, \mathcal{H}^{\beta\theta}, \Omega^{\beta\theta}) \frac{d\theta}{2\pi} \tag{6}$$

where the GNS-triples correspond to ω^β and $\omega^{\beta\theta}$, respectively. The cyclic vector Ω^β is also separating for $\mathcal{M}^\beta := \pi^\beta(\mathcal{A})''$ and $(\mathcal{M}^\beta, \mathcal{H}^\beta, \Omega^\beta)$ is standard in the sense of [3] with the antilinear isomorphism $j^\beta : \mathcal{M}^\beta \to \mathcal{M}^{\beta\prime}$.

Introducing the "renormalized" local Hamiltonians $K_a := H_a - j^\beta(H_a) \in B(\mathcal{H}^\beta)$ (π^β being dropped) we can show that

$$\text{strong resolvent} \lim_{a\to\infty} K_a = K^\beta \quad , \tag{7}$$

where the selfadjoint limiting operator is locally given by ($A \in \mathcal{A}_a$)

$$K^\beta A \Omega^\beta = [(\varepsilon(|\mathcal{J}_a| - \bar{b}^2_a) - 2g(\bar{b} b^*_a + \bar{b}^* b_a)), A]_- \Omega^\beta . \tag{8}$$

Here we have denoted

$$\bar{b} := s\text{-}\lim_a (b_a / |\mathcal{J}_a|) = w(\beta) \int_0^{2\pi} e^{-i\theta}\, 1^{\beta\theta} \frac{d\theta}{2\pi} \in \mathcal{M}^\beta \cap \mathcal{M}^{\beta\prime} , \tag{9}$$

and the last equality follows from the form of ω^β. From (9) we see that $\bar{b}$ behaves under gauge transformations as b_k and in the same way does $v := (1/2)^{\frac{1}{2}}(1-\varepsilon/E)^{\frac{1}{2}} \int_0^{2\pi} e^{-i\theta} 1^{\beta\theta} d\theta/2\pi$, $E := g\,\hat{z}\,\hat{y}_1$. The scalar $u := (1/2)^{\frac{1}{2}}(1+\varepsilon/E)^{\frac{1}{2}}$ is, of course, gauge invariant. The quasi-particle annihilation operators

$$\begin{aligned} \gamma_\varkappa &:= u\, c_\varkappa - v\, c^*_{-\varkappa} \\ \gamma_{-\varkappa} &:= v\, c^*_\varkappa + u\, c_{-\varkappa} \end{aligned} \tag{10}$$

are no longer in $\mathcal{A}$, but in $\mathcal{M}^\beta$, and transform as c_κ. The appropriate number operator in a nonrelativistic theory with one particle species is the generator of the unitaries which implement the gauge automorphisms in the representation algebra (here $\mathcal{M}^\beta$) and can directly be obtained by

$$\text{strong resolvent} \lim_a \Big(\sum_{\substack{k \in \mathcal{J}_a \\ \sigma}} c^*_\varkappa c_\varkappa - j^\beta\Big(\sum_{\substack{k \in \mathcal{J}_a \\ \sigma}} c^*_\varkappa c_\varkappa \Big)\Big) =: N^\beta \tag{11}$$

The vectors $(\bar{b}^*)^n \prod_\kappa \gamma^*_\kappa \prod_{\kappa'} \gamma_{\kappa'} \Omega^\beta$ constitute a total set of common eigenvectors to K^β and N^β. Beside the quasi-particle and -hole excitations they contain the condensed Cooper pairs $\bar{b}^*$, which are duly counted with the particle number two.

1) W. Thirring, Commun. Math. Phys. 7, 181 (1968)

2) O. Bratteli and D.W. Robinson, Operator Algebras and Quantum Statistical Mechanics I, II, Springer, Berlin 1979, 1981

3) M. Takesaki, in Lecture Notes in Physics 20, p. 205, ed. A. Lenard, Springer, Berlin 1973

ON THE ALGEBRAIC PROPERTIES OF THE LUTTINGER MODEL

K.A.Rustamov
Institute of Physics,Academy of Sciences of the Azerbaijan SSR, Baku, USSR

ABSTRACT

The system of differential equations for the elements of the tangent vector field corresponding to the Lie group of equivalence of the Schrödinger equations family with the Luttinger type Hamiltonian depending on an arbitrary potential function are analyzed. The absense of any operator of the first order in linear momentum operator in the Lie algebra of invariance of the problem has been proved. The obtained equations in the limit $\sigma = 0$ of the parameters enable one to get the Lie groups of equivalence of a various problems of quantum mechanics. In the case of $\sigma = 0$ the kernel of the basic groups of the mentioned family of equations is demonstrated to be physically trivial.

I. Introduction

The Luttinger model is the point of departure of the most of recent works devoted to the energy spectra problems of excitons and acceptors in cubic solids. According to the symmetry considerations Luttinger /1/ wrote the corresponding Hamiltonian as the general expression in second order in linear momentum operator $\vec{p}$ and with cubic symmetry

$$\mathcal{H} = \left(\gamma_1 + \frac{5}{2}\gamma_2\right)\frac{p^2}{2m_o} - \frac{\gamma_2}{m_o}\left(p_1^2 J_1^2 + p_2^2 J_2^2 + p_3^2 J_3^2\right) + \frac{2\gamma_3}{m_o}\Big\{2p_1p_2J_1J_2 - i\hbar p_1p_2J_3 + \quad (1)$$
$$+ 2p_2p_3J_2J_3 - i\hbar p_2p_3J_1 + 2p_3p_1J_3J_1 - i\hbar p_3p_1J_2\Big\} + \rho f(x_1, x_2, x_3),$$

where m_o is the free electron mass; γ_1, γ_2, γ_3 are the parameters responsible for description of the quasiparticle dispersion relation near the center of the Brillouin zone; $\rho f(x_1, x_2, x_3)$ is the potential function, where ρ is the constant depending on the electron charge and crystal dielectric constant; $\vec{J}$ is the angular momentum operator corresponding to spin 3/2. On the strength of this model the satisfactory solution of the energy spectra problems of excitons and acceptors have been obtained for a large number of cubic solids.

However the latest studies based on this model have shown that obtaining of valid wave functions being important for the solution of various physical problems in certain class of cubic solids is not an obvious question /10/.

Besides that and the traditional interest to the algebraic properties of the differential equations of the theoretical physics /2/ the present work is stimulated by the fact that certain physical quantities calculated independently by two methods phenomenologically and using the model Hamiltonians reveals different symmetry properties necessitating the additional requirements for the approachs based on the model Hamiltonians /9/.

2. On the method of investigation and results

Thus it is the purpose of this work to study the algebraic properties of the equation giving rise to the following linear differential operator

$$\hat{L}(p_\nu, x_\nu) = p_4 + \alpha p_i p_i + \beta p_i^2 J_i^2 + \gamma\Big\{2p_1p_2J_1J_2 - i\hbar p_1p_2J_3 + \quad (2)$$
$$+ 2p_2p_3J_2J_3 - i\hbar p_2p_3J_1 + 2p_3p_1J_3J_1 - i\hbar p_3p_1J_2\Big\} + \delta f(x_1, x_2, x_3),$$

where we use the Einstein convention on repeated indices; $i=1,2,3$; $\nu=1,2,3,4$; $x_4=t$; $P_4=\partial/\partial t$.

The direct method of solving of the formulated above problem requires the diagonalization of the symbol /5/ of the operator (2), which is rather difficult in discussing case because of noncommutativness of the apparent matrices /4/.

Therefore we have chosen the more simple method of investigation and leave the matrix representation of $\mathcal{J}_i$ till the concretization of the representation of the obtained algebra of invariance.

Let us sketch the details of the used method.

Assume that $\hat{A}(x_\nu)$ is the element of the algebra of invariance of the differential eq.

$$\hat{L}(P_\nu,x_\nu)\Psi(x_\nu)=0 \tag{3}$$

and let $\hat{A}(x_\nu)$ be a linear differential operator of the form

$$\hat{A}(x_\nu)=b_\mu(x_\nu)P_\mu+a(x_\nu)\,;\,(\mu=1,\dots,4)\,. \tag{4}$$

It will be recalled that the manifold of the operators is said to be the Lie algebra of invariance of eq.(3) if this manifold closed under commutation and if for an arbitrary operator $\hat{A}$ from this algebra and an arbitrary vector Ψ from the space of solutions of the eq.(3) $\Psi'=\hat{A}\Psi$ is again a solution of the eq.(3) /2-8/, so the space of solutions of the eq.(3) is the carrier space for the representation of the group of equivalence of the problem.

Since $\hat{L}(P_\nu,x_\nu)$ is the only annihilator of an arbitrary solution of eq.(3) one can get the following ehuation fot the operator

$$\left[\hat{L}(p_\nu,x_\nu),\hat{A}(x_\nu)\right]=\hat{L}(p_\nu,x_\nu)\hat{A}(x_\nu)-\hat{A}(x_\nu)\hat{L}(p_\nu,x_\nu)=\varphi(x_\nu)\hat{L}(p_\nu,x_\nu)\,, \tag{5}$$

where $\varphi(x_\nu)$ is an arbitrary function.

In case when $\hat{L}(p_\nu, x_\nu)$ contains an arbitrary function $V(x_\nu)$ the problem on the kernel of the basic groups of equivalence can be solved by the classical method /3/ of expanding the equations for the $b_{\mu}(x_\nu)$, $a(x_\nu)$ and collecting terms that multiply $V(x_\nu)$ and its derivatives.

The similar prosedure has been carried out for the eq. (3). From the equations obtained for the $b_\mu(x_\nu)$ and $a(x_\nu)$ or what is the same, for the elements of the tangent vector field of the connected Lie group, it is easy to conclude that no operators linear in the linear momentum operator are admitted by the eq.(3) excluding the case of $\beta-\gamma=\sigma=0$. The mentioned equations are too room consuming, that is why we cannot give their explicite form. In the case of $\sigma=0$ the obtained equations allow to cary out the algebras of invariance of various problems of nonrelativistic quantum mechanics and free particle problem admitting the 12-parameter Lie group of equivalence is the simplest example of the mentioned /7/. From the equations for $b_\mu(x_\nu)$ and $a(x_\nu)$ it also follows that the kernel of the basic algebras of invariance of the differential equations family $\hat{L}(p_\nu, x_\nu, f(x_i), \sigma=0)\Psi(x_\nu)=0$ is the physically trivial.

3. References

/1/. J.M.Luttinger, Phys.Rev., 102, 1030, 1956.

/2/. I.A.Malkin, V.I.Man'ko, Dinamicheskie simmetrii i cogerentnye sostoiania quantovykh sistem, Nauka, 1979.

/3/. L.V.Ovsiannikov, Gruppovoi analiz differencialnykh uravnenii, Nauka, 1978.

/4/. V.I.Fuschich, Teoretiko-gruppovie metody v matematicheskoi fisike, Kiev, 1978.

/5/. M.A.Shubin, Psevdodifferencialnye operatory i spectralnaia teoria, Nauka, 1978.

/6/. W.Miller, Jr., Symmetry and Separation of Variables, 1977.

/7/. U.Niederer, Helvetica Phys.Acta, 45, 802, 1972; 47, 119, 1974.

/8/. R.L.Anderson, Revista Mexicana de Fisica, 21, 1, 1972; J.Math.Phys., 14, N 11, 1973.

/9/. F.M.Gashimzade, K.A.Rustamov, Doklady Akademii Nauk Azerbaijanskoi SSR, N 3, 21, 1978.

/10/. M.A.Mekhtiev (to be published).

DYNAMICAL GROUPS AND COEXISTENCE PHENOMENA

Allan I. Solomon
Faculty of Mathematics
The Open University
Milton Keynes, U.K.

ABSTRACT

This note describes an application of dynamical Lie groups to many body systems exhibiting phase transitions. The specific model exemplified is that of a three-phase many fermion system for which the appropriate groups is SO(6).

1. Foreword

The following talk describes what I think is a fairly unusual application of Lie algebras, to the solution of problems involving collective phenomena. I was first introduced to Lie algebras, and more specifically spectrum-generating Lie algebras in the context of particle physics, while visiting Professor Yuval Ne'eman at the Department of Physics of Tel Aviv University in 1968-69. It gives me great pleasure to acknowledge the influence that he exercised on my subsequent work, and I am very happy to be able to contribute to the Proceedings of this year's Group Theory Conference, the year in which Professor Ne'eman received the Wigner Medal for his outstanding contributions in the field.

2. Dynamical Groups

The association of symmetry groups with phase transitions is well-founded[1]. The system in question in its disordered state, above a critical temperature T_C, is described by a hamiltonian H having symmetry group G_S. In the ordered state, below T_C, the system is conventionally described by a reduced hamiltonian H_{red} which is invariant under a smaller symmetry group $G_B \subset G_S$.

Dynamical groups - non-symmetry groups of the system, arise in the following way: the reduced hamiltonian, usually a mean-field approximation, is exactly solvable (diagonalizable), and is a representation of an element of a Lie algebra, the so-called Spectrum Generating Algebra (SGA). A working definition of the dynamical group would be the Lie group of this SGA. The best-known example is the so(4,2) SGA of the hydrogen atom[2]; the name of the algebra derives from its property of generating the spectrum of the system. This valuable property is shared by the SGA's of solvable many-body models; such SGA's usually have the form of a direct sum of Lie algebras indexed by momentum, $g = \oplus_k g_k$, where each g_k is isomorphic to a fixed Lie algebra (which we shall occasionally loosely refer to as the SGA of the system). An example is the SGA for an exactly solvable model[3] of superfluid Helium Four; here $g \sim \oplus_k \, su(1,1)_k$.

3. Quadratic Hamiltonians

A general example of such an exactly-solvable system will be given by a reduced hamiltonian H_{red} which is expressed as a direct sum of (essentially one-particle) commuting hamiltonians indexed by momentum k, where each H_k is quadratic[4] in some set of creation and annihilation operators given by a column d-vector A_k. Thus

$$H_{red} = \sum_k H_k \quad \text{with} \quad H_k = A_k^+ M_k A_k$$

where M_k is a hermitian $d \times d$ matrix over $\mathbb{C}$. This is sufficient to ensure that H_{red} is a representative of an element x_{red} of a Lie algebra $g \sim \bigoplus_k g_k$. If we assume for definiteness that we have a fermion system

$$\{A_{ik}, A_{jk}^+\} = \delta_{ij}$$

then each g_k is isomorphic to a subalgebra of u(d).

4. Bogoliubov Transformation

In general, we may choose a Cartan basis for a rank-ℓ, n-dimensional semi-simple complex Lie algebra g

$$\{h_1, h_2, \ldots, h_\ell; e_1, e_2, \ldots, e_{n-\ell}\} \tag{1}$$

where
$$[h_i, h_j] = 0, \quad [h_i, e_\alpha] = \alpha_i e_\alpha \tag{2}$$

$$[e_\alpha, e_\beta] = N_{\alpha+\beta} e_{\alpha+\beta} \quad (\text{if } \alpha + \beta \neq 0),$$

$$[e_\alpha, e_{-\alpha}] = \sum \alpha^i h_i$$

The diagonalizable H_{red} is a representative of a semi-simple element $x_{red} \in g$; as a consequence, there exists an automorphism $\phi : g \to g$ which sends x_{red} into the commuting Cartan Subalgebra (CSA) generated by h_i [5]. This automorphism will be implemented in the d-dimensional representation by a rotation R, and in Fock space by a unitary transformation U. Thus the matrix $M' = RMR^+$ is diagonal; and if we define new operators $\mathcal{A}_k$ by $\mathcal{A}_k = U^+ A_k U$ then, expressed in terms of the new operators $\mathcal{A}_k$, the reduced hamiltonian H_{red} is diagonal, $H_k = \mathcal{A}_k^+ M_k' \mathcal{A}_k$. The transformation $A_k \mapsto \mathcal{A}_k = U^+ A_k U = RA_k$ is known as the Bogoliubov transformation[6] from its first use in solving the Helium Four problem.

5. Order Parameters

There is a certain arbitrariness in the choice of the automorphism ϕ. In the case where the symmetry group G_S of the original hamiltonian H (in the disordered state, about T_C) is abelian, we may choose the corresponding commutative algebra g_s to be a sub-algebra of the CSA of g. In this case, the element $x_{sym} \in g$ defined by $x_{sym} = \phi(x_{red}) = \sum_{i=1}^{\ell} \mu_i h_i$ 'recovers' the original symmetry of H. If we take its Fock space representative H_{sym} as a 'good' (in the group theoretical sense) description of the original system (above T_C), we see that the elements e_α of the Cartan basis (1) behave as order parameters for the system[7]. This follows from the commutation relations (2)

$$[x_{sym}, e_\alpha] = [\sum_i \mu_i h_i, e_\alpha] = (\sum_i \mu_i \alpha_i) e_\alpha .$$

Taking the expectation value of this commutator in eigenstates of H_{sym}, we have

$$<[H_{sym}, e_\alpha]>_{sym} = \sum_i \mu_i \alpha_i <\hat{e}_\alpha>_{sym} = 0.$$

If $\sum_i \mu_i \alpha_i \neq 0$, this shows that the Hilbert space representative $\hat{e}_\alpha$ of e_α vanishes in eigenstates of H_{sym} - representating the system above T_C - and is thus a good candidate for an order parameter. In the case of the BCS model of superconductivity, this leads to the well-known local complex order parameter; a similar result is obtained for a u(2) model of charge-density waves[8] (since the considerations above for semi-simple algebras also hold good for the reductive u(2)).

6. so(6) Coexistence Model

We now illustrate the preceding ideas by a many-fermion model which exhibits coexisting phases. Our starting point is thus the anti-commutation relations

$$\{a_{k\sigma}, a^+_{k'\sigma'}\} = \delta_{kk'} \delta_{\sigma\sigma'}$$

for fermions of wave vector k,k' and spin σ,σ'. We define a four-component operator A_i by

$$(A_1, A_2, A_3, A_4)_{(k)} = (a_{k\uparrow}, a^+_{-k\downarrow}, a_{\bar{k}\uparrow}, a^+_{-\bar{k}\downarrow})$$

where $\bar{k} = k + Q$. Here Q is a characteristic wave vector for the physical problem; in the case we shall discuss, that of charge-density waves (CDW) and anti-ferromagnetic order (AF), $Q = 2k_F$, where k_F is the Fermi surface wave vector. Defining $X_{ij} = A^+_i A_j$ (suppressing the implicit k-dependence) we see that

$$[X_{ij}, X_{i'j'}] = \delta_{ji'} X_{ij'} - \delta_{ij'} X_{i'j} .$$

Thus the X_{ij} generate $g\ell(4,R)$; or, as we have only hermitian combinations of operators, u(4). We take as our mean field hamiltonian $H_{red} = \sum_k H_k$, where $H_k = A^+ MA$ for some (hermitian) matrix M. Identification of the kinetic energy term as $\epsilon(k) = \epsilon(-k)$ leads to a traceless M; we are therefore dealing with the SGA su(4) ~ so(6). Typical terms occurring in the model are

$X_{12} = a^+_{k\uparrow} a^+_{-k\downarrow}$; superconductor (SC) pairing,

$X_{13} = a^+_{k\uparrow} a_{\bar{k}\uparrow}$; CDW term,

$X_{14} = a^+_{k\uparrow} a^+_{-\bar{k}\downarrow}$; AF anomalous pairing(9).

A Cartan basis for this 15-dimensional rank-3 Lie algebra has the form

$$\{h_1,h_2,h_3;\ e_1,e_2,\ldots,e_{12}\}.$$

We may identify each of the h_i with an operator conserved above T_C; thus

$h_1 \sim N$	Number operator, (sum of k and $\bar{k}$ numbers)
$h_2 \sim P$	Linear (one-dimensional) momentum
$h_3 \sim A$	Anomalous number, (difference of k and $\bar{k}$ numbers).

Physically, it is clear we can recover subphases by considering those operators commuting with h_1,h_2 and h_3 in turn. Thus, the SC terms will certainly not commute with $h_1 \sim N$. Mathematically, this process of taking the centralizer $C(h_i)$ of each h_i in turn leads to a sub-algebra, which we may identify with the SGA of a subphase(10). Each centralizer is given by

$$C(h_i) \sim s(u(2) \oplus u(2)) \sim u(1) \oplus so(4).$$

We therefore obtain a model, which, in the absence of magnetic terms, has been previously treated in some detail(11).

According to our previous remarks, the order parameters are e_α; and they will determine the presence or absence of phases. Following the root-vector diagram analysis of Van der Waerden (1933) as quoted by Wybourne(12), we see that the root-vectors for so(6) are given by the 12 combinations $\pm\underline{u}_i \pm \underline{u}_j$, where $\underline{u}_1 = (1,0,0)$, $\underline{u}_2 = (0,1,0)$ and $\underline{u}_3 = (0,0,1)$.

One immediate conclusion that we may draw is that no root-vector can have non-vanishing components along all 3 axes. For each α, the corresponding local order parameter $\Phi_\alpha(x)$ defined by

$$\Phi_\alpha(x) = \sum_k e_\alpha(k)e^{ikx}$$

must commute with at least one of the h_i(N,P or A) and hence cannot serve as a simultaneous order parameter for more than two of the subphases. This implies the impossibility of the simultaneous coexistence of all three phases.

References

(1) See, for example, the review article by Louis Michel, Rev.Mod. Phys. 52, 617 (1980).

(2) A.O. Barut: Phys. Rev. 135, B839 (1964).

(3) A.I. Solomon: J. Maths. Phys. 12, 390 (1971).

(4) P. Broadbridge and C.A. Hurst: Physica 108A, 39 (1981), and P. Broadbridge: Hadronic Journal 4, 879 (1981).

(5) Standard results on semi-simple complex Lie algebras will be found in most texts, for example; "Lie Groups, Lie Algebras and their Representations", V.S. Varadarajan (Prentice-Hall, 1974).

(6) N. Bogoliubov: J. Phys. USSR 11, 23(1947).

(7) A.I. Solomon: Annals of the New York Academy of Sciences (to be published).

(8) A.I. Solomon and J.L. Birman: Physics Letters 88A, 413 (1982).

(9) M.J. Nass, K. Levin and G.S. Grest: Phys. Rev. Letters 46, 614 (1981).

(10) A.I. Solomon: Proceedings of IX International Colloquium on Group Theoretical Methods, Mexico. Lecture Notes in Physics 135, page 42 (Springer-Verlag 1980).

(11) J. L. Birman and A.I. Solomon: Phys. Rev. Letters 49, 230 (1982).

(12) B. G. Wybourne: "Classical Groups for Physicists", page 70, (John Wiley, 1974).

LIST OF PARTICIPANTS

H.AKÇAY
Ortadoğu Teknik Üniversitesi
Fizik Bölümü
Ankara, Turkey

Y.C.AKGÖZ
Ortadoğu Teknik Üniversitesi
Fizik Bölümü
Ankara, Turkey

G.AKTAŞ
Boğaziçi Üniversitesi
Fizik Bölümü
P.K.2 Bebek
İstanbul, Turkey

Ö.AKYÜZ
Boğaziçi Üniversitesi
Fizik Bölümü
P.K.2 Bebek
İstanbul, Turkey

D.ALPHANDARİ
Boğaziçi Üniversitesi
Fizik Bölümü
P.K.2 Bebek
İstanbul, Turkey

G.I.AMUNATEGUI
Universidad Catolica de
Valparaiso
Instituto Fisica
Casilla 4059
Valparaiso, Chile

E.ARIK
Boğaziçi Üniversitesi
Fizik Bölümü
P.K.2 Bebek
İstanbul, Turkey

M.ARIK
Boğaziçi Üniversitesi
Fizik Bölümü
P.K.2 Bebek
İstanbul, Turkey

T.ARMAĞAN
İstanbul Üniversitesi
Veteriner Fakültesi
İstanbul, Turkey

M.I.AROYO
Sofia University
Faculty of Physics
5 Anton Ivanov Blvd. BG-1126
Sofia, Bulgaria

A.ARŞIK
Boğaziçi Üniversitesi
Fizik Bölümü
P.K.2 Bebek, İstanbul, Turkey

N.M.ATAKISHIYEV
Azerbaijan Academy of Sciences
Institute of Physics
Baku, USSR

S.ATAĞ
Ankara Üniversitesi
Fen Fakültesi
Fizik Bölümü
Beşevler, Ankara, Turkey

Z.AYDIN
Ankara Üniversitesi
Fen Fakültesi
Fizik Bölümü
Beşevler, Ankara, Turkey

N.BALKAN
Boğaziçi Üniversitesi
Fizik Bölümü
P.K.2 Bebek
İstanbul, Turkey

A.BARAN
Diyarbakır Üniversitesi
Fen Fakültesi
Diyarbakır, Turkey

B.R.BARRET
Department of Physics
University of Arizona
Tucson, Arizona 85721, USA

I.BARS
Yale University
Physics Department
New Haven, Conn. 06520, USA

A.BARUT
Department of Physics
University of Colorado
Campus Box 390
Boulder, Colorado 80309, USA

J.BECKERS
Université de Liége
Institut de Physique
Sart-Tilman B-4000
Liége 1, Belgium

H.BEKER
Boğaziçi Üniversitesi
Fizik Bölümü
P.K.2 Bebek
İstanbul, Turkey

C.BEZMEN
Boğaziçi Üniversitesi
Fizik Bölümü
P.K.2 Bebek
İstanbul, Turkey

A.H.BİLGE
Ortadoğu Teknik Üniversitesi
Fizik Bölümü
Ankara, Turkey

A.BÖHM
University of Texas
Department of Physics
Austin, Texas 78712, USA

L.L.BOYLE
University Chemical Laboratory
Canterbury, Kent GB-CT2 7NH
England

M.P.BOŽIĆ
Institute of Physics
11001 Beograd, Yugoslavia

E.CELEGHINI
Instituto Fisica Teorica
Largo E. Fermi 2
50123 Firenze, Italy

D.CHUDNOVSKY
Department of Mathematics
Columbia University
New York, NY 10027, USA

P.COTTA-RAMUSINO
Universita degli Studi di Milano
Istituto di Scienze
Via Celoria 16
Milano 20133, Italy

P.CVITANOVIĆ
Nordita, Blegdolmsvej 17
DK-2100, KØbenhavn Ø
Denmark

J.CZYZ
Polish Academy of Sciences
Institute of Mathematics
Warsaw 00-950 PO Box 137
Poland

O.ÇAĞLAYAN
İ.D.M.M.A.
İzmit, Turkey

A.ÇELİKEL
Ankara Üniversitesi
Fen Fakültesi
Fizik Bölümü
Beşevler, Ankara, Turkey

M.DAMNJANOVIĆ
Faculty of Natural and
Mathematical Sciences
Department of Physics and
Meteorology
11001 Beograd, Yugoslavia

Ö.F.DAYI
Ankara Üniversitesi
Fen Fakültesi
Fizik Bölümü
Beşevler, Ankara, Turkey

J.A.de AZCÁRRAGA
Departamento de Fisica Teorica
Facultad de Ciencias Fisicas
Burjasot, Valencia, Spain

M.B.de LEONE GAYDUCATI
Centre de Recherches Nucleaires
Physique Theorique des Hautes
Energies, PB 20
67037 Strasbourg, Cedex, France

H.de MEYER
Seminarie Wiskundige Natuurk.
Rijksuniversiteit-Gent
Krijgslaan 281-S9
B-900 Gent, Belgium

G.DENARDO
Istituto di Fisica Teorica
Universita degli Studi di Trieste
34014 Trieste, Italy

T.DERELİ
Ortadoğu Teknik Üniversitesi
Fizik Bölümü
Ankara, Turkey

D.DİBEKÇİ
İ.D.M.M.A.
İzmit, Turkey

M.DİKMEN
Boğaziçi University
Fen Edebiyat Fakültesi
Bebek P.K.2.
İstanbul, Turkey

R.DIRL
Institut für Theoretische Physik
TU Wien, Karlsplatz 13
A-1040 Wien, Austria

L.DOLAN
Physics Department
Rockefeller University
New York, NY10021, USA

M.DURGUT
Ortadoğu Teknik Üniversitesi
Fizik Bölümü
Ankara, Turkey

İ.H.DURU
Diyarbakır Üniversitesi
Fizik Bölümü
Diyarbakır, Turkey

H.EICHENHERR
CERN Theory Division
CH-1211 Genevê 23
Switzerland

H.EIZENBERG
Department of Physics and Astronomy
Tel-Aviv University
Ramat-Aviv, Tel-Aviv, Israel

D.ERCAN
Othm. Kirchenweg 105
2000 Hamburg 50
W.Germany

A.ERİŞ
Ortadoğu Teknik Üniversitesi
Fizik Bölümü
Ankara, Turkey

N.EVİN
T.E.K.
Ankara, Turkey

H.N.FICEK
Instytut Fizyki Molekularnej
Polskiej Akademii Nauk
ulica Smoluchowskiego 17-19
Poznan, PL 60179, Poland

M.GADELLA
Universidad de Santander
Facultad de Ciencias
Av. los Castros, Santander, Spain

H.GOLDSCHMIDT
The Institute for Advanced Study,
Princeton
New Jersey 8540, U.S.A.

H.S.GREEN
University of Adelaide
Department of Mathematical Physics
Adelaide, South Australia

Y.GÜLER
Ortadoğu Teknik Üniversitesi
Fizik Bölümü
Ankara, Turkey

M.GÜNAYDIN
Ecole Normale Supérieure
24, rue Lhomond
75231 Paris Cedex 05-France

Y.GÜNDÜÇ
Hacettepe Üniversitesi
Fizik Bölümü
Beytepe, Ankara, Turkey

M.GÜRSES
Ortadoğu Teknik Üniversitesi
Fizik Bölümü
Ankara, Turkey

F.GÜRSEY
Yale University
Physics Department
New Haven Conn. 06520, USA

Z.B.GÜVENÇ
Ankara Üniversitesi
Fen Fakültesi
Fizik Bölümü
Beşevler, Ankara, Turkey

A.HACINLIYAN
Boğaziçi Üniversitesi
Fizik Bölümü
P.K.2 Bebek
İstanbul, Turkey

M.HORTAÇSU
Boğaziçi Üniversitesi
Fizik Bölümü
P.K.2 Bebek
İstanbul, Turkey

J.W.B.HUGHES
University of London
Queen Mary College
Department of Applied Math.
Mile end Road
London E1 4NS, England

E.İNÖNÜ
Boğaziçi Üniversitesi
Fizik Bölümü
P.K.2 Bebek
İstanbul, Turkey

H.JAKOBSEN
Mathematics Institute
Universitatsparken 5
DK-2100, Copenhagen Ø Denmark

A.JANNER
Instituut voor Theoretische Fysica
Katholieke Universiteit
Toernooiveld, 6525 ED
Nijmegen, Holland

T.JANSSEN
Institute for Theoretical Physics
Toernooiveld, 6525 ED
Nijmegen, Holland

M.JARIĆ
Frei Universitat Berlin FB20
WE2, Arnimallee 3,
1000 Berlin 33
West Germany

B.JULIA
Laboratoire de Physique Theorique
Ecole Normale Supérieure
24 Rue Lhomond, 75231 Paris
Cedex 05, France

J.KALAYCI
Boğaziçi Üniversitesi
Fizik Bölümü
P.K.2 Bebek
İstanbul, Turkey

A.KALKANLI
Ortadoğu Teknik Üniversitesi
Fizik Bölümü
Ankara, Turkey

C.KARACA
Ege Üniversitesi Fen Fakültesi
Fizik Bölümü
Bornova, İzmir, Turkey

A.KARASU
Ortadoğu Teknik Üniversitesi
Fizik Bölümü
Ankara, Turkey

P.KASPERKOVITZ
Institut für Theoretische Physik
der Technische Universitat Wien
Karls-platz 13
A-1040 Vienna, Austria

P.W.KASTELEYN
Instituut Lorentz
voor Theoretische Natuurkunde
Niewsteeg 18, Leiden
Netherlands

B.KENDİRLİ
Ortadoğu Teknik Üniversitesi
Matematik Bölümü
Ankara, Turkey

A.KILIÇ
İstanbul Üniversitesi Fen Fakültesi
Fizik Bölümü
İstanbul, Turkey

R.C.KING
University of Southampton
Faculty of Mathematical Studies,
Southampton SO 95 NH
England

A.KLEIN
University of Pennsylvania
Physics Department
Philadelphia 19104, USA

M.KOCA
Çukurova Üniversitesi
Fizik Bölümü
Adana, Turkey

H.W.KUNERT
Institute of Physics
Technical University
Piotrovo 3, 60-965 Poznan
Poland

LING-LIE CHAU
Physics Department
Broohaven National Laboratory
Upton NY 11973, USA

J.LUKIERSKI
Institute of Theoretical Physics
University of Wroclaw, Cybulskiego 36
Wroclaw, Poland

K.T.MAHANTHAPPA
University of Colorado
Department of Physics
Boulder, Colorado 80309, USA

K.B.MARATHE
Department of Mathematics
Brooklyn College
Brooklyn, NY 11210, USA

L.MICHEL
I.H.E.S.
35 Route de Chartres
91440 Bures-sur-Yvette, Paris, France

M.MOSHINSKY
UNAM
Instituto de Fisica
Apdo.Postal 20-364
Mexico DF 01000, Mexico

P.MOYLAN
The University of Texas
Department of Physics
Austin, Texas 78712, USA

W.NAHM
Max-Planck Institut für Mathematik
Gottfried-Claren str. 26
5300 Bonn 3, Germany

F.NEYZİ
Boğaziçi Üniversitesi
Fizik Bölümü
P.K.2 Bebek
İstanbul, Turkey

M.ÖNDER
Ortadoğu Teknik Üniversitesi
Fizik Bölümü
Ankara, Turkey

O.ÖZALTIN
Diyarbakır Üniversitesi
Fen Fakültesi
Diyarbakır, Turkey

M.PERROUD
Departement de Mathematique
Appliquées
Ecole Polytechnique Montréal,
Quebec H3C 3A7, Canada

J.W.PHIPPEN
Weber State College
Department of Physics
3750 Harrison Blvd.
Ogden, Utah 84408, USA

J.F.POMMARET
Departement de Mathématique
Ecole Nationale des Ponts et
Chaussées, Paris F-75007, France

C.QUESNE
Université Libre de Bruxelles
Physique Théorique et Mathematique
CP 229, B-1050 Bruxelles, Belgium

M.RASETTI
Instituto di Fisica
Politecnico di Torino
Corso Abruzzi 10129 Torino, Italy

A.RIECKERS
Institut für Theoretische Physik
der Universitat Tübingen
D-7400 Tübingen, West Germany

H.RÖMER
Albert Ludwigs Universitat
Fakultat für Physik
Herman-Hedrer Str. 3
7800 Freiburg, Germany

H.RUEGG
Department de Physique Théorique
32 Boulevard d'Yvoy
CH-1211, Geneve 4, Switzerland

K.RUSTAMOV
Azerbaijan Academy of Sciences
Institute of Physics
370143 Baku, USSR

C.SAÇLIOĞLU
Boğaziçi Üniversitesi
Fizik Bölümü
P.K.2 Bebek
İstanbul, Turkey

H.SALLER
Max-Planck Institute für Physik
und Astrophysik
Foehringer Ring 6
8000 München 40, West Germany

T.SCHÜCKER
Université de Genevé
Sektion de Physique
CH-1211, Geneve 4, Switzerland

A.SCIARRINO
Instituto Fisica Teorica
Mostra d'oltremare, PAD 19
Napoli 80125, Italy

S.SCIUTO
Instituto Fisica Teorica
C.Massimo d'Azeglio 46
10125 Torino, Italy

M.SERDAROĞLU
Boğaziçi Üniversitesi
Fizik Bölümü
P.K.2 Bebek
İstanbul, Turkey

A.SEVGEN
Boğaziçi Üniversitesi
Fizik Bölümü
P.K.2 Bebek
İstanbul, Turkey

R.M.SHANKAR
Yale University
Physics Department
New Haven Conn. 06520, USA

R.T.SHARP
McGill University
Rutherford Physics Bldg.
3600 University Str.
Montreal, Quebec H3A2T8, Canada

R.SHAW
University of Hull
Department of Applied Mathematics
Hull, HU6 7RX, England

W.M.SILVA
Libera Universita degli Studi
di Trento,
Dpt. di Mathematica
38050 Povo, Trento, Italy

B.R.SITARAM
Department of Physics
Delhi University
Delhi 110007, India

Y.SKARLATOS
Boğaziçi Üniversitesi
Fizik Bölümü
P.K.2 Bebek
İstanbul, Turkey

J.SOKOLOFF
Physics Department
Northeastern University
Boston, Mass.02115, USA

A.SOLOMON
The Open University
Faculty of Physics
Milton Keynes MK6AA, England

P.SORBA
L.A.P.P.
Chemin de Bellevue
Boite Postale 909
74019 Annecy-le-Vieux, Cedex
France

S.SÖZÜER
Boğaziçi Üniversitesi
Fizik Bölümü
P.K.2 Bebek
İstanbul, Turkey

M.TARLINI
Istituto Fisica Teorica
Largo E.Fermi 2
50123 Firenze, Italy

S.TİLAV
Boğaziçi Üniversitesi
Fizik Bölümü
P.K.2 Bebek
İstanbul, Turkey

S.TOURE
Institut de Recherches
Mathematiques
08BA2030 ABİDJAN 08
Ivory Coast

S.TWAREQUE-ALİ
Concordia University
Math Department
Loyola Campus, Montreal
Quebec H4 B 1R6, Canada

B.ÜN
İstanbul Üniversitesi
Fen Fakültesi, Fizik Bölümü
İstanbul, Turkey

N.ÜNAL
Diyarbakır Üniversitesi
Fen Fakültesi, Fizik Bölümü
Diyarbakır, Turkey

G.VAN DEN BERGHE
Seminarie Wiskundige Natuurkunde
Rijksuniversiteit-Gent
Krijgslaan 281-S9
B-9000 Gent, Belgium

P.VAN DER BROEK
Twente University of Technology
Department of Applied Mathematics
P.O.Box 217
7500 AE Enschede, Netherlands

J.VAN DER JEUGT
Seminarie Wiskundige Natuurkunde
Rijksuniversiteit-Gent
Krijgslaan 281-S9
B-9000 Gent, Belgium

P.VAN ISACKER
Instituto de Fisica, UNAM
Apdo.Postal 20-364
01000 Mexico DF, Mexico

P.VAN NIEUWENHUIZEN
CERN-Theory Division
CH-1211, Geneva 23
Switzerland

H.VENABLES
Portsmouth Polytecnic
Department of Mathematics
Portsmouth PO1 2EG, England

Y.A.VERDİYEV
Azerbaijan Academy of Sciences
Institute of Phyics
370143 Baku, USSR

J.L.VERDIER
Centre de Mathematiques
Ecole Normale Superieure
45 rue d'Ulm, 75230 Paris,
Cedex 05, France

G.VITIELLO
Universita di Salerno
Istituto di Fisica
84100 Salerno, Italy

P.WINTERNITZ
CRMA
Université de Montreal
CP 6128 Montreal
Quebec H3C3J7, Canada

C.WULFMAN
University of Pacific
Department of Physics
Stockton, California 95211, USA

A.U.YILMAZER
Ankara Üniversitesi
Fen Fakültesi
Beşevler-Ankara, Turkey

A.YALAMA
Boğaziçi Üniversitesi
Fizik Bölümü
P.K.2 Bebek
İstanbul, Turkey

H.YÜKSEL
Ankara Üniversitesi,
Fen Fakültesi, Fizik Bölümü
Beşevler, Ankara, Turkey

U.YURTSEVER
Ortadoğu Teknik Üniversitesi
Fizik Bölümü
Ankara, Turkey

G.ZUCKERMAN
Yale University
Department of Mathematics
New Haven, Conn. 06520, USA

APPENDIX I

The Wigner Medal for 1980

The Wigner Medal for 1980 was awarded to I. M. Gelfand in recognition of the numerous mathematical contributions having profound applications in physics. I. M. Gelfand was honored in a ceremony at the IX International Colloquium on Group Theoretical Methods in Physics held at Cocoyoc, Mexico in June 1980. He could not attend the meeting and the Medal was presented to him by A. Bohm, Chairman of the Group Theory Foundation, in Moscow in July 1981. At this occasion, I. M. Gelfand read the following statement which is reproduced verbatim below:

> "I am honored to receive the Wigner medal. It is a great honor for a mathematician to learn that his works are used in Physics. I am also very touched that the medal has the name of Prof. Wigner - a man whose work I respect very highly and whom personally I love and admire. I am also very touched that the medal was handed to me by my friend Prof. A. Bohm whom I like very much, and this is a consolation for being unable to come to the conference"

I. M. Gelfand, Moscow

21-VII-81

APPENDIX II

PRESENTATION OF THE WIGNER MEDAL

November 11, 1982

Thompson Conference Center
University of Texas

"Only the introductory talks are reproduced here as the full proceedings on the Ceremony will be published elewhere".

THE WIGNER MEDAL FOR 1982

Introductory Remarks by A. Bohm

As Chairman of the Group Theory Foundation I would like to welcome you to this brief, improvised, almost secret session in which the 1982 Wigner Medal will be presented to Yuval Ne'eman.

The Wigner Medal is not connected with the Solvay Conferences and is usually presented at the International Colloquia on Group Theoretical Methods in Physics. Unfortunately this year an unforeseen development in world events has forced the Group Theory Foundation to look for another host conference. The Solvay Conference was the earliest opportunity for the presentation of the Medal and we are grateful to Professors Prigogine, Van Hove and the organizers of the conference to provide this distinguished setting.

The Wigner Medal is unique, because it is a truly international award not connected with any state or national society. It is awarded approximately every two years for outstanding contributions to the understanding of physics through group theory. Previous recipients were Eugene Wigner, Valentine Bargmann and Izrail' Moiseevich Gel'fand. The recipient is chosen by an international selection committee whose members serve for four years, except for Wigner who is a lifetime member.

The tribute to Yuval Ne'eman will be given by the chairman of the selection committee, Larry Biedenharn.

TRIBUTE TO YUVAL NE'EMAN

L.C. Biedenharn

It is a great pleasure to take part in this award of the Wigner medal to our esteemed colleague, Yuval Ne'eman. This is an especial pleasure for me since the award simultaneously honors two men whose influence on my own work has been profound: Eugene Wigner--the 'patron saint', as it were, of symmetry in quantum physics--and Yuval Ne'eman --one of the originators of unitary symmetry in particle physics.

First let me say a few words of a biographical nature. Yuval is a native Israeli, the scion of one of the older Jewish families in the country; he attended Herzliya high school in Tel Aviv and the Technion in Haifa for his undergraduate work. Yuval grew up in a time of turmoil, taking part, out of strong conviction, in the wars of independence in 1948, 1956, 1967. It was only in 1958 that he was able to begin graduate work at Imperial College (London) under Salam. Science transcends national origins, and national struggles, uniting us all in a common endeavor, but I have mentioned these facets of Yuval's past only to emphasize how remarkable it is that he could have made any scientific contributions at all in view of his frequently interrupted, postponed, and abbreviated graduate training in the essentials of our discipline.

The many contributions of Ne'eman to particle physics are all, in one way or another, related to his extraordinary ability to discern symmetry, and pattern, in the confused welter of experimental results. Probably his most spectacular result was the discovery of unitary symmetry; with Murray Gell-Mann he was one of the discoverers of the "eight fold way" characterizing $SU3^{flavor}$ (as it is now known). Yuval noticed that all reactions allowed by isospin and strangeness conservation seemed actually to occur: from this he concluded--intuitively-- that a simple rank 2 Lie group was required and systematically tested and eliminated, all but SU3. There were difficulties (such as the $\Delta I=1/2$ rule, still unsettled) but Salam encouraged him to publish: the result is his famous Nuclear Physics paper (submitted 13 February 1961) which--with Gell-mann's equally famous, and slightly earlier (20 January 1961), Cal Tech report CTSL-20--laid the foundations of the octet model, radically (and permanently) altering the course of particle physics.

Yuval's approach to particle physics may be summarized under the rubric "pattern, structure, then dynamics". The *pattern* he discovered in hadron physics was, initially, the octet of SU3 for both baryons and hadrons (Nucl. Phys. 26, 222 (1961)). The *structure* which he induced shortly there-after was that of triplets carrying fractional (1/3) baryon number (Nuovo Cim. 27, 1, (1963))--precursors of the quark--, the *dynamics* he perceived to be that of a gauged SU3-symmetry carried by an octet of vector mesons (a pre-figuring for flavor of the currently accepted QCD).

One of the great powers of symmetry techniques is that it enables one to make meaningful statements even without a detailed knowledge of the underlying dynamics. Ne'eman knew this well and exploited it by: *classification* (the decuplet representation completing the 3-3 resonance); *branching and intensity rules* (SU3 Wigner coefficients and the independent discovery of triality); even an early version of SU6 (in January 1964!).

I have gone into Yuval's contributions to unitary symmetry in the 'heroic' phase of particle physics in some detail, but one mustn't conclude these to be his only contributions! On the contrary, his gift of discerning patterns and structure have enriched many fields, not least relativistic astrophysics, even gravitational theory. Thus Yuval has originated "white holes" (1965), "lagging cores", hyper-condensed nuclear matter (1969), supergravity (with Gell-Mann), and metric affine gravity (This latter found a use for the 'quarkel' as a world spinor).

It is clear from this brief sketch that Yuval Ne'eman has made major contributions to physics, especially in discerning, applying and extending symmetry techniques and group theoretical methods. He is accordingly a worthy recipient of the Wigner medal, and it is a pleasure and honor for me--on behalf of the Selection Committee--to present this medal to him.

Selection Committee:

H.Bacry
V.Bargmann
L.C.Biedenharn, Chairman
M.Hamermesh
E.P.Wigner

Texts and Monographs in Physics

Editors:
W. Beiglböck, M. Goldhaber, E.H. Lieb, W. Thirring

A Selection

Springer-Verlag
Berlin
Heidelberg
New York
Tokyo

R. D. Richtmyer

Principles of Advanced Mathematical Physics I

1978. 45 figures. XV, 422 pages.
ISBN 3-540-08873-3

Contents: Hilbert Spaces. – Distributions; General Properties. – Local Properties of Distributions. – Tempered Distributions and Fourier Transforms. – L^2-Spaces. – Some Problems Connected with the Laplacian. – Linear Operators in a Hilbert Space. – Spectrum and Resolvent. – Spectral Decomposition of Self-Adjoint and Unitary Operators. – Ordinary Differential Operators. – Some Partial Differential Operators of Quantum Mechanics. – Compact. Hilbert-Schmidt, and Trace-Class Operators. – Probability: Measure. – Probability and Operators in Quantum Mechanics. – Problems of Evolution: Banach Spaces. – Well-Posed Initial-Value Problems: Semigroups. – Nonlinear Problems: Fluid Dynamics. – References. – Index.

R.D. Richtmyer

Principles of Advanced Mathematical Physics II

1981. 60 figures. XI, 322 pages.
ISBN 3-540-10772-X

Contents: Elementary Group Theory. – Continuous Groups. – Group Representations I: Rotations and Spherical Harmonics. – Group Representations II: General; Rigid Motions; Bessel Functions. – Group Representations and Quantum Mechanics. – Elementary Theory of Manifolds. – Covering Manifolds. – Lie Groups. – Metric and Geodesics on a Manifold. – Riemannian, Pseudo-Riemannian, and Affinely Connected Manifolds. – The Extension of Einstein Manifolds. – Bifurcations in Hydrodynamic Stability Problems. – Invariant Manifolds in the Taylor Problem. – The Early Onset of Turbulence. – References. – Index.

G. Gallavotti

The Elements of Mechanics

1983. 53 figures. Approx. 530 pages.
ISBN 3-540-11753-9

Contents: Phenomenic Reality and Models. – Qualitative Aspects of the One-Dimensional Motion. – Systems with Many Degrees of Freedom. Theory of the Constraints. Analytical Mechanics. – Special Mechanical Systems. – Stability Properties for Dissipative and Conservative Systems. – Appendices A–P. – Definition and Symbols. – References. – Index.

Lecture Notes in Physics

Vol. 144: Topics in Nuclear Physics I. A C
Review of Recent Developments. Edited by
S.S.M. Wong. XX, 567 pages. 1981.

Vol. 145: Topics in Nuclear Physics II. A C
Review of Recent Developments. Proceed
Edited by T. T. S. Kuo and S. S. M. Wong. V
pages. 1981.

Vol. 146: B. J. West, On the Simpler Aspect
Fluctuating. Deep Gravity Waves. VI, 341

Vol. 147: J. Messer, Temperature Dependent
Theory. IX, 131 pages. 1981.

Vol. 148: Advances in Fluid Mechanics. Proce
Edited by E. Krause. VII, 361 pages. 1981.

Vol. 149: Disordered Systems and Localiza
ings, 1981. Edited by C. Castellani, C. Castr
XII, 308 pages. 1981.

Vol. 150: N. Straumann, Allgemeine Relativit
relativistische Astrophysik. VII, 418 Seiten. 19

Vol. 151: Integrable Quantum Field Theory.
1981. Edited by J. Hietarinta and C. Montonen
1982.

Vol. 152: Physics of Narrow Gap Semiconduc
ings, 1981. Edited by E. Gornik, H. Heinrich a
hofer. XIII, 485 pages. 1982.

Vol. 153: Mathematical Problems in Theore
Proceedings, 1981. Edited by R. Schrader,
D.A. Uhlenbrock. XII, 429 pages. 1982.

Vol. 154: Macroscopic Properties of Disor
Proceedings, 1981. Edited by R. Burridge, S.
G. Papanicolaou. VII, 307 pages. 1982.

Vol. 155: Quantum Optics. Proceedings, 19
C.A. Engelbrecht. VIII, 329 pages. 1982.

Vol. 156: Resonances in Heavy Ion Reactions.
1981. Edited by K.A. Eberhard. XII, 448 pages

Vol. 157: P. Niyogi, Integral Equation Metho
Flow. XI, 189 pages. 1982.

Vol. 158: Dynamics of Nuclear Fission and Related Collective Phenomena. Proceedings, 1981. Edited by P. David, T. Mayer-Kuckuk, and A. van der Woude. X, 462 pages. 1982.

Vol. 159: E. Seiler, Gauge Theories as a Problem of Constructive Quantum Field Theory and Statistical Mechanics. V, 192 pages. 1982.

Vol. 160: Unified Theories of Elementary Particles. Critical Assessment and Prospects. Proceedings, 1981. Edited by P. Breitenlohner and H.P. Dürr. VI, 217 pages. 1982.

Vol. 161: Interacting Bosons in Nuclei. Proceedings, 1981. Edited by J.S. Dehesa, J.M.G. Gomez, and J. Ros. V, 209 pages. 1982.

Vol. 162: Relativistic Action at a Distance: Classical and Quantum Aspects. Proceedings, 1981. Edited by J. Llosa. X, 263 pages. 1982.

Vol. 163: J. S. Darrozes, C. Francois, Mécanique des Fluides Incompressibles. XIX, 459 pages. 1982.

Vol. 164: Stability of Thermodynamic Systems. Proceedings, 1981. Edited by J. Casas-Vázquez and G. Lebon. VII, 321 pages. 1982.

1982. Edited by W. von Oertzen. VIII, 258 pages. 1983.

Vol. 179: Dynamical Systems and Chaos. Proceedings, 1982. Edited by L. Garrido. XIV, 298 pages. 1983.

Vol. 180: Group Theoretical Methods in Physics. Proceedings, 1982. Edited by M. Serdaroğlu and E. Ínönü. XI, 569 pages. 1983.